Navigation: Science and Technology

Volume 13

This series *Navigation: Science and Technology (NST)* presents new developments and advances in various aspects of navigation - from land navigation, marine navigation, aeronautic navigation to space navigation; and from basic theories, mechanisms, to modern techniques. It publishes monographs, edited volumes, lecture notes and professional books on topics relevant to navigation - quickly, up to date and with a high quality. A special focus of the series is the technologies of the Global Navigation Satellite Systems (GNSSs), as well as the latest progress made in the existing systems (GPS, BDS, Galileo, GLONASS, etc.). To help readers keep abreast of the latest advances in the field, the key topics in NST include but are not limited to:

- Satellite Navigation Signal Systems
- GNSS Navigation Applications
- Position Determination
- Navigational instrument
- Atomic Clock Technique and Time-Frequency System
- X-ray pulsar-based navigation and timing
- Test and Evaluation
- User Terminal Technology
- Navigation in Space
- New theories and technologies of navigation
- Policies and Standards

This book series is indexed in **SCOPUS** and **EI Compendex** databases.

Lei Yan · An Li · Wanfeng Ji · Yang Li

Mathematical and Physical Fundamentals of Navigation and Positioning with Earth's Natural Fields

Springer

Lei Yan
Institute of Remote Sensing and GIS
Peking University
Beijing, China

Wanfeng Ji
Naval Aeronautical University
Yantai, Shandong, China

An Li
Naval University of Engineering
Wuhan, Hubei, China

Yang Li
Naval University of Engineering
Wuhan, Hubei, China

ISSN 2522-0454 ISSN 2522-0462 (electronic)
Navigation: Science and Technology
ISBN 978-981-97-5523-3 ISBN 978-981-97-5524-0 (eBook)
https://doi.org/10.1007/978-981-97-5524-0

This Springer imprint is published by the registered company Springer Nature Singapore Pte Ltd.
The registered company address is: 152 Beach Road, #21-01/04 Gateway East, Singapore 189721, Singapore

Foreword by Jiancheng Fang

Earth observation and navigation positioning, as the most important high-technology national strategic area in the twenty-first century, has attracted remarkable attention. Although the inertial navigation system has high instantaneous positioning accuracy and strong anti-jamming ability, the positioning error accumulates over time, and there are major limitations. Alternatively, the gravity field, the Earth's magnetic field, the terrain elevation field, the polarization field of solar incidence, and even the gravitational field, the natural or artificial electrostatic field, and other Earth's natural vector fields can be used for navigation and positioning in different ways. The navigation and positioning of the Earth's natural field belong to the passive navigation and positioning technology; it is not affected by weather and other external conditions, making them ideal for deep space, deep earth, and deep sea exploration, as well as in national defense applications. With the development of national military strategy, natural field-based positioning and navigation technologies are expected to become one of the important methods to enhance national security and be widely used by the public.

This book systematically elaborates the theoretical models and technical methods of various natural field navigation and positioning, covering navigation with the Earth's gravitational field, the Earth's magnetic field, topographic elevation, sky polarization field, electrostatic field, and relativistic air and space field effect, forming a more complete theoretical and technical system. The mathematical and physical essence of the various aspects of the Earth's vector field navigation positioning system is abstracted, achieving outstanding results in physical model construction and the numerical algorithms of navigation filtering, with strong theoretical, fundamental, and practical significance. The academic level of the book is advanced in China, and it has outstanding value for publication and dissemination.

Fang jiancheng

Beijing, China
May 2024

Jiancheng Fang
Academician of Chinese Academy of Sciences
Professor of Beihang University (BUAA)

Foreword by Yuntai Chen

Navigation is an ancient and young science. Since the 1990s, the Global Positioning System (GPS) has provided continuous, high-precision three-dimensional positioning information. However, GPS also has its limitations, such as satellite signals being susceptible to interference, poor reception in obscured areas, and certain difficulties in underwater positioning. Moreover, GPS is under the control of the military of the country hosting the system, leading to potential limitations in its use during critical moments. Although the Inertial Navigation System offers high instantaneous positioning accuracy, errors can accumulate over time. Therefore, developing a new, independent navigation and positioning technology that can function in environments where satellite signals are unavailable is crucial. This book introduces the mathematical and physical fundamentals of passive navigation and positioning, using the natural vector field as the main source.

Over the past two decades, the navigation of Earth's natural fields has garnered increasing interest. However, the research in this area is insufficient both domestically and internationally. With the support of national 863 projects, national defense 973 projects, and the special projects of the industry department, the author's team has made systematic theoretical and technical breakthroughs over more than ten years. Focusing on the principles of gravity field, geomagnetic field, topographic field, polarization vector light field navigation and positioning, and Einstein's theory of relativity, the book forms a comprehensive theoretical and technical system and abstracts the mathematical and physical essence of each sector of the Earth's vector field navigation and positioning system, serving as a crucial foundation for original innovation in the field of navigation and positioning in China and an important reference for engineering implementation.

The book holds significant academic merit and practical importance, offering insights into algorithm principles and engineering applications, and providing essential support for precision navigation and deep space exploration.

陈运泰

Beijing, China
May 2024

Yuntai Chen
Academician of Chinese Academy of Sciences
University of Chinese Academy of Sciences

Preface

Earth observation is regarded as the most important high-tech national strategic area of the twenty-first century, which includes two key factors. One is the observation object and its molecular physical and chemical characteristics, known as remote sensing. The other is the four-dimensional time and space, also known as the navigation positioning. Essentially, the qualitative and quantitative descriptions of the two aspects are closely related to the intrinsic properties of Earth's natural fields. Earth's natural fields refer to vector fields that can be naturally formed globally or in certain specific regions of the Earth, such as the gravity field, the geomagnetic field, the topographic and geomorphic elevation field, the solar incident polarization field, and even the universal gravity field, natural or artificial electrostatic fields. These fields can be utilized in various forms for passive navigation positioning, and are not affected by objective conditions such as weather, so they have significant potential applications in deep space and deep earth exploration. With the rapid development of national military strategies, the natural field navigation and positioning technology based on passive Earth observation will become one of the most important methods to enhance national security capabilities and be widely applied to the public.

In the past two decades, Earth's natural field navigation technology has received widespread attention. The author's team of the book has conducted over ten years of research supported by national 863 projects, the national defense 973 projects, and special projects from the industry, achieving systematic theoretical and technological breakthroughs, resulting in the publication of this book. The book systematically elaborates on various theoretical models and technical methods of natural field navigation and positioning. It consists of 11 chapters, covering topics such as navigation with Earth's gravity field, Earth's magnetic field, topographic elevation, sky polarization field, electrostatic field, and relativistic sky field effect. It presents a comprehensive theoretical and technical system, abstracting the mathematical and physical essence of various stages of Earth's vector field navigation and positioning systems. The book has made prominent and distinctive achievements in the construction of physical models, and navigation filtering mathematical computation algorithms, and possesses strong theoretical, foundational, and practical value. It can serve as an important foundation for original innovation in the field of navigation and positioning in China and an essential reference for engineering implementation.

The publication of this book has been greatly supported by the leaders and colleagues from the School of Earth and Space Sciences of Peking University (PKU), the College of Electrical Engineering of the Naval University of Engineering (NUE), and the Department of Aviation Foundation of the Naval Aeronautical University (NAU). We would like to express our gratitude. In particular, Prof. Yan Lei's team at Peking University provided a wealth of material for the book. Professor Li An's team at the Naval University of Engineering contributed significantly to the writing, translation, proofreading, and error correction of this book. Associate Professor Ji Wanfeng's team at the Naval Aeronautical University also provided strong support for the compilation of this book. In addition, we sincerely thank our group members, Prof. Qin Fangjun, Dr. Shang Ke, Dr. Li Yang, Dr. Sun Jiaqing, and the excellent graduate students Qian Leiyuan, Han Yu, Mao Ning, Ma Jinxiu, Xu Rui, Zhu Tiangao, and Zhu Jiupeng, for their efforts on this book. Additionally, special thanks to Ms. Dou Jing from Springer for her assistance in the publication of this book.

Due to the rapid development of navigation technology and the fast-paced advancement of theories and technologies in related fields, coupled with the limited knowledge of the authors, there may be omissions in the content of this book. We sincerely invite readers and colleagues to provide valuable feedback on this book.

Beijing, China — Lei Yan
Wuhan, China — An Li
Yantai, China — Wanfeng Ji
Wuhan, China — Yang Li
May 2024

Contents

Chapter 1
Introduction

Abstract As the most important high technology national strategic field in the twenty-first century, earth observation consists of two elements, one is the physical and chemical characteristics of the observed object (What), namely remote sensing: the other is four-dimensional time and space, namely navigation and position (Where, When). Thus, navigation is an old and young science. Before the appearance of satellite position systems, radio navigation systems were mainly used for remote navigation and position. Radio navigation position has played a very important role in the history of navigation through methods such as Roland-C, Omega and Doppler systems. However, it covers a small working area. This has largely limited its development and application, because radio wave propagation is affected by the atmosphere, the position accuracy and other factors. It was not until the development of global navigation position in the 1990s, which can provide continuous and high-precision three-dimensional position velocity and time information to users at sea, on land, in air and space in around the clock on a global scale and can lead to epoch-making changes in the navigation and position of carriers such as ships, aircraft and automobiles. However, globle position system also has some shortcomings, such as satellite signals which are susceptible to be interfered, poor reception in hidden areas, no in-water capability, and some difficulty in underwater position. And being controlled by the military of the country where the system is located, the application can be very limited at critical times. On the other hand, although the instantaneous position accuracy of inertial navigation system is high, the error will accumulate over time. Based on this consideration, it is particularly important to study new technologies for independent navigation and position in environments where satellite signals do not cover or are subject to electromagnetic interference. This monograph is the mathematical-physical basis of passive navigation and positioning based on natural vector fields as the main source.

L. Yan et al., *Mathematical and Physical Fundamentals of Navigation and Positioning with Earth's Natural Fields*, Navigation: Science and Technology 13,
https://doi.org/10.1007/978-981-97-5524-0_1

1.1 Earth Natural Vector Field

The field, we mean the stable field naturally formed by earth. Because of the stability, it can be used as a spatial reference frame for human activities. The extension can also be expanded to serve artificial fields for natural space observation. For example, artificial electrostatic field gyroscopes can serve for the verification of the detection of space-sky relativistic effects.

A vector field is generally the combination of two elements, field axis and the force line. The field axis generally refers to the energy-emitting source. The force line is the diffusion of the "force" perpendicular to the field axis. The "lines" of the same size and value spread around the field axis are closed together to form a group of force lines, collectively called field force lines. The field axis and the force line interact with each other, which cannot be missing. The magnetic field, for example, has N and S poles as field axes, and magnetic lines of force of different sizes perpendicular to the poles, which has the same value of the line of force. We describe it in terms of sagittal force lines. The tangents at different points are the direction of the sagittal path of the point, with the head and tail running through. Therefore, it is called a vector field.

The natural vector field of the earth is a vector field that can be formed naturally in the entire area of earth or an equivalent region. For example, with solar incidence as the field axis, the dispersion of light is generated under the atmosphere and particle medium, namely the polarization effect of light. Different polarization values form lines of polarization force perpendicular to the incident light. Because the sun irradiates the globe, a solar polarization vector field with a 24-h period is formed. Biological using polarized light to navigation has proven its objectivity, periodicity, availability. Chinese scholar Yan Lei's team systematically proposed the definition, mechanism and application system of sky polarization vector field in the world, and was recognized by international peers. Thus, the observational evidence so far suggests that the sky polarization vector field can become the third natural vector field of earth after the geomagnetic and gravity fields known to man.

Due to the presence of gravity in nature and the rotation of earth, the earth's gravitational field and the earth's magnetic field are distributed in the space around earth. Influenced by the Newtonian mechanics brought about by the rotation of earth, different topographical formations appear on the surface of earth under the influence of the inhomogeneity of the earth's internal medium. Thus, gravity fields, geomagnetic fields, topographic elevation "fields", solar incident polarization fields and electrostatic fields can all be used in different forms for navigation and position. Since these navigation methods are passive navigation, they are not affected by objective conditions such as weather, and have strong potential for application in deep space and deep earth exploration.

The earth's gravitational field refers to the earth's gravitational effect, which is generally considered to be the earth's universal gravity acting on the earth, and usually refers to the earth's gravitational field near the earth's surface. In the earth's gravitational field, the magnitude and direction of gravity at each point is related only to the position of that point. Like the earth's magnetic and electrostatic fields, the earth's gravity field has elements such as gravity, gravity lines, gravity levels, and isotropic planes.

The earth's magnetic field is a vector field that is formed due to the earth's rotation and is located in the space within and around the earth. Many animals in nature use the earth's magnetic field for navigation and position. The geomagnetic field is spread over the earth's near-earth space and is a vector field. The geomagnetic field is composed of a superposition of magnetic fields from various sources, which can be distinguished into two major parts according to their origins. One part is a stable magnetic field mainly originating from the interior of the earth. The other part is a rapidly changing magnetic field mainly originating from the exterior of the earth. The geomagnetic field and its space and time variations contain important information about the solid earth and the geospace environment.

Due to the inhomogeneity of the earth's internal medium, the surface of the earth appears in various topographical features under the effect of the earth spinning. This 3-dimensional terrain that changes with height can also be considered as a kind of earth vector field, and its various properties are used for navigation.

An electrostatic field is also an electrostatic field that exists within the earth's space. The support system can research electrostatic gyroscopes with very high precision. Therefore, electrostatic gyroscopes are also used for navigation.

1.2 Gravity Natural Vector Field Navigation and Positioning

Gravity navigation is a graphical tracking navigation technique implemented using gravity-sensitive instrument measurements. Making a gravity map beforehand, and leading each route in the map has a special gravity distribution. The gravity distribution map is stored in the navigation system, and then gravity-sensitive instruments are used to determine the gravity field properties to search for the desired route. Artificial neural networks and statistical characteristic curve recognition methods enable carriers to confirm, track or traverse a route, reaching a certain destination point. This method is called passive gravity navigation because it does not radiate and does not use external coordinates. Passive gravity navigation has high accuracy, no time limit, no need to reach out of the water, no radiation, and can finally solve the stealth of submarines, which meets the demand of "high precision, long time, autonomy, and passive" navigation of underwater vehicles in the twenty-first century, and is the direction of future navigation development.

1.2.1 History of Foreign Research in Gravity-Assisted Inertial Navigation Science

Foreign research on gravity navigation systems can be basically divided into three stages [1]. Before the mid-1980s, research efforts focused on the motion base gravity gradiometer, gravity navigation principles, and matching theory. Around the 1990s, research efforts focused on passive navigation systems with gravity gradients as the matching object. High-precision passive gravity navigation and seafloor topography estimation with gravity anomalies and gravity gradients as matching objects in the late 1990s. The gravity matching theory and algorithm are based on Kalman filtering theory, and some systems use Iterative Closest Contour Point (ICCP) algorithm.

Prior to the mid-1980s, research on gravity navigation systems was represented by Warren G. Heller of Analytic Sciences Corporation. Two gravity navigation schemes were proposed, namely Gradiometer as an External Navigation Aid (GAEA) and Reference Ellipsoid Formula As an External Navigation Aid (REFAEA). Reference [2] laid the foundation of gravity navigation research. Since gravity maps were not available at that time, the scheme focused on estimating position, velocity, and gravity gradient errors through Kalman filter using statistical models of gravity fields and gravity gradients as observables. The state vector of the Kalman filter is $X = \left[g\xi \ \mu \ \delta V_n \ \delta R_n \right]^T$. Where, g is the normal gravity value, ξ is the angle of deviation of the northward plumb line, δV_n is the northward speed error, δR_n is the northward position error. The research method is mainly based on theoretical analysis and simulation experiments.

In the early 1990s, with the successful development of Bell Aerospace's rotating Gravity Gradiometer Instrument (GGI), the new concept of using gravity pattern matching techniques to improve the performance of inertial navigation systems was introduced. Bell Aerospace, Lockheed-Martin Federal System Inc and other organizations had conducted special research on gravity pattern matching technology and had achieved the expected results. Bell Labs researchers, represented by Albert Jircitaano, developed the Gravity Aided Inertial Navigation System (GAINS) and applied for a U.S. patent (Patent No. 5339684) in 1994. The system consists of up to three gravity gradiometers, a three-axis Inertial Navigation System (INS) with a vertical accelerometer (GMA, as a gravimeter), a bathymetry, gravity maps, and an optimal filter. By matching with the gravity map to provide position coordinates to reduce and limit the inertia error in a passive manner, a planar position error (CEP) of about 6 to 305 m and a horizontal velocity error of 0.004 to 0.5144 m/s can be obtained through simulation tests.

In the late 1990s, Lockheed Martin Navigation and Gravity System (NGS, which Bell Aerospace has now merged into NGS) developed the Universal Gravity Module (UGM). The system, which provides Gravity Passive Navigation (PN) and Terrain Estimation (ET) capabilities, was patented in the United States in 2000 (Patent No. 6014103). UGM is not only low cost, but also can be directly applied to existing navigation systems to greatly enhance the cruise capability of present and future shipborne missiles and submarines. The UGM was demonstrated by the U.S. Navy on board surface ships in 1998 and submarines in 1999. Experimental data showed that the using of gravity pattern matching technology can reduce the longitude and latitude error of the navigation system to 10% of the nominal error of the navigation system. The use of high precision Full Tensor Gradiometer (FTG) gravity navigation on the AUV system enabled navigation accuracy up to 30 m.

1.2.2 History of Domestic Research in Gravity-Assisted Inertial Navigation Science

Domestic research on gravity navigation systems began at the beginning of this century, and the main units include the Naval Equipment Research Institute, Peking University, Southeast University, Harbin Engineering University, the Institute of Surveying and Geophysics of the Chinese Academy of Sciences, and the Naval University of Engineering. Among them, Peking University, in cooperation with the Naval Equipment Research Institute, has made some progress in systematic research from global gravity field model applicable to gravity navigation, gravity real-time measurement theory and method, two-dimensional random field interpolation theory of gravity anomaly map and gravity anomaly matching theory. Harbin Engineering University mainly focused on gravity matching localization based on ICCP algorithm and Kriging interpolation algorithm of gravity field. The Institute of Surveying and Geophysics of the Chinese Academy of Sciences, with the support of the National Natural Science Foundation of China, has researched on the application of SITAN algorithm for gravity matching positioning and the inversion of gravity field anomalies using satellite altimetry data, and made some progress. Southeast University, with the support of the "Tenth Five-Year Plan" national defense pre-research project, has researched on the implementation of combined navigation system based on rate orientation inertial platform and gravity matching, using Kalman filter. The main research units carrying out gravity sensors and real-time gravity measurement systems for motion bases include the 707 Research Institute of China Shipbuilding Industry Corporation, Huazhong University of Science and Technology, and Naval University of Engineering.

In response to the limitations of a single matching algorithm, emerging a combined matching algorithm combining SITAN algorithm and TERCOM algorithm [3], which can reduce the number of iterations and improved the matching accuracy. Combining the correlation analysis algorithm and the nonlinear filtering algorithm can achieve the optimal estimation of the true position of the carrier [4]. The correlation between adjacent points in the inertial guidance indication trajectory is added to the point set matching by extracting vector information of the underwater carrier's navigation trajectory [5]. Then, a vector matching method that combines the velocity of the carrier's motion with the heading information and corrects the matching results by the position correlation between the sampled points in the sequence. For the limitation of low matching accuracy with large error, there is the gravity background map alignment technique based on the triangular constraint model [6]. In addition, the accuracy of gravity matching-assisted navigation has been analyzed by studying the characteristics of ocean gravity anomaly maps [7]. A hierarchical neighborhood thresholding approach to can be used to improve the matching efficiency of underwater gravity navigation [8]. In 2019, Wang et al. proposed a feature parameter matching algorithm to improve the matching accuracy by calculating the particle filtering range for different navigation trajectories of underwater carriers and considering the gravity field feature parameters into the particle weights [9]. With the continuous development of artificial intelligence, artificial intelligence matching algorithms based on neural networks, support vector machines [10] and swarm search [11] are also gradually applied, but it is still the traditional algorithms that dominate.

In summary, the domestic research in gravity navigation is mainly focused on the research of navigation algorithms, while the research on gravity sensors is relatively small, and there is still a big gap compared with foreign countries.

1.3 Geomagnetic Natural Vector Field Navigation and Positioning

The geocentric dipole magnetic field and the anomalous magnetic field are collectively referred to as the earth's primary magnetic field, also known as the fundamental magnetic field. It is a stable magnetic field originating from the earth's interior and constitutes the main component of the geomagnetic field, which changes extremely slowly. The earth's variable magnetic field is a variety of short-period geomagnetic variations that originate externally and are superimposed on the main magnetic field. The earth's fundamental magnetic field changes very slowly, with an annual variation rate of less than 1 in 1,000. The time scale of the exogenous field variation is relatively small and the variation is more complex. The intensity of the exogenous field when it is calm is less than one thousandth of the endogenous field, and the exogenous field when it is strongly disturbed is less than one hundredth of the endogenous field. However, the varying magnetic field component is very sensitive to the physical conditions in near-earth space and is a very effective physical quantity for characterizing space environmental perturbations.

1.3.1 History of Foreign Research in Geomagnetic Navigation Science

Research in geomagnetic navigation technology originated from aircraft navigation and missile guidance. By the 1980s, Boeing had made geomagnetic navigation systems standard on its aircraft, the core of which is the Global Geomagnetic Field

Model software. Russia's 1276 transport aircraft are equipped with fluxgate sensors, which unlike gyro-inertial guidance systems that require a 15-min lead time, are ready to take off. In the field of missile guidance, Russia has studied geomagnetically matched guidance technology for a long time and has set up specialized institutes. Using the intensity of the geomagnetic field as the characteristic quantity, the fluxgate sensor is used to match the guidance method with the geomagnetic field contour, and extensive experiments have been conducted. Russia test-fired its SS-19 intercontinental missile with a maneuvering variable-track multiple warhead. It reportedly uses geomagnetic field matching guidance technology. France has worked on a new geomagnetic field-based artillery guidance system since 1997 and is currently in the third phase of research, focusing mainly on the development of a Kalman filter. Test firing experiments of the shells fitted with filters were conducted in 2004 to test the self-correcting calibration capability of the shells during flight. China's Space 35 Institute, through the 863–2 aerospace field project funding, has now developed the GPS/three-axis magnetometer combination of autonomous navigation system software. The system is validated by simulation experiments using simulated data and actual satellite data in the laboratory of the Institute, Kangxi Grassland in the western suburbs of Beijing, and Guan Hall Reservoir, respectively, with position accuracy of 12 m, velocity accuracy of 0.03 m/s and fixing accuracy of 0.05° when using high precision gravity field model correction. In pure geomagnetic navigation, the position accuracy is 0.3–2 km, the velocity accuracy is 0.3–2 m/s, and the fixing accuracy is $<1°$.

For the development of geomagnetic navigation technology, the U.S. Department of Defense has included the Geomagnetic Data Reference Navigation System in its list of military critical technologies in August 2003. It is hoped that the navigation accuracy of the pure geomagnetic navigation system developed in the next 5–10 years will be better than 30 m Circular Error Probability (CEP) for ground and air positioning, and better than 500 m for underwater positioning. Geomagnetic navigation has been studied by NASA Goddard Space Center and related universities in the United States, and numerous ground experiments have been conducted.

1.3.2 History of Domestic Research in Geomagnetic Navigation Science

In China, the main units studying underwater geomagnetic navigation technology are Tianjin Institute of Navigation Instruments, Peking University, National University of Defense Technology, Harbin Engineering University, Naval University Engineering, et al. Aisino Science and Technology Group, National University of Defense Technology, Northwestern Polytechnical University, and College of Ordnance Engineering have conducted preliminary discussions on application issues. Peking University has conducted a long research on underwater terrain and gravity-assisted inertial navigation during the "Tenth Five-Year Plan" period with the support of the National Defense 973 project. Due to the similarity between gravity field and geomagnetic field, many research results can be used in geomagnetic navigation systems, and basic research about underwater geomagnetic navigation has been conducted for more than 2 years. Tianjin Institute of Marine Instruments is at the forefront of research on inertial navigation systems and combined navigation systems on water, and has undertaken national defense projects related to underwater geomagnetic/inertial guidance systems. China National Land and Resources Exploration and Remote Sensing Center has completed the preliminary general mapping of China's geomagnetic field. This includes airborne magnetic surveys in coastal areas, and has accumulated very rich experience in processing geomagnetic maps. For the geomagnetic matching positioning technology, Aisino 35 has conducted verification tests of geomagnetic matching positioning in the laboratory of the institute, Kangxi Grassland in the western suburbs of Beijing, and Guan Hall Reservoir, respectively, to initially verify the feasibility and accuracy of static geomagnetic matching positioning. For underwater carriers, navigation using a combination of geomagnetic information and inertial systems is undoubtedly a good solution. With the gradual establishment of our precise marine geomagnetic map and the emergence of high-precision magnetic sensors in China, the research of underwater geomagnetic navigation system has been paid more and more attention.

1.4 Natural Vector Field Navigation and Positioning of Terrain and Landscape Elevation

Topographic features can be characterized by elevation fields. Because the change of surface elevation is continuous, we can take the change of vertical gradient of elevation at different regional scales as the "field axis", and the smooth areas of different elevations as different "force lines" of elevation. As a result, stable terrain and landscape can be standardized with natural elevation vector fields, forming natural vector fields of terrain and landscape elevations, and enabling navigation and positioning. Subsequently, for simplicity, terrain navigation is used to represent the full description of topographic elevation natural vector field navigation and positioning.

The principle of terrain matching navigation and positioning technology is shown in Fig. 1.1. The method starts with surveying the underwater topography of the submersible mission area and constructing a database of underwater 3D reference digital topographic maps of the area according to the mapping standards. During the submersible mission the sensors are used to obtain information on the topography of the seafloor at the location of the submersible and compare it with the baseline digital topographic map in the database. After continuous measurements and suitable algorithms for matching operations, the exact position information of the submersible is finally determined. The matching position and the inertial guidance information for information fusion processing, which can better correct the positioning error due to gyroscope, accelerometer, other inertial sensors drift and data extrapolation, to assist

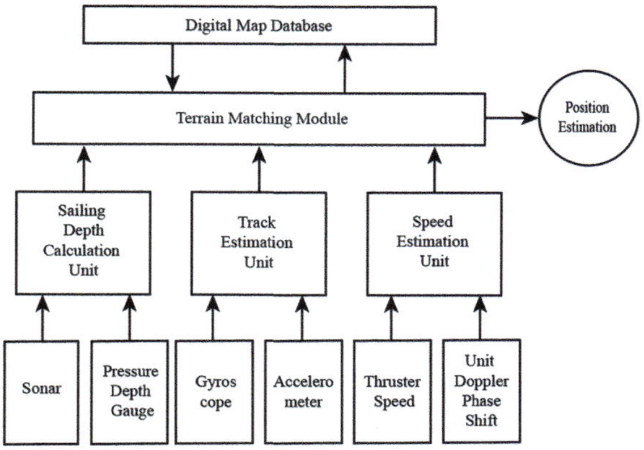

Fig. 1.1 Block diagram of underwater terrain-assisted navigation principle

the inertial navigation system to complete underwater autonomous remote accurate navigation and position. Therefore terrain-assisted navigation is generally not used as a stand-alone navigation system, but as an adjunct to inertial navigation systems.

1.4.1 History of Foreign Research in Terrain Navigation Science

The research of terrain navigation started more than forty years ago and achieved obvious results in the late 1970s and early 1980s, followed by the completion of automatic control of aircraft doing terrain tracking and terrain avoidance in 88 years. The first is the gradual development of automatic navigation and control of underwater robots, and the University of Sydney in particular is the most prominent research in this area. The Robotics Research Centre in the School of Mechanical Engineering at the University of Sydney was the first to engage in automatic navigation and control of underwater robots based on terrain-assisted navigation, and proposed the special localization and mapping algorithm SLMA (Simultaneous Localization and Mapping Algorithm) technology. The first step is to perform the underwater unmanned vehicle (UUV) position estimation, using the UUV position and the feature structure information of the seafloor environment acquired by the seafloor topography scanning sonar and bathymetry sensors as the input data of the Extended Kalman Filter EKL (Extended Kalman Filter), and estimate the position of the UUV and the position of the feature features of the seafloor to determine the UUV position relative to the feature topography of the seafloor. Translated with www.DeepL.com/Translator (free version). The next step is to perform seafloor feature extraction and identify seafloor features to provide more accurate seafloor topographic observation data for EKL. The algorithm is not only able to determine the UUV location, but also validate and update the seafloor topography data in a timely manner. Another commonly used terrain matching algorithm is the multi-model adaptive algorithm, which uses a set of parallel Kalman filters to search for the best matching location. Using the position information of INS, the information acquired by the bathymetry sensor and the digital seafloor topographic information as input data, the output information is the best match point position information and the correction information for INS. British Aerospace developed Terrain Contour Matching (TERCOM, Terrain Contour Matching), based on extended Kalman filtering to accurately model errors in barometric/inertial altitude, stored digital map data and inertial guidance systems, corrected with true radar altimeter measurements.

The new Sandia Inertial Terrain-aided Navigation (SITAN, Sandia Inertial Terrain-aided Navigation), designed by the United States, uses the Kalman filter principle to continuously combine data measured by INS with data measured by radar altimeter to estimate best the position, velocity, attitude, and other information of the vehicle. The ISS terrain navigation system, developed with the investment of the German Ministry of Defense and Downier, consists of three main components: the LiDAR, the data processor, and the real-time positioning through the LiDAR information and the corresponding reference information.

1.4.2 History of Domestic Research in Terrain Navigation Science

In China, since the 1990s, some research institutes have done a lot of research on the algorithm. Cheng Hui et al. 2003 of Huazhong University of Science and Technology applied wavelet packet decomposition for fast iterative topographic 2D matching. Yuan Shuming et al. of Harbin Engineering University designed a geographic information entropy minimum variance matching algorithm by introducing topographic information entropy in 2004. In recent years, the terrain-aided navigation system has gradually developed towards the establishment of 3-dimensional visualization and high-precision digital terrain models of the seafloor. According to the accuracy requirements of UUV diving, a Digital Elevating Model (DEM) based on grid model and TIN model (irregular triangular network) suitable for underwater navigation is established, and accurately reflects various topographic features underwater. During the UUV dive, the grid control point information of the DEM can be directly used as the input information of the terrain matching algorithm to determine the longitude and latitude position and forward state of the UUV, and it can carry out the simulation of the seafloor topography to realize the 3D visualization of the seafloor terrain matching navigation.

In recent years, terrain navigation systems have also evolved in the direction of combinations such as TERPROM/Doppler, TERPROM/GPS and newer forms of systems. Meanwhile, foreign countries vigorously develop the application of terrain matching technology on ships, submarines and underwater carriers. Among them, the technical index of U.S. underwater carriers was to improve the navigation accuracy of underwater carriers by 10 times in 2000 and to 100 times in 2005, and the main realization technology used was terrain matching technology.

Navigation algorithms are a core research component of terrain navigation algorithms, and there are certain similarities in the principles of navigation by air and underwater vehicles using terrain information. Therefore, terrain navigation algorithms are proposed either for aerial applications or for underwater applications without differentiation. The navigation algorithms that are more mature and concentrated in related fields can be divided into two main categories: batch-related navigation algorithms and iterative recursive navigation algorithms. Batch-related navigation algorithms include Terrain Contour Matching (TERCOM) and Iterated Closest Contour Point (ICCP). The iterative recursive navigation algorithms mainly include SITAN (Sandia Inertial Terrain-Aided Navigation) algorithm, VITERBI maximum a posteriori estimation algorithm and BITAN (Baihang Inertial Terrain-Aided Navigation) algorithm, etc. The latter two algorithms are the improvement and development of SITAN algorithm.

1.5 Polarized Light Natural Vector Field Navigation and Positioning

Polarized light navigation is one of the natural methods of navigation in nature. The polarization field is another important natural field in addition to the geomagnetic and gravitational fields. The causes of the existing natural field in the whole domain are Newtonian mechanics, i.e., gravity and the rotation effect of earth, without considering the energy incidence effect of celestial bodies, i.e., the energy incidence effect of sunlight. The polarization field is a full-domain natural field formed by the sun irradiating the entire earth and distributed in space around the earth, which spans the inertial space of Newtonian mechanics. Entering the category of light quantum motion, i.e., the speed of light, becomes the underlying source of excitation for the relativistic effects of space–time.

Many animals, including ants, bees, crickets, and migratory birds, have polarized vision systems that use the polarization properties of sunlight scattered in the atmosphere for navigation. Polarized light navigation has the ability to navigate accurately. According to the August 11, 2006 issue of Science, migratory birds calibrate the accumulated errors in their geomagnetic compasses by the polarized light patterns of sunrise and sunset, allowing them to accurately return to their nests or reach their destinations despite the thousands of kilometers of seasonal migration each year.

1.5.1 History of Foreign Research in Polarized Light Field Navigation Science

As a novel navigation technology, polarized light navigation has attracted a lot of attention from researchers at home and abroad. Currently, scholars in the United States, Switzerland, Sweden, Australia, Russia and other countries are conducting relevant research on polarized light navigation in biology.

Insects such as bees, crickets, sand ants, spiders[12] birds [13], and some aquatic vertebrates are able to use polarized light for navigation. Migrating birds are extremely sensitive to the direction of sky polarized light, and nocturnal migrating birds use the sky polarized light distribution at dusk to correct their heading. Rudiger Wehner, who began his research on polarized visual navigation in insects in the 1960s, found that insects such as sand ants and crickets are able to navigate using polarized light because there is a relatively stable pattern map of polarized light in the sky and because they have their own visual nervous system that is extremely sensitive to the direction of sunlight polarization. Squid possess polarized vision and use polarization information for intraspecific identification and communication; some animals can use the polarization of moonlight for navigation and orientation. For both bees and sand ants, they obtain directional information from the pattern of polarized light distribution in the sky, but the process of extrapolating distances is different. Bees use the light flow perceived during flight to project flight distances, while sand ants may use proprioceptive information to project distances.

People is also gradually focusing on the study of polarized light fields. Talbot H.Waterman(1954)conducted a measurement study on the distribution of underwater polarized light and showed that the distribution of underwater polarized light has some similarities with the distribution of sky polarized light, as shown in Fig. 1.2. A and B in Fig. 1.2 show the polarization patterns observed in the observer's line of sight from the horizontal rise to the zenith as the Sun rises from due east. C and D indicate the polarization pattern diagrams observed by the observer's line of sight descending from horizontal to zenith at noon when the Sun passes the zenith.

Animal applications of polarized light are limited to the relatively stable relationship between the E-vector direction of sky polarized light and solar altitude. In fact, additional navigation information can be obtained from the dynamic properties of the sky polarized light distribution, such as the observer latitude, the local solar time angle, and the true compass direction. Inspired by the use of polarized light for navigation by animals in nature, humans have also begun to investigate polarized light navigation methods. Inspired by the navigation capabilities of insects, [14] proposed a biological model of path integration and visual piloting and developed a corresponding platform that was successfully used on the mobile robot Sahabot 2, as shown in Fig. 1.3. The experimental measurements show that the average error of navigation using polarized light is 13.5 cm with a standard deviation of 6.3 cm, as shown in Fig. 1.4.

Reference [15] conducted an experimental study of the distribution of sky polarized light in the red, green and blue bands under different weather conditions using

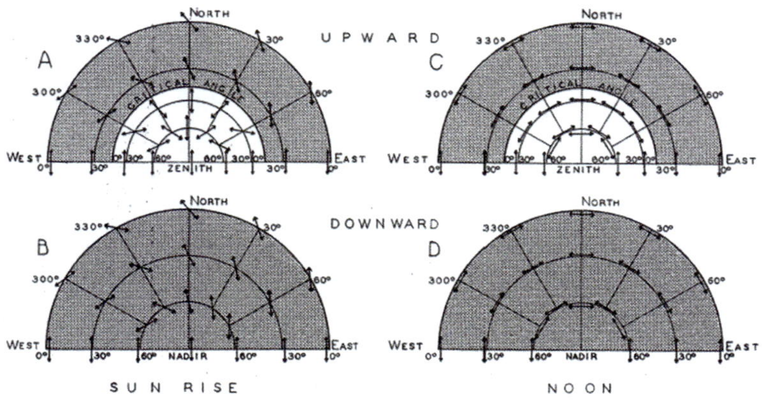

Fig. 1.2 Underwater polarized light distribution pattern map

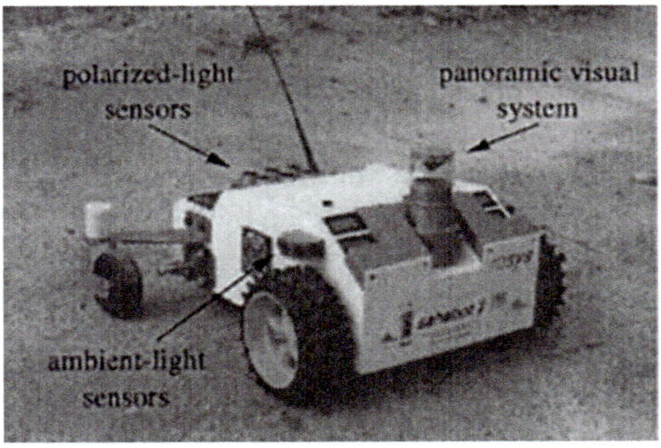

Fig. 1.3 Sahabot, a mobile robot using polarized light sensors

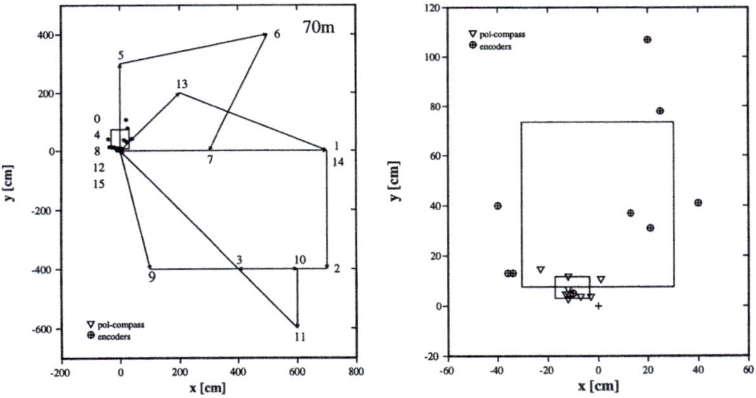

Fig. 1.4 Measurement results using polarized light sensors

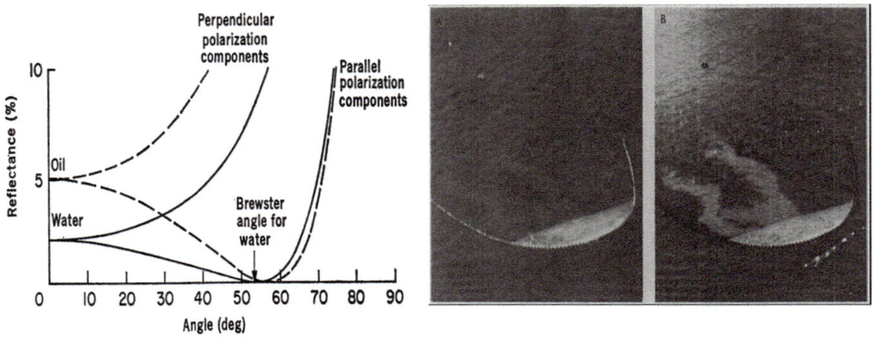

Fig. 1.5 Reflection of polarized light by the surface of the water body

all-sky polarimetry. The results show that the shorter the wavelength of visible light during cloudy days, the stronger the applicability of sky polarization information. Reference [16] discussed polarized vision of underwater organisms and its role in target recognition, contrast enhancement, and signal detection, as shown in Fig. 1.5.

The polarization and polarization angle distribution of the sky polarized light are closely related to the position of the sun. References [17] and [18]quantified the distribution of line polarization and radiance in the ultraviolet and blue-green bands and concluded that the line polarization was greatest in the ultraviolet band in cloudy or vegetation-shaded areas. Therefore, polarization is most effectively measured at the UV band line in cloudy weather or where vegetation is obscured, as shown in Fig. 1.6.

Reference [19] studied the measurement of the polarized light distribution at 220 m underwater and concluded that the underwater polarized light distribution is related to the water depth and wavelength in addition to the solar altitude angle, as shown in Fig. 1.7.

Numerous studies have shown that the sky polarization decreases sharply (Pmax $\leq$ 14%, Paverage $\leq$ 8%) due to smog, which will have a great impact on insects navigating with polarized light. The Swiss biological neurologist [20] showed that the sand ant retina is composed of hundreds of optic nerve sensory rods facing different directions, each sensitive only to the polarized light in the same direction as it. The single sensor rod is composed of eight unidirectional sensors that cross perpendicular to each other (as shown in Fig. 1.8). This structure makes the stimulus response of each sensor rod to polarized light in a specific direction a sinusoidal curve and a strict direction analyzer (as shown in Fig. 1.10). In turn, all the polarization-sensitive neurosensory rods simulate the distribution of polarized light in the sky to form a polarization-sensitive array map on the retina, as shown in Fig. 1.9. The outer ring of short rods indicates the distribution of polarized light in different directions in the sky, and the inner ring of short rods indicates the array of nerve-sensitive rods.

In this way, a systematic change in neural response occurs as the sand ant rotates its body around its head. When the sand ant body axis coincides with the solar meridian, the response of the retinal neurosensory rods reaches an extreme value.

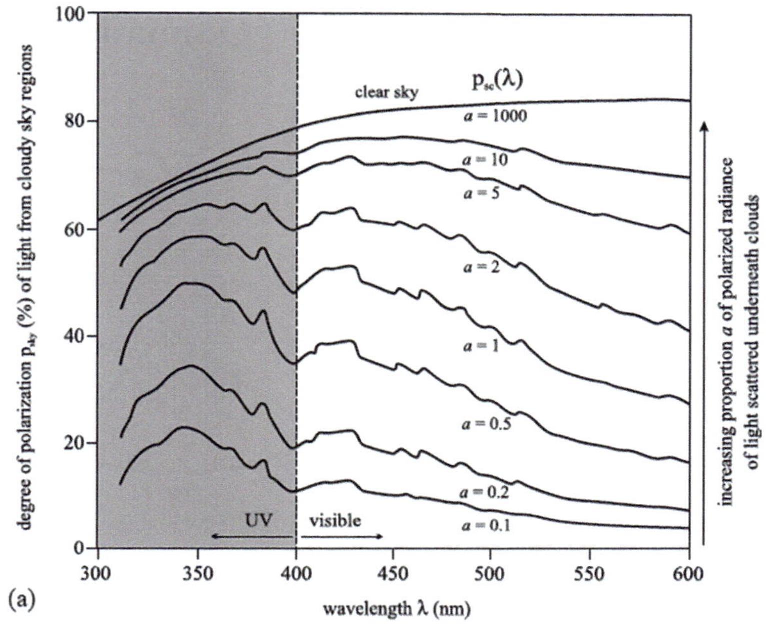

Fig. 1.6 Variation of polarization under different conditions

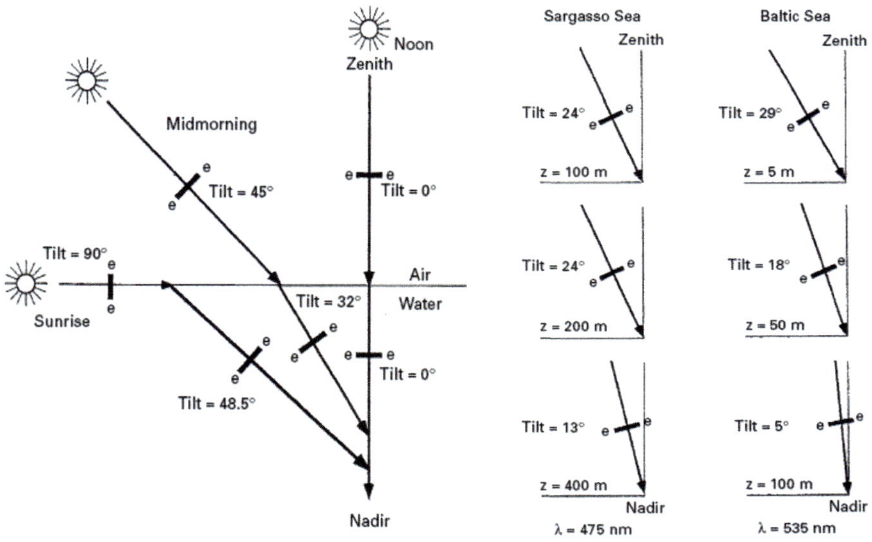

Fig. 1.7 Underwater polarized light distribution as a function of solar altitude angle

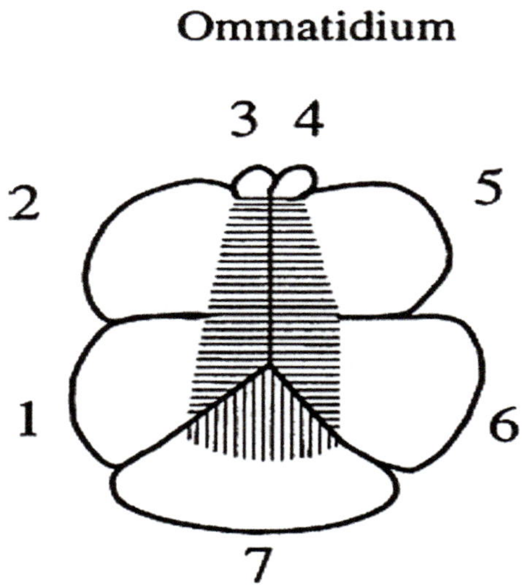

Fig. 1.8 Polarization sensitive nerve sensing rod

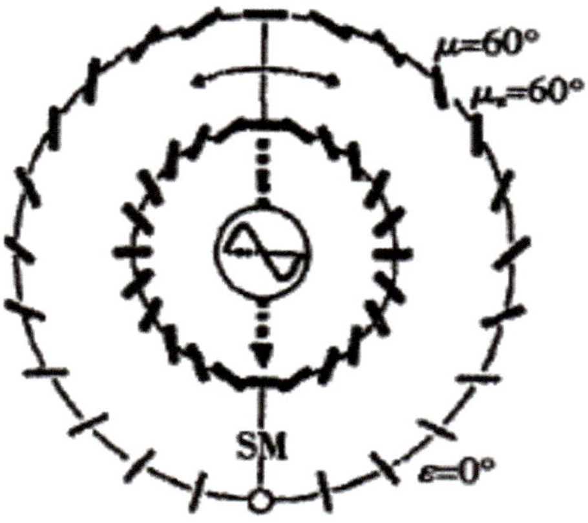

Fig. 1.9 Schematic diagram of the retinal rod-sensing array

The responses of retinal polarization-sensitive nerve rods converge as inputs to the optic nerve lobes of the central nervous layer. Intermediate neurons in the lobes of the optic nerve are divided into three types, which are polarized orthogonal neurons that receive input from optic nerve sensory rods that are perpendicular to each other in the direction of maximum sensitivity in the retina. The response of these three interneurons to retinal input was also sinusoidal, reaching maximum response values at angles of 0°, 60°, and 120° between the sand ant body axis and the solar meridian, respectively (as shown in Fig. 1.11). The response results of the intermediate neurons are used as inputs to the compass neurons in the brain of the sand ants, which are calculated and decoded to derive the angle between the body axis of the sand ant and the solar meridian, thus realizing its navigation function.

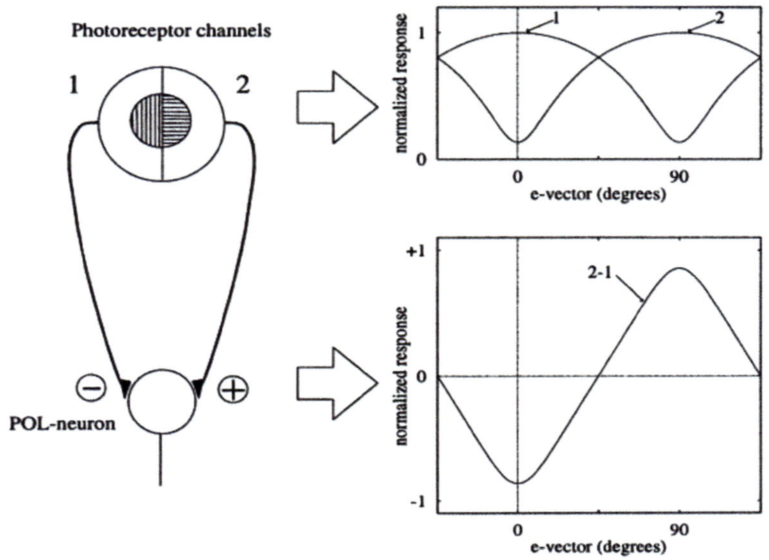

Fig. 1.10 Response curve of neurosensory rod to E-vector

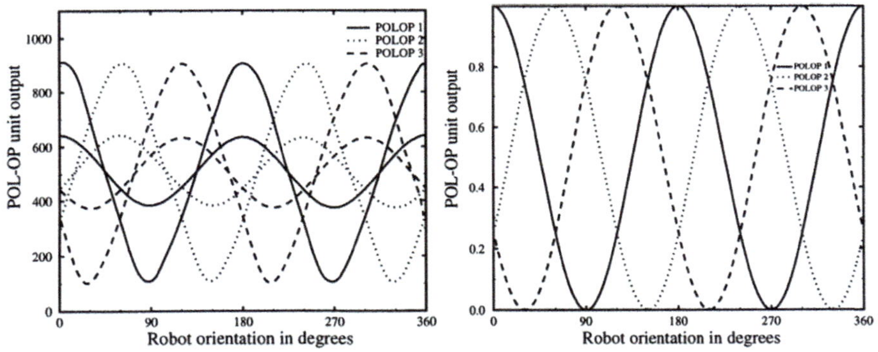

(a) Neuronal response to polarized light (b) Normalized response curve

Fig. 1.11 Response curve of neurons to polarized light

1.5.2 *History of Domestic Research in Polarized Light Field Navigation Science*

The domestic research in this area is relatively few and late. Xiaobing Sun, Jin Hong and Yanli Qiao processed and analyzed the atmospheric spectral polarization data in different spatial and temporal time to obtain the information on the spatial and temporal distribution of atmospheric polarization characteristics. This work provides material for inversion of atmospheric parameters and modeling of the polarization characteristics of atmospheric scattered radiation. Chu Jinkui and Zhao Kaichun of Lien Polytechnic University conducted an in-depth analysis and study of foreign achievements in polarized light navigation and proposed an improved method of navigation model. Hongqian Lu, Hang Yin and Xianlin Huang proposed a combined polarized light/geomagnetic/GPS/SINS navigation method, derived the principle of applying polarized light/geomagnetic-assisted attitude measurement in three-dimensional space, proved that the system has full energy observability, and pointed out that the observation structure has degradation phenomena and degradation conditions. The combined navigation algorithm is implemented by the federal Kalman filter method, and the effect of attitude correction and energy perception improvement of the combined navigation system using polarized light alone and polarized light/geomagnetic assistance is examined and compared by using Matlab simulation. Fan Zhiguo et al. proposed a polarization information detection method imitating the POL neurons of sand ants based on the special structure of polarization-sensitive neurons in the compound eyes of sand ants, and designed a detection system. Xin Cheng and Xiangzhao Wang derived a mathematical expression for the polarization of scattered light caused by water droplets based on Mie theory, and calculated and analyzed the maximum polarization in different wavelengths of visible light for a given scattering angle and water droplet size. Jinkui Chu, Kaichun Zhao, Qiang Zhang and Tichang Wang developed a sensitive polarization goniometric sensor and analyzed and tested its performance, as shown in Fig. 1.12.

The polarization properties of atmospheric and sky light were demonstrated by theoretical and experimental studies of multi-angle polarization remote sensing by Yan Lei, Wu Taixia and Guan Guixia of Peking University. Previous studies suggest that the navigation of the pavement ant may be based on recording its step length as the mode (length) of its movement vector while foraging and determining the direction of its movement vector based on the sky polarization pattern map to obtain an accurate vector relationship with its nest [21–23]. The sky polarization light distribution

Fig. 1.12 Goniometric stage with sensitive polarized light

was analyzed by the all-sky polarization measurement method, and it was initially determined that the sky polarization distribution was distributed in a circular pattern around several polarization neutrals. The distribution of polarized light in the sky at different wavelengths of visible light has been measured and studied. The results show that the polarization of clear sky increases with the increase of the observed wavelength, while the long wavelength band has a greater influence on the decay of polarization under cloudy weather conditions, and the short wavelength band is relatively stable. This indicates that the blue band is the most suitable detection band for polarized light navigation. Meanwhile, it is proposed to correct the angle between the solar meridian and the long axis of the carrier by the solar azimuth to obtain a constant navigation reference direction, which makes the polarization navigation more practical [24, 25].

1.6 Electrostatic Vector Field Navigation and Localization with Deep Space Relativistic Effect Verification

The first few sections explore the principles of establishing gravitational field, geomagnetic field, and topographic and geomorphic elevation field navigation and positioning, corresponding to the Newtonian mechanics of earth's universal gravity and earth's rotation. Based on sunlight quantum incidence polarization field bionic navigation positioning across inertial space, electrostatic field navigation can be combined with relativistic null-sky effect verification. This results in a more complete theoretical and technical system and abstracts the mathematical and physical essence of each link of the earth vector field navigation and positioning system, which is an important basis for China's source innovation in the field of navigation and positioning and an important reference for engineering implementation. It also lays the foundation of electrostatic field ultra-precision navigation and positioning gyroscope for China's deep space exploration across inertial space and pioneering relativistic effect exploration.

As a kind of inertial measurement element, gyroscope has been widely used in aviation, aerospace, navigation, measurement and other fields. As the accuracy of gyroscope directly affects the accuracy of navigation, guidance and measurement systems, etc. Therefore, since the introduction of gyroscope, people have been making unremitting efforts to improve the accuracy of gyroscope. Rotor gyroscope is in the continuous reduction of support friction torque, improve the support structure, and seek new support methods, has been developed from the initial ball bearing mechanical gyro to liquid floating and hydrostatic (floating along the output axis) air floating gyroscope, flexible gyroscope and electrostatic gyroscope. At the same time, scientists from various countries also break through the box of the original definition of gyroscope and develop a variety of new gyroscopes, such as ring laser gyroscope, fiber optic gyroscope, nuclear magnetic resonance gyroscope, superconducting gyroscope, etc. Some of these new gyroscopes are already in the practical stage, while others are still in the laboratory stage.

Electrostatic field, refers to the electric field observed when the observer is at rest relative to a charge whose charge does not change with time. It is a special form of matter that exists in the space around an electric charge, and its basic characteristic is the powerful action on the static charge placed in it. The electrostatic field is artificial, but its navigational positioning role is significant beyond Newtonian mechanics inertial space. Electrostatic field navigation and positioning is by far the highest accuracy navigation system in the world. Electrostatic gyro ESG is the world's highest precision inertial device with an accuracy of 4.4*10–2 arc second/year, electrostatic field based electrostatic gyro can be used for deep space exploration navigation and positioning. Using the gyroscopic drift, it is possible to verify two predictions of Einstein's theory of relativity—the short-range line effect and the coordinate system dragging effect, about which will be described in detail in Chap. 11.

1.6.1 Electrostatic Gyroscope and Electrostatic Support Principle

Electrostatic suspended gyro (ESG), also known as the electrostatically supported free rotor gyroscope, consists of a spherical rotor, ceramic housing assembly, reading system, electrostatic support system, titanium ion pump, starting and centering coils, and shielding cover. Since electrostatic suspended gyro the electrostatic force of an electrostatic field to support a high-speed rotating spherical rotor in an ultra-high vacuum spherical cavity, and use photoelectric sensors or mass imbalance modulation to pick up attitude information, the frictional and elastic moments and liquid or gas disturbances inherent to mechanical support are eliminated. Compared with other gyros have the following characteristics.

1. High precision.
2. Simple structure, only one moving part, less basic precision parts compared with conventional gyro.
3. Full attitude measurement.
4. High reliability, small size, light weight and low power consumption.
5. Complex and demanding support lines and reading system lines.
6. Requireing complex super-finishing process, difficult and costly to manufacture.

Electrostatic suspended gyro is currently recognized as the most accurate gyroscope available. Many experts believe that light-guided luminescence technology, including laser gyro, cannot completely replace the traditional spin-rotor-based gyro. Currently, very low drift rates (10–4 /h or less) are also guaranteed by electrostatic suspended gyro. Navigation devices composed of electrostatic suspended gyro, such as electrostatic gyroscopic monitors and electrostatic gyroscopic navigators, have been widely used on nuclear submarines, survey ships, long-range ballistic missiles, strategic bombers and space shuttles since the 1980s.

The Electrostatic suspension system (ESS) consists of two major parts: the gyro mechanical structure and the electrostatic support circuit. The gap between the spherical rotor and the spherical electrode is very small. When a high voltage is connected to the spherical electrode and the spherical rotor is kept at zero potential, due to electrostatic induction, charges of opposite polarity will be generated on the corresponding surface of the rotor, so electrostatic suction is generated between the rotor and the spherical electrode. Since the magnitude of electrostatic suction is inversely proportional to the square of the gap and proportional to the square of the voltage applied to the pole plate, the rotor is in equilibrium when the gaps on both sides are equal. If there are three pairs of spherical electrodes outside the spherical rotor along three orthogonal axes, when the electrostatic suction of each pair of electrodes on the spherical rotor are balanced, the spherical rotor is suspended by the electrostatic suction and stabilized in the center of space, as shown in Fig. 1.13.

However, if the voltage applied to the electrodes is a non-adjustable, fixed value, it still does not support the rotor. This is because the gap between the corresponding electrode and the rotor will change when the rotor is displaced relative to the electrode by a force in the gravitational field or in the acceleration field. Referring to Fig. 1.14, when the rotor is displaced Δx along the axis direction of the pair of electrodes A and B, i.e. the x-axis direction, the gap d_A between electrode A and the rotor becomes smaller and the gap d_B between electrode B and the rotor increases. If the voltage applied to electrode A and electrode B is still the same, then the electrostatic suction F_A of electrode A on the rotor increases due to the gap reduction, and the electrostatic suction F_B of electrode B on the rotor decreases due to the gap increase, thus attracting the rotor to the side of electrode A and destroying the electrostatic support. Therefore,

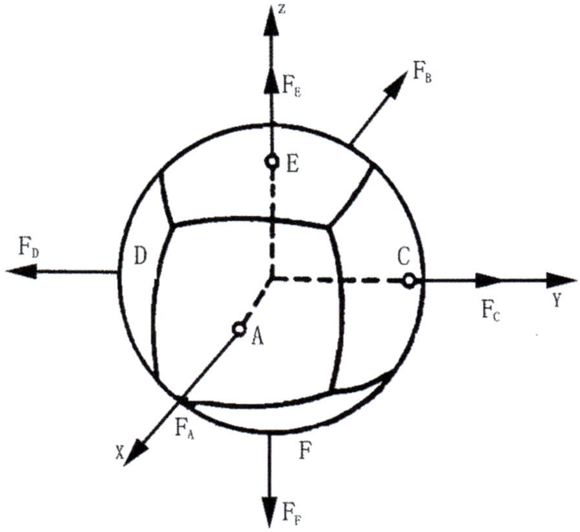

Fig. 1.13 Three pairs of electrostatic power-supported suspended ball rotors

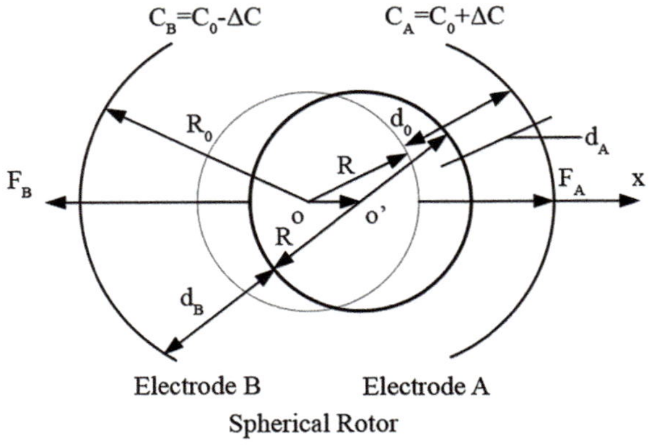

Fig. 1.14 Rotor ball center deviates from electrode ball center

when the rotor is displaced, the size of the voltage applied to the corresponding two electrodes must be automatically adjusted so that the voltage applied to the electrode at the end of the gap becomes smaller to reduce the electrostatic suction, and the voltage at the electrode at the end of the gap becomes larger to increase the electrostatic suction. And the speed of voltage change should be greater than the speed of gap change, so as to suppress the displacement of the rotor, so that the rotor back to the middle position and play the role of supporting the rotor. Only when this automatic adjustment of the electrostatic suction force is achieved can the stable balance of the rotor be guaranteed.

In order to automatically adjust the voltage on the electrodes and thus the electrostatic suction when the rotor is displaced by a small amount, it is necessary to first be sensitive to this small displacement. However, it is very difficult to measure this small displacement directly with instruments in an ultra-high vacuum spherical cavity. However, from the electric capacity $C = \varepsilon_0 \frac{S}{\delta}$ between the two electrodes ($\varepsilon_0 = 8.86 \times 10^{-12}\,F/m$ is the vacuum dielectric constant, δ is the gap between the electrodes and S is the area of the electrodes), the electric capacity between the electrodes and the rotor will change when the rotor is displaced relative to the electrodes. And when the displacement is very small, the amount of change in capacitance is proportional to the displacement of the rotor. Therefore, the use of capacitive sensors to accurately and sensitively measure this capacitance change can be indirectly sensitive to the small displacement of the rotor relative to the electrode, thus achieving automatic regulation of the electrostatic suction force. Figure 1.15 shows the schematic block diagram of the electrostatic support system.

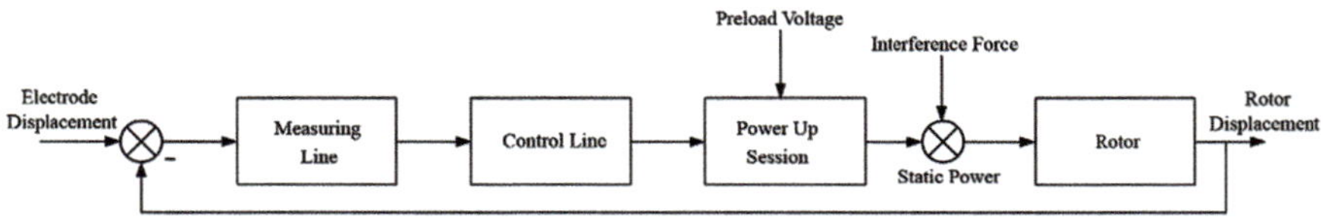

Fig. 1.15 Block diagram of the electrostatic support system

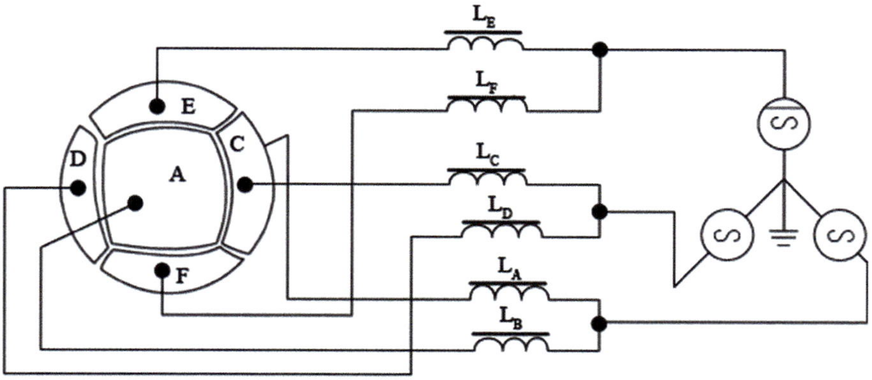

Fig. 1.16 Schematic diagram of the principle of passive type triaxial electrostatic support circuit

1.6.2 Suspension Support Loop of a Gyro in an Electrostatic Field

The support circuit of electrostatic gyroscope is divided into two types: active electrostatic support and passive electrostatic support, and active support can also be divided into active square wave control system, active sine wave control system and active time-sharing control system.

1. Passive electrostatic support

Figure 1.16 shows the schematic diagram of the passive type triaxial electrostatic support circuit. The six capacitors C_{A}、$C_{\mathrm{B}}\ldots C_{\mathrm{F}}$ formed between the positive hexahedral electrodes and the rotor and the six inductors L_{A}、$L_{\mathrm{B}}\ldots L_{\mathrm{F}}$ form six series resonant circuits and are supplied with three-phase AC power. The use of three-phase power supply ensures the virtual grounding of the ball rotor without lead wires.

Figure 1.17 shows the schematic diagram of a single axis (i.e., one channel) passive support loop. Figure 1.18 shows the curve of uniaxial support voltage versus frequency and gap variation. When the rotor is in the middle position of the two electrodes, the capacitance between the two electrodes and the rotor is equal, i.e. $C_{\mathrm{A}} = C_{\mathrm{B}} = C_0$, and if the two inductances are also equal, i.e. $L_{\mathrm{A}} = L_{\mathrm{B}} = L$, the resonant frequency of the two resonant circuits is

$$\omega_A = \omega_B = \omega_0 = \frac{1}{\sqrt{LC_0}} \tag{1.1}$$

Appropriate parameters are chosen so that the resonant frequency ω_0 is slightly less than the power supply frequency ω_r, while making the voltage U_0 applied to the two Electrodes $1/\sqrt{2}$ times the resonant voltage U_{m}. That is, the second half-power point of the resonant voltage curve is chosen as the operating point of the support

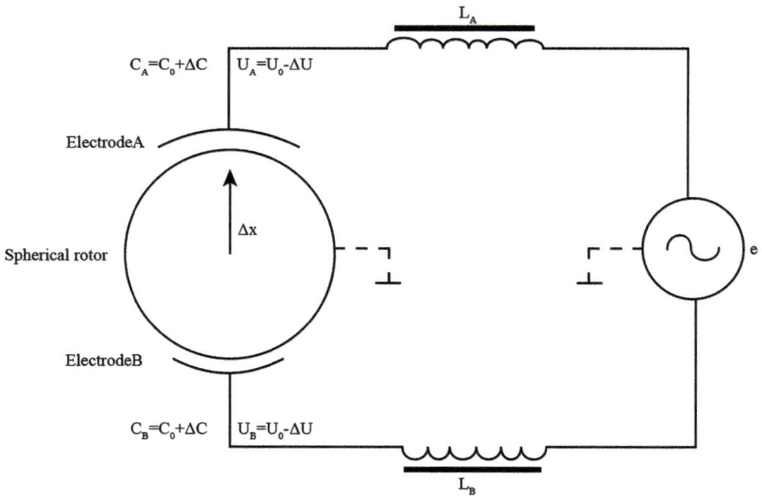

Fig. 1.17 Single-axis passive support circuit schematic

Fig. 1.18 Single-axis support voltage versus frequency and clearance curve

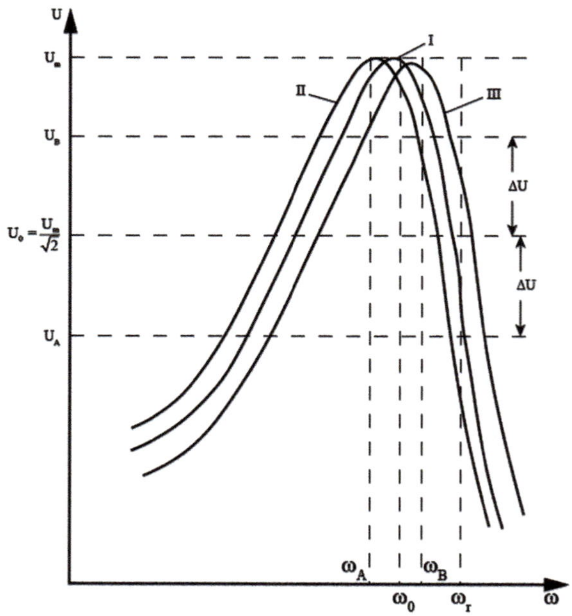

loop. At this point, the voltage applied to the two electrodes is equal, acting as a preload voltage.

When the rotor is displaced by the disturbing force by a small displacement, e.g. Δx to one side of electrode A, the capacitance between electrode A and the rotor is $C_A = C_0 + \Delta C$ and the capacitance between electrode B and the rotor is C_B

$= C_0 - \Delta C$. The capacitance C_A increases so that the resonant frequency ω_A of the circuit decreases, and the capacitance C_B decreases so that the resonant frequency ω_B increases, namely.

$$\omega_A = \frac{1}{\sqrt{L(C_0 + \Delta C)}} < \omega_0 \tag{1.2}$$

$$\omega_B = \frac{1}{\sqrt{L(C_0 - \Delta C)}} > \omega_0 \tag{1.3}$$

As the resonant frequency of the two loops changes, the resonant voltage curve of the electrode loop shifts to the left, i.e., from curve I to curve II. The resonant voltage curve of the electrode circuit shifts to the right, i.e., from curve I to curve III. Since the power supply frequency ω_r remains constant, the operating point of the electrode circuit is shifted downward, so that the voltage applied to the electrode decreases. And the operating point of the electrode circuit is shifted upward, so that the voltage applied to the electrode increases, there is

$$U_A = U_0 - \Delta U \tag{1.4}$$

$$U_B = U_0 + \Delta U \tag{1.5}$$

Thus, the electrostatic suction of electrode A to the rotor decreases and the electrode B to the rotor increases. In this way, a support force is generated to overcome the disturbance force, which acts as a support for the rotor. Since the ball rotor is in vacuum, a saturation reactance is also connected to the series circuit to introduce damping so that the rotor avoids violent oscillations.

The advantage of using passive type electrostatic support is that the line is simpler and the device is more compact. The disadvantage is that the support capacity and support stiffness is low, it is suitable for ships and other moving bodies where the impact acceleration is not very large [26].

2. Active electrostatic support

The principle of the active electrostatic support control loop is shown in Fig. 1.19. It is an axial support control loop, and the other two axial support control loops are the same.

The capacitors C_A and C_B between the electrodes A and B and the ball rotor, with two fixed capacitors C_1 and C_2 form the bridge arm of the sensitive bridge. e_1 is the low-voltage high-frequency power supply for this sensitive bridge, which operates at a typical frequency of 1 ~ 2MHz. C_3 and C_4 are high voltage isolation capacitors that isolate the low frequency high voltage added to the two electrodes from the sensitive bridge circuit.

When the ball rotor is in the center of the two electrodes, the capacitance between the two electrodes and the rotor is equal, i.e. $C_A = C_B = C_0$, there is no signal output from the sensitive bridge and no control voltage output from the electronic circuit of

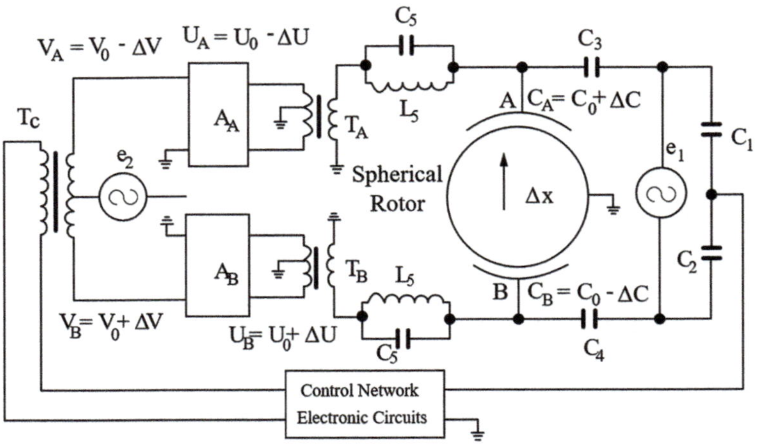

Fig. 1.19 Active electrostatic support control circuit schematic

the control loop. The low-frequency power supply e_2 of the control loop gives a fixed voltage V_0 added to the center tap of the secondary pole of the synthesis transformer T_C, which is amplified by the power amplifiers A_A and A_B to drive the high-voltage output transformers T_A and T_B. The low frequency high voltage U_0 output from transformers T_A and T_B is added to electrode A and electrode B as preload voltage via high frequency isolation networks C_5, $L_5$and C_6, L_6, respectively. Networks C_5, L_5 and C_6, L_6 are resonated in parallel on the sensitive bridge power supply frequency in order to isolate the high frequency signal from the sensitive bridge so that it does not affect the operation of the high voltage output transformers T_A and T_B.

When the rotor is displaced by a small disturbing force, for example Δx to one side of electrode A, the capacitance between electrodes A and B and the rotor becomes $C_A = C_0 + \Delta C$ and $C_B = C_0$-ΔC, when the sensitive bridge is out of balance and a signal is output. This high frequency signal is amplified, demodulated, filtered, DC amplified, network corrected, low frequency modulated, band pass filtered and low frequency AC amplified by the electronic circuit of the control loop, and then a low frequency control voltage ΔV is added to the primary of the synthesis transformer T_C. ΔV and V_0 have the same frequency, and these two voltages are synthesized in the synthesis transformer T_C, and the output voltages of the secondary poles of T_C are

$$\mathrm{V_A} = \mathrm{V_0} - \Delta\mathrm{V}, \quad \mathrm{V_B} = \mathrm{V_0} + \Delta\mathrm{V} \tag{1.6}$$

The voltages V_A and V_B are amplified by power amplifiers A_A and A_B, respectively, to drive the high-voltage output transformers T_A and T_B. The outputs of their poles are

$$\mathrm{U_A} = \mathrm{U_0} - \Delta\mathrm{U}, \quad \mathrm{U_B} = \mathrm{U_0} + \Delta\mathrm{U} \tag{1.7}$$

They are added to A and electrode B, respectively, so that the voltage on electrode A decreases and the electrostatic suction on the rotor decreases accordingly. The voltage on the electrode B increases and the electrostatic suction on the rotor increases accordingly, thus playing a role in stabilizing the support rotor.

The advantage of active electrostatic support system is the wide operating band, support stiffness and load-bearing capacity. The disadvantage is that the wiring is more complex, and the use of high-voltage output transformers increases the size of the structure. This form is suitable for moving bodies with high impact acceleration [26].

1.7 Summary

Earth's natural field navigation and positioning occupies an important position in deep space and deep earth and sea scientific research, and the basic elements in space applications, i.e., location and orientation (Where), are expressed in four-dimensional orientation in time and space. This chapter illustrates the principles and application examples of various natural field navigation and positioning methods, aiming to provide a more comprehensive theoretical and technical approach for related scientific research and applications.

This chapter emphasizes the theoretical, fundamental and practical aspects of navigation and positioning by using natural vector fields. It mainly includes: (1) the principles of navigation and positioning for gravitational fields, geomagnetic fields, and topographic and geomorphic elevation fields, corresponding to the Newtonian mechanics of the earth's universal gravity and earth's rotation; (2) Polarization field navigation and localization combined with solar incident light wave energy field effect; (3) Electrostatic field navigation and localization and combined with relativistic air-sky effect verification. This results in a more complete theoretical and technical system and abstracts the mathematical and physical essence of each link of the earth vector field navigation and positioning system, which is an important basis for China's source innovation in the field of navigation and positioning and an important reference for engineering implementation. It also provides an important basis for the scientific progress of deep space and deep earth and sea exploration in China.

For China's deep-sea exploration strategy, the use of ocean waters to achieve navigation on the ground course is particularly important for the lack of reference over water and ocean navigation. In particular, the use of underwater gravity, geomagnetic and topographic navigation is of particular importance to our independent marine research and marine homeland security protection. The authors' team has studied the natural field navigation methods of earth more systematically, and obtained more outstanding results in the physical model construction and navigation filtering mathematical computation algorithms.

For China's deep space exploration strategy, bionic polarization navigation and electrostatic gyroscopic navigation are new methods and new means that span Newtonian mechanics related to the role of natural light fields and the verification of Einstein's relativity effect, with China's unique intellectual property rights, and are also an important physical foundation link for China's future deep space exploration.

References

1. Xu, Z. (2007). *Simulation study on the principle, model and matching algorithm of underwater gravity-assisted inertial navigation.* Jiangsu:China University of Mining and Technology.
2. Heller, W. G. (1975). Gradiometer-Aided Inertial Navigation. Report TR-312–5. Analytic Sciences Corporation.
3. Wang, S., Zhang, H., Zhao, J., et al. (2011). Joint geomagnetic matching navigation using TERCOM and ICCP. *Journal of Wuhan University (Information Science Edition), 36*(10), 1209–1212.
4. Zhao, L., Gao, N., Huang, B. Q., et al. (2015). A novel terrain-aided navigation algorithm combined with the TERCOM algorithm and particle filter. *IEEE Sensors Journal, 15*(2), 1124–1131.
5. Yu, L. (2016). *Research on underwater carrier weight matching algorithm.* Beijing University of Technology.
6. Zhu, Z. S., Guo, Y. Y., & Yang, Z. L. (2016). Study oninitial gravity map matching technique based on triangle constraint model. *The Journal of Navigation, 69*(2), 353–372.
7. Wang, H. B., Wu, L., Chai, H., et al. (2017). Characteristics of marine gravity anomaly reference maps and accuracy analysis of gravity matching-aided navigation. *Sensors, 17*(8), 1851.
8. Li, Z., Zheng, W., Wu, F., et al. (2019). Improving matching efficiency of gravity matching navigation for underwater submersibles based on a novel hierarchical neighborhood threshold search method. *Journal of Geophysics, 62*(7), 2405–2416.
9. Wang, B., Zhu, J. W., Deng, Z. H., et al. (2019). A characteristic parameter matching algorithm for gravity-aided navigation of underwater vehicles. *IEEE Transactions on Industrial Electronics, 66*(2), 1203–1212.
10. Huang, P., Cheng, Y. (2011). Application of improved BP neural network algorithm in track matching. *Computer Engineering, 37*(11), 218–219+222.
11. Dai, T., Miao, L. J., Shao, H. J., et al. (2019). Solvinggravity anomaly matching problem under large initial errors in gravity aided navigation by using an affine transformation based artificial bee colony algorithm. *Frontiers in Neurorobotics, 13*, 19.
12. Labhart, T., & Meyer, E. P. (1999). Detectors for polarized skylight in insects: A survey of ommatidial specializations in the dorsal rim area of the compound eye. *Microscopy Research and Technique., 47*, 368–379.
13. Moore, F. R. (1986). Sunrise, skylight polarization and the early morning orientation of night-migrating warblers. *The Condor., 88*(4), 493–498.
14. Cronin, T. W., & Shashar, N. (2003). Polarization vision and its role in biological signaling. *Integrative and Comparative Biology.*
15. Pomozi, I., Horvath, G., & Wehner, R. (2001). How the clear-sky angle of polarization pattern continues underneath clouds: Full-sky measurements and implications for animal orientation. *The Journal of Experimental Biology., 204*, 2933–2942.
16. Cronin, T. W., Shashar, N., Caldwell, R. L., Marshall, J., Cheroske, A. G., & Chiou, T.-H. (2003). Polarization vision and its role in biological signaling. *Integrative and Comparative Biology, 43*, 549–558.
17. Barta, A., & Horvath, G. (2004). Why is it advantageous for animals todetect celestial polarization in the ultraviolet? Skylight polarization under clouds and canopies is strongest in the UV. *Journal of Theoretical Biology., 226*, 429–437.

18. Horvath, G., & Varju, D. (2004). *Polarized light in animal vision*. Berlin: Springer.
19. Waterman, T. H. (2006). Reviving a neglected celestial underwater polarization compass for aquatic animals. *Biological Research, 81*, 111–115.
20. Wehner, R., & Menzel, R. (1969). Homing in the ant cataglyphis bicolor. *Science, 164*, 192–194.
21. Taixia, W., & Yunsheng, Z. (2005). The bidirectional polarized reflectance model of soil. *IEEE Transactions on Geoscience and Remote Sensing, 43*(12), 2854–2858.
22. Hongying, Z., Hu, Z., Lei, Y., & Yunsheng, Z. . Model of reflection spectrum of rock surface in 2π space, ACTA GEOLOGICA SINICA (English Edition).
23. Hao, F., Lei, Y., Zhongliang, D., & Kedong, W. (2004). Data map characteristic in aided navigation. In *IEEE position, location and navigation symposium. PLANS 2004 26–29:771–774*.
24. Guan, G., Yan, L., Chen, J., & Wu, T. (2008). Vehicle tracking algorithm based on GPS and map-matching. In *Measurement theory and systems and aeronautical equipment*, 2008(EI).
25. Yan, L., Guan, G., & Chen, J., et al. (2009). A preliminary investigation on the mechanism of bionic navigation and orientation based on sky polarization light distribution pattern. *Journal of Peking University (Natural Science Edition), 45*(4).
26. Zhang, Y. (1985). *Gyroscopic support systems*. Shanghai Jiao Tong University Press.

Chapter 2
The Basis and Theoretical Model of Earth Gravity Field Navigation and Positioning

Abstract The irregularity of the earth's shape and the non-uniformity of density lead to different gravity fields at various points of the earth, which are expressed as a function of spatial position (longitude, latitude, height). Therefore, during navigation, the submersible can collect the gravity data of the route through the gravity measuring instrument, match it with the pre stored gravity data, obtain the current positioning information of the submersible, and then correct the position error accumulated by the inertial navigation system. Gravity assisted navigation system in the process of measuring gravity field data, the submersible does not need to be exposed to or close to the water surface, and the measuring instrument does not need to send or receive external signals to the outside. The system can carry out passive and covert navigation and positioning, and the submersible can still achieve the purpose of autonomous covert navigation in the special case of satellite and radio positioning system failure or damage. With the development of gravity measurement instruments and space measurement technology [Chen et al. in Advances in Earth Sciences. 16:1–13, 2001], it has become a reality to obtain gravity data quickly and accurately all over the world, which makes the mapped gravity aided navigation system have the ability to correct ins accumulated position error. There are mainly two representative systems of the charted gravity assisted navigation system, both developed by Bell aerospace, namely, the gravity gradient navigation system developed in 1990 and the gravity assisted inertial navigation system developed in 1991. The gravity assisted positioning system can be divided into four main parts: inertial navigation system, gravity measurement instrument, digital gravity reference map and matching positioning algorithm. The performance difference of each part has an important impact on the performance of the gravity assisted positioning system. Gravity model (benchmark map) is the basis of gravity assisted positioning. Whether the description of gravity model is accurate, whether the gravity features contained are rich, and whether the resolution meets the requirements will affect the performance of gravity assisted positioning system. Based on the external gravity model J2 gravity field model, this chapter discusses the calculation method of normal gravity

L. Yan et al., *Mathematical and Physical Fundamentals of Navigation and Positioning with Earth's Natural Fields*, Navigation: Science and Technology 13,
https://doi.org/10.1007/978-981-97-5524-0_2

on the surface of rotating ellipsoid, and establishes a small depth underwater three-dimensional gravity field model and the corresponding underwater inertial navigation system INS combined model, which provides a basic theoretical method for the application of gravity assisted positioning system.

2.1 Standard Ellipsoid Gravitational Field Model and Its Numerical Analysis of Normal Gravity

The gravity models used in inertial navigation include the central gravity field and the general gravity model. The central gravity field model simplifies the earth into a standard sphere for general explanation of navigation principles. Common normal gravity models include EGM96 (earth gravity model 96) and J2 gravity models based on wgs1984 (World geodetic system 1984). J2 model is a normal gravity model obtained by solving Laplace equation with ellipsoid as boundary [2, 3].

In 1996, David y. Hsu compared the normal gravity model above the level, and in 1998, he proposed an approximate calculation model of normal gravity. In 1997, Xiaoguang Hao studied the effects of different earth parameters (density parameters, interface depth parameters and ellipsoid parameters) on the normal gravity field. The normal gravity field is an artificial setting in understanding the earth's gravity field. Various models have certain rationality, and the research on it is still of practical significance [4].

The J2 gravity model is described by using the earth constant and coordinate system defined by WGS84. The conventional gravitational potential energy is universal gravitational potential energy V and centrifugal potential energy Φ, seeing formula (2.1)

$$W = V + \Phi \tag{2.1}$$

The gravitational potential energy of the outer space is a harmonic field, which satisfies the Laplace equation.

$$\nabla V = \frac{\partial^2 V}{\partial x^2} + \frac{\partial^2 V}{\partial y^2} + \frac{\partial^2 V}{\partial z^2} = 0 \tag{2.2}$$

(x, y, z) represents the coordinate points in the ECEF coordinate system (earth central earth fixed). The boundary conditions are determined by the conventional gravity of the earth's surface, and V can be expanded in the form of sum of lower order numbers.

$$V = \frac{GM}{r}\left[1 + \sum_{n=0}^{\infty}(a/r)^{2n} C_{2n,0}\overline{P}_{2n,0}\left(\frac{z}{r}\right)\right] \tag{2.3}$$

G: : Gravitational constant, 6.672 × 10-11m3/(kg·s2);M: : Earth mass, gm = (3,986,004.418 × 0.008) × 108 m^3/s^2, is the gravitational constant of the earth; a: Radius of the earth's long axis, 6,378,137.0 m; $C_{2n,0}$ is the spherical harmonic coefficien:$r = \sqrt{x^2 + y^2 + z^2}$, is the distance from the point to the center of the earth.

J2 gravity model only uses the first two terms of formula (2.3) to obtain a truncated conventional gravity model:

$$V_{J2} = \frac{GM}{r}\left[1 + \frac{J_2 a^2}{2r^2}\left(1 - \frac{3z^2}{r^2}\right)\right] \tag{2.4}$$

where, $J_2 = -\sqrt{5}C_{2,0}$, $C_{2,0} = -0.484165371736 \times 10^{-3}$.

Centrifugal potential Φ. It can be obtained from the following formula.

$$\Phi = \frac{\Omega^2}{2}\left(x^2 + y^2\right) \tag{2.5}$$

where, Ω is earth rotation rate, which is 7.2921150 × 10–5 rad/s。

J2 gravity model is obtained by adding formula (2.4) and formula (2.5).

$$W_{J2} = V_{J2} + \Phi = \frac{GM}{r}\left[1 + \frac{J_2 a^2}{2r^2}\left(1 - \frac{3z^2}{r^2}\right)\right] + \frac{\Omega^2}{2}\left(x^2 + y^2\right) \tag{2.6}$$

Taking the gradient of W_{J2} to get the gravity vector.

$$g_x = \frac{\partial W_{J2}}{\partial x} = -\frac{xGM}{r^3}\left[1 + \frac{3J_2 a^2}{2r^2} - \frac{15 J_2 a^2 z^2}{2r^4}\right] + x\Omega^2 \tag{2.7}$$

$$g_y = \frac{\partial W_{J2}}{\partial y} = -\frac{yGM}{r^3}\left[1 + \frac{3J_2 a^2}{2r^2} - \frac{15 J_2 a^2 z^2}{2r^4}\right] + y\Omega^2 \tag{2.8}$$

$$g_z = \frac{\partial W_{J2}}{\partial z} = -\frac{zGM}{r^3}\left[1 + \frac{9J_2 a^2}{2r^2} - \frac{15 J_2 a^2 z^2}{2r^4}\right] \tag{2.9}$$

The expression of gravity in the local geodetic vertical (LGV) coordinate system requires an additional transformation.

$$g^N = C_E^N \begin{pmatrix} g_x \\ g_y \\ g_z \end{pmatrix} \tag{2.10}$$

where, C_E^N is the conversion matrix from ECEF to LGV coordinate system.

$$C_E^N = \begin{bmatrix} -\cos\lambda\sin\varphi & -\sin\lambda\sin\varphi & \cos\varphi \\ \sin\lambda & -\cos\lambda & 0 \\ \cos\lambda\cos\varphi & \sin\lambda\sin\varphi & \sin\varphi \end{bmatrix} \tag{2.11}$$

λ and φ are longitude and latitude of the point.

The above gravity model is based on the gravity value obtained from the coordinates of ECEF points. In inertial navigation, the position of the carrier is expressed by the geodetic coordinate system (λ, φ, h). The conversion matrix from geodetic coordinate system to ECEF coordinate system is

$$\begin{bmatrix} x \\ y \\ z \end{bmatrix} = \begin{bmatrix} (N+h)\cos\varphi\cos\lambda \\ (N+h)\cos\varphi\sin\lambda \\ \left(\frac{b^2N}{a^2}+h\right)\sin\varphi \end{bmatrix} \tag{2.12}$$

b: Minor axis radius of ellipsoid;h: Depth of underwater motion carrier;N: The radius of curvature of this position (the relationship between H and N is shown in Fig. 2.1)

$$N = a^2 \cdot \left(a^2\cos^2\varphi + b^2\sin^2\varphi\right)^{-\frac{1}{2}} \tag{2.13}$$

When $h = 0$, the ECEF coordinate value of each point on the earth's surface is obtained, and the gravity value of the earth's surface is obtained by substituting into formulas (2.7, 2.8 and 2.9). The results are shown in Fig. 2.2.

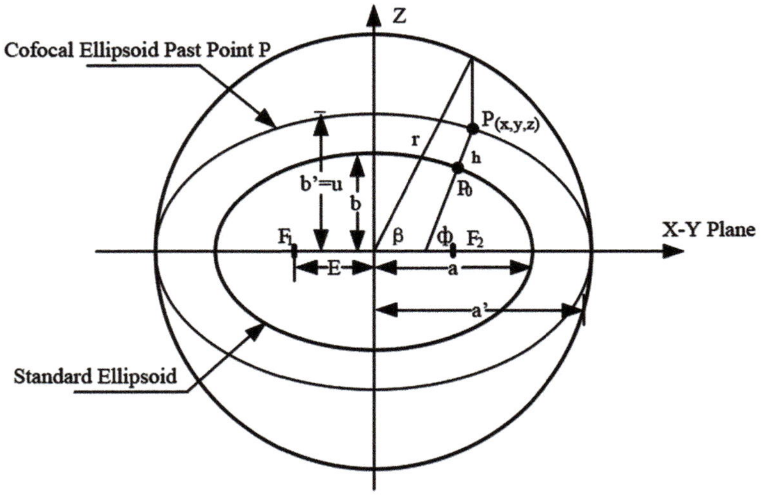

Fig. 2.1 Schematic diagram of ECEF coordinate system and ellipsoidal coordinate system

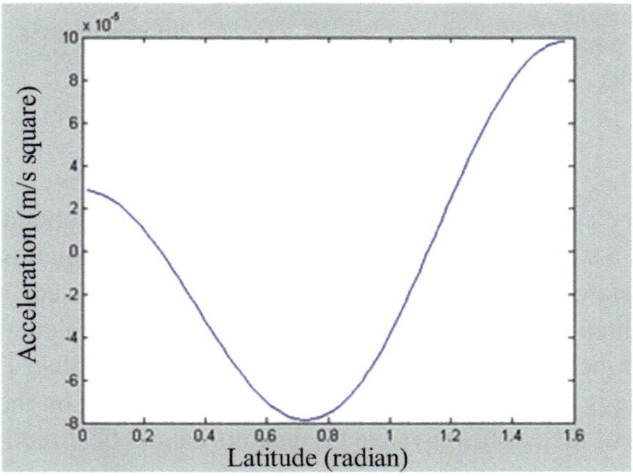

(a) J2 model and Helmut gravity vertical component difference latitudinal variation curve

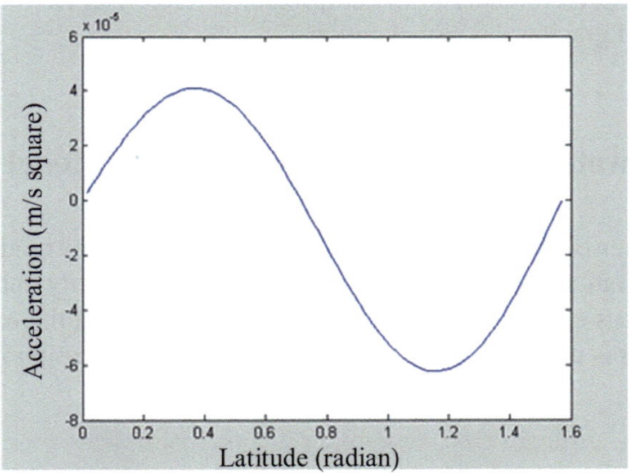

(b) J2 model latitudinal variation curve of the north-south component of the Earth's surface

Fig. 2.2 Calculation results of conventional gravity model on the earth's surface

The surveying and mapping department uses the Helmert formula to obtain the conventional gravity value of the earth's surface, as shown in formula (2.14).

$$g_\varphi = 9.78030\left(1 + 0.005302\sin^2\varphi - 0.000007\sin^2 2\varphi\right)m/s^2 \tag{2.14}$$

The earth surface gravity value obtained by Helmert formula for surveying and mapping is compared with the vertical component of the local geodetic vertical coordinate system (LVG) of J2 gravity model. The comparison results are shown in Fig. 2.2a, b. The maximum difference value is shown in Table 2.1

Table. 2.1 Comparison results of Helmert formula and J2 gravity model

	Difference(m/s^2)	Latitude (°)
Maximum difference	9.812×10^{-5}	90
Minimum difference	-7.862×10^{-5}	42

The boundary conditions used in J2 gravity model are obtained by assuming that the earth is a standard ellipsoid with uniform density, which is the same as the assumption of Helmert formula. The ellipsoid parameters used in the calculation vary slightly, which is the main reason for the difference in calculation results.

The physical meaning of J2 gravity model is more clear than Helmert formula. Helmert formula is an approximation of the standard gravity on the earth's surface, and the changes of the earth's physical parameters cannot be directly reflected in the formula.

In the LVG coordinate system, in addition to the vertical component, the ellipsoidal surface gravity of J2 gravity model also has a small component upward in the north, that is, it is not perpendicular to the ellipsoidal surface. The Helmert formula only gets the vertical gravitational acceleration.

2.2 Underwater Three-Dimensional Gravity Model

For the discussion of the three-dimensional conventional gravity model of the seabed, the direct integration method is often used to obtain the gravity and its potential in the homogeneous ellipsoid. However, the average density of the earth's density at different depths is different, and the error caused by simplification is large [5–7].

2.2.1 Basic Assumptions of the Earth's Gravity Model

The earth model with a depth of less than 1000 m refers to the J2 gravity model. The different depths of the carrier indicate that the carrier moves on ellipsoids with different major axis radii, which have the same center point and oblateness.

At present, the change of the earth's density is not fully understood, but Bren gave a depth distribution model of the earth's density. When the depth is less than 1000 m, it can be considered that the average density of this layer of ellipsoidal shell is 2.9×103 kg/m^3.

2.2.2 Underwater Three-Dimensional Conventional Gravity Model

The basic idea of the three-dimensional conventional gravity model of the seabed: calculate the surface gravity acceleration of the imaginary ellipsoid whose position of the underwater carrier is in focus with the earth's reference ellipsoid; Investigate the influence of the external ellipsoidal shell on the underwater carrier [8].

It can be verified that when the ellipsoid shrinks inward and changes very little, because the structural parameters of the ellipsoid, such as the oblateness of the ellipsoid, do not change, only the radius of the ellipsoid changes slightly, and the influence of this change on the gravitational potential can be approximately reflected by the change of the earth's mass. According to the standard ellipsoid radius of the earth, when the variation range is very small (<1000 m), the gravitational potential of the ellipsoid surface and its external space can be calculated directly by quoting formula (2.6). However, the mass part of the formula should be changed accordingly. The change of ellipsoid radius and gravitational potential function can be expressed by the change of mass.

The gravity of the remaining ellipsoidal shell on its internal point is calculated by direct integration. The calculation results show that when the depth change is less than 1000 m, the gravity of the shell to the internal point is very small, and the order of magnitude is 10-12 m/s2.

Therefore, directly using formulas (2.15, 2.16 and 2.17), we can obtain the normal gravity distribution on the surface of an ellipsoid that is in the same focus with the earth's standard ellipsoid at a certain depth.

$$g_x = \frac{\partial W' J_2}{\partial x} = -\frac{xGM'}{r^3}\left[1 + \frac{3J_2 a'^2}{2r^2} - \frac{15J_2 a'^2 z^2}{2r^4}\right] + x\Omega^2 \tag{2.15}$$

$$g_y = \frac{\partial W_{J2'}}{\partial y} = -\frac{yGM'}{r^3}\left[1 + \frac{3J_2 a'^2}{2r^2} - \frac{15J_2 a'^2 z^2}{2r^4}\right] + y\Omega^2 \tag{2.16}$$

$$g_z = \frac{\partial W_{J2'}}{\partial z} = -\frac{zGM'}{r^3}\left[1 + \frac{9J_2 a'^2}{2r^2} - \frac{15J_2 a'^2 z^2}{2r^4}\right] \tag{2.17}$$

where, M' and a' are the mass and major axis radius of the internal confocal ellipsoid.

2.2.3 Approximate Calculation Model

Using formulas (2.15), (2.16) and (2.17) to calculate the gravity value on grid points with latitude of 0–90 °, latitude interval of 1 min, depth of 0–1000 m and depth interval of 20 m. The gravity values are converted to LGV coordinates. The results are shown in Fig. 2.3.

At each latitude point, the normal gravity value of the corresponding depth is linearly fitted. The fitting results are shown in Fig. 2.4.

The linear fitting formula can be expressed in the following form.

$$g_m = g_s + k \cdot h \tag{2.18}$$

g_s: ellipsoid surface gravity, k: slope.

The maximum linear fitting variance of the northward component is 4.6018 × 10–20, the maximum linear fitting variance of the vertical component is 1.8931 × 10–14。

The conventional three-dimensional gravity model of the seabed includes: J2 gravity model, north component depth correction slope array, and vertical component depth correction array.

The average slope of the north component is 7.587 × 10-4mgal/m, that is, the maximum depth change of the calculated area does not exceed 1mgal. In order to further simplify the model, the north component correction of the latitude area with a slope less than the mean value is omitted, that is, the correction area is Latitude 20 °–70 °.

The slope of the vertical component varies from -0.37393 to -0.37388mgal/m. It can be seen from the figure that the slope changes monotonously with latitude, so the corresponding latitude region can be determined according to the slope range. The results are listed in Table 2.2.

2.3 Free Air Gravity Anomaly Calculation Model

Gravity anomaly is the difference between the gravity on the geoid (a closed geopotential level that extends to the interior of the continent and surrounds the whole earth) and the reference ellipsoid (an idealized sphere, which can completely use mathematical formulas to express the points on the sphere). It is caused by the mass distribution that causes the geoid to deviate from the reference ellipsoid. The general method of calculating gravity anomaly is to correct the observed value to the equivalent gravity value at sea level, and then subtract it from the normal gravity of the reference ellipsoid. Different corrections result in corresponding gravity anomalies. The existing correction models include free air correction, Bouguer correction and equilibrium correction. The gravity anomaly at sea is usually the free air gravity anomaly with only free air correction [9–11].

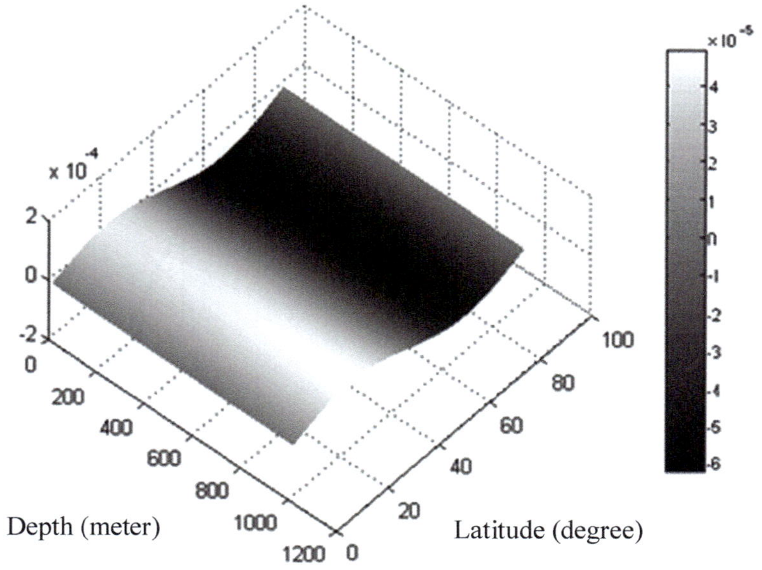

(a) Gravity northward component depth-latitude distribution

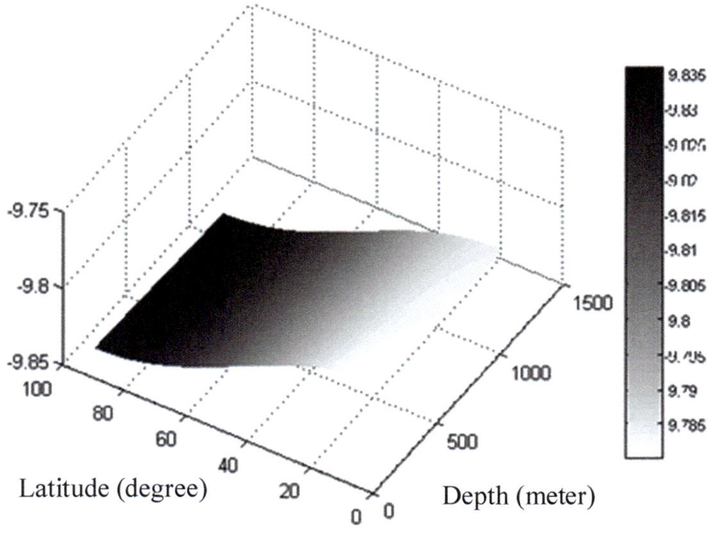

(b) Vertical component of gravity depth-latitude distribution

Fig. 2.3 Depth latitude distribution of North and vertical components of seafloor conventional gravity

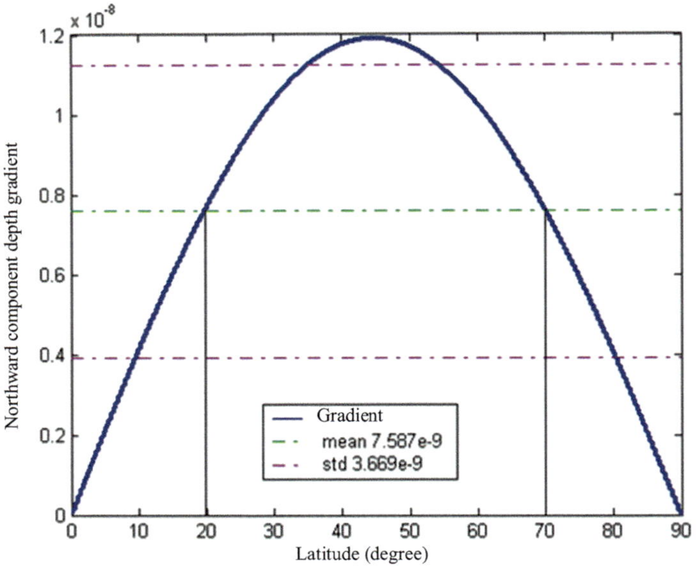

(a) Latitudinal variation of the northward component depth gradient

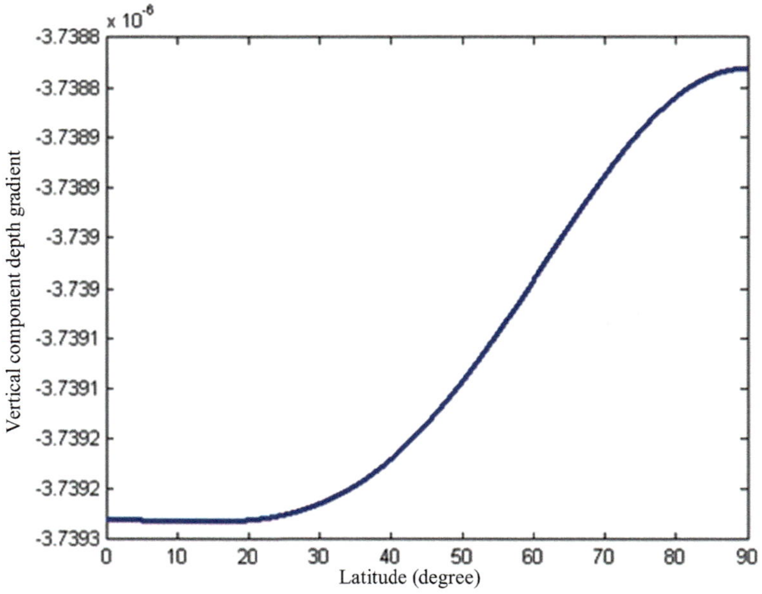

(b) Vertical component depth gradient latitudinal variation

Fig. 2.4 Linear fitting constants of each gravity component in the depth direction

Table. 2.2 Analysis list of calculation results of vertical component linear fitting slope

k	-0.37392	-0.37391	-0.37390	-0.37389	-0.37388
φ	< 34.3	< 49.33	< 59.30	< 68.83	90

2.3.1 Free Air Gravity Anomaly Calculation Model

The free air gravity anomaly is obtained only by spatial correction. The derivation process of free air correction (correction of gravity observation value at height h) is briefly described as follows.

$$g_0 = \frac{GM}{r^2} \tag{2.19}$$

where, r is the radius of the earth. The normal gravity at height h is:

$$g_h = \frac{GM}{(r+h)^2} \tag{2.20}$$

Taylor expansion for g_h near r:

$$\delta_F = (g_0 - g_h) = 2g_m\left(\frac{h}{r}\right)\left[1 - \frac{3}{2}\left(\frac{h}{r}\right) + \cdots\right] \tag{2.21}$$

where g_m is the global average gravity. When the height is less than 700 m, the second term of the formula is ignored.

$$\delta_F = 0.3086h\ mGal \tag{2.22}$$

When the carrier is sailing underwater, the gravity measurement point is lower than the geoid, and its free air correction is obviously different from the calculation method mentioned above.

It is generally believed that when the observation point is on the underwater moving carrier, the mass of the seawater layer above the observation point reduces its gravity, and when the observation point moves up to the sea surface. The mass of the sea water layer is below the observation point, which increases the gravity of the observation point on the sea surface, so the correction number of this part is twice the mass gravity of the sea water layer. The gravity of seawater layer mass is calculated by the interlayer correction formula, and the seawater density is taken as 1.03×103 kg/m^3, the obtained correction formula:

$$g_\sigma = 0.0861h\ mGal \tag{2.23}$$

The sea level is close to the geoid. When the carrier is sailing at a certain depth, the data collection point is considered to be in the earth's mass. As mentioned earlier,

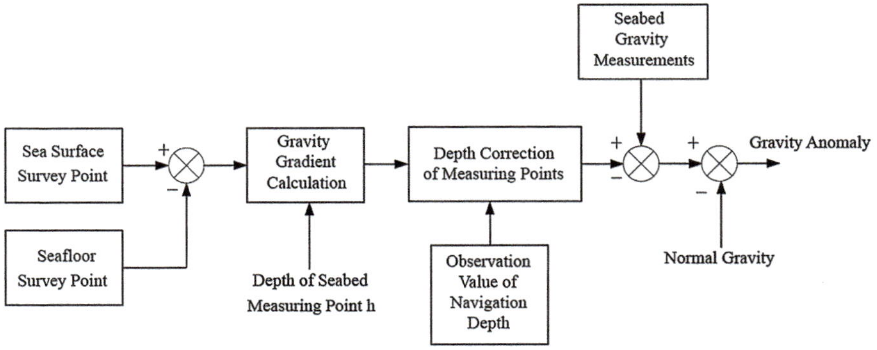

Fig. 2.5 Calculation process of seabed gravity anomaly

the description of the gravity field in outer space is not applicable to the gravity field inside the earth in theory.

Another gravity anomaly calculation principle based on gradient calibration is to obtain the distribution of terrestrial gravity anomaly according to the gravity measurement results on the leveling line, so as to obtain the gravity data of coastal waters, and calculate the vertical gradient of gravity according to the seabed gravity measurement values. A gravity reference point is set on the seabed, which is used as the reference point for offshore gravity measurement for experimental measurement.

The basic process of gravity anomaly reduction to the level is shown in Fig. 2.5. When the depth difference between the measuring points is small, the relative value of the seabed measurement can be compared with the offshore measurement results. The accuracy of gravity anomaly obtained by this method is greatly affected by the change of the sailing depth of the moving carrier (the change of depth will change the relative change value of gravity).

2.3.2 Free Air Gravity Anomaly Model Based on Underwater Normal Gravity Field

For the calculation of gravity anomaly of measuring points in the interior of the earth, it is not appropriate to directly quote the correction formula of external measuring points in theory.

From the definition of gravity anomaly, underwater gravity anomaly δg can be obtained by first calculating the normal gravity of the measuring point, and then subtracting it from the observed gravity value of the measuring point. Using the three-dimensional normal gravity model proposed in Chapter 2, the normal gravity of this point can be calculated from the spatial position. The process is shown in Fig. 2.6.

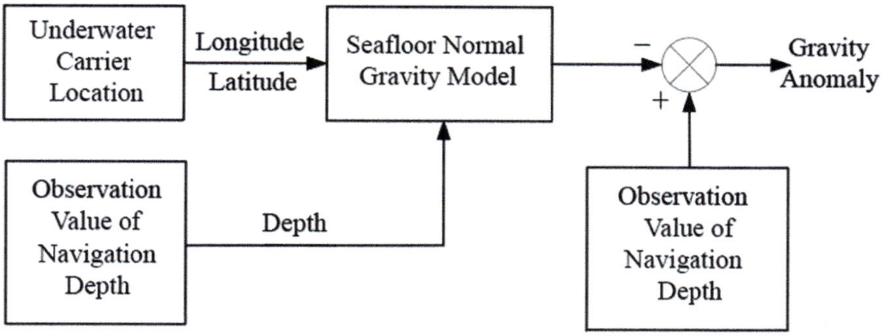

Fig. 2.6 Calculation of free air gravity anomaly on the seafloor for navigation

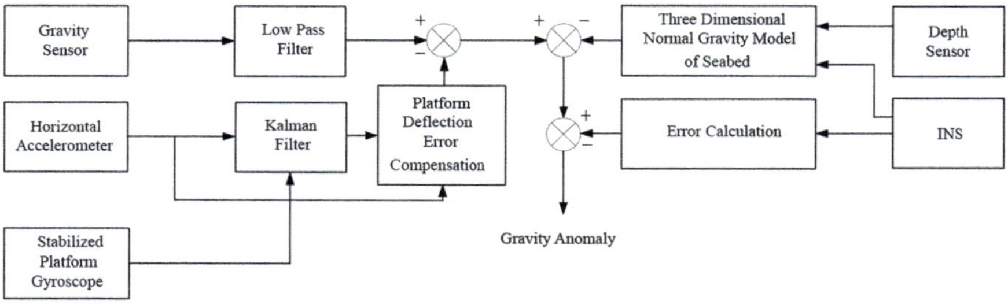

Fig. 2.7 Principle block diagram of gravity data processing

$$\delta g = g_{obs} - g_{pn} \tag{2.24}$$

where g_{obs} is the gravity observation value, g_{pn} is the normal gravity value at this point.

Sections 2.2, 2.3 and 2.4 of this chapter are connected to form the overall framework of gravity data processing, as shown in Fig. 2.7.

2.4 Fitting Function Approximation Method of Gravity Anomaly Random Field

There are many methods of fitting function approximation, commonly used are, surface interpolation approximation, polynomial fitting method, multi-surface function method, regression approximation method, moving surface method, fast Fourier transform method, etc. They can both locally approximate the data and fit the whole map in its entirety. The difference in calculation methods lies in the construction of different fitting functions [12–15].

2.4.1 Mathematical Description of the Fitting Function Approximation Problem

For the gravity anomaly data map, the corresponding local coordinates are established, the minimum latitude and longitude points of the map are used as the local coordinate origin, and the northeast direction corresponds to the positive direction of the x and y axes of the local coordinates.

Assume that the minimum cell grid of the map is a square, and take its unit spacing as the unit of length. Any point on the map is denoted by $X(x, y)$, and the corresponding gravity anomaly data is g_x.

The fitting function method assumes that in the fitted region

$$g_x = F(X, \alpha) \tag{2.25}$$

where, $F(\cdot)$ is the gravity anomaly fitting function. $\alpha = \begin{bmatrix} \alpha_1 \; \alpha_2 \ldots \alpha_n \end{bmatrix}$ is the coefficients to be determined for the function.

Different forms of $F(\cdot)$, corresponding to different methods.

1. Typical fitted functional form
(1) Moving Surface Method

The moving surface is approximated by a binary polynomial as a fitting function to a numerical field. The generic form of $F(\cdot)$ in Eq. (2.25) is

$$F(x, y, \alpha) = \sum_{i=1}^{n} \alpha_i x^k y^l \tag{2.26}$$

where $k + 1 = 0, 1, \ldots, m; n = (m + 1)(m + 2)/2$.

It can also be expressed in the form of a matrix

$$F(x, y, \alpha) = \begin{bmatrix} 1 \; x^1 y^0 \cdots x^k y^l \; x^0 y^m \end{bmatrix}_{1\times n} \begin{bmatrix} \alpha_1 \\ \alpha_2 \\ \vdots \\ \alpha_i \\ \alpha_n \end{bmatrix}_{n\times 1} \tag{2.27}$$

(2) Multifaceted function method

The multi-surface function fit was proposed by Hardy in 1977 and is based on the theory that any rounded mathematical surface can always be approximated with arbitrary accuracy by the sum of a series of regular mathematical surfaces.

The $F(\cdot)$ constructed by the polyhedral function method can be uniformly expressed as

$$F(X, \alpha) = \sum_{j=1}^{m} \alpha_j f\left(X, X_j\right) \tag{2.28}$$

X_j constructs the node coordinates $\left(x_j, y_j\right)$ of the polyhedral function. $f(\cdot)$ is the kernel function. The kernel function is constructed with the distance between two points as variables, and for convenience note $f(x_k, y_k)$ as f_{kj}.

Different forms of kernel functions can be selected for different numerical field characteristics, and typical kernel functions are:

Hardy's hyperbolic form

$$f(d) = \sqrt{d^2 + \delta^2} \tag{2.29}$$

where, $d = \sqrt{\left(x - x_j\right)^2 + \left(y - y_j\right)^2}$, δ is the rounding factor.

Arthur's exponential form

$$f(d) = e^{-\frac{2.5d^2}{a^2}} \tag{2.30}$$

The common form used in the gravity anomaly fit is

$$f(d) = (1 + d^2)^{-\frac{1}{2}} \tag{2.31}$$

Chunxi Guo et al. proposed the corresponding improved kernel function form

$$f(d, \Delta h) = (d^2 + (h_x - h_j)^2 + s_{jmin}^2)^{-\frac{1}{2}} \tag{2.32}$$

$s_{j\min}$ is the plane distance between point j and the nearest point x; h_x and h_j are the topographic elevation values of the interpolated point and the reference point, respectively.

In the absence of elevation data, the elevation-related term in Eq. (2.32) is omitted.

$$f(d) = (d^2 + s_{\text{jmin}}^2)^{-\frac{1}{2}} \tag{2.33}$$

(3) Two-tone and spline interpolation

The bimodal and spline interpolation can be considered as a form of polyhedral function fitting. The application of splines in modeling is an important area of computer graphics, and this paper does not explore its underlying theory. Its kernel function is the Green function.

The double harmonic spline assumes that the fourth-order derivative of the displacement at all points except at the nodes under the point action is zero.

$$\frac{d^4\phi}{dx^4} = 6\delta(x) \tag{2.34}$$

The n-node m-dimensional surface fitting function w.

$$w(x) = \sum_{j=1}^{n} \phi_m(x - x_j) \tag{2.35}$$

φ_m is the m-dimensional Green function. The different dimensional b_i-adjustment and Green functions are shown in Table 2.3.

Thus, a two-dimensional surface fitting function can be obtained

$$F(d, \alpha) = \sum_{j=1}^{N} \alpha_j |d|^2 (2\ln|d| - 1) \tag{2.36}$$

Obviously, it is formally consistent with the polyhedral function fit, corresponding to Eq. (2.37), where the kernel function $f(\cdot)f(\cdot)$

$$f(d) = |d|^2 (2\ln|d| - 1) \tag{2.37}$$

2. Determination of the coefficient to be determined

The coefficients to be determined by the gravity data are known at the point $X(x, y)$. There are m known points of gravity data, which in this study are all points on an equally spaced grid in the determined area. Let the boundaries of the region be:$xb < xi < xs$, $yb < yi < ys$; the map unit grid spacing is Δ. The set of gravity data points A is thus:

$$A = \{XX_i - X_j = [n\Delta\ m\Delta];\ x_b < x_i < x_s, y_b < y_i < y_s n, m \in N\} \tag{2.38}$$

where N is the set of integers.

The gravity data corresponding to the point $Xk(xk, yk)$ is denoted as g_k. The M equations obtained from Eq. (2.39) are expressed in the form of a matrix.

Table. 2.3 Bi-tonal and green functions

Dimension (m)	Green function φm(x)	Green function gradient ∇φm(x)
1	$\|x\|^3$	$x\|x\|$
2	$\|x\|^2(\ln\|x\|-1)$	$x(2\ln\|x\|-1)$
3	$\|x\|$	$x\|x\|^{-1}$
4	$\ln\|x\|$	$x\|x\|^{-2}$
5	$\|x\|^{-1}$	$-x\|x\|^{-3}$
6	$\|x\|^{-2}$	$-2x\|x\|^{-4}$
m(>4)	$\|x\|^{4-m}$	$(4-m)x\|x\|^{2-m}$

$$G = FA \tag{2.39}$$

where,

$$G = \begin{bmatrix} g_1 \ g_2 \cdots g_k \cdots g_m \end{bmatrix}'$$

$$F = \begin{bmatrix} f_{11} \ f_{12} \cdots f_{1m} \\ f_{21} \ f_{22} \cdots f_{2m} \\ \vdots \ \ \vdots \ \ \vdots \ \ \vdots \\ f_{m1} \ f_{m2} \cdots f_{m3} \end{bmatrix}$$

$$A = \begin{bmatrix} \alpha_1 \ \alpha_2 \cdots \alpha_m \end{bmatrix}'$$

where $k = 1, 2, \ldots m$. All control points are used as centroids of some kernel function, i.e., containing m kernel functions with m coefficients to be determined.

Then, the pending coefficients can be found:

$$A = F^{-1}G \tag{2.40}$$

If a moving surface fit is used, such as a quadratic surface, the corresponding matrix F becomes

$$F = \begin{bmatrix} 1 & x_1 & y_1 & x_1^2 & x_1y_1 & y_1^2 \\ 1 & x_i & y_i & x_i^2 & x_iy_i & y_i^2 \\ \vdots & \vdots & \vdots & \vdots & \vdots & \vdots \\ 1 & x_m & y_m & x_m^2 & x_my_m & y_m^2 \end{bmatrix}$$

Usually $m > 6$, so that Eq. (2.41) is modified to

$$A = (F^TF)^{-1}F^TG \tag{2.41}$$

The gravity anomaly can be obtained by interpolating the coordinates of the location of the point by substituting the corresponding coefficients to be determined into Eq. (2.42).

$$F(x_k) = \begin{bmatrix} 1 \ x_k \ y_k \ x_k^2 \ x_ky_k \ y_k^2 \end{bmatrix} F^{-1}G \tag{2.42}$$

Or

$$F(x_k) = \begin{bmatrix} f_{k1} \ f_{k2} \cdots f_{kn} \end{bmatrix} F^{-1}G \tag{2.43}$$

2.4.2 *Physical Significance of the Fitted Function Approximation Method*

To further understand its mathematical meaning, Eq. (2.44) is treated as follows.

Letting $F_k = \left[f_{k1}\, f_{k2}\, \cdots\, f_{km} \right]$,

$$F(x_{\dot{k}}) = F_{\dot{k}} F^{-1} G \tag{2.44}$$

Then,

$$P = F_k F^{-1} \tag{2.45}$$

Thus

$$F(x_k) = PG \tag{2.46}$$

Expressing Eq. (2.46) as a polynomial

$$F(x_k) = \sum_{i=1}^{m} P_i G_i \tag{2.47}$$

Equation (2.47) shows that: the value of the interpolated points is the "weighted average" of the data at each control point, and the weight P_i is related to the kernel function, the distribution of control points, and the location of the interpolated points.

The kernel function is a function with distance as the independent variable, so its distance selection characteristics will affect the interpolation accuracy. The kernel function curve is shown in Fig. 2.8.

Comparing Eqs. (2.29, 2.30, 2.33, and 2.37) shows that the first three Eqs. (2.29 and 2.30) contain a smooth factor a, and Eq. (2.33) has corresponding to the smooth factor, and Eq. (2.48) can be changed to the following form.

$$f(d) = (d^2 + a^2)^{-\frac{1}{2}} \tag{2.48}$$

Letting $s_d = d/a$, then the corresponding equations are

$$f(s_d, a) = a\sqrt{s_d^2 + 1} \tag{2.49}$$

$$f(s_d) = e^{-2.5 s_d^2} \tag{2.50}$$

$$f(s_d, a) = \frac{a}{\sqrt{s_d^2 + 1}} \tag{2.51}$$

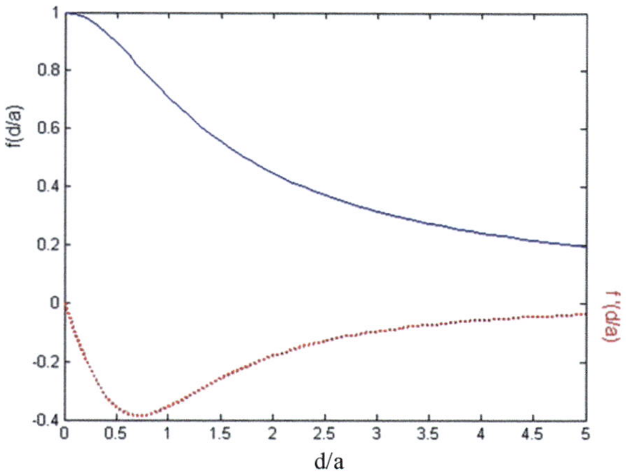

(a) Gravity model functions and their derivative curve

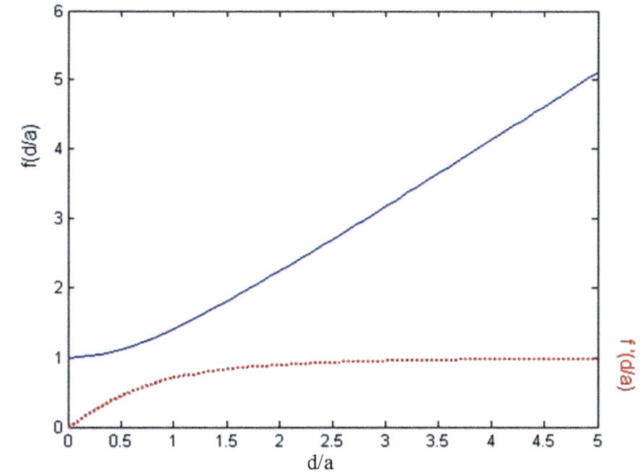

(b) Hardy kernel functions and their derivative curve

Fig. 2.8 Kernel function and its derivative curve

Figure 2.8a and b represent the $\left(s_d^2+1\right)^{\frac{1}{2}}$ and $\left(s_d^2+1\right)^{-\frac{1}{2}}$ function properties.

Figure 2.9 represents the functional characteristics of Eqs. (2.49 and 2.51), respectively.

As can be seen from the two plots, with sd as the variable, Fig. 2.8a and c are similar to the low-pass filter, where the function value decays when sd takes large values (high frequencies), and when the smooth factor is fixed, the effect of the points at the corresponding distance from it on the interpolated points (the weight factor also decays).

Figure 2.8b and d can then be considered as high-pass filtering.

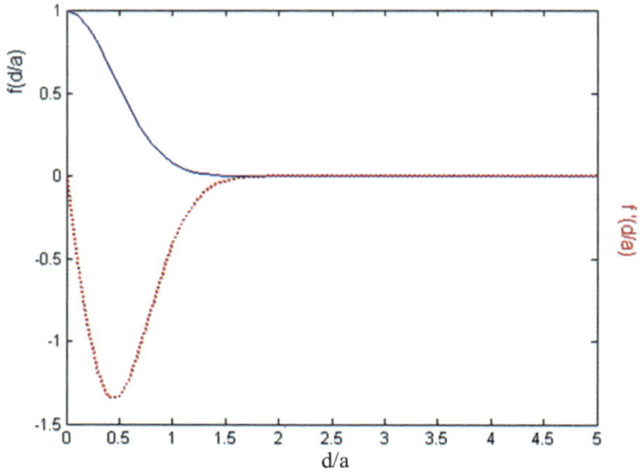

(c) Exponential model kernel functions and their derivative curve

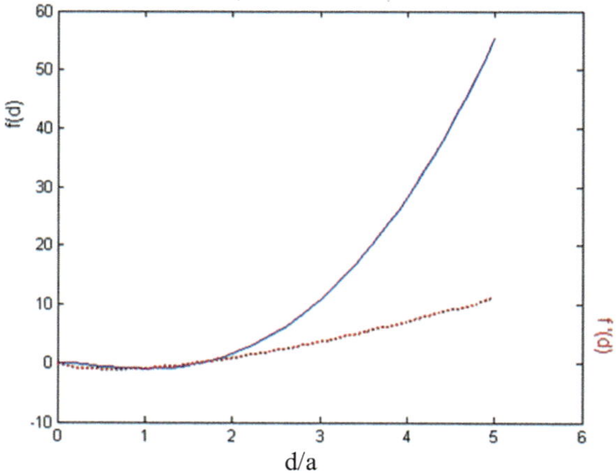

(d) Biharmonic and spline functions and their derivative curve

Fig. 2.8 (continued)

2.5 Kriging Interpolation Method for Gravity Anomaly-Oriented Reference Maps

Kriging interpolation method considers physical observations of spatial points (gravity anomalies) as random variables, and it is an unbiased and optimal interpolation method to use known observations of one or several variables in space to infer and recognize its(their) laws and characteristics that vary with position in space, especially their values at points outside the observation. It describes the statistical characteristics of two-dimensional random fields with spatial variational structure, keeping the overall statistical characteristics of the interpolated data unchanged while

Fig. 2.9 Relationship between kernel function and distance and photoconformity factor

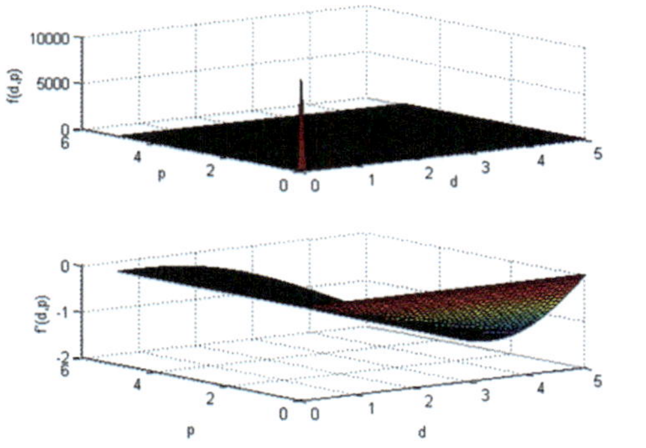

(a) Gravity model kernel functions and their derivatives for surface

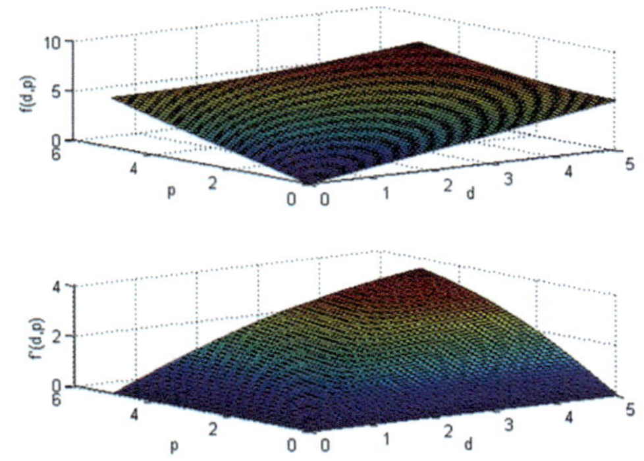

(b) Hardy model kernel functions and their derivatives for surface

describing the local details of the surface, and it can be used as a reference map interpolation method for gravity anomalies under moderate scale conditions [16].

Translated with www.DeepL.com/Translator (free version) The Krige interpolation method was first proposed by D. G. Krige, a South African gold mining engineer, in the 1950s, and was theoretically enhanced by Matheron, a French mathematician, in the 1960s, to form the geological statistics. With the deepening of theoretical research in geostatistics, a variety of Kriging interpolation methods have been developed, such as ordinary Kriging, pan-Kriging, cooperative Kriging, indicated Kriging and Bayesian Kriging. The study in this section is carried out based on the ordinary Kriging method.

2.5.1 Basic Principles of Kriging Interpolation

Let the gravity anomaly map under study be in region A and the coordinates (x, y) of point P, then:

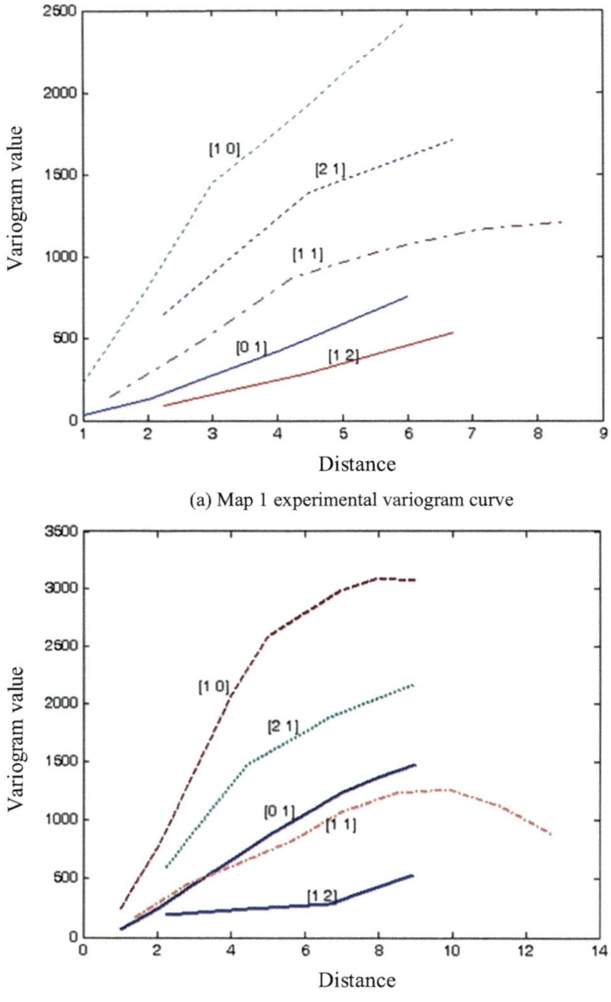

(a) Map 1 experimental variogram curve

(b) Map 2 experimental variogram curve

Fig. 2.10 Results of experimental variation function of gravity anomaly map

$$A = \{ P(x, y) | x_a \le x \le x_b; y_a \le y \le y_b \} \tag{2.52}$$

If $P_0(x_a, x_b)$ is defined to be the point (0, 0), then

$$P' = P - P_0 \tag{2.53}$$

$z(P_i)$ denotes the gravity anomaly stochastic function at control point $P_i(i = 1, 2, ..., n)$, and $z(P_i)$ is a realization of the gravity anomaly at P_i. The gravity anomaly estimates the interpolation point P is $z^*(P)$.

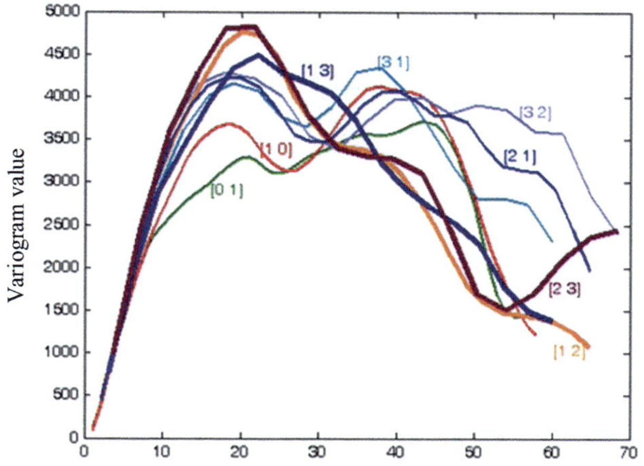

(c) Map 3 experimental variogram curve

Fig. 2.10 (continued)

$$z^*(P) = \lambda_0 + \sum_{i=1}^{n} \lambda_i z(P_i) \quad (2.54)$$

λ_i are the weighting coefficients to be determined and they make the estimates satisfy the minimum variance criterion, i.e.

$$E\left\{\left[z(P) - z^*(P)\right]^2\right\} = \min \quad (2.55)$$

Therefore, Kriging interpolation is a linear minimum variance method. From the estimation theory, it is known that its weighting coefficients can be expressed in terms of the first-order moments and second-order moments of the gravity anomaly random function $z(P_i)$ at the control point. That is, in the scope of linear least variance estimation, it is not necessary to examine the probability distribution functions of the random variables as long as their first-order moments and second-order moments are involved.。

Different forms of Kriging interpolation are derived from the different properties of the first- and second-order moments of the random variable. There are three cases of first-order moments of the random function as follows.

Mathematical expectation of the gravity anomaly $m(P) = E[z(P)]$ for all P known and allowed $m(P) = m(P')$.

Mathematical expectation independent of the position P, but unknown, considering the random function to be first-order smooth.

Mathematical expectations are both unknown and non-smooth.

The above three conditions correspond to, respectively, simple Kriging, ordinary Kriging and pan-Kriging. considering the characteristics of the gravity anomaly reference map compound condition II, the ordinary Kriging method is used as the basis for the interpolation study.

Kriging is a linear minimum variance estimator, then it is also an unbiased estimator, in that the random function satisfies the condition A, with

$$E\big[Z(P) - Z^*(P)\big] = 0 \tag{2.56}$$

Substituting (2.54) into (2.56), after a simple transformation, we get

$$\lambda_0 = m_0 - \sum_{i=1}^{n} \lambda_i m_i \tag{2.57}$$

Letting $\lambda 0 = 0$ in Eq. (2.54), and because $m_0 = m_1 = m_2 = \ldots = m_n = m$. From Eq. (2.57) it follows that

$$\sum_{i=1}^{n} \lambda_i = 1 \tag{2.58}$$

$$Z^*(P) = \sum_{i=1}^{n} Z(P_i) \tag{2.59}$$

The ordinary Kriging method objective function is expanded as before

$$E\Big\{\big[Z(x_0) - Z^*(x_0)\big]^2\Big\} = E\left\{\left[\sum_{i=0}^{n} \alpha_i (Z(x_i) - m_i)\right]^2\right\} = \sum_{i=0}^{n} \sum_{i=0}^{n} \alpha_i \alpha_j C\big(P_i, P_j\big) \tag{2.60}$$

It can be seen that this is a problem of finding the extremum under the equation constraint. Constructing a new objective function using Lagrange multiplier method

$$F = E\Big\{\big[Z(P) - Z^*(P)\big]^2\Big\} - 2\mu\left(\sum_{i=1}^{n} \lambda_i - 1\right) \tag{2.61}$$

From extreme value theory, it is obtained that the system of equations for solving $\lambda_i \lambda_i$

$$\sum_{i=1}^{n} \lambda_i C\big(P_i, P_j\big) - \mu = C\big(P, P_j\big)$$

$$\sum_{i=1}^{n} \lambda_i = 1 \tag{2.62}$$

Namely, the set of ordinary Kriging equations.
Ordinary Kriging estimation variance:

$$\sigma_{ok} = C(P, P) - \sum_{i=1}^{n} \lambda_i C(P, P_i) + \mu \tag{2.63}$$

2.5.2 *Spatial Variation Function for Gravity Anomaly Reference Maps*

A. Definition of variation function

In the set of Kriging equations, it is necessary to calculate the covariance, which is defined as:

$$C(P_i, P_j) = E\{[Z(P_i) - E(Z(P_i))][Z(P_j) - E(Z(P_j))]\} \tag{2.64}$$

In gravity anomaly map interpolation, what is known is the gravity anomaly values at the map grid nodes, and there is no series of values at the nodes reflecting the stochastic function $Z(P)$, i.e., only one realization of this stochastic function at the grid nodes. Therefore, a gravity anomaly covariance in the strict sense cannot be obtained.

Ordinary Kriging assumes that $Z(P)$ is second-order smooth, namely, there is a covariance $C(D)$ between each pair of random variables $Z(P)$ and $Z(P + D)$.

$$C(D) = E\{[Z(P + D) - E(Z(P + D))][Z(P) - E(Z(P))]\} = E[Z(P + D)Z(P)] - m^2 \tag{2.65}$$

where, D is the vector of distances between two points.

The system of ordinary Kriging equations satisfying the second-order smoothness condition can be written in the following form.

$$\begin{cases} \sum_{i=1}^{n} \lambda_i C(P_i - P_j) - \mu = C(P_j - P), j = 1, 2, \ldots, n \\ \sum_{i=1}^{n} \lambda_i = 1 \end{cases} \tag{2.66}$$

However, (2.66) still requires that E[Z(P)] is known and therefore defines the variogram $\gamma\left(P_i, P_j\right)$.

$$\gamma\left(P_i, P_j\right) = \frac{1}{2}E\left\{\left[Z(P_i) - Z\left(P_j\right)\right]^2\right\} \tag{2.67}$$

It is related to $C\left(P_i, P_j\right)$ in the second-order smooth condition as

$$\gamma\left(P_i, P_j\right) = C(0) - C\left(P_i, P_j\right) \tag{2.68}$$

From (2.68), we can obtain that the set of ordinary Kriging equations expressed in terms of variational functions

$$\begin{cases} \sum_{i=1}^{n} \lambda_i \gamma\left(P_i - P_j\right) + \mu = \gamma\left(P_j - P\right), j = 1, 2, \ldots, n \\ \sum_{i=1}^{n} \lambda_i = 1 \end{cases} \tag{2.69}$$

The basis of such ordinary Kriging is the determination of the variation function. The variation function is determined in two steps: first, the experimental variation function is found using the known gravity anomaly data from the map; second, the theoretical variation function is obtained using the experimental variation function.

B. Determination of Experimental Variation Function

Definition of the experimental variance function.

Let h be a vector with mode $r = |h|$ and direction a. If there exist N_h pairs of observed data points separated by vector h, then the experimental variation function $\gamma^*(h)$ corresponding to vector h in the direction a can be approximated as:

$$\gamma^*(h) = \gamma^*(r, a) = \frac{1}{2N_h}\sum_{i=1}^{N_h}[Z(P_i + h) - Z(P_i)]^2 \tag{2.70}$$

The map data are finite, and when the mode of h (i.e., the distance between pairs of data points) is close to the length of the map interval, the number of data pairs Nh is small and the confidence of the obtained experimental variance function becomes low. For this reason, the concept of distance tolerance $\varepsilon(r)$ is introduced to increase the number of data pairs.

If the distance between two data points $D_{ij} = \left|P_i - P_j\right|$, and the difference between h is less than $\varepsilon(r)$.

$$\left|D_{ij} - h\right| \leq \varepsilon(r) \tag{2.71}$$

Then the data pair can be used in the statistics of the data pair with interval (Fig. 2.10)h.

C. Fitting of the theoretical variance function

The experimental variation function is a discrete value obtained on the basis of a grid data point pair, and a continuous variation function can be obtained by fitting the method. A typical theoretical model of the variation function is shown in Table 2.4.

Figure 2.16 shows the experimental variance function result curves for a map. (c) plot is the result of a 60 × 60 array map; (a) and (b) are the calculated results of its two sub-maps, which are 8 × 8 and 11 × 11 arrays, respectively.

It can be seen that the gravity anomaly map exhibits an isotropic variance function structure, so the average variance function is found before the theoretical variance function is obtained.

$$\gamma^*(h) = \left[\sum_{k=1}^{K} N_k(h)\gamma_k^*(h, \alpha_k)\right] \Big/ \left[\sum_{k=1}^{K} N_k(h)\right] \tag{2.72}$$

where the subscript k denotes the k_{th} variation function (distance pair is h and direction is α_k). K is the number of variation functions that can be averaged.

Due to the limitation of the map grid, different unit spacing can be obtained in different directions, and the unit spacing in some directions is shown in Table 2.5.

The variance functions of pairs of points with the same distance can only be averaged. Figure 2.11 shows the results of averaging the variance functions. Figure 2.12 also gives the results of the fit with polynomials.

Figure 2.12 and Table 2.6 show the results of the fits obtained using different theoretical models.

Table 2.4 Theoretical model of variation function

Spherical model	$\gamma(h) = \begin{cases} 0 & h = 0 \\ C_0 + C\left(\frac{3}{2}\cdot\frac{h}{a} - \frac{1}{2}\cdot\frac{h^3}{a^3}\right) & 0 < h \le a \\ C_0 + C & h > a \end{cases}$
Index Model	$\gamma(h) = \begin{cases} 0 & h = 0 \\ C_0 + C\left(1 - e^{-\frac{h}{a}} \right. & h > 0 \end{cases}$
Gaussian model	$\gamma(h) = \begin{cases} 0 & h = 0 \\ C_0 + C\left(1 - e^{-\frac{h^2}{a^2}}\right) & h > 0 \end{cases}$
Linear ab-initio model	$\gamma(h) = \begin{cases} C_0 & h = 0 \\ Ah & 0 < h \le a \\ C_0 + C & h > a \end{cases}$
Linear ab-initio-free model	$\gamma(h) = \begin{cases} C_0 \; h = 0 \\ Ah \; h > 0 \end{cases}$

Table 2.5 Partial direction unit spacing

Direction	Unit spacing	Direction	Unit spacing	Direction	Unit spacing
[0 1]	1	[1 2]	2.236	[3 1]	3.162
[1 0]	1	[1 3]	3.162	[2 3]	3.6055
[1 1]	1.4147	[2 1]	2.236	[3 2]	3.6055

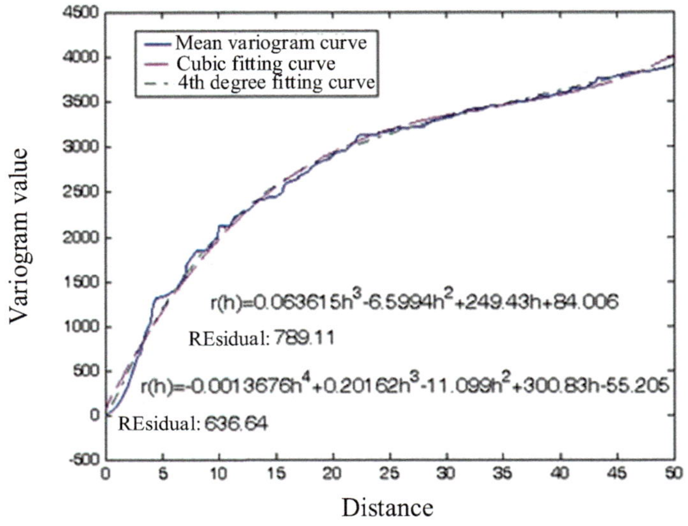

Fig. 2.11 Mean variance function and its curve fitting results

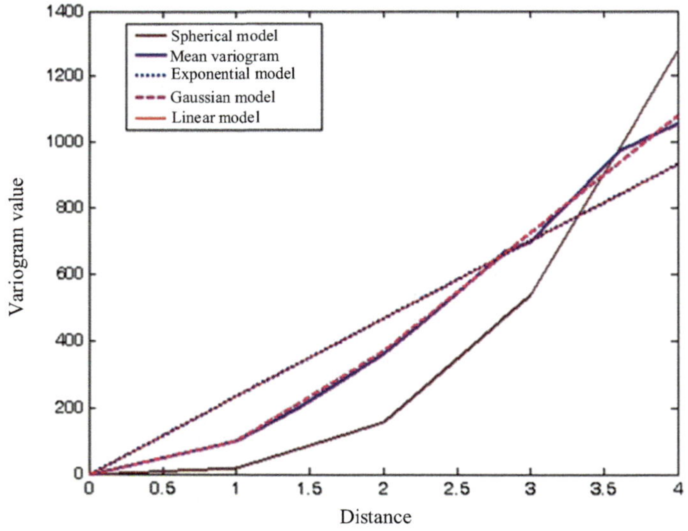

Fig. 2.12 Variance function theoretical model fit

Table 2.6 Parameters of the theoretical model of variation function

	Spherical model	Index Model	Gaussian model	Straight line model
a	0.14933	2.087×10^8	4.2093	–
c	−0.13382	4.8764×10^{10}	1824.1	233.66

D. Theoretical variational function fitting for gravity anomaly-oriented reference maps

In the traditional fitting of the theoretical variance function, the average experimental variance function is a further treatment of the isotropic spatial data, and the average variance function obtained on the basis of the experimental variance function is limited by the experimental variance function. This is because the isotropic variance function only needs to consider the statistical characteristics of data pairs with the same distance, while the experimental variance function defined in (2.70) includes the directional restrictions of the spatial data.

Redefining from isotropic characteristics, the (average) experimental variance function

$$\gamma^*(h) = \frac{1}{2N_h}\sum_{i=1}^{N_h}[Z(P_j) - Z(P_i)]^2 \|P_j - P_i\| - h| \leq \varepsilon \tag{2.73}$$

N_h is the number of all point pairs on the map grid that satisfy a distance of $h \pm \varepsilon$.

In Fig. 2.13, method 1 is the average variance function obtained by the traditional method; method 2 is the improved method. The theoretical fitting method for the mean variation function is the same. Table 2.7 shows the calculation results of the theoretical variation function.

E. Conclusion

Conclusion I.

The improved mean variance function curve is smoother when the distance is greater than a certain value ($h\varepsilon$) when the map data behave isotropically.

The minimum distance difference on the grid is shown in Fig. 2.14 as:

$$\Delta_{h\min} = \sqrt{h^2 + 1} - h \tag{2.74}$$

The grid spacing is normalized. When $\Delta_{h\min} \leq \varepsilon$, the number of point pairs increases and the variance function curve is smooth, so the lower limit of the smoothing distance hε satisfies

$$h_\varepsilon \geq \frac{1}{2\varepsilon} \tag{2.75}$$

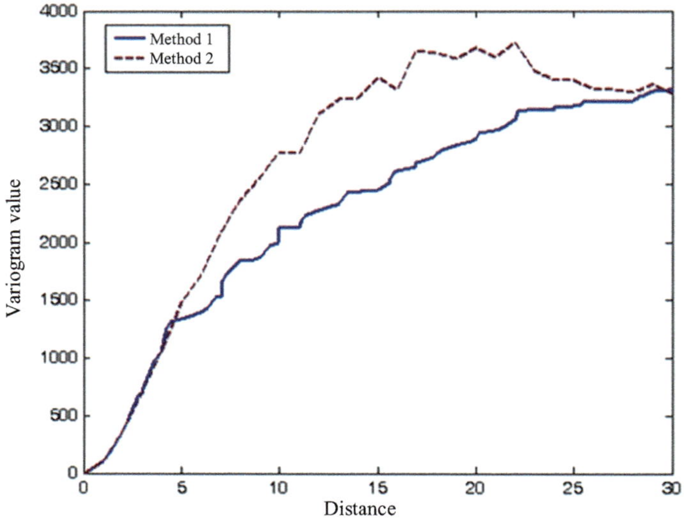

Fig. 2.13 Improvement results of the mean variance function

Table 2.7 Parameters of the theoretical model of variation function

	Spherical model	Index Model	Gaussian model	Straight line model
a	0.44706	2.9374×10^8	-6.1219	–
c	−0.68278	8.0011×10^{10}	2903	272.38

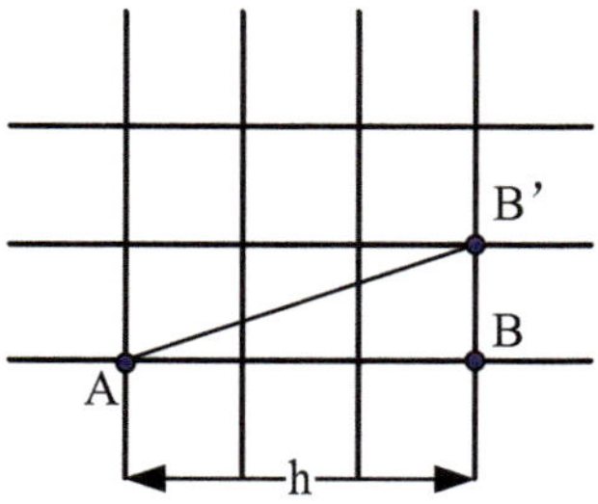

Fig. 2.14 Schematic diagram of grid point-to-distance

when the mean variance function is improved. The higher terms of the denominator of Eq. (2.75) are omitted.

Conclusion II.

The theoretical model fitting results show that the Gaussian model fits best near the zero point, but the Gaussian model is sensitive to changes in the data; as can be seen from the model parameters in Tables 2.6 and 2.7, the Gaussian model has the most significant parameter changes.

The parameters of the exponential model are too large, and near the zero point its theoretical model is close to the linear model, and the difference between the theoretical curve obtained by the spherical model and the average variance function curve is the largest.

The following conclusions were drawn from the experimental results.

The smaller the interpolation radius, the smaller the influence of the variance function model on the interpolation results. This is because the fit of the theoretical variance function model to the experimental variance function is close near the zero point; while other experiments show that the variance function model parameters have a more pronounced effect on the interpolation accuracy when the interpolation radius is small.

The interpolation radius increases while the map size remains the same, then the resulting edge effect (not enough points at the edge of the map to participate in the interpolation) leads to a decrease in the interpolation accuracy; at this point, changes in the model parameters do not significantly improve the interpolation accuracy.

The interpolation accuracy is not only related to the choice of the interpolation method, but also to the statistical characteristics and the change law of the map. In the interpolation process, it is found that the location where the maximum value of the interpolation error appears does not change due to the interpolation method, and even in small interpolation radius, its value does not change.

2.6 Fractal Reconstruction of Gravity Anomalies with Airports Under Large Scale Conditions

In regions where gravity anomaly data are scarce, the accuracy of both the fitting function and Kriging method is greatly reduced as the number of interpolations between adjacent points increases. In this paper, we consider the reconstruction of the gravity anomaly random data field under such conditions using the fractal approach at large scale conditions. This reconstructed simulation provides a closer approximation to the real gravity environment for the underwater GAINS system [17].

From the beginning, fractals conveyed information about the description of irregular and fragmented natural shapes. In 1980, Goodchild considered the use of fractal theory to analyze geomorphic surfaces; in 1991, Clark and Schweitzer defined fractal estimates of topographic surfaces with better robustness; starting in 1991, Kenichi Arakawa and Eric Krotkov studied the "reconstruction of natural terrain using fractals", using FBM (Fractal Brownian Motion) to estimate the fractal structure of natural terrain on an irregular set of sampling points and construct natural terrain based on it (1996). Naokazu Yokoya et al. gave an interpolation method based on the FBM model, on the basis of which it has been continuously refined.

In the terrain navigation research, Jixian Zhang and Kang Su introduced fractal dimensions to systematically evaluate terrain in their studies, respectively; subsequently, Ning Zou and Jian Liu et al. (1999) and Jun Li and Xin Yang et al. (2001) studied and explored fractals in terrain classification techniques and terrain environment understanding, respectively.

This section discusses the method of FBM fractal interpolation for reconstructing gravity anomaly random fields under large scale conditions, and the normalized length units discussed earlier are followed in the study.

2.6.1 *Definition of Fractional Dimension and Its Estimation*

Brownian motion (BM, Brownian Motion) was discovered by the British botanist R. Brown in 1982, and the mathematical model of BM was established by the German mathematician N. Wiener in 1923.BM is a stochastic process $B(t)$ satisfying the following two conditions (2.76 and 2.77).

$$B(0) = C \tag{2.76}$$

$$B(b) - B(a) \sim N\left(0, (b - a)\sigma^2\right), a \leq b \tag{2.77}$$

where C is a constant; the increments between any two points of the stochastic process ($B(b) - B(a)$ and $B(c) - B(b)$,$a \leq b \leq c$) are independent of each other.

$B(t)$ is written in the following form

$$B(rt) = r^{\frac{1}{2}} B(t) \tag{2.78}$$

Obviously, it has a self-consistent nature.

Mandelbrot proposed the FBM function, from which the BM process is generated.

$$B_H(0) = C \tag{2.79}$$

$$B_H(t) - B_H(0) = C\left[\int_{-\infty}^{0} \left((t - s)^{H-0.5} - (-s)^{H-0.5}\right) dB(S) + \int_{0}^{t} (t - s)^{H-0.5} dB(S) \right] \tag{2.80}$$

Penland defines the statistical self-similarity feature of the FBM function

$$P_r\left\{\frac{f(t + \Delta t) - f(t)}{\Delta t^H} < x\right\} = g(x) \tag{2.81}$$

$f(t)$ is the FBM function; $g(x)$ is the cumulative distribution function ($g(x) \sim N(0, \sigma 2)$); H is the self-affine parameter, which is related to the fractional dimension

D of $f(t)$ as

$$D = n + 1 - H \tag{2.82}$$

n is the geometric topological dimension, and in the gravity anomaly map,$n = 2$.

From Eq. (2.81), it follows that

$$E\left[\frac{f(t+\Delta t) - f(t)}{\Delta t^H}\right] = E\left[g(x)\right] \tag{2.83}$$

Then

$$E\left[|g(x)|\right] = 2\int_0^{\infty} x \frac{1}{(2\pi)^{\frac{1}{2}}\sigma} e^{-\frac{x^2}{2\sigma^2}} dx = \frac{2\sigma}{\sqrt{2\pi}} = C \tag{2.84}$$

$$E\left[|f(t+\Delta t) - f(t)|\right] = C \cdot \Delta t^H \tag{2.85}$$

Taking the logarithm of both sides

$$\log E\left[|f(t+\Delta t) - f(t)|\right] = H \log\|\Delta t\| + \log C \tag{2.86}$$

With $\log E\left[|f(t + \Delta t) - f(t)|\right]$ as the dependent variable and $\log ||\Delta t||$ as the independent variable, Eq. (2.94) is the linear equation (both H and C are constants).

In the estimation of the fractional dimensionality of a two-dimensional random data field, Δt is replaced by the distance ΔX ($\sqrt{\Delta x^2 + \Delta y^2}$) between two points on the map.

$$\log E\left[|f(X + \Delta X) - f(X)|\right] = H \log\|\Delta X\| + \log C \tag{2.87}$$

The $E\left[|f(X + \Delta X) - f(X)|\right]$ of the increments at different distances can be obtained statistically, and finally, the linear equations can be fitted by least squares to obtain the H and σ values of the map by simple transformation.

2.6.2 Gravity Anomaly Map Fractional Dimension Estimation

The estimation of the fractional dimension of the gravity anomaly map is also performed based on Eq. (2.94).

The $E\left[|f(X + \Delta X) - f(X)|\right]$ at different ΔX needs to be obtained in the calculation. Obviously, on the basis of the gravity anomaly map array, the ΔX that can be obtained is discrete, and its value is related to the map grid unit distance $\left[\Delta x_u, \ \Delta y_u\right]$ and the scale of the map array $\left[m_x, \ m_y\right]$, as shown in Fig. 2.15.

The distance that can be taken is ΔX

Fig. 2.15 Schematic diagram of the map grid

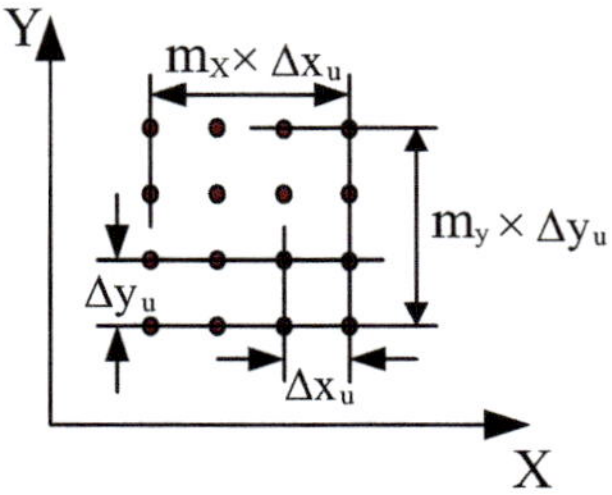

Fig. 2.16 Fractal characteristic curve of gravity anomaly map

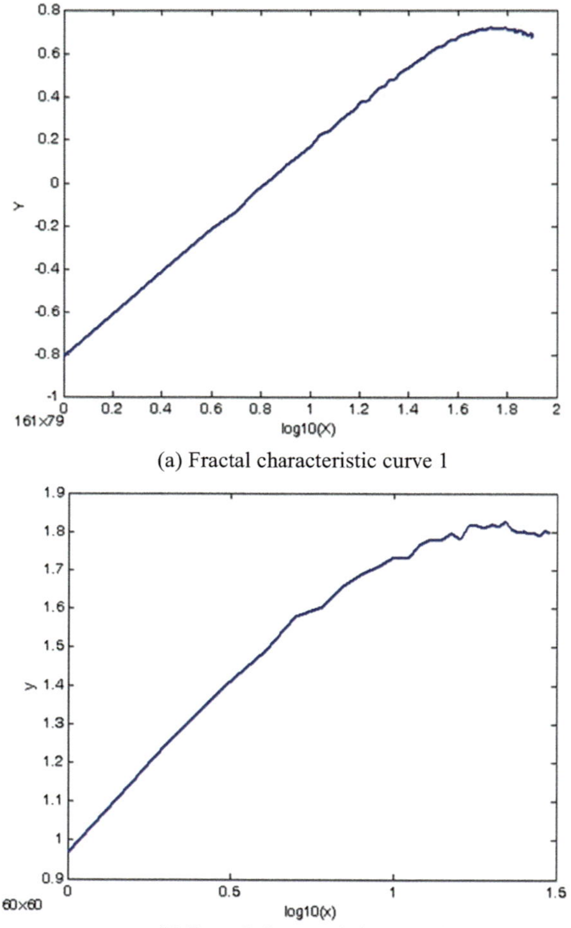

(a) Fractal characteristic curve 1

(b) Fractal characteristic curve 2

Table 2.8 Calculated results of fractal characteristics

	Fitting distance	H	D_f	Log C	σ	Residuals
161 × 79	15	0.97609	2.02391	−0.80349	0.1970	0.018934
60 × 60	5	0.86986	2.13014	0.97461	11.8214	0.016769

$$\Delta X = \sqrt{(i \cdot u_x)^2 + (j \cdot u_y)^2} \tag{2.88}$$

$i = 1, 2, \ldots, m_x; j = 1, 2, \ldots, m_y$; Only the case $u_x = u_y = u$ is discussed here.

Obviously, similar to the spatial variational structure analysis of Kriging mentioned before, ΔX increases, and the number of statistical points that can be involved in the calculation of $E\left[|f(X + \Delta X) - f(X)|\right]$ decreases sharply. Based on the same idea, setting the distance tolerance can better reflect the statistical characteristics of the map.

That is, let ε_x be the distance tolerance, when the distance between two points on the map satisfies

$$|\Delta X_p - \Delta X| < \varepsilon_x \tag{2.89}$$

Then, the increment of gravity anomaly at two points can be accounted for in the calculation of $E\left[|f(X + \Delta X) - f(X)|\right]$ corresponding to ΔX where ΔX_p is the actual grid distance between the two points.

According to this method, the number of sub-dimensions was calculated for the two gravity anomaly maps, and the results are shown in Fig. 2.16.

The results obtained by fitting a straight line with the least squares method are shown in Table 2.8.

2.6.3 FBM Fractal Interpolation

The basic steps for generating a new small-scale gravity anomaly map using FBM fractal interpolation are shown in Fig. 2.17.

The fractal features can be fully represented only within a limited scale range, which is called the scale-free interval $[\Delta X_{min}, \Delta X_{max}]$.

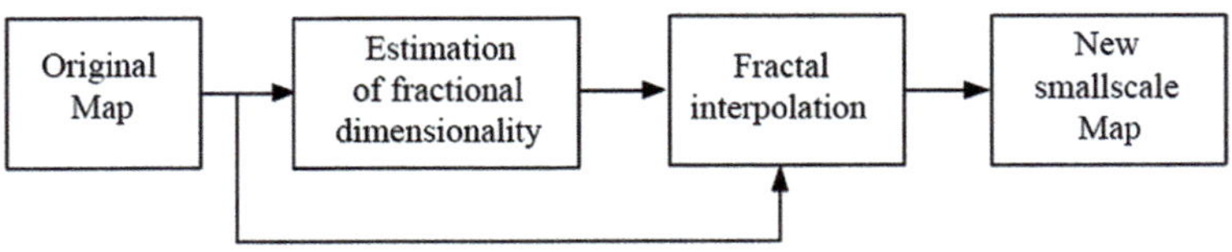

Fig. 2.17 Basic flow of fractal interpolation method

The lower limit of the scale-free interval of fractal features obtained from the map is determined by the map interval.

$$\Delta X_{min} = u \tag{2.90}$$

The upper limit is the upper limit of the linear fit region of the fractal characteristic curve.

The minimum scale uc of the map obtained after fractal interpolation is smaller than u. Therefore, the minimum scale that can be obtained by fractal interpolation cannot be determined by the original map. The fractal interpolation of gravity anomaly maps discussed in this paper can only be used for coarse interpolation of gravity anomaly maps.

The fractal interpolation uses the random midpoint displacement method, where the points in the region near the insertion point are understood as a sum of random data conforming to (0, σ2) and the mean of the region, and it is implemented as shown in Fig. 2.18.

In Fig. 2.18, the filled points are the control points of the original map, and the dashed lines are the interpolation points; the interpolation points are divided into two categories, and the interpolation point marked as 1 (x, y) is interpolated by the four control points of the original map around it as reference.

$$\begin{aligned} g(x, y) = \frac{1}{4}\big[g(x - u_c, y - u_c) + g(x + u_c, y - u_c) + g(x - u_c, y + u_c) + g(x + u_c, y + u_c)\big] \\ + \sqrt{1 - 2^{2H-2}} u_c^H \sigma Gauss \end{aligned} \tag{2.91}$$

The interpolation points marked as 2, the reference points used consist of two original map control points and two interpolation points marked as 1

$$\begin{aligned} g(x, y) = \frac{1}{4}\big[g(x - u_c, y) + g(x, y - u_c) + g(x + u_c, y) + g(x, y + u_c)\big] \\ + \sqrt{1 - 2^{2H-2}} \cdot \left(\frac{u_c}{\sqrt{2}}\right)^H \cdot \sigma \cdot Gauss \end{aligned} \tag{2.92}$$

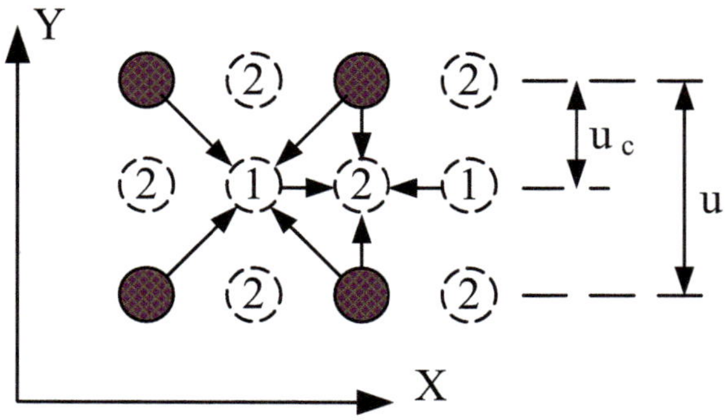

Fig. 2.18 Midpoint displacement interpolation method

where Gauss is a random signal consistent with a mean of 0 and a standard deviation of 1.

2.6.4 *Fractal Interpolation Experiments and Analysis*

A Experiment I.

For a certain (60 × 60) gravity anomaly map to be interpolated, a 30 × 30 map is obtained by equally spaced sampling on the original map, and then the above fractal interpolation method is used to interpolate and compare the interpolated map with the original map.

The fractal characteristics of the sampled maps are shown in Table 2.9.

The comparison of the interpolated gravity anomaly map with the original map is shown in Fig. 2.19.

The parameters used for the accuracy analysis after interpolation were defined using (2.2. 2.52, 2.2. 2.53, 2.2. 2.54, 2.2. 2.55, and 2.2. 2.56).

B Experiment II.

The interpolation was performed on the original gravity anomaly map and the changes in the fractal characteristics of the map after the interpolation were

Table 2.9 Table of analysis characteristics parameters

	Fitting distance	H	Df	Log C	σ	Residuals
30 × 30	6	0.7514	2.2486	1.0261	13.3094	0.015019
119 × 119	5	0.83967	2.16033	0.99224	12.3112	0.028032

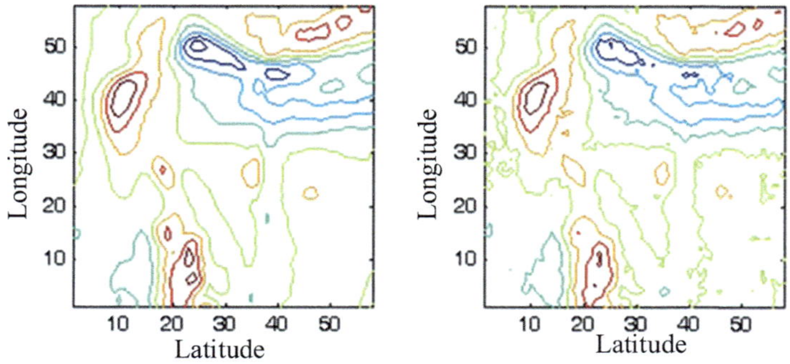

(a) Contour map of original data (b) Contour map after fractal interpolation

Fig. 2.19 Fractal interpolation results after sampling on the original data analyzed. Figure 2.20 shows the comparison of the interpolated contour maps. Figure 2.21 shows the fractal characteristic curves of the map after interpolation. The characteristic parameters are shown in the data in row 2 of Table 2.10.

Obviously, the fractal features of the interpolated map change very little from those of the pre-interpolated map.

Fractal interpolation has the following characteristics.

The accuracy of fractal interpolation is related to the standard deviation; the larger the standard deviation, the lower the accuracy.

Fractal interpolation introduces random signals and therefore uncertainty in the gravity anomaly values at the interpolation points.

Interpolation using the midpoint displacement method with a fixed position of the interpolation point, namely,$uc = u/2$.

Fractal interpolation as a reconstruction method of the natural surface of terrain has a strong sense of realism; the correlation between gravity anomalies and terrain

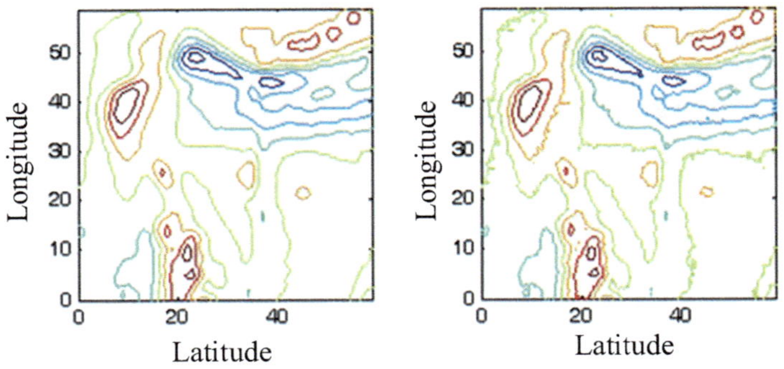

(a) Contour map of original data (b) Contour map after fractal interpolation

Fig. 2.20 Results of direct fractal interpolation of raw data

Fig. 2.21 Map fractal feature curve after interpolation

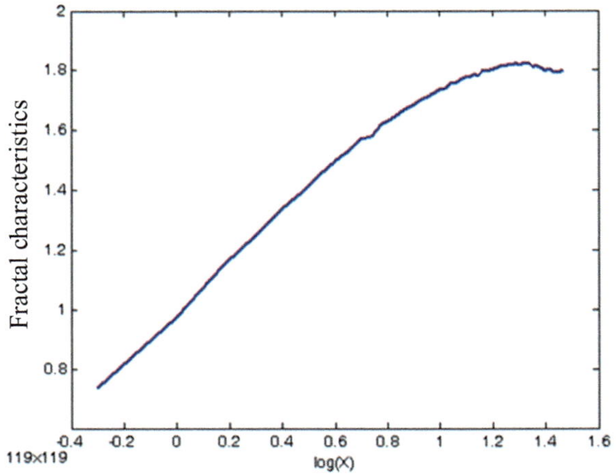

Table 2.10 Analysis of fractal interpolation accuracy

$E(g_e)$	$E(\|g_e\|)$	$Max(\|g_e\|)$	$E(g_e^2)$	$D(g_e)$
0.013683	6.4704	53.419	98.316	7.5144

gives us the association of using fractals to interpolate gravity anomalies. Due to the limitation of conditions, no actual measurement of interpolation results is verified, but from the results of computer simulation experiments, fractal interpolation as a coarse interpolation is feasible.

2.7 Summary

This chapter introduces the physical parameters related to gravity-aided navigation, describes the normal gravity field, gravity perturbation and other related physical concepts, studies the establishment of a three-dimensional normal gravity model for underwater, and constructs the INS base model that can be used for underwater GAINS and the method of fusing multi-source gravity anomaly data to generate a gravity anomaly reference map, which is the theoretical basis for the implementation of gravity-aided navigation system.

References

1. Chen, J., Wen, S., & Cheng, P. (2001). Some issues on the development of geodetic science in China. *Advances in Earth Sciences., 16*(5), 1–13.
2. Kai Wang, Chuanding Zhang, Xing Wu, Qiang Hou. (2014). Topography/equilibrium gravity field model constructions and their proposed stability analysis. *Journal of Geophysics, 57*(03):760-769
3. Qingbin Wang, Dongming Zhao. (2016). Fundamentals of Earth's gravity field. *Survey and Mapping Press.* 75–80
4. Motao Huang. (2005). Ocean gravity field determination and its applications. *Survey and Mapping Press.* 290–294
5. Jian Ding, Houze Xu, Chuanyin Zhang. (2016). Evaluation of the Earth's gravity field model based on the classical definition of the geoid. *Journal of the University of Chinese Academy of Sciences, 33*(04):528-536
6. Lane, R. (1972). Johnson and Joe. *J. Litehiser, A method for Computing the Gravitational Attraction of Three-Dimensional Bodies in a Spherical or Ellipsoidal Earth, Journal of Geophysical Research, 77*(35), 6999–7009.
7. Forsberg, R. (1987). A new covariance model for inertial gravity and gradiometry. *Journal of Geophysical Research, 92*(B2), 1305–1310.
8. Lei Song. (2015). Global and regional gravity field approximation theory methods. *Shandong University Press.* 56–60
9. Han Xu, Qiangbo Zhou. (2016) Accuracy analysis of gravity anomalies from satellite gravity gradient data. *Survey and Mapping Science. 41*(11):17-24
10. Li Yan, Hualing Wu. (2009). Research on multi-scale analysis of gravity field datum map. *Mapping Bulletin,* (02):10-13

11. Zhang, S., Li, J., & Kong, X. (2020). Laplace equation-based inversion of gravity anomalies in global seas by the vertical deviation method. *Journal of Surveying and Mapping., 49*(04), 452–460.
12. Qiujie Chen, Yunzhong Shen, Xingfu Zhang, Wu Chen, and Houze Xu. (2016). A high-precision global static gravity field model based on GRACE satellite data. 45(04): 396–403
13. Hao Feng. (2004). Theory, modeling and simulation in future underwater gravity-assisted inertial navigation microsystems[D]. *Beijing University of Posts and Telecommunications: Ph.D. Dissertation (joint training by Beifu and Peking University)*
14. Linjuan Nie, Nengfang Chao, Taoyong Jin, Dingbo Chao. (2016). Accuracy analysis of satellite altimetry inversion of ocean gravity anomalies. *Surveying and Mapping Science*, (9), 41(09):1–7
15. Jinyu Yang, Xunhua Zhang, Feifei Zhang, Bo Han, Zhenxing Tian. (2014). Application of gravity anomalies from multiple sources to prepare a spatial gravity anomaly map and gravity field interpretation for China's sea, land and adjacent areas. *Journal of Geophysics, 57*(12):3920-3931
16. Zunyi Xu, Yuxiang Jiang, Liang Zhao, Fuxing Yang. (2010). Improved Shepard algorithm and its application in gravity anomaly interpolation. *Journal of Wuhan University (Information Science Edition). 35*(04):477–480
17. Pan, Z., Wang, W., & An, R. (2013). Multi-source gravity field information fusion processing method. *Journal of Surveying and Mapping Science and Technology., 30*(03), 236–240.

Chapter 3
Navigation and Positioning Method of Earth Gravity Field

Abstract At present, the inertial navigation system (hereinafter referred to as "inertial navigation system" and "INS") is the core navigation system of the underwater vehicle. However, due to the problem that the positioning error of the inertial navigation system accumulates with the increase of the running time, it is necessary to use other external information to regularly correct the inertial navigation system, that is, auxiliary navigation technology. Gravity-assisted navigation technology uses the sensors installed on the carrier that can accurately measure the earth's gravity field and the accurate earth's gravity field map, and uses modern optimal control theory and methods to regularly determine the position of the carrier. It is one of the technical ways to modify the inertial navigation system, and is especially suitable for underwater vehicles (Zhao, X., & Chen, G. (2020). Development status and trend of ship navigation. Navigation and control, 19(z1), 82–87; Wei, B., Lv, W., Fan, X., Zhu, Y., & Guo, Y. (2019). AUV navigation technology development status and prospects. *Journal of Underwater Unmanned Systems*, *27*(01), 9; Zhou, B., & Liu, S. (2012). Research on underwater navigation technology. *Modern navigation*, *3*(01), 19–23; Xiong, Z., Cai, K., & Fengde, W., et al. (2002). A summary of the development of navigation technology of U.S. strategic submarines in the 21st century. *Ship science and technology*, *24*(3), 30−37; Feizhou, Z. (2003). *Research on underwater passive navigation technology [postdoctoral report]*. Beijing Peking University.). The basic principle of the gravity aided navigation system is that the gravity sensor measures the gravity characteristic data in real time during the movement of the carrier; At the same time, the gravity data is read from the gravity map according to the position information of the inertial navigation system. The two data are sent to the matching solution computer, and the matching solution software is used to solve the problem to obtain the best matching position. Using this information to correct the inertial navigation system can suppress the inertial navigation error and improve the navigation accuracy (Moryl, J., Rice, H., & Shinners, S. (1998). The universal gravitymodule for enhanced submarine navigation. In *Position Location and Navigation Symposium 1998* (pp. 324−331). IEEE; Feng, H., Yan, L., Ge, Y. et al. (2004). Real time gravity correction of underwater vehicle based on INS. *Journal of Wuhan University (Engineering Edition)*, *37*(3), 135−138.). To sum up, the gravity assisted navigation system mainly includes gravimeter, gravity field model and gravity data

L. Yan et al., *Mathematical and Physical Fundamentals of Navigation and Positioning with Earth's Natural Fields*, Navigation: Science and Technology 13,
https://doi.org/10.1007/978-981-97-5524-0_3

processing, gravity map matching theory and matching algorithm, and correction theory and method of various system errors. The second chapter focuses on the gravity field model and gravity data processing method. The underwater application is the most important application scenario of gravity assisted navigation. When introducing the specific methods of gravity assisted navigation, the application is gravity. From the perspective of application, this chapter focuses on the moving base gravity sensor, the moving base marine gravity real-time measurement, providing relevant theoretical basis and technical support for the final gravity assisted inertial navigation system (Zhang, H. (2013). *Research on key technologies of underwater gravity field aided navigation and positioning* (pp. 56–60). Doctoral dissertation of Harbin Engineering University; Javed, W., Ghani, S., & Elmqvist, N.: GravNav conference (pp. 217–224). In *Advanced Visual Interfaces*; Zhu, Z., & Zhou, P. (2011). Research on multi-scale characteristics of gravity field in gravity assisted inertial navigation. *Progress in geophysics*, *26*(05), 1868–1873.).

3.1 Auxiliary Inertial Navigation Model Based on Passive Gravity Field

The navigation coordinate system is the foundation of any navigation system. Firstly, this chapter gives a brief introduction to the commonly used navigation coordinate system at present. Aiming at the main navigation coordinate system adopted in this paper, the mutual conversion relationship between other navigation coordinate systems and the main navigation coordinate system is derived. Then, starting from the ins dynamic equation, the influence of the earth's gravity field on INS error is analyzed in detail, and the configuration scheme of gravity navigation system based on gravimeter and gravity gradiometer is proposed. Finally, various existing INS error models are analyzed. According to the requirements of the gravity navigation system, an INS error model for the gravity navigation system is established [11].

3.1.1 Common Navigation Coordinate System

The coordinate system can be established by software or hardware. The coordinate systems established by the software are all orthogonal coordinate systems. Due to the imperfection of the instrument, the coordinate system established by hardware is mostly non orthogonal coordinate system. Determining a coordinate system must indicate its origin position and three-axis direction. The coordinate systems used are established in the navigation computer and are all software based coordinate systems.

1. Main coordinate system used in this book.

The coordinate system used in this paper mainly includes geocentric inertial coordinate system, geocentric earth fixed coordinate system, local geodetic vertical coordinate system and carrier coordinate system.

(1) Geocentric inertial (ECI) coordinate system

It is referred to as i system, and the three coordinate axes are represented as x_i, y_i, z_i. Inertial coordinate system can be considered as a coordinate system that can be established by Newton physics. In this coordinate system, there is no acceleration, no gravitational mass, and the direction is fixed relative to the average background of all objects in the universe. Therefore, as long as the above conditions are met, the orientation of the coordinate axis can be arbitrary. In this book, since all motions are limited to the earth and its vicinity, the origin of the geocentric inertial coordinate system is the center of the earth's mass, the z_i-axis is along the earth's rotation direction, x_i, y_i is in the equatorial plane, and the direction is fixed with respect to the distant stars (pointing to the stars). Because of the earth's rotation, the x_i-axis is not parallel to the zero meridian plane of the earth. This coordinate system is the reference datum for inertial instrument measurement. In the equation, the coordinate system is marked by superscript i.

(2) Geocentric earth fixed (ECEF) coordinate system

It is abbreviated as e system, and the three coordinate axes are represented as x_e, y_e, z_e. The coordinate origin is at the center of the earth's mass, the z_e axis coincides with the earth's rotation axis, the x_e axis points to the Greenwich meridian in the equatorial plane, and the y_e axis is in the equatorial plane. $x_e y_e z_e$ constitutes a right-handed coordinate system. Since the coordinates remain fixed relative to the rotating earth, the direction of each axis is specified to remain fixed throughout the navigation period. The ECEF coordinate system rotates relative to the ECI coordinate system at the earth's rotation rate (earth rotation angle speed) ω_{ie}. In the equation, the coordinate system is marked by superscript e.

(3) Local geodetic vertical (LGV) coordinate system.

Sometimes it is also called geographical coordinate system, abbreviated as n system (some data abbreviated as g system), and the three coordinate axes are respectively represented as N, E, U. The coordinate origin is at the mass center of the carrier, the N-axis points to the north of the earth, the E-axis is perpendicular to the meridian plane where the carrier is located, the direction is moving, the U-axis is opposite to the direction of gravity g, and passes through the carrier (IMU mass center) along the local geodetic vertical line. In the equation, the coordinate system is marked by superscript n.

(4) Carrier (VB) coordinate system

Referred to as b system, the three coordinate axes are represented as x_b, y_b, z_b. The coordinate origin is at the center of gravity of the aircraft, x_b is along the longitudinal axis of the aircraft (pointing to the east), y_b is consistent with the horizontal axis of the aircraft (pointing to the north), z_b is along the vertical axis of the aircraft, and $x_b y_b z_b$ constitutes a right-hand coordinate system. In the equation, the coordinate system is marked by superscript b. The azimuth of the carrier coordinate system relative to the geographic coordinate system is the attitude and heading of the aircraft.

2. Mutual Conversion Relationship Between Main Navigation Parameters and Coordinate System of Carrier

The most basic main navigation parameters of the carrier are the instantaneous geographic position P, ground speed V and attitude heading ϕ information of the carrier. In this book, the local geodetic vertical coordinate system (geographic coordinate system) is selected as the navigation coordinate system, then the instantaneous geographic position P is expressed as $P(\lambda, \varphi, h)$, where λ, φ is the geographic longitude and latitude, and h is the height of the carrier above the geoid, as shown in Fig. 3.1.

(1) Relationship between geographic coordinate system and ECI coordinate system

$$\begin{bmatrix} E \\ N \\ U \end{bmatrix} = \begin{bmatrix} -\cos\lambda^i \sin\varphi & -\sin\lambda^i \sin\varphi & \cos\varphi \\ \sin\lambda^i & -\cos\lambda^i & 0 \\ \cos\lambda^i \cos\varphi & \sin\lambda^i \cos\varphi & \sin\varphi \end{bmatrix} \begin{bmatrix} x_i \\ y_i \\ z_i \end{bmatrix} = C_i^n \begin{bmatrix} x_i \\ y_i \\ z_i \end{bmatrix} \tag{3.1}$$

where, $\lambda^i = \lambda + \omega_{ie} t$.

(2) Transformation relationship between geographical coordinate system and ECEF coordinate system

$$\begin{bmatrix} E \\ N \\ U \end{bmatrix} = \begin{bmatrix} -\cos\lambda \sin\varphi & -\sin\lambda \sin\varphi & \cos\varphi \\ \sin\lambda & -\cos\lambda & 0 \\ \cos\lambda \cos\varphi & \sin\lambda \cos\varphi & \sin\varphi \end{bmatrix} \begin{bmatrix} x_e \\ y_e \\ z_e \end{bmatrix} = C_e^n \begin{bmatrix} x_e \\ y_e \\ z_e \end{bmatrix} \tag{3.2}$$

(3) Relationship between geographic coordinate system and carrier coordinate system

$$\begin{aligned} \begin{bmatrix} x_b \\ y_b \\ z_b \end{bmatrix} &= \begin{bmatrix} \cos\psi\cos\gamma + \sin\psi\sin\theta\sin\gamma & -\sin\psi\cos\gamma + \sin\gamma\sin\theta\cos\psi & -\cos\theta\sin\gamma \\ \sin\psi\cos\theta & \cos\psi\cos\theta & \sin\theta \\ \cos\psi\sin\gamma - \sin\psi\sin\theta\cos\gamma & -\sin\psi\sin\gamma - \cos\psi\sin\theta\cos\gamma & \cos\theta\cos\gamma \end{bmatrix} \begin{bmatrix} E \\ N \\ U \end{bmatrix} \\ &= C_n^b \begin{bmatrix} E \\ N \\ U \end{bmatrix} \end{aligned} \tag{3.3}$$

where, ψ is the heading angle (calculated clockwise from the geographical north, $0 \le \psi < 360°$), θ is the pitch angle (calculated from the longitudinal horizontal axis, positive upward and negative downward, $-90° \le \theta \le 90°$), and γ is the

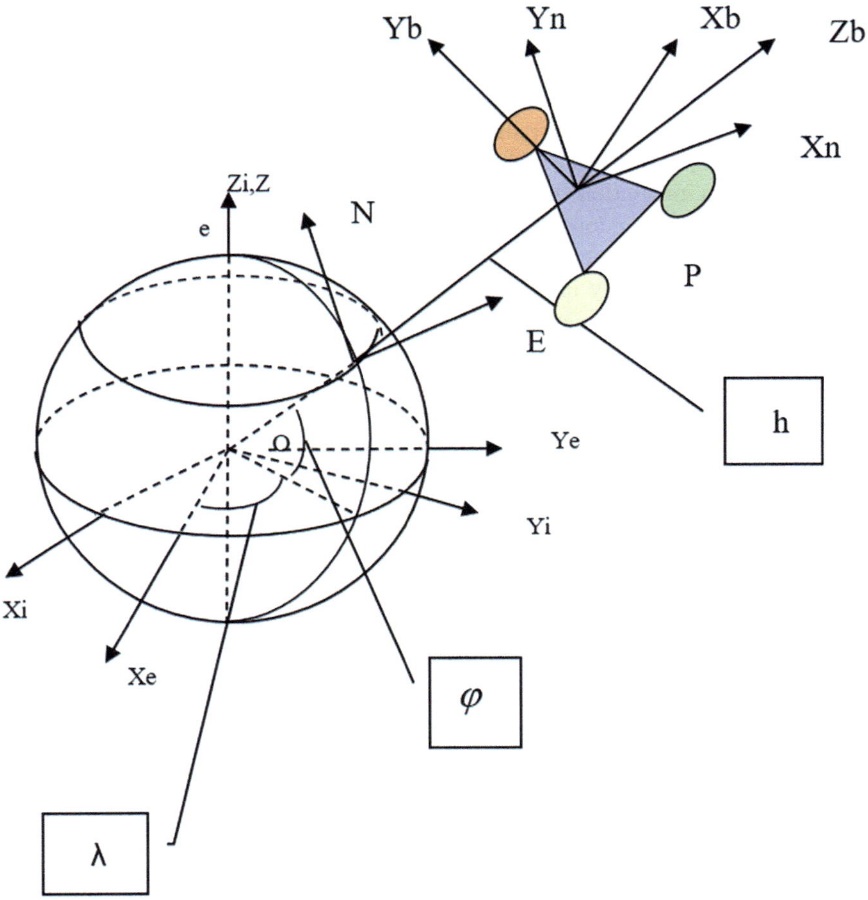

Fig. 3.1 Schematic diagram of various coordinate systems used

inclination angle (calculated from the vertical plane, positive right and negative left, $-90° \leq \theta \leq 90°$).

(4) Transformation relationship between ECI coordinate system and ECEF coordinate system

$$\begin{bmatrix} x_e \\ y_e \\ z_e \end{bmatrix} = \begin{bmatrix} \cos\omega_{ie}t & \sin\omega_{ie}t & 0 \\ -\sin\omega_{ie}t & \cos\omega_{ie}t & 0 \\ 0 & 0 & 1 \end{bmatrix} \begin{bmatrix} x_i \\ y_i \\ z_i \end{bmatrix} = C_i^e \begin{bmatrix} x_i \\ y_i \\ z_i \end{bmatrix} \tag{3.4}$$

where t is the time from the start of navigation.

3.1.2 Basic Principle of Inertial Navigation System

The basic principle of inertial navigation is based on Newton's law of mechanics. The acceleration of the carrier motion is measured by the navigation accelerometer in the carrier body, and the velocity and position information of the carrier are obtained through integration. The principle of strapdown inertial navigation system is shown in Fig. 3.2. Strapdown inertial navigation system is composed of Strapdown accelerometer group, gyroscope group and navigation computer. The accelerometer and gyroscope are directly installed on the carrier. All navigation information is processed and given by the navigation computer.

In inertial space, two forces determine the carrier's motion, namely gravity and inertial force. The ratio of inertial force to carrier mass is called specific force, namely.

$$f = \frac{F_I}{m} \tag{3.5}$$

The unit of specific force vector is m/s^2. The specific force of inertial navigation system is usually caused by thrust, lifting force, resistance and other forces. It is a basic concept in inertial navigation. The accelerometer measures the specific force f^b of the carrier relative to the inertial space, and the gyroscope measures the angular velocity ω_{ib}^b of the carrier relative to the inertial space. According to the coordinate system selected by the navigation system, f^b and ω_{ib}^b need to be transformed accordingly.

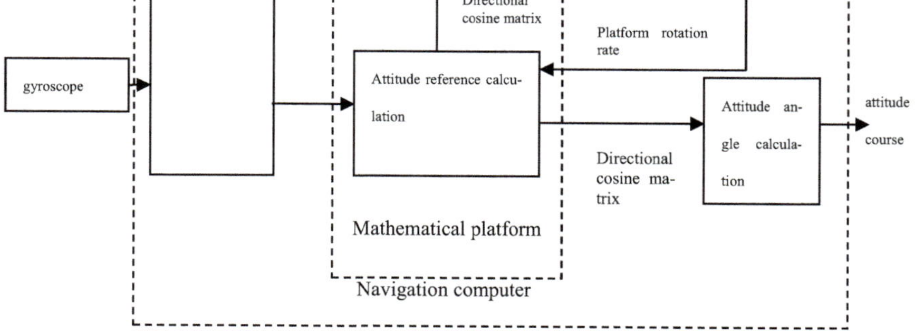

Fig. 3.2 Principle block diagram of strapdown inertial navigation system

When the navigation coordinate system adopts the geographic coordinate system (n system, E, N, U), the basic equation of the inertial navigation system is:

$$\dot{\vec{V}}^{n} = \vec{f}^{n} - (2\overrightarrow{\omega_{ie}^{n}} + \overrightarrow{\omega_{en}^{n}}) \times \vec{V}^{n} - \overrightarrow{g^{n}} \tag{3.6}$$

where, $\dot{\vec{V}}^{n}$ is the acceleration vector of the carrier. $\vec{f}^{n}$ is the specific force vector, which is obtained from the measured value of accelerometer through attitude matrix transformation. $\vec{\omega}_{ie}^{n}$ is the rotation angle velocity vector of the earth in the geographic coordinate system $\vec{\omega}_{en}^{n}$ is the rotational angular velocity vector of the carrier relative to the earth. $\vec{V}^{n}$ is the carrier speed vector. $\overrightarrow{g^{n}}$ is the gravitational acceleration vector. For the expansion of formula (3.6), there are

$$\begin{bmatrix} \dot{V}_E^n \\ \dot{V}_N^n \\ \dot{V}_U^n \end{bmatrix} = \begin{bmatrix} f_E^n \\ f_N^n \\ f_U^n \end{bmatrix} + \begin{bmatrix} 0 & (\lambda + 2\omega_{ie}) \sin\varphi & -(\lambda + 2\omega_{ie}) \cos\varphi \\ -(\lambda + 2\omega_{ie}) \sin\varphi & 0 & -\varphi \\ (\lambda + 2\omega_{ie}) \cos\varphi & \varphi & 0 \end{bmatrix}$$
$$\begin{bmatrix} V_E^n \\ V_N^n \\ V_U^n \end{bmatrix} + \begin{bmatrix} 0 \\ 0 \\ -g \end{bmatrix} \tag{3.7}$$

Inertial navigation is usually divided into four stages: calibration, alignment, state initialization and current state calculation.

Calibration: due to the errors of the accelerometer and gyroscope, the relationship between the sensitive specific force and rotation rate of the accelerometer and gyroscope and the actual specific force and rotation rate shall be determined by calibrating the accelerometer and gyroscope respectively before navigation. The calibration shall be carried out at a very precise location determined by the apparent gravity vector value and the geographical position so as to provide a coefficient that correctly expresses the output of the accelerometer and gyroscope.

Alignment: at the beginning of navigation, determine the direction of the input axis of each accelerometer, that is, align the accelerometer gyroscope assembly to the local apparent gravity vector (the astronomical vertical direction of the pendulum) and the astronomical North (the direction close to the earth's rotation axis), so that the navigation computer can determine coordinate transformation and convert the output of the accelerometer into the coordinates of the navigation computer.

State initialization: provide the initial position and speed value (if not zero) to the computer for initialization of computer integration operation. The initial position can be obtained from the previous geodetic survey or from GPS, and the ground speed can be obtained by some measuring device outside the inertial equipment.

3.1.3 Gravity/Inertial Navigation Integrated Navigation System

According to the basic principle of inertial navigation system, the accelerometer measures the specific force, that is, the difference between inertial acceleration and gravity vector. In order to distinguish the inertial acceleration and the gravitational acceleration of the carrier movement, the inertial navigation instrument must have a mathematical model of the gravity field. The traditional ins gravity calculation is shown in Fig. 3.3. At present, gravity anomaly is the largest residual error of high-performance ins. For ship navigation system, the gravity anomaly component has a deviation of about 30–50 mGal. Therefore, in order to improve the accuracy of INS, in addition to continuously improving the mathematical model of gravity field, it is also hoped that the gravity field value of the ship's position can be obtained in real time.

Based on the above analysis, we can get the following main reasons for INS error:

(1) Gravity anomaly on ship track.
(2) Random drift error of inertial sensor.
(3) INS initial alignment error.
(4) INS is the cumulative error caused by reasons (1), (2) and calculation.

The planned track of the underwater carrier passes through the gravity anomaly matching area, and the gravity anomaly is significant, and the standard deviation is more than 10 mGal (at present, the error of the gravity anomaly map is 5–8 mGal). When the underwater carrier enters the gravity anomaly matching area, it moves in a straight line (east–west or north–south) with approximately constant depth and uniform speed. Use the measurement results of gravimeter or gravity gradiometer to correct the gravity field model adopted by ins and the random drift error of INS sensor. INS position error (initial alignment error, cumulative error) is corrected by gravity map matching.

Since the INS system is usually installed at the center of mass of the underwater carrier, the gravity navigation equipment and INS can share a stable platform, be installed adjacent to each other, or be installed at other suitable positions using INS platform parameters according to the actual situation.

As shown in Fig. 3.4, during the movement of the carrier, the gravity sensor measures the gravity characteristic data in real time. At the same time, the gravity

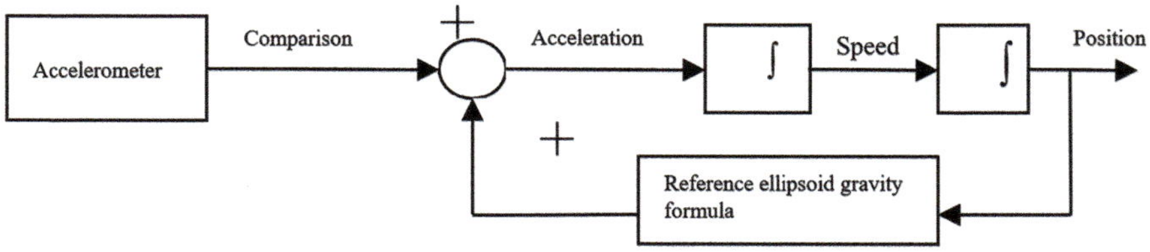

Fig. 3.3 Calculation of gravity in traditional inertial navigation system

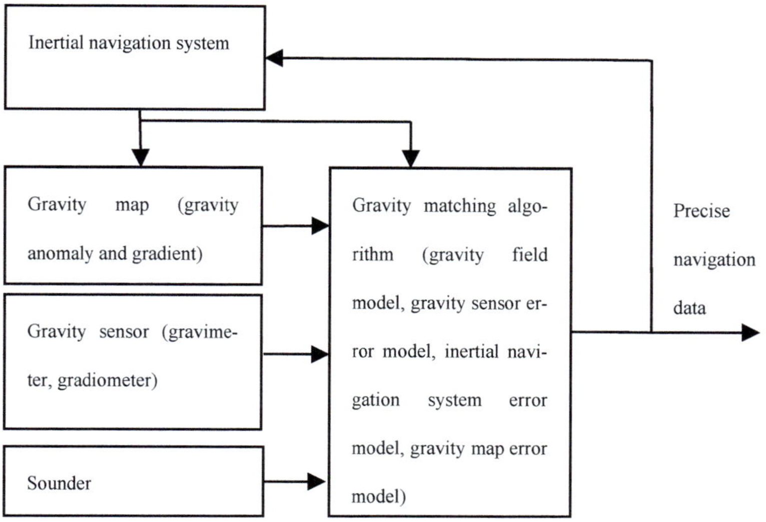

Fig. 3.4 Gravity assisted inertial navigation system

data is read from the gravity map according to the position information of INS. The two data are sent to the matching solution computer, and the matching solution software is used to solve the problem to obtain the best matching position. Using this information to correct ins can restrain INS error and improve navigation accuracy.

3.2 Gravity Aided Navigation Dynamic Measurement System

Gravity measurement is divided into dynamic measurement and static measurement. Dynamic gravity measurement is a gravity measurement technology on the motor seat, and its measurement methods can be divided into scalar and vector gravity measurement. Vector gravity measurement is still in the development stage. This book focuses on the dynamic scalar gravity measurement system [12].

The scalar gravity measurement only measures the vertical gravity acceleration, and the gravity acceleration sensor is only sensitive to the axial acceleration. It is installed on the stable platform, which isolates the rotation and vibration of the carrier and ensures that the sensitive axis of the gravity acceleration sensor is always vertical to the ground.

Typical dynamic scalar gravity measurement systems include kss31 of Graf company, S-type of lacoste & Romberg and bgm-3 marine gravimeter produced by Bell area space company. Seeing Table 3.1 for their comparison.

Table. 3.1 Performance comparison of different types of gravimeters

Instrument type	LaCoste & Romberg S	KSS31	BGM-3
Resolving power	0.01 mGal	0.01 mGal	0.1 mGal
Measuring range	12,000 mGal	10,000 mGal	978,000–983000 mGal
Drift	3 mGal/month	3 mGal/month	1.2 mGal/month
Dynamic accuracy	1.0 mGal	1.0 mGal	1.0 mGal
Response time		36 s	
Recording frequency	1 Hz		1 Hz
Working temperature	46–53 °C	50 °C	15–32 °C
Volume	152 × 67 × 117 mm	Φ195 × 357 mm	120 × 170 × 130 mm
Weight	3.2 kg	12 kg	3 kg

As mentioned in the introduction, vector gravity measurement technology is still in the development stage, and it is not discussed as a gravity measurement prototype in gains.

3.2.1 Signal Processing of Gravity Dynamic Measurement

The accuracy of dynamic gravity measurement data is defined by two aspects: one is the accuracy of gravity sensitive data. The second is the position accuracy of the moving carrier. The underwater navigation gravity measurement is different from the gravity mapping in that the position estimation of the carrier is obtained by matching the real-time measured gravity data with the reference gravity anomaly map. Therefore, the signal processing of underwater navigation gravity measurement is mainly to improve the accuracy of gravity sensitive data and extract the gravity anomaly data from the sensitive data in real time.

As mentioned above, the scalar dynamic gravity measurement is to obtain the vertical acceleration information, place the gravity sensitive element on the stable platform, and ensure that the axis of the sensitive element is always perpendicular to the local geoid. Secondly, it is necessary to ensure that the sensing element only outputs the axial acceleration information of the sensing axis.

As shown in Fig. 3.5, the bgm-3 marine gravity sensor is provided with a transverse spring at the horizontal position to restrict the movement of the sensitive mass in the horizontal direction, forcing it to only move vertically, eliminating the cross coupling effect of acceleration.

The gravity sensor on the moving carrier is sensitive to the vertical acceleration generated by the movement of the carrier while obtaining the gravitational acceleration. These accelerations can even reach tens of Gal, which is the main error source. These errors must be eliminated before the gravity anomaly value is obtained.

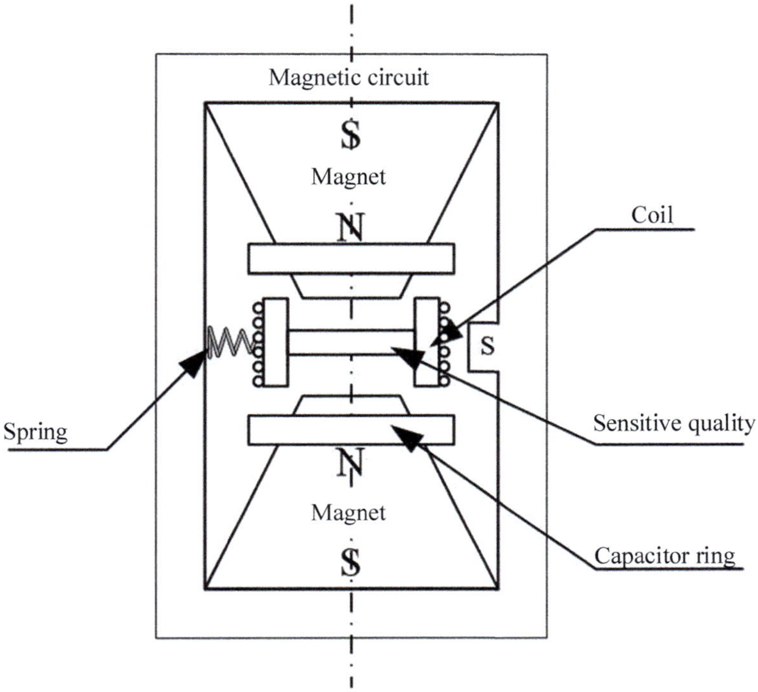

Fig. 3.5 Structural diagram of BGM force balance accelerometer

The interference of the movement of the carrier to the gravimeter can be divided into three aspects: vertical acceleration, deflection of the stable platform and Eötvös effect. Eötvös effect refers to the influence of the movement of the carrier relative to the earth on the gravity measurement.

A lot of work has been done at home and abroad on the research of gravity signal processing to obtain high-precision measurement values. The research of foreign dynamic gravimeter manufacturers and relevant laboratories shows that low-pass filtering is effective for most noise. Jiandi Wang, Institute of Surveying and Geophysics, Chinese Academy of Sciences, discussed the problem of eliminating the distortion of dynamic gravity measurement by Kalman inverse filtering. Gauss-Markov。In establishing the equation of state, he used a second-order Gauss Markov process to represent the covariance model of the earth's gravity anomaly. Chihang Zhao of Southeast University discussed the method of processing Gravimeter Data with wavelet analysis. His main idea is that low-pass filtering will lead to distortion of output signal. Therefore, wavelet is used to decompose the collected gravity signal. If the wavelet coefficient is less than a certain threshold value, it will be set to 0. Then the signal is reconstructed to obtain the original gravity signal.

These methods are proposed for the purpose of gravity anomaly mapping. Gravity sensitive data were recorded and then processed in the laboratory. However, the general average filtering has a small linear range, and the gravity signal itself will be distorted while filtering. There are defects in real-time signal processing. In this chapter, a method of using finite impulse response digital filter (FIR) to ensure the stability and phase linearity of the filter and realize real-time signal processing is proposed.

Other errors can be compensated only by external observation. The deflection of the stable platform makes the horizontal acceleration of the carrier coupled to the output of the gravimeter, which requires real-time estimation of the deflection angle of the stable platform. Correction of the Eötvös effect requires estimation of the carrier speed.

In gravity surveying and mapping, Eötvös effect correction relies on reliable continuous positioning, such as GPS, to provide the speed observation of the carrier. It is also possible to correct the Eötvös effect by using the gravity measurement difference at the intersection of the survey lines. There is zero speed correction method in land survey. But these methods can not meet the real-time requirements of underwater navigation gravity measurement. Based on the analysis of the Eötvös effect model, the Eötvös effect correction model based on the gravity anomaly reference map is proposed.

3.2.2 Gravity Anomaly Calculation of Gravity Dynamic Measurement

Gravity anomaly-the difference in gravity between the geoid and the reference ellipsoid-is caused by the mass distribution that causes the geoid to deviate from the reference ellipsoid. The published gravity anomaly values are almost all based on the sea level. Similarly, in underwater gravity navigation, the given reference map is gravity anomaly map. The gravity data must be converted into gravity anomaly data to match with the reference map.

The basic calculation flow of gravity anomaly is shown in Fig. 3.6.

The terrestrial gravity survey is usually carried out above the sea level. At this time, before it subtracts the normal gravity value on the ellipsoid to obtain the gravity

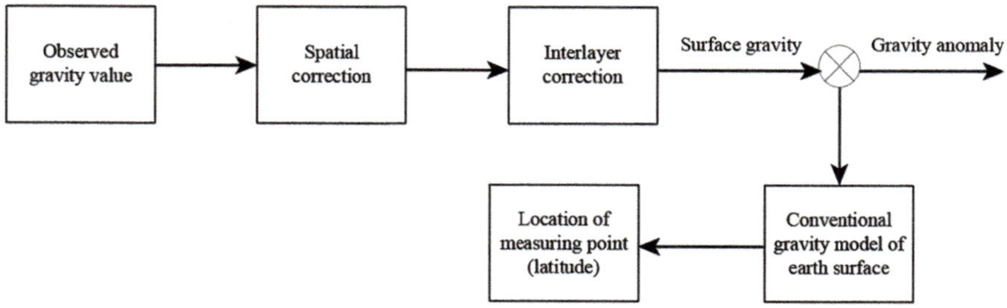

Fig. 3.6 Gravity anomaly calculation process

anomaly, the observed gravity value must be corrected to the equivalent gravity value of the sea level.

In order to obtain the equivalent value at sea level, the observation value is corrected on the basis of a variety of assumptions, although none of these assumptions can truly reflect the actual situation. Free air correction: only the height of the observation point is corrected to correct the observation value to the sea level. Bouguer correction: the influence of mass above the sea level is subtracted from the observed gravity value to correct the gravity value to the sea level based on the assumption that these masses do not exist. Equilibrium correction: remove the mass influence above sea level, and then reintroduce it at the depth directly downward, so as to eliminate the "root" of the mountain range (or "mountain root" of the ocean basin) that compensates for the terrain.

Free air anomaly can be obtained by free air correction; Bouguer anomaly can be obtained after Bouguer correction; The equalization correction corresponds to the equalization anomaly.

Because people's understanding of submarine terrain is not as good as that of land terrain, the existing maps of marine gravity anomalies are all free air anomalies. The measuring point of underwater gravity survey is lower than the geoid, and the calculation model of gravity anomaly is different from that of terrestrial gravity anomaly. One of the existing calculation models is to add the interlayer correction calculation of sea water on the basis of the land free air calculation model. The other is to calibrate the gravity gradient to obtain the depth correction value. The first method is not complete in theory, and the second method cannot be realized in underwater navigation. This book starts from the definition of gravity anomaly, corrects the normal gravity model to the observation point, and calculates the gravity anomaly value. The method is based on the establishment of a three-dimensional normal gravity model of the seabed.

3.3 Interference Processing Method of Gravity Measurement

In Sect. 3.2.1, it is introduced that the interference sources of underwater dynamic gravity measurement are mainly vertical acceleration, horizontal acceleration and Eötvös effect. This section focuses on the generation mechanism, model establishment and treatment methods of these three types of interference [13, 14].

3.3.1 Processing Method of Vertical Acceleration Interference in Gravity Dynamic Measurement

The vertical acceleration of the carrier is directly sensitive by the gravity acceleration sensor, which is an important error source in gravity measurement.

The vertical acceleration AV of the carrier can be divided into constant value AVC and periodic signal.

$$a_v = a_{ve} + \sum a_{vo}^{(i)} \cos 2\pi f_{vo}^{(i)} t \tag{3.8}$$

a_{vo} and f_{vo} respectively represent the amplitude and frequency of the periodic vertical acceleration, and the superscript i represents the ith frequency component.

The constant value is mainly generated by the accelerated ascent and submergence of the underwater carrier. Under normal circumstances, the measured value of the gravimeter is not used for auxiliary navigation at this time.

When the underwater vehicle is sailing at a fixed depth, it will move periodically in the vertical direction due to the ocean current. This part of motion is different from the vibration of the moving carrier itself and cannot be removed by the mechanical low-pass filter of the gravimeter base. The observation shows that its amplitude can reach 5–10 Gal and the frequency range is 0.05–0.5 Hz.

In order to effectively filter the periodic acceleration interference in the vertical direction, it is necessary to explore the difference between the interference signal and the gravity acceleration signal in the frequency domain. The frequency domain range of the interference signal is between 0.05–0.5H through observation.

The frequency component of gravity acceleration is related to the distribution of gravity field, and its spatial frequency component is fixed. For a gravity measurement system, what is needed is the frequency component of gravity with respect to time. The conversion from the spatial frequency f_{sm} of the gravity field to the time frequency f_{tm} is defined by formula (3.9).

$$f_{tm} = v f_{sm} = v / n_d \tag{3.9}$$

where n_d is the grid spacing of the gravity anomaly reference map.

The frequency is related to the speed and the resolution of the reference map, and the relationship is shown in Fig. 3.7.

In order to distinguish the interference acceleration and the gravity acceleration in the filtering, it is necessary to distinguish the frequency components of the two, so that the interference signal frequency f_{tm} and the gravity signal frequency f_{vo} must meet.

$$\max(f_{tm}) < \frac{1}{2} \min(f_{vo}) = 0.025 \tag{3.10}$$

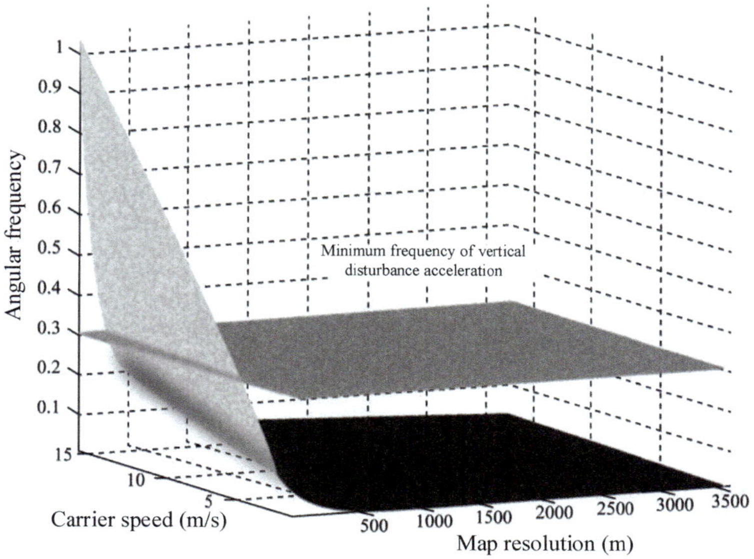

Fig. 3.7 Relationship between gravity signal frequency, carrier speed and reference map resolution

The coefficient 1/2 is to prevent aliasing between frequencies during filtering (according to Shannon's sampling theorem). In practical applications, the denominator value is usually 2.5.

The navigation speed is limited by the map resolution. The unit of the map resolution is meters, and the speed is expressed in international standard units (m/s) and knots (nautical miles/hour) (1 nautical mile = 1852 m).

$$v < 0.025n_d,\ v < 0.04859n_d,\ \text{(knot)}$$

The realization of fir (fine impulse response) filter is to approximate the characteristics of ideal filter by some methods, including interpolation method, least square approximation method and uniform approximation method. They represent different approximation criteria in numerical approximation.

The existence and uniqueness of the model are proved by Chebyshev approximation theory, and the construction method of $p(x)$ is proposed. Later, McClellan J. H, Parks T. W and Rabiner L. R, etc. proposed a computer-aided design method for FIR digital filters by applying Chebyshev's theory. The method has the property of equal ripple in both pass band and stop band, and can accurately control the ripple peak. The design parameters of the low-pass filter include the cut-off frequency of the low-pass filter ω_s. Lower cut-off frequency of stopband ω_p. Passband ripple peak δ_1. Stop band ripple peak δ_2 and unit sampling response time length N.

According to the analysis of error source characteristics, the frequency of error is between 0.1 π-π. During navigation, the highest frequency component of gravity anomaly is 0.05 π by controlling the speed of the carrier. When designing a low-pass filter, its cut-off frequency is usually between 0.05 and 0.1 π.

3.3.2 Compensation Method of Horizontal Acceleration Disturbance in Gravity Dynamic Measurement

If the acceleration sensitive axis deviates from the direction of the geodetic vertical line θ, The horizontal acceleration ah generates a corresponding component on the gravity acceleration sensitive axis and superimposes it on the vertical acceleration g_v to obtain the output g_{sen} of the gravimeter sensor.

$$g_{sen} = g_v \cos\theta - a_h \sin\theta \tag{3.11}$$

The corresponding acceleration disturbance can be obtained.

$$\delta_{g_p} = g_v - g_{sen} = g_v(1 - \cos\theta) + a_h \sin\theta \approx \frac{1}{2} g_v \theta^2 + a_h \theta \tag{3.12}$$

The gyro stabilized platform is used to isolate the rotation movement of the carrier and ensure that the axis of the gravimeter is perpendicular to the local geoid.

By formula (3.12), the error of platform deflection angle will increase the calculation error of horizontal acceleration disturbance. The estimatison of platform deflection angle is the key of horizontal acceleration compensation.

The control method of gyro stabilized platform is one of the key technologies of inertial technology. The stabilization platform can provide relative inertial coordinate stabilization for airborne equipment, thus providing a reference for the alignment of various equipment. In this paper, a stable platform is introduced, and the variation mechanism of its misalignment angle is introduced, which provides a theoretical basis for the correction of horizontal acceleration disturbance.

It can be seen from formula (3.12) that the stable platform is forced to vibrate under the influence of horizontal acceleration, and its frequency response function is.

$$H(s) = -\omega_0 * \frac{2f_s + \omega_0}{s^2 + 2f\omega_0 s + \omega_0^2} \tag{3.13}$$

where the input signal is a_h / g.

Any signal can be understood as the form of periodic signal sum according to Fourier transformation. In the process of gravity navigation, usually the carrier does not make maneuvering navigation, and the main form of disturbance of horizontal motion is periodic signal. Therefore, this paper only analyzes the disturbance acceleration caused by periodic horizontal acceleration.

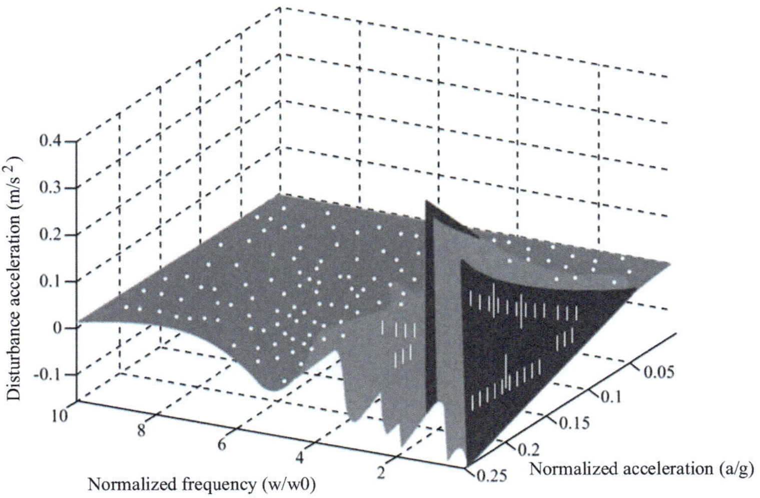

Fig. 3.8 Constant value component of gravimeter measurement disturbance caused by periodic acceleration

From Eqs. (3.13) and (3.12), the calculation formula of disturbance acceleration g_θ caused by periodic horizontal acceleration can be obtained.

$$\begin{aligned} g_\theta = & \frac{1}{2}A(\omega)a_h^2(\cos\vartheta(\omega) + \frac{g}{2}A(\omega) + \cos(2\omega t + \vartheta(\omega)) \\ & - \frac{g}{2}A(\omega) + \cos(2\omega t + 2\vartheta(\omega)) \end{aligned} \tag{3.14}$$

The relationship between horizontal acceleration and disturbance acceleration amplitude at different frequencies and amplitudes is shown in Fig. 3.8.

It can be proved that this effect is minimized when damping $f = 1/\sqrt{2}$.

The periodic component is the remaining part, and its frequency is twice that of the original horizontal acceleration signal. The disturbance to the gravimeter is shown in Fig. 3.9.

Obviously, the error caused by the periodic horizontal acceleration is mainly reflected in the periodic disturbance to the gravimeter. However, its frequency component is twice that of the original horizontal acceleration signal, so as long as low-pass filtering is adopted in signal processing, its influence can be eliminated.

If the stable platform deflection angle in formula (3.14) is directly determined by the gyroscope reading, the compensation obtained must include the random error of the gyroscope. In this paper, based on the established relationship model between horizontal acceleration and the inclination angle of the stabilized platform, the Kalman filtering method is used to obtain the optimal estimation of the inclination angle of the stabilized platform. And error compensation is performed based on the estimated value.

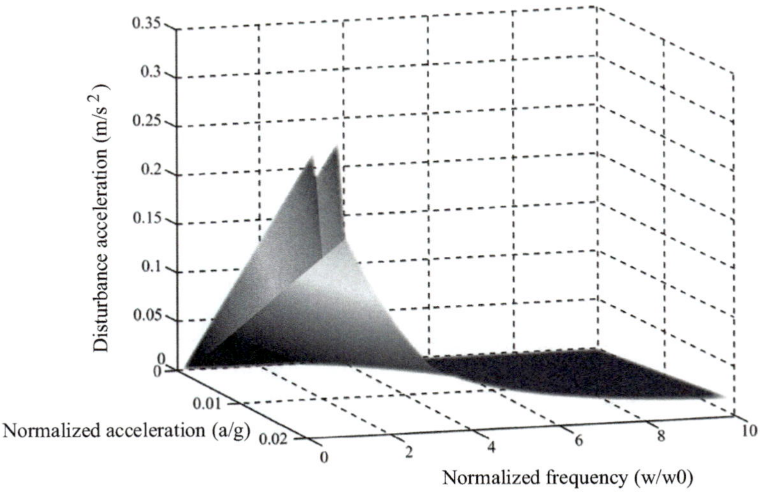

Fig. 3.9 Periodic disturbance caused by periodic horizontal acceleration

The linear state equation of the system can be obtained from (3.15) platform motion differential equation.

$$\dot{x} = \begin{bmatrix} -2f\omega_0 & -\omega_0^2 \\ 1 & 0 \end{bmatrix} x + \begin{bmatrix} -\frac{\omega_0^2}{g} & -\frac{2f}{g} \\ 0 & 0 \end{bmatrix} \begin{bmatrix} \frac{a_h}{g} \\ \frac{\dot{a}_h}{g} \end{bmatrix} \tag{3.15}$$

where, $x = \begin{bmatrix} \dot{\theta} \ \theta \end{bmatrix}^T$.

The input ah of the system can be decomposed into the sum of the real signal and the white noise $a_h = \widehat{a} + w$, where w is the measurement noise of the horizontal accelerometer. Then formula (3.15) can be written in the following form.

$$\dot{x} = \begin{bmatrix} -2f\omega_0 & -\omega_0^2 \\ 1 & 0 \end{bmatrix} x + \begin{bmatrix} -\frac{\omega_0^2}{g} & -\frac{2f}{g} \\ 0 & 0 \end{bmatrix} \begin{bmatrix} \frac{\widehat{a}_h}{g} \\ \frac{\widehat{a}_h}{g} \end{bmatrix} + \begin{bmatrix} -\frac{\omega_0^2}{g} & -\frac{2f}{g} \\ 0 & 0 \end{bmatrix} \begin{bmatrix} w \\ \dot{w} \end{bmatrix} \tag{3.16}$$

The measurement equation of the system is as follows.

$$y = Cx + v \tag{3.17}$$

where, $C = [0 \ \ 1]$, v is the gyro measurement noise.

The measurement errors of calibrated accelerometers and gyroscopes can be regarded as white noise w and v and the statistical characteristics are known, that is to say

$$E(w) = E(v) = 0, E[(w(t)v^T(j)] = 0, \forall t, j$$
$$E[(w(t)w^T(j)] = \sigma_w \delta_{tj}, E(v(t)v^T(j)) = \sigma_v \delta_{tj}$$

3.3.3 Real Time Correction Model of Eötvös Effect

Eötvös effect refers to the influence of the movement of the carrier relative to the earth on the gravity measurement. The disturbance acceleration produced by it in gravity measurement δg_E can be defined by formula (3.18).

$$\delta_{g_E} = \left(1 - \frac{h}{R_\varphi}\right)\left(2\omega_{ie} v \cos\varphi \sin\wedge + \frac{v^2}{R_\varphi}\right) \tag{3.18}$$

where R_φ Is the local earth radius. φ is latitude. ω_{ie} is the earth rotation rate. h is the navigation depth of the carrier. v is the speed of the carrier. $\wedge$ is the course of the carrier.

$$R_\varphi \approx (1 - f \sin^2 \varphi)a \tag{3.19}$$

where f is the earth oblateness. a is the radius of the earth's major axis.

The navigation depth of the lower carrier varies from 100 to 1000 m, and the carriers that can navigate below 300 m are mostly unmanned carriers. According to the maximum sailing depth, the value of h/R_φ in formula (3.18) is 1.573 × 10–4。 Therefore, the first term in the formula can be simplified to 1.

Similarly, for underwater carriers, the maximum speed is about 18 knots (1 knot = 1852 m/hour), and the normal speed is 4–6 knots. The maximum value of v^2/R_φ is 1.3489 × 10−5m/s2,1.3mGal。 About 0.1mgal during normal navigation.

The $2\omega_{ie}v \cos \varphi \sin \Lambda$ term in the formula can be up to 30mgal when the carrier is calculated according to the normal sailing speed of 4 knots. Therefore, formula can be simplified to the following form (Fig. 3.10).

$$\delta g_E = 2\omega_{ie} v \cos\varphi \sin\Lambda \tag{3.20}$$

Formula defines the influence of E ö TV ö s effect on the output of gravimeter, and the compensation for it needs to obtain the speed, heading (or eastward speed) and latitude information of the carrier. The observation of speed and heading of the underwater carrier introduced in this chapter mainly depends on INS.

The influence of velocity, heading and latitude errors on the calculation of Eötvös effect is discussed here. In the formula, the product of the sine of the heading and the velocity is the eastward velocity v_e ($v_e = v \sin \Lambda$), which can be obtained by substituting it into the formula and performing partial differentiation on the function.

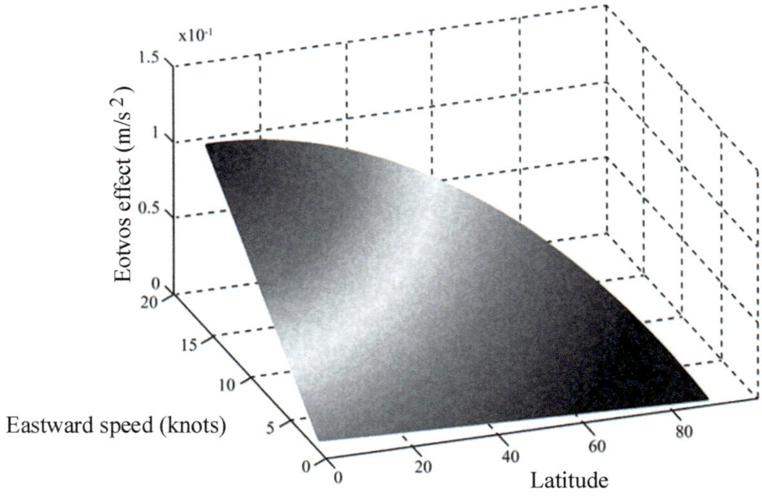

Fig. 3.10 Relationship between latitude, eastward velocity and Eötvös effect

$$\frac{\partial \delta_{gE}}{\partial v_e} = 2\omega_{ie} \cos \varphi \tag{3.21}$$

$$\frac{\partial \delta_g}{\partial \varphi} = -2\omega_{ie} v_e \sin \varphi \tag{3.22}$$

Obviously, the Eötvös effect correction accuracy of the gravimeter should reach 1mGal, and the velocity observation error δv, latitude error $\delta\varphi$ and heading error $\delta\wedge$ shall meet:

$$\begin{aligned} &\delta_{v_e} < \frac{1}{2\omega_{ie} \times 10^5} = 0.068\ m/s \\ &\delta_\varphi < 0.06856/v \\ &\delta\wedge < 0.0686/v \end{aligned} \tag{3.23}$$

When the underwater vehicle travels at the maximum speed of 18 knots, the latitude and heading angle tolerance can reach 0.424°. It can be seen that Eötvös effect correction requires high speed observation, but is not sensitive to latitude and heading errors.

As mentioned above, the navigation and positioning conditions of underwater navigation gravity measurement are limited, and there is a large error in Eötvös effect correction by the velocity observation provided by INS.

Since the Eötvös effect is not sensitive to latitude and heading errors, it is assumed that the underwater carrier keeps moving at a constant speed, and the Eötvös effect is approximately a constant value. This error remains in the gravity observation data gE after the high-frequency noise is eliminated.

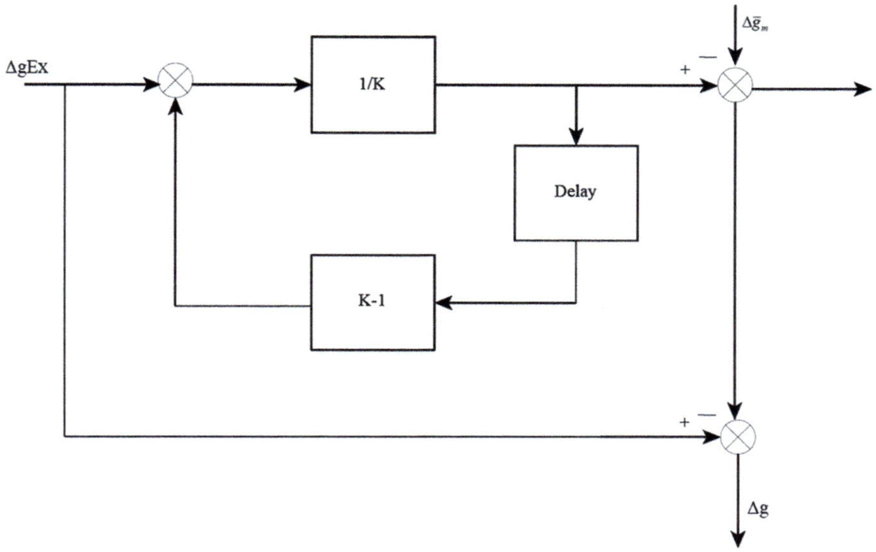

Fig. 3.11 Eötvös effect correction based on gravity anomaly reference map

$$g_E = g + \delta g_E \tag{3.24}$$

Gravity anomaly value calculated by g_E Δg_E includes this part of the error, using the gravity anomaly sequence measured in real time. The Eötvös effect can be corrected by comparing the expectation of Δg_{Ei} value with the local average value of the gravity anomaly reference map. The process is shown in Fig. 3.11.

Real time the expected calculation model of Δg_E is shown in.

$$\Delta \bar{g}_{EK} = \frac{k-1}{k} \Delta \bar{g}_{E(k-1)} + \frac{1}{k} \Delta g_{EK} \tag{3.25}$$

The calculation range of the average value $\Delta \bar{g}_m$ of gravity anomaly map is determined by the matching period Tm of gravity navigation and INS error. In a matching period, the positions of gravity anomaly sampling points are arranged in time:PK, $PK + 1$, $PK + 2, \ldots,$ $PK + n$. The calculation area of the average value of gravity anomaly covers all possible ranges.

3.4 Navigation and Positioning Algorithm of Gravity Field

The gravity field navigation algorithm is the core of the whole gravity aided navigation system. Its output is used to correct ins and limit the accumulation or even divergence of INS errors over time. The low-speed characteristics of underwater navigation and the low resolution of gravity anomaly map are the key factors affecting the accuracy and stability of map matching. This section focuses on the two positioning algorithms of correlation analysis and Kalman filtering according to the characteristics of underwater gravity assisted navigation [15–19].

3.4.1 Correlation Analysis Algorithm

The correlation analysis algorithm is also known as batch processing algorithm. Its basic idea is to use a series of real-time gravity anomaly data collected by the carrier along the track to conduct correlation analysis with the reference gravity anomaly map data, find a series of points with the largest correlation on the map and output them, as shown in Fig. 3.12 [20–22].

The most typical and successfully applied correlation analysis algorithm is the American TERCOM system, which is a terrain correlation matching system. Its basic flow is shown in Fig. 3.13.

The dashed line in Fig. 3.14 shows the flight height measurement system. Similarly, gravity correlation matching system can be constructed.

The correlation analysis algorithm can be described by the following mathematical model. Definition of data map set A:

$$A = \{S(i,j)|(i,j) \in \Omega\} \tag{3.26}$$

where $S(i,j)$ is the physical value (elevation, gravity anomaly, etc.) at the grid coordinates (i,j). Ω is the matching area for navigation.

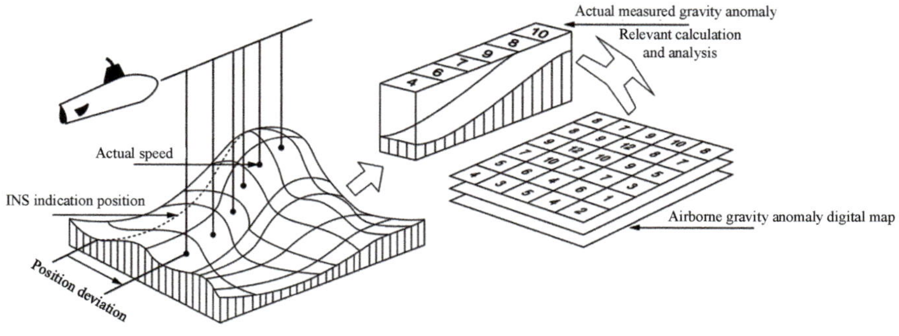

Fig. 3.12 Basic principle of correlation analysis algorithm

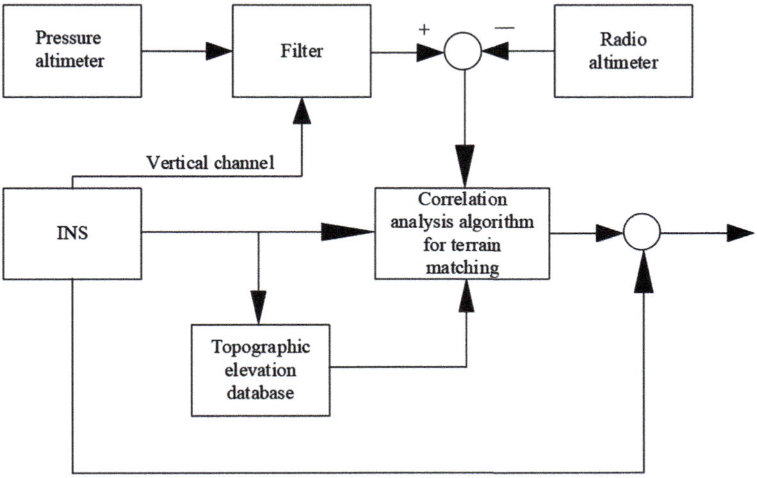

Fig. 3.13 TERCOM system flow

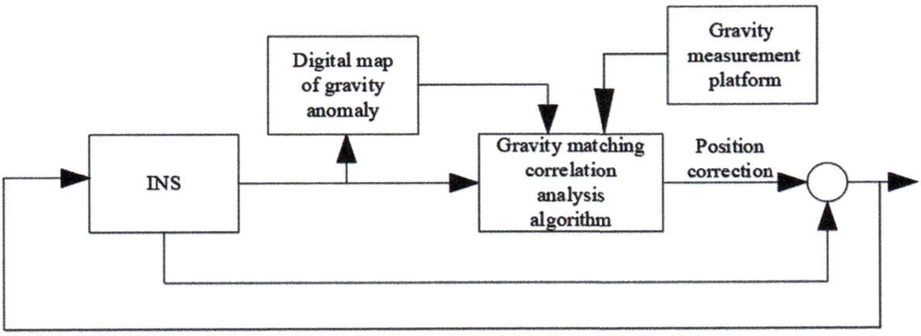

Fig. 3.14 Matching process of gravity correlation analysis

$$\Omega = \{(i,j)|1 \le i \le H, 1 \le j \le W\} \tag{3.27}$$

The data $S(T_k)$ acquired in real time is a one-dimensional array based on time sequence K. The position estimation P_k shall meet the requirement of maximum correlation between $S(T_k)$ and $S(P_k)$, namely:

$$\max\{\delta_i = f(S(P_k), S(T_k))\} \tag{3.28}$$

where $f(\cdot)$is the correlation function.

In the implementation of the algorithm, it is necessary to define the correlation function and determine the timing length.

The gravity anomaly sequence $S(T_k)$ (abbreviated as S_{tn}) measured in real time and the gravity anomaly sequence $S(P_k)$ (abbreviated as S_{pn}) with the same length along the track direction on the map can be regarded as two vectors, and their distance indicates their similarity. The distance can be expressed by a universal number. The correlation functions often used in the correlation analysis map matching algorithm are as follows:

1. Cross correlation algorithm

By using the inner product of vectors, the cross-correlation calculation formula can be constructed.

$$J_{cor} = \frac{1}{n}\sum_{i=1}^{n} S_{ti} \cdot Spi \tag{3.29}$$

Relevant literature points out that the maximum value of the product correlation algorithm may not be unique even under ideal conditions, so it is normalized.

$$J_{\text{cor}} = \frac{1}{n}\sum_{i=1}^{n} (S_{ti}/\|S_{tn}\|^2)(S_{pi}/\|S_{pn}\|^2) \tag{3.30}$$

After normalization, J_{cor} is the inner product of two unit vectors. Only when the included angle is 0, J_{cor} reaches the maximum value. This algorithm has a large amount of computation.

2. Absolute difference algorithm

Their similarity is defined by the abstract universal number $||x||$ 1.

$$J_{ad} = \|S_{tn} - S_{pn}\| = \sum_{i=1}^{n} |S_{ti} - S_{pi}| \tag{3.31}$$

The mean absolute difference algorithm is further improved.

$$J_{mad} = \frac{1}{n}\sum_{i=1}^{n} |S_{ti} - S_{pi}| \tag{3.32}$$

When J_{mad} is the smallest, the correlation between the two vectors is the largest.

3. Square difference algorithm.

The difference is defined by the square of Euclidean universal number $||x||$ 2.

$$J_{sd} = \|S_{tn} - S_{pn}\|^2 = \sum_{i=1}^{n} (S_{ti} - S_{pi})^2 \tag{3.33}$$

The actual algorithm is the mean square difference algorithm.

$$J_{msd} = \frac{1}{n}\sum_{i=1}^{n}\left(S_{ti} - S_{pi}\right)^2 \tag{3.34}$$

When J_{msd} is the smallest, their correlation is the largest.

4. Modulo two addition (XOR) algorithm

The Hamming distance of binary codes is used to define their similarity.

S_{tn} and S_{pn} are quantized as 0 or 1 values, respectively, defined by Boolean algebra.

$$S_{ti} \oplus S_{pi} = S_{ti} \cdot \overline{S}_{pi} + \overline{S} \cdot S_{pi} \tag{3.35}$$

The "." and " + " operations respectively represent "and" and "or", the top horizontal line "–" represents "Negation", and " ⊕ " is a modulo two addition operator. Hamming distance is defined as:

$$J_{xor} = \sum_{i=1}^{n} S_{ti} \oplus S_{pi} \tag{3.36}$$

In the above methods, XOR algorithm requires a large number of data within each correlation analysis length for analysis, and correlation matching based on gravity anomaly is difficult to meet the requirements. Cross correlation algorithm, MAD algorithm and MSD algorithm are compared (Literature). The comparison results show that the calculation results of MAD algorithm and MSD algorithm are close and the accuracy is similar, and MSD algorithm is slightly better than mad algorithm. The cross correlation algorithm has the lowest accuracy.

3.4.2 Gravity Field Matching Location Algorithm Based on Extended Kalman Filter Theory

Kalman filter is a method proposed by R. E. Kalman in 1960, which is applied to the optimal estimation of stationary and nonstationary multi-dimensional stochastic processes. It is the core content of modern control theory. The meaning and method of Kalman filtering are completely different from that of conventional filtering, and it should be called optimal estimation theory. Kalman filter estimates the required signal through algorithm from the measurement related to the extracted signal. The estimated signal is a random response caused by white noise excitation. The transfer structure (system equation) between the excitation source and the response is known, and the functional relationship (measurement equation) between the measurement and the estimated signal is known. The following information is used in the estimation process: system equation, measurement equation, statistical characteristics of white noise excitation Statistical characteristics of measurement error. From the implementation form, the Kalman filter is essentially a set of recursive algorithms implemented by computers. Each recursive cycle includes two processes of time update and measurement update of the estimated quantity. The time update is determined by the measurement update result of the previous step and the prior information when designing the Kalman filter. The measurement update is determined based on the measured value obtained in real time on the basis of the time update. The measurement is the input of the Kalman filter, and the estimation is the output of the Kalman filter. The two are connected by the time update and measurement update algorithms. The Kalman filtering theory was applied to practical engineering as soon as it was put forward. Apollo lunar landing program and C-5A aircraft navigation system were the most successful examples of its early application; At present, Kalman filtering theory, as the most important optimal estimation theory, is widely used in inertial navigation, guidance system, GPS, target tracking, communication and signal process, stochastic optimal control, fault diagnosis and other application fields. The design of integrated navigation system is the most important aspect of its successful application [23].

Let the state equation (without considering the control action) and the measurement equation of the random linear discrete system be:

$$\begin{cases} X_k = \Phi_{k,k-1}X_{k-1} + \Gamma_{k,k-1}W_{k-1} \\ Z_k = H_kX_k + V_k \end{cases} \tag{3.37}$$

where X_k is the n-dimensional state vector of the system. $\Phi_{k,k-1}$ is the $n \times n$-dimensional state transition matrix of the system. $\Gamma_{k,k-1}$ is an $n \times p$-dimensional noise input matrix. W_k is an p-dimensional noise input matrix. Z_k is the m-dimensional observation sequence of the system. H_k is an $m \times n$-dimensional observation matrix. V_k is the m-dimensional observation noise sequence.

If the statistical characteristics of system process noise and observation noise meet:

$$\begin{cases} E[W_k] = 0, E\left[W_kW_j^T\right] = Q_k\delta_{kj} \\ E[V_k] = 0, E\left[V_kV_j^T\right] = R_k\delta_{kj} \\ E\left[W_kV_j^T\right] = 0 \end{cases} \tag{3.38}$$

where Q_k is the $p \times p$-dimensional symmetric nonnegative definite variance matrix of the system process noise W_k. R_k is the $m \times m$-dimensional symmetric positive definite variance matrix of the systematic observation noise V_k.δ_{kj} is the Kronecker-δ function. If the observed value at time k is Z_k and the estimated value of state vector X_k is $\hat{X}_k$, the Kalman filtering equation is:

Status one step prediction (time update):

$$\hat{X}_{k,k-1} = \Phi_{k,k-1}\hat{X}_{k-1} \tag{3.39}$$

State estimation (observation update):

$$\hat{X}_k = \hat{X}_{k,k-1} + K_k\left[Z_k - H_k\hat{X}_{k,k-1}\right] \tag{3.40}$$

Filter gain matrix:

$$K_k = P_{k,k-1}H_k^T[H_kP_{k,k-1}H_k^T + R_k]^{-1} \tag{3.41}$$

One step prediction error variance matrix (time update):

$$P_{k,k-1} = \Phi_{k,k-1}P_{k-1}\Phi_{k,k-1}^T + \Gamma_{k,k-1}Q_{k-1}\Gamma_{k,k-1}^T \tag{3.42}$$

Estimation error variance matrix.

$$P_k = [I - K_kH_k]P_{k,k-1} \tag{3.43}$$

Initial value.

$$\hat{X}_0 = E[X_0]\,P_0 = E\left[X_0X_0^T\right]\,and\,E\left[X_0W_k^T\right] = 0,\,E\left[X_0V_k^T\right] = 0 \tag{3.44}$$

The above equations are the final equations used by various optimal estimation based on Kalman filtering theory.

The standard Kalman filtering problem is based on the physical system. The mathematical model is linear, but the physical system mathematical model encountered in engineering practice is often nonlinear, that is, the system equation is nonlinear, or the system equation and the measurement equation are nonlinear. For general nonlinear systems, there is no strict recursive filtering method in theory at present, and the approximate method is usually used to solve the nonlinear filtering problem. The Kalman filtering of nonlinear system shall meet the following conditions: the theoretical solution of nonlinear equation must exist, and the difference between the theoretical solution and the actual solution can be expressed by a linear differential equation, which is called "linear interference equation", "small deviation equation" or "perturbation equation". This condition is basically satisfied in engineering practice.

There are two main forms of linearization of nonlinear filtering: linearized Kalman filter and extended Kalman filter (EKF). The linearized Kalman filter is to expand the system state equation or measurement equation into Taylor series around the nominal state of the system (the theoretical solution of the nonlinear equation when the system noise or measurement noise is constant zero). However, because it cannot ensure that the deviation between the true trajectory and the nominal trajectory is small enough, it is rarely used in engineering practice. Instead, the EKF method is widely used, that is, the linearization method around the optimal state estimation is adopted.

1. Extended Kalman filter model

Consider the following continuous stochastic nonlinear systems:

$$\begin{cases} \dot{X}(t) = f[X(t), t] + G(t)w(t) \\ Z(t) = h[X(t), t) + v(t) \end{cases} \tag{3.45}$$

where $f[\cdot]$ is a continuous function of n-dimensional nonlinear vector. $h[\cdot]$ is an m-dimensional nonlinear vector continuous function. $w(t)$ and $v(t)$ are zero mean white noise sequence. Its statistical characteristics are as follows:

$$\begin{cases} E[w(t)] = 0,\ E[w(t) \cdot w^T(\tau)] = q(t)\delta(t-\tau) \\ E[v(t)] = 0,\ E[v(t) \cdot v^T(\tau)] = r(t)\delta(t-\tau) \\ E[w(t) \cdot v^T(\tau)] = 0,\ E[X(0) \cdot w^T(\tau)] = 0,\ E[X(0) \cdot v^T(\tau)] = 0 \end{cases} \tag{3.46}$$

For Eq. (3.46), the linear interference equation can be obtained by expanding the optimal state estimate $\hat{X}(t)$ into Taylor series and taking its first approximation.

$$\begin{cases} \delta\dot{X}(t) = F(t)\delta X(t) + G(t)w(t) \\ \delta Z(t) = H(t)\delta X(t) + v(t) \end{cases} \tag{3.47}$$

where,

$$F(t) = \left.\frac{\partial f[X(t), t]}{\partial X(t)}\right|_{X(t)=\hat{X}(t)} = \begin{bmatrix} \frac{\partial f_1[X(t),t]}{\partial x_1(t)} & \frac{\partial f_1[X(t),t]}{\partial x_2(t)} & \cdots & \frac{\partial f_1[X(t),t]}{\partial x_n(t)} \\ \frac{\partial f_2[X(t),t]}{\partial x_1(t)} & \frac{\partial f_2[X(t),t]}{\partial x_2(t)} & \cdots & \frac{\partial f_2[X(t),t]}{\partial x_n(t)} \\ \vdots & \vdots & & \vdots \\ \frac{\partial f_n[X(t),t]}{\partial x_1(t)} & \frac{\partial f_n[X(t),t]}{\partial x_2(t)} & \cdots & \frac{\partial f_n[X(t),t]}{\partial x_n(t)} \end{bmatrix}_{X(t)=\hat{X}(t)} \tag{3.48}$$

$$H(t) = \left.\frac{\partial h[X(t), t]}{\partial X(t)}\right|_{X(t)=\hat{X}(t)} = \begin{bmatrix} \frac{\partial h_1[X(t),t]}{\partial x_1(t)} & \frac{\partial h_1[X(t),t]}{\partial x_2(t)} & \cdots & \frac{\partial h_1[X(t),t]}{\partial x_n(t)} \\ \frac{\partial h_2[X(t),t]}{\partial x_1(t)} & \frac{\partial h_2[X(t),t]}{\partial x_2(t)} & \cdots & \frac{\partial h_2[X(t),t]}{\partial x_n(t)} \\ \vdots & \vdots & & \vdots \\ \frac{\partial h_m[X(t),t]}{\partial x_1(t)} & \frac{\partial h_m[X(t),t]}{\partial x_2(t)} & \cdots & \frac{\partial h_m[X(t),t]}{\partial x_n(t)} \end{bmatrix}_{X(t)=\hat{X}(t)} \tag{3.49}$$

The $F(t)$ and $H(t)$ matrices are referred to as "Jacobian matrices". Continue to discretize the basic solution matrix and the discrete linear interference equation is:

$$\begin{cases} \delta X_k = \Phi_{k,k-1}\delta X_{k-1} + \Gamma_{k,k-1}W_{k-1} \\ \delta Z_k = H_k\delta X_k + V_k \end{cases} \tag{3.50}$$

When the sampling period T is small.

$$
\begin{aligned}
\Phi_{k,k-1} &\simeq I + F(t_{k-1}) \bullet T \\
&= I + T \bullet \begin{bmatrix} \frac{\partial f_1[X(t_{k-1}),t_{k-1}]}{\partial x_1(t_{k-1})} & \frac{\partial f_1[X(t_{k-1}),t_{k-1}]}{\partial x_2(t)} & \cdots & \frac{\partial f_1[X(t_{k-1}),t_{k-1}]}{\partial x_n(t)} \\ \frac{\partial f_2[X(t_{k-1}),t_{k-1}]}{\partial x_1(t_{k-1})} & \frac{\partial f_2[X(t_{k-1}),t_{k-1}]}{\partial x_2(t_{k-1})} & \cdots & \frac{\partial f_2[X(t_{k-1}),t_{k-1}]}{\partial x_n(t_{k-1})} \\ \vdots & \vdots & & \vdots \\ \frac{\partial f_n[X(t_{k-1}),t_{k-1}]}{\partial x_1(t_{k-1})} & \frac{\partial f_n[X(t_{k-1}),t_{k-1}]}{\partial x_2(t_{k-1})} & \cdots & \frac{\partial f_n[X(t_{k-1}),t_{k-1}]}{\partial x_n(t_{k-1})} \end{bmatrix}_{X(t_{k-1})=\hat{X}_{k-1}}
\end{aligned} \tag{3.51}
$$

$$
\begin{aligned}
H_k &= \left.\frac{\partial h[X(t_k),t_k]}{\partial X(t_k)}\right|_{X(t_k)=\hat{X}_{k,k-1}} \\
&= \begin{bmatrix} \frac{\partial h_1[X(t_k),t_k]}{\partial x_1(t)} & \frac{\partial h_1[X(t_k),t_k]}{\partial x_2(t)} & \cdots & \frac{\partial h_1[X(t_k),t_k]}{\partial x_n(t)} \\ \frac{\partial h_2[X(t_k),t_k]}{\partial x_1(t)} & \frac{\partial h_2[X(t_k),t_k]}{\partial x_2(t)} & \cdots & \frac{\partial h_2[X(t_k),t_k]}{\partial x_n(t)} \\ \vdots & \vdots & & \vdots \\ \frac{\partial h_m[X(t_k),t_k]}{\partial x_1(t_k)} & \frac{\partial h_m[X(t_k),t_k]}{\partial x_2(t_k)} & \cdots & \frac{\partial h_m[X(t_k),t_k]}{\partial x_n(t_k)} \end{bmatrix}_{X(t_k)=\hat{X}_{k,k-1}}
\end{aligned} \tag{3.52}
$$

$$
\begin{cases} \Gamma_{k,k-1} = G(t) \cdot T \\ W_{k-1} = q(t)/T \,,\ V_k = r(t)/T \end{cases} \tag{3.53}
$$

Referring to the basic equation of discrete Kalman filtering, it is easy to obtain the deviation Kalman filtering equation:

$$
\begin{cases} \delta\hat{X}_{k,k-1} = \Phi_{k,k-1}\delta\hat{X}_{k-1} \\ \delta\hat{X}_k = \delta\hat{X}_{k,k-1} + K_k[\delta Z_k - H_k\delta\hat{X}_{k,k-1}] \\ K_k = P_{k,k-1}H_k^T[H_kP_{k,k-1}H_k^T + R_k]^{-1} \\ P_{k,k-1} = \Phi_{k,k-1}P_{k-1}\Phi_{k,k-1}^T + \Gamma_{k,k-1}Q_{k-1}\Gamma_{k,k-1}^T \\ P_k = (I - K_kH_k)P_{k,k-1}(I - K_kH_k)^T + K_kR_kK_k^T \end{cases} \tag{3.54}
$$

where

$$
\delta Z_k = Z_k - h[\hat{X}_{k,k-1}, k] \tag{3.55}
$$

The discrete nonlinear EKF equation can be obtained.

$$
\begin{cases} \hat{X}_{k,k-1} = \hat{X}_{k-1} + f[\hat{X}_{k-1}, k-1] \cdot T \\ \hat{X}_k = \hat{X}_{k,k-1} + \delta\hat{X}_k \\ \delta\hat{X}_k = K_k[Z_k - h[\hat{X}_{k,k-1}, k] \\ K_k = P_{k,k-1}H_k^T[H_kP_{k,k-1}H_k^T + R_k]^{-1} \\ P_{k,k-1} = \Phi_{k,k-1}P_{k-1}\Phi_{k,k-1}^T + \Gamma_{k,k-1}Q_{k-1}\Gamma_{k,k-1}^T \\ P_k = (I - K_kH_k)P_{k,k-1}(I - K_kH_k)^T + K_kR_kK_k^T \end{cases} \tag{3.56}
$$

Initial condition.

$$\hat{X}_0 = E[X_0] = m_{X0}\,,\ P_0 = C_{X0} \tag{3.57}$$

Finally, it should be pointed out that the above nonlinear filtering method is not optimal in fact, but is approximately optimal. Therefore, this kind of filter is often called "suboptimal Kalman filter". EKF is often called "generalized Kalman filter" or "extended Kalman filter", which is widely used in engineering practice.

2. Establishment of state equation of gravity/INS combined system.

When the Kalman filter is used as a closed-loop control system, it can be used not only to correct the error of the inertial navigation system itself, but also to correct the error of gyroscopes and accelerometers that cause the error of the inertial navigation system, that is, to correct the error source of the inertial navigation system. The errors of gyroscope mainly include scale coefficient error and drift, both of which are random errors. Generally, the calibration coefficient error of gyroscope is generally described by random constant, and the drift error is described by random constant, first-order Markov process and white noise process. The error model of the accelerometer is similar to that of the gyroscope. In the integrated navigation system, the scale coefficient error is described by a random constant, and the drift error is described by a first-order Markov process.

Thus, the continuous state equation of gravity/INS combined system can be established as follows:

$$\dot{X}(t) = F(t)X(t) + G(t)w(t) \tag{3.58}$$

where,

$$X(t) = [\delta\varphi \quad \delta\lambda \quad \delta h \quad \delta V_E \quad \delta V_N \quad \delta V_U \quad \phi_E \quad \phi_N \quad \phi_U \quad \varepsilon_{bx} \quad \varepsilon_{by} \quad \varepsilon_{bz} \quad \varepsilon_{rx} \quad \varepsilon_{ry} \quad \varepsilon_{rz} \\ \nabla_x \quad \nabla_y \quad \nabla_z \quad \delta K_{Gx} \quad \delta K_{Gy} \quad \delta K_{Gz} \quad \delta K_{Ax} \quad \delta K_{Ay} \quad \delta K_{Az}]^T$$

where,$\varepsilon_{bi}(i = x, y, z)$ is the constant drift of the gyroscope. $\varepsilon_{ri}(i = x, y, z)$ is the gyro related drift. $\nabla_{ri}(i = x, y, z)$ is the zero bias of the accelerometer.$\delta K_{Gi}(i = x, y, z)$ is the gyroscope scale factor error. $\delta K_{Ai}(i = x, y, z)$ is the calibration coefficient error of the accelerometer.

$$w(t) = \left[000000\,w_{gx}\,w_{gy}\,w_{gz}\,000\,w_{rx}\,w_{ry}\,w_{rz}\,w_{ax}\,w_{ay}\,w_{ax}\,000000\right]^T$$

where, $w_{gi}(i = x, y, z)$ is the gyro random walk coefficient. $w_{ri}(i = x, y, z)$ is the gyro related drift intensity variance. $w_{ai}(i = x, y, z)$ is the variance of the relevant drift intensity of the accelerometer.

The non-zero element of $F(t)$ is.

$$F_{1,3} = -\frac{V_N}{R^2},\ F_{1,5} = \frac{1}{R^2}$$

$$F_{2,1} = \frac{V_E \tan\varphi}{R\cos\varphi}, F_{2,3} = -\frac{V_E}{R^2\cos\varphi}, F_{2,4} = \frac{1}{R\cos\varphi}, F_{3,6} = 1$$

$$F_{4,1} = 2\omega_{ie}\cos\varphi V_N + \frac{V_E V_N}{R\cos^2\varphi} + 2\omega_{ie}\sin\varphi V_U, F_{4,3} = \frac{V_E V_U - V_E V_N \tan\varphi}{R^2}$$

$$F_{4,4} = \frac{V_N\tan\varphi}{R} - \frac{V_U}{R}, F_{4,5} = 2\omega_{ie}\sin\varphi + \frac{V_E\tan\varphi}{R}, F_{4,6} = -\left(2\omega_{ie}\cos\varphi + \frac{V_E}{R}\right)$$

$$F_{4,8} = -f_U, F_{4,9} = f_N, F_{4,16} = C_b^n(1,1), F_{4,17} = C_b^n(1,2),$$
$$F_{4,18} = C_b^n(1,3), F_{4,22} = C_b^n(1,1)f_x, F_{4,23} = C_b^n(1,2)f_y, F_{4,24} = C_b^n(1,3)f_z$$

$$F_{5,1} = -(2\omega_{ie}\cos\varphi V_E + \frac{V_E^2}{R\cos^2\varphi}), F_{5,3} = \frac{V_E^2\tan\varphi + V_N V_U}{R^2}$$

$$F_{5,4} = -2(\omega_{ie}\sin\varphi + \frac{V_E\tan\varphi}{R}), F_{5,5} = -\frac{V_U}{R}, F_{5,6} = -\frac{V_N}{R}$$

$$F_{5,7} = f_U, F_{5,9} = -f_E, F_{5,16} = C_b^n(2,1), F_{5,17} = C_b^n(2,2), F_{5,18} = C_b^n(2,3),$$
$$F_{5,22} = C_b^n(2,1)f_x, F_{5,23} = C_b^n(2,2)f_y, F_{5,24} = C_b^n(2,3)f_z, F_{6,1} = -2\omega_{ie}\sin\varphi V_E$$

$$F_{6,3} = -\frac{V_E^2 + V_N^2}{R^2}, F_{6,4} = 2\omega_{ie}\cos\varphi + \frac{V_E}{R}, F_{6,5} = \frac{2V_N}{R}$$

$$F_{6,7} = -f_N, F_{6,8} = f_E, F_{6,16} = C_b^n(3,1), F_{6,17} = C_b^n(3,2), F_{6,18} = C_b^n(3,3),$$
$$F_{6,22} = C_b^n(3,1)f_x, F_{6,23} = C_b^n(3,2)f_y, F_{6,24} = C_b^n(3,3)f_z, F_{7,3} = \frac{V_N}{R^2}$$

$$F_{7,5} = -\frac{1}{R}, F_{7,8} = \omega_{ie}\sin\varphi + \frac{V_E}{R}\tan\varphi, F_{7,9} = -\left(\omega_{ie}\cos\varphi + \frac{V_E}{R}\right)$$

$$F_{7,10} = C_b^n(1,1), F_{7,11} = C_b^n(1,2), F_{7,12} = C_b^n(1,3), F_{7,13} = C_b^n(1,1), F_{7,14} = C_b^n(1,2), F_{7,15}$$
$$= C_b^n(1,3), F_{7,19} = C_b^n(1,1)\omega_x, F_{7,20} = C_b^n(1,2)\omega_y, F_{7,21} = C_b^n(1,3)\omega_z.$$

$$F_{8,1} = -\omega_{ie}\sin\varphi, F_{8,3} = -\frac{V_E}{R^2},$$

$$F_{8,4} = \frac{1}{R}, F_{8,7} = -\left(\omega_{ie}\sin\varphi + \frac{V_E\tan\varphi}{R}\right), F_{8,9} = -\frac{V_N}{R}$$

$$F_{8,10} = C_b^n(2,1), F_{8,11} = C_b^n(2,2), F_{8,12} = C_b^n(2,3), F_{8,13} = C_b^n(2,1), F_{8,14} = C_b^n(2,2),$$
$$F_{8,15} = C_b^n(2,3), F_{8,19} = C_b^n(2,1)\omega_x, F_{8,20} = C_b^n(2,2)\omega_y, F_{8,21} = C_b^n(2,3)\omega_z.$$

$$F_{9,1} = \omega_{ie}\cos\varphi + \frac{V_E}{R\cos^2\varphi}, F_{9,3} = -\frac{V_E\tan\varphi}{R^2}$$

$$F_{9,4} = \frac{\tan\varphi}{R}, F_{9,7} = \omega_{ie}\cos\varphi + \frac{V_E}{R}, F_{9,8} = \frac{V_N}{R}.$$

$$F_{9,10} = C_b^n(3,1), F_{9,11} = C_b^n(3,2), F_{9,12} = C_b^n(3,3), F_{9,13} = C_b^n(3,1), F_{9,14} = C_b^n(3,2),$$
$$F_{9,15} = C_b^n(3,3), F_{9,19} = C_b^n(3,1)\omega_x, F_{9,20} = C_b^n(3,2)\omega_y, F_{9,21} = C_b^n(3,3)\omega_z.$$

$$F_{13,13} = -\frac{1}{\tau_G}, F_{14,14} = -\frac{1}{\tau_G}, F_{15,15} = -\frac{1}{\tau_G}.$$

where, C_b^n is the attitude matrix of strapdown inertial navigation. $\omega_i(i = 1, 2, 3)$ is the angular velocity of the download body in the carrier coordinate system.$f_i(i = 1, 2, 3)$ is the projection of the ideal specific force of the carrier in the carrier coordinate system.

The system noise distribution matrix is:

$$G(t) = \begin{bmatrix} 0_{6\times3} & 0_{6\times3} & 0_{6\times3} \\ C_b^n & 0_{3\times3} & 0_{3\times3} \\ 0_{3\times3} & 0_{3\times3} & 0_{3\times3} \\ 0_{3\times3} & I_{3\times3} & 0_{3\times3} \\ 0_{3\times3} & 0_{3\times3} & I_{3\times3} \\ 0_{6\times3} & 0_{6\times3} & 0_{6\times3} \end{bmatrix}$$

3. Establishment of observation equation of gravity/INS combined system

The measured value obtained from the gravity real-time measurement system installed on the carrier is the actual gravity anomaly value at the position of the carrier, which is a function of the carrier position and expressed in the inertial coordinate system as

$$\Delta g(\vec{r}) = -\frac{\vec{r}}{r}\left(\vec{n} - \frac{d\vec{v}}{dt} - \frac{GM\vec{r}}{r^3}\right) = f(\varphi_{\mathrm{T}}, \lambda_{\mathrm{T}}) \tag{3.59}$$

where, $\vec{r}$ is the radial vector of the position of the carrier. $\vec{n}$ is the state space accelerometer output. $\vec{v}$ is the moving speed of the carrier. (φ_T, λ_T) is the true position of the carrier in the geodetic coordinate system.

The gravity anomaly ∇g_M corresponding to the indicated position $(\varphi_{INS}, \lambda_{INS})$ of the inertial navigation system on the gravity anomaly map is:

$$\Delta g_M = f(\varphi_{INS} + \delta\varphi, \lambda_{INS} + \delta\lambda) \tag{3.60}$$

where, $\delta\varphi = \varphi_T - \varphi_{INS}$. $\delta\lambda = \varphi_T - \lambda_{INS}$ is ins position error, then.

$$\delta g_a = \Delta g_M - \Delta g_S = f(\varphi_{INS} + \delta\varphi, \lambda_{INS} + \delta\lambda) - f(\varphi_{INS}, \lambda_{INS}) \tag{3.61}$$

When $\delta\varphi$ and $\delta\lambda$ are small quantities, expand Eq. (3.61) by Taylor series and take the first approximation value

$$\delta g_a = \frac{\partial f}{\partial \varphi}\delta\varphi + \frac{\partial f}{\cos\varphi\partial\lambda}\delta\lambda + \varepsilon_P$$

Therefore, the continuous observation equation of the system is:

$$Z(t) = \delta g_a = \frac{\partial f}{\partial \varphi}\delta\varphi + \frac{\partial f}{\cos\varphi\partial\lambda}\delta\lambda + v = \begin{bmatrix} \frac{\partial f}{\partial \varphi} & \frac{\partial f}{\cos\varphi\partial\lambda} & 0_{1\times 22} \end{bmatrix}^T X(t) + v(t) \quad (3.62)$$

where $v(t)$ is composed of the real-time measurement error ε_S of gravity anomaly, the linearization error ε_P of gravity map, the mapping error ε_M of gravity anomaly and other random errors ε_O, which can be uniformly expressed as the measurement error of gravimeter.

Since the gravity anomaly is generally nonlinear and random in nature with respect to the distance, the mean tangent line (MTL) technique is adopted in this paper with reference to the terrain stochastic linearization technique in the terrain navigation system. MTL technology uses the gravity anomaly data of all grid points on the gravity anomaly map to determine $\frac{\partial f}{\partial \varphi}$ and $\frac{\partial f}{\cos\varphi\partial\lambda}$. the area of the gravity anomaly map is about $5\sigma_x \times 5\sigma_y$, and its center is on the point (m, n) where $\frac{\partial f}{\partial \varphi}$ and $\frac{\partial f}{\cos\varphi\partial\lambda}$ need to be calculated, as shown in Fig. 3.15.

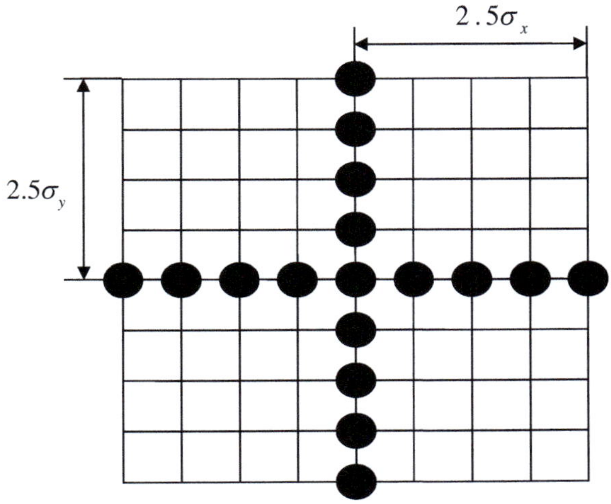

Fig. 3.15 Plot geometry for average tangent technique

$$\begin{cases} \dfrac{\partial f}{\partial \varphi} = \dfrac{1}{2Nd} \sum\limits_{\substack{k=-N \\ k \neq 0}}^{N} \dfrac{\nabla g(m, n+k) - \nabla g(m, n)}{|k|} \\ \dfrac{\partial f}{\partial \lambda} = \dfrac{1}{2Md} \sum\limits_{\substack{k=-M \\ k \neq 0}}^{M} \dfrac{\nabla g(m+k, n) - \nabla g(m, n)}{|k|} \end{cases} \tag{3.63}$$

For the estimation of inertial navigation position error, Farrell error estimation formula is adopted here:

$$\begin{cases} \overline{x}_E(t) = x_E(0) + V_E(0)\dfrac{\sin Wt}{W} + R_e(\phi_N + \dfrac{\nabla_y}{g})(1 - \cos Wt) \\ + R_e(\varepsilon_{bx} - \overline{\omega}_2\phi_U)(t - \dfrac{\sin Wt}{W}) + R_e\overline{\omega}_2\varepsilon_{bz}(\dfrac{1 - \cos Wt}{W^2} - \dfrac{1}{2}t^2) + \sigma_E \\ \overline{x}_N(t) = x_N(0) + V_N(0)\dfrac{\sin Wt}{W} + \sqrt{gR_e}\phi_E(\dfrac{\cos Wt - 1}{W}) + \nabla_x\dfrac{\sin Wt}{W^2} \\ - R_e(\varepsilon_{by} + \overline{\omega}_1\phi_U)(\dfrac{\sin Wt}{W} - t) + R_e\overline{\omega}_1\varepsilon_{bz}(\dfrac{\cos Wt - 1}{W^2} - \dfrac{1}{2}t^2) + \sigma_N \end{cases} \tag{3.64}$$

where, $x_E(0)$ and $x_N(0)$ are initial position errors. σ_E and σ_N are position random errors. $V_E(0)$ and $V_N(0)$ are initial velocity errors. R_e is the average radius of the earth. g is the standard gravitational acceleration, and $W = \sqrt{\frac{g}{R_e}}$(rads/s). ϕ_E, ϕ_N and ϕ_U are misalignment angles. ε_{bx}, ε_{by} and ε_{bz} are constant drift of gyroscope. ∇_x, ∇_y and ∇_z are the zero bias of the accelerometer, and $\begin{cases} \overline{\omega}_1 = \omega_{ie}\cos\varphi \\ \overline{\omega}_2 = \omega_{ie}\sin\varphi \end{cases}$.

In addition, in order to prevent mismatching caused by Kalman filtering, the book uses average weighted square residual (*AWRS*) to judge:

$$AWRS = \frac{1}{L}\left\{\sum_{k=1}^{L}\left[\Delta g_s - \Delta\hat{g}_s\right]^2 \Big/ \left[H_k P_k H_k^T + r_k\right]\right\} \tag{3.65}$$

If the *AWRS* value is still greater than the given threshold after L times of filter correction, it indicates that the filter matching fails.

3.4.3 Gravity Field Matching Algorithm Based on Unbiased Kalman Filter Theory

In the above two sections of this chapter, the traditional Kalman filtering theory, the extended Kalman filtering theory and their applications in underwater gravity ins matching navigation are introduced in detail, and the underwater gravity anomaly matching algorithm of EKF filter is given; At the same time, EKF was also the most widely used sub optimal approximation method to solve nonlinear problems. However, EKF still has the following disadvantages: (1) when the higher order term of Taylor expansion of nonlinear function cannot be ignored, the linearization will cause large errors in the system and even make the filter difficult to stabilize; (2) In many practical problems, it is difficult to obtain the derivative of Jacobian matrix of nonlinear function; (3) EKF requires derivation, so it must clearly understand the specific form of nonlinear function, and cannot be packaged in black box, so it is difficult to apply in modularization. At present, although there are many improved methods for EKF, such as high-order truncated EKF and iterative EKF, these defects are still difficult to overcome.

Because it is easier to approximate the probability density distribution of nonlinear functions than to approximate nonlinear functions, the approach to solve nonlinear problems by using sampling methods to approximate nonlinear distributions has recently attracted extensive attention. Among them, unscented Kalman filter (UKF) is a large class of methods to approximate nonlinear distributions with sampling strategies. UKF is based on ut transform and adopts Kalman linear filtering framework. The specific sampling form is deterministic sampling. The number of particle points sampled (generally referred to as sigma points) is small. The specific number depends on the selected sampling strategy. The most commonly used is 2n + 1 sigma point symmetric sampling. UKF algorithm has the following characteristics: (1) it approximates the probability density distribution of the nonlinear function instead of approximating the nonlinear function; (2) The calculation accuracy of nonlinear distribution statistics shall reach at least second order, and higher order accuracy can be achieved for special sampling strategies, such as Gaussian distribution fourth-order sampling and skewness sampling; (3) The Jacobian matrix does not need to be derived; (4) It can deal with non additive noise and discrete systems, expanding the application range; (5) The calculation amount is of the same order as that of EKF; (6) Due to the deterministic sampling strategy, the particle decay problem is avoided. In a word, the amount of computation of UKF is basically the same as that of EKF algorithm, but its performance is better than that of EKF. It has shown a wider application prospect than the standard EKF estimation method in the field of nonlinear state estimation.

This section will introduce the gravity/SINS integrated navigation method based on UKF. The basic error equation of INS is adopted for the state equation, and the difference between the measured gravity anomaly of gravimeter and the gravity anomaly map at the position indicated by sins is still adopted for the measurement. Because the earth gravity anomaly is a nonlinear function of the earth's plane position, the position error, velocity error and attitude error of SINS system also show nonlinear characteristics.

1. Basic principle of UT (unscented transformation)

UT transformation is based on prior knowledge: it is easier to approximate the probability density of a nonlinear function than to approximate its function. Assuming that the random variable x is an n-dimensional vector, the mean value is $\overline{x}$, and the

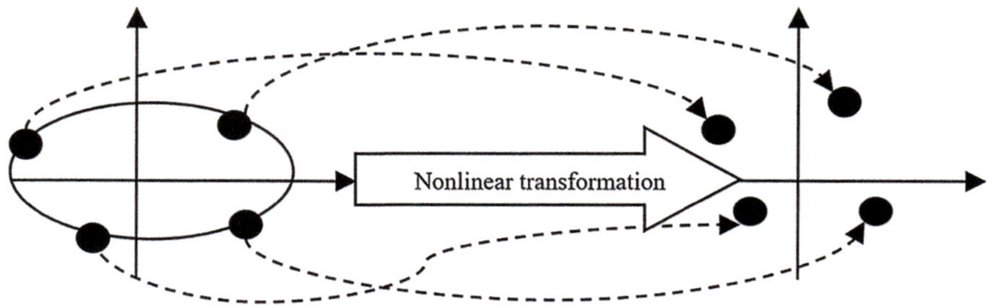

Fig. 3.16 Ut conversion principle

covariance is P_{xx}. To predict the mean and covariance $\overline{y}$ of m-dimensional random variable y, the relationship between y and x is defined by the following nonlinear transformation:

$$y = f(x)$$

The specific transformation method of UT is shown in Fig. 3.16. On the premise of ensuring that the sampling mean and covariance are $\overline{x}$ and P_{xx}, a set of points (called sigma point set) is selected, and the nonlinear transformation is applied to each sigma point sampled to obtain the non-linear transformed point set. $\overline{y}$ and P_{yy} are the statistics of the transformed sigma point set.

In general, the steps of the UT transform algorithm (applicable to any sigma sampling strategy) framework are as follows:

(A) According to the statistics $\overline{x}$ and P_{xx} of the input variable x, a sigma point sampling strategy is selected to obtain the sigma point set $\{\chi_i\}$, $i = 1, 2, \cdots, L$ of the input variable and the corresponding weights W_i^m and W_i^c. Where, L is the number of sampling sigma points of the sampling strategy adopted, W_i^m is the weight used for mean weighting, and W_i^c is the weight used for covariance weighting.

(B) Each sigma point in the sampled input variable sigma point set $\{\chi_i\}$ is subjected to $f(\cdot)$ nonlinear transformation to obtain the transformed sigma point set $\{y_i\}$.

$$y_i = f(\chi_i), i = 1, 2, \cdots, L \tag{3.66}$$

(C) The transformed sigma point set is weighted to obtain the statistics $\overline{y}$ and P_{yy} of the output variable y. The specific weight is still based on the corresponding weight of each sigma point that samples the input variable x.

$$\begin{cases} \overline{y} = \sum_{i=0}^{L-1} W_i^m y_i \\ P_{yy} = \sum_{i=0}^{L-1} W_i^c (y_i - \overline{y})(y_i - \overline{y})^{\mathrm{T}} \end{cases} \tag{3.67}$$

The approximate values of Y and P obtained after Ut change are:

$$\begin{cases} \overline{y} = f(\overline{x}) + E[f^{(2)}(\varepsilon^2)] \\ P_{yy} = f^{(1)}(\varepsilon) \cdot P_{xx} \cdot \left[f^{(1)}(\varepsilon)\right]^{\mathrm{T}} \end{cases} \tag{3.68}$$

where ε is the neighborhood partial value of x at $\overline{x}$, and $f^{(i)}(\cdot)$ is the i-order partial derivative value of x at $\overline{x}$.

In the UT transform algorithm, the most important thing is to determine the sigma point sampling strategy, that is, to determine the number, position and corresponding weight of the sigma points. At present, the existing sigma point sampling strategies include symmetric sampling, simplex sampling, third-order moment skew sampling and Gaussian distribution fourth-order moment symmetric sampling. In this paper, proportional symmetric sampling using proportional correction framework is selected.

2. UKF matching algorithm

In the Kalman filtering algorithm, for the one-step prediction equation, the UT transform is used to deal with the nonlinearity of the mean and covariance, which becomes the UKF algorithm. The gravity field matching algorithm based on UKF is divided into four parts:

(1) Selecting the weight of system state and covariance matrix.

$$\begin{cases} W_0^m = \lambda / (n_s + \lambda) \\ W_0^c = (1 - \alpha^2 + \beta) + 0.5\lambda / (n_s + \lambda) \\ W_i^m = W_i^c = 0.5\lambda / (n_s + \lambda), i = 1, 2, \cdots, 2n_s \end{cases} \tag{3.69}$$

where n_s is the number of system states. β is a parameter introducing the information of the higher-order term of $f(\cdot)$. $\beta = 2$ is optimal for Gaussian distribution when the higher-order term information of $f(\cdot)$ is not used. α is a scaling factor greater than zero. The distance between sigma point and $\overline{x}$ can be adjusted by adjusting the value of α, that is, to determine the distribution of sigma points around $\overline{x}$, which is usually set to a small positive number ($1e - 004 \leq \alpha < 1$). λ is the proportional parameter related to the constant specified by two users. The calculation formula is.

$$\lambda = \alpha^2(n_s + \kappa) - n_s \tag{3.70}$$

where κ is the kurtosis of measuring sigma point distribution, usually set to 0 or $3 - n_s$.

(2) Establishing sigma point set $\{\chi_i\}$.

$$\begin{cases} \chi_0[k|k-1] = \hat{X}[k|k-1] \\ \chi_i[k|k-1] = \hat{X}[k|k-1] + (\sqrt{(n_s+\lambda)\hat{P}[k|k-1]})_i, i = 1, 2, \cdots, n_s \\ \chi_i[k|k-1] = \hat{X}[k|k-1] - (\sqrt{(n_s+\lambda)\hat{P}[k|k-1]})_i, i = n_s+1, n_s+2, \cdots, 2n_s \end{cases} \tag{3.71}$$

(3) System prediction equation.

$$\begin{aligned} & X_i[k|k-1] = f(\chi_i[k-1]) \\ & \hat{X}[k|k-1] = \sum_{i=0}^{2n_s} W_i^m \chi_i[k|k-1] \\ & Z_i[k|k-1] = h(\chi_i[k]) \\ & \hat{Z}[k|k-1] = \sum_{i=0}^{2n_s} W_i^m Z_i[k|k-1] \\ & \hat{P}[k|k-1] = \sum_{i=0}^{2n_s} W_i^c (X_i[k|k-1] - \hat{X}[k|k-1])(X_i[k|k-1] - \hat{X}[k|k-1])^{\mathrm{T}} + Q \\ & P_{ZZ}[k|k-1] = \sum_{i=0}^{2n_s} W_i^c (Z_i[k|k-1] - \hat{Z}[k|k-1])(Z_i[k|k-1] - \hat{Z}[k|k-1])^{\mathrm{T}} + R \\ & P_{XZ}[k|k-1] = \sum_{i=0}^{2n_s} W_i^c (X_i[k|k-1] - \hat{X}[k|k-1])(Z_i[k|k-1] - \hat{Z}[k|k-1])^{\mathrm{T}} \end{aligned} \tag{3.72}$$

(4) Update equation.

$$\begin{cases} K[k] = P_{XZ}[k|k-1](P_{ZZ}[k|k-1])^{-1} \\ \hat{X}[k|k] = \hat{X}[k|k-1] + K[k](Z[k] - \hat{Z}[k|k-1]) \\ P[k|k] = P[k|k-1] - K[k]P_{ZZ}[k|k-1](K[k])^{\mathrm{T}} \end{cases} \tag{3.73}$$

The calculation flow chart of UKF algorithm is shown in Fig. 3.17.

3. Establishment of system state equation and measurement equation.

Select the matching area where the carrier track and INS possible error region (3 σ) are located to conduct gravity anomaly modeling. After repeated experiments, the point mass model can be obtained as follows. Point mass point 3. The positions are:λ_j = 122.918751, φ_1 = 24.142241, φ_2 = 24.275575, φ_3 = 24.408908. The corresponding point masses are ($\times 10^{16}$kg): M_1 = −2.4774, M_2 = 2.4836, and

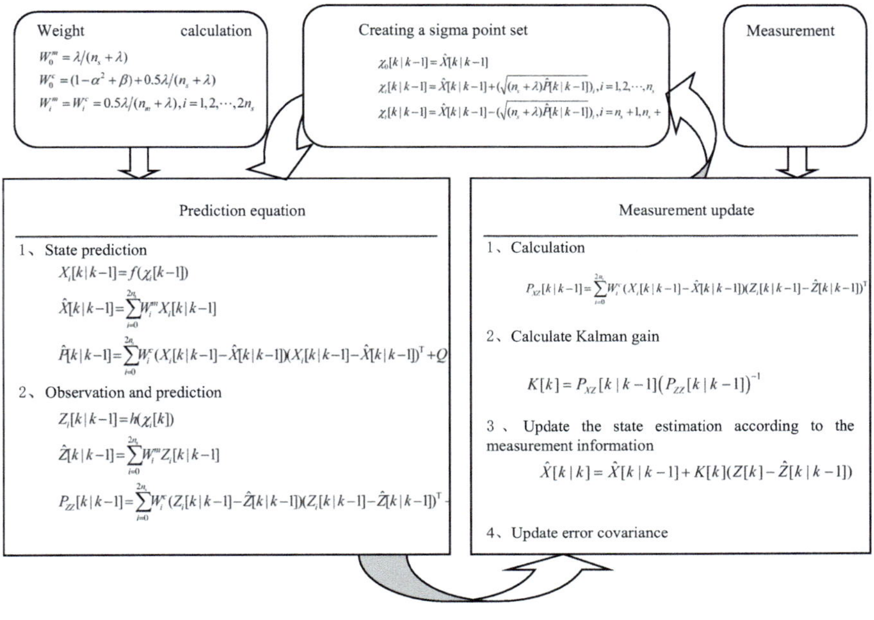

Fig. 3.17 UKF algorithm calculation flow chart

$M_3 = 0.4568$. Depth is $D_j = 40$km. Substitute the above data into the point quality model, namely.

$$\begin{aligned}
&\Delta g(\lambda, \varphi) = A_1 M_1 + A_2 M_2 + A_3 M_3 + \delta g_a \\
&A_1 = \frac{R - (R - D_j)(\sin\varphi \sin\varphi_1 + \cos\varphi \cos\varphi_1 \cos(\lambda - \lambda_j))}{l_1^3} - \frac{2}{l_1 R} \\
&l_1 = \sqrt{R^2 + (R - D_j)^2 - 2R(R - D_j)(\sin\varphi \sin\varphi_1 + \cos\varphi \cos\varphi_1 \cos(\lambda - \lambda_j))} \\
&A_2 = \frac{R - (R - D_j)(\sin\varphi \sin\varphi_2 + \cos\varphi \cos\varphi_2 \cos(\lambda - \lambda_j))}{l_2^3} - \frac{2}{l_2 R} \\
&l_2 = \sqrt{R^2 + (R - D_j)^2 - 2R(R - D_j)(\sin\varphi \sin\varphi_2 + \cos\varphi \cos\varphi_2 \cos(\lambda - \lambda_j))} \\
&A_3 = \frac{R - (R - D_j)(\sin\varphi \sin\varphi_3 + \cos\varphi \cos\varphi_3 \cos(\lambda - \lambda_j))}{l_3^3} - \frac{2}{l_3 R} \\
&l_3 = \sqrt{R^2 + (R - D_j)^2 - 2R(R - D_j)(\sin\varphi \sin\varphi_3 + \cos\varphi \cos\varphi_3 \cos(\lambda - \lambda_j))}
\end{aligned} \tag{3.73}$$

where δg_a is the error of gravity real-time measurement system. $R = R_e + h_i$, and R_e is the radial diameter of the matching area or the average radius of the earth. h_i is the carrier elevation.

The system measurement equation can be obtained as follows:

$$\Delta g(t) = \Delta g_s(\lambda, \varphi) - \Delta g_M(\lambda + \delta\lambda, \varphi + \delta\varphi) + \delta g_a \tag{3.75}$$

where $\Delta g_s(\lambda, \varphi)$ is the measured gravity anomaly at the inertial navigation indication position. $\Delta g_M(\lambda+\delta\lambda, \varphi+\delta\varphi)$ is the gravity anomaly obtained on the gravity anomaly map corresponding to the inertial navigation indication position. $(\delta\lambda, \delta\varphi)$ is the inertial navigation position error.

3.5 Summary

This chapter mainly introduces the methods of marine gravity assisted inertial navigation, including the composition of the real-time gravity measurement system and the implementation of the gravity field navigation method, which provides useful experience for the further research and engineering application of the gravity assisted navigation matching algorithm.

References

1. Zhao, X., & Chen, G. (2020). Development status and trend of ship navigation. *Navigation and Control, 19*(z1), 82–87.
2. Wei, B., Lv, W., Fan, X., Zhu, Y., & Guo, Y. (2019). AUV navigation technology development status and prospects. *Journal of Underwater Unmanned Systems, 27*(01), 9.
3. Zhou, B., & Liu, S. (2012). Research on underwater navigation technology. *Modern Navigation, 3*(01), 19–23.
4. Xiong, Z., Cai, K., Fengde, W., et al. (2002). A summary of the development of navigation technology of U.S. Strategic submarines in the 21st century. *Ship Science and Technology, 24*(3), 30−37.
5. Feizhou, Z. (2003). *Research on underwater passive navigation technology [postdoctoral report]*. Beijing Peking University.
6. Moryl, J., Rice, H., & Shinners, S. (1998). The universal gravity module for enhanced submarine navigation. In *Position Location and Navigation Symposium 1998* (pp. 324−331). IEEE.
7. Feng, H., Yan, L., Ge, Y., et al. (2004). Real time gravity correction of underwater vehicle based on INS. *Journal of Wuhan University (Engineering Edition), 37*(3), 135−138.
8. Zhang, H. (2013). *Research on key technologies of underwater gravity field aided navigation and positioning* (pp. 56–60). Doctoral dissertation of Harbin Engineering University.
9. Javed, W., Ghani, S., & Elmqvist, N. GravNav conference. In *Advanced Visual Interfaces* (pp. 217–224).
10. Zhu, Z., & Zhou, P. (2011). Research on multi-scale characteristics of gravity field in gravity assisted inertial navigation. *Progress in Geophysics, 26*(05), 1868–1873.
11. Feng, H. (2004). *Research on theory, model and simulation offuture underwater gravity assisted inertial navigation micro system [D]*. Beijing University of Posts and Telecommunications: Doctoral Dissertation (Beiyou\Peking University joint training).
12. Wei, X. (2017). Review of research progress of atomic interference gravity gradiometer. *Optics and Optoelectronic Technology, 15*(02), 99–104.
13. Guo, Q., & Xu, Z. (2012). Progress in geophysics of gravity measurement error correction in underwater gravity assisted navigation, *27*(3), 1274−1279.

14. Zhao, C. (2003). Application of wavelet multi-scale analysis in gravimeter data processing. *Chinese Journal of Inertial Technology, 11*, 1–5.
15. Dong, C. (2017). *Research on underwater gravity field aided inertial navigation matching algorithm and matching area selection* (pp. 6–12). Master's thesis of Wuhan University.
16. Zhu, Z., & Yang, Z. (2012). Triangle matching algorithm and simulation of passive gravity navigation. *Journal of instrumentation, 33*(10), 2387–2394.
17. Ouyang, M., Sun, Y., Kuang, Y., Hfan, H., & Wang, H. (2021). Underwater gravity matching navigation method using robust estimation Sitan algorithm. *Chinese Journal of inertial technology, 29*(02), 214–220.
18. Wang, B., Fu, M., Li, X., Wu, T., & Deng, Z. (2020). Overview of underwater gravity matching positioning algorithm. *Navigation and control, 19*(z1), 170–178.
19. Meng, W. U., Jie, M. A., Jinwen, T., & Jin, L. I. U. (2011). Fusion navigation algorithm based on three-axis magnetometer and gravity measurement. *Information and control, 40*(6), 760–766.
20. Wu, L., Wang, H., Chai, H., Zhang, L., Hsu, H., & Wang, Y. (2017). Performance evaluation and analysis for gravity matching aided navigation. *Sensors, 17*(4), 769.
21. Wang, H., Wu, L., Chai, H., Xiao, Y., Hsu, H., Wang, Y. (2017). Characteristics of marine gravity anomaly reference maps and accuracy analysis of gravity matching-aided mavigation. *Sensors, 17*(8), 1851.
22. Wang, H., Wu, L., Chai, H., Hsu, H., & Wang, Y. (2016). Technology of gravity aided inertial navigation system and its trial in South China Sea. *IET Radar, Sonar & Navigation, 10*(5), 862–869.
23. Xu, Z., Yan, L., & Zhao, L. (2010). Simulation research of gravity assisted navigation based on H ∞ filtering technology. *Ship Science and Technology, 31*(1), 124–128ff.

Chapter 4
Fundamentals and Theoretical Models of the Geomagnetic Navigation and Positioning

Abstract The geomagnetic field is a fundamental geophysical field (M.-L. Zhang, in Introduction to geophysics. Petroleum Industry Press, pp. 23–25, 2019), which consists of the Earth's internal and external magnetic field. Among them, the internal magnetic field mainly consists of the main magnetic field formed by the Earth's core and the crustal anomalous field formed by the lithosphere, which accounts for more than 99% of the geomagnetic field. The external magnetic field is mainly composed of the magnetic field superposition, which is generated by the solar activity acting on the atmospheric ionosphere and magnetosphere (Y. Kamide, A. C. L. Chian, in Handbook of the solar-terrestrial environment. J.-L. Jian. Science Press, pp. 91–96, 2010). The geomagnetic field has the following characteristics: (1) It continuous distribution in the near-Earth space; (2) It differences in the geomagnetic field in various regions due to the distribution of the Earth's magnetic materials;(3) It is strong near the ground and gradually decays with the increase of height; (4) The change period of the main magnetic field originating from the Earth's core is measured in centuries, and the change of the anomalous field originating from the Earth's crust is recorded in geological ages, so the main magnetic field and the anomalous field are stable; (5) The changing magnetic field mainly originates from outside the Earth, and the trend of change is consistent within hundreds of kilometers of geography. The geomagnetic field has been widely used for orientation navigation of ships and airplanes, and the geomagnetic field orientation information is measured using compasses and magnetic compasses to obtain a stable north pointing of the Earth (K. Liu, in Research on long-distance geomagnetic navigation based on magnetotactic search. NorthWestern Polytechnical University, pp. 45–50, 2019). Geomagnetic positioning navigation technology, unlike orientation measurement technology, uses the characteristic that the intensity of the geomagnetic field has different distributions with geospatial location to achieve the positioning of the motion platform. Through magnetic field sensors installed on the motion platform, the geomagnetic field data on the motion track is measured in real-time. Features are extracted and matched with the geomagnetic field model or geomagnetic reference map obtained and stored in advance to determine the real-time position of the motion platform (K. Chen et al., in Journal of Zhejiang University-Science A (Applied Physics & Engineering), 22(5):357–368, 2021; Feng-Min et al. in Journal of Projectiles, Rockets,

L. Yan et al., *Mathematical and Physical Fundamentals of Navigation and Positioning with Earth's Natural Fields*, Navigation: Science and Technology 13,
https://doi.org/10.1007/978-981-97-5524-0_4

Missiles and Guidance 41:10–14, 2021; G.-G. Wang et al. in A geomagnetic localization method based on road magnetic field characteristics. Beijing: CN107621263B, 2019–12-27; X. Li et al., in Chinese Journal of Sensors and Actuators, 30(12):1869–1875, 2017; Zhuang-Sheng et al. in Progress in Geophysics 26:1473–1477, 2011). Ultimately, a high-precision local three-dimensional spatial reference geomagnetic field that meets the requirements of geomagnetic positioning and navigation can be constructed.

4.1 Composition of the Geomagnetic Field

The geomagnetic field is a fundamental physical field inherent of the Earth, which is generated by the superposition of various magnetic field components generated by magnetic rocks inside the Earth and the current system distributed inside and outside the Earth. More than 95% of the geomagnetic field originates from the magnetic fluid generating process of the Earth's core, which becomes the core field or main magnetic field; 4% of the geomagnetic field is generated by magnetic rocks in the Earth's crust, which is called the local field, crustal field or anomaly field. The main magnetic field and anomalous field change very slowly with time, so they can be collectively called stable magnetic field. The exogenous field, which accounts for 1% of the geomagnetic field, originates from the space current system above the earth's surface and varies more rapidly with time. It's periods ranging mainly from a few tenths of a second to several days, and is called the earth's changing magnetic field. Studies have shown that although the geomagnetic field has a complex temporal and spatial distribution, it has the following characteristics: (1) the time-varying properties of the magnetic field change only gently with space; and (2) the spatial variation of the magnetic field has little dependence on time [9–11].

The magnetic field component for navigation heading fixation is the main magnetic field, and the magnetic field component of interest for geomagnetic positioning navigation is the anomaly field. The anomalous field is regionally distributed on the Earth's surface, with a typical range of about hundreds of meters to hundreds of kilometers, and it hardly changes with time, which is the basis for the anomalous field to be used for positioning and navigation.

4.2 Basic Theory and Analytical Representation of the Main Magnetic Field

Although geomagnetic positioning navigation is referenced to the Earth's anomalous field, the main magnetic field accounts for more than 94% of the geomagnetic field, so attention to the main magnetic field is also required for geomagnetic positioning navigation applications [12, 13].

From the set of Maxwell equations for the geomagnetic field, the geomagnetic field magnetic potential satisfies the Laplace equation

$$\nabla^2 U = 0 \tag{4.1}$$

Suppose the Earth is a sphere with radius a, and in a spherical coordinate system with the center of the sphere O as the origin, the polar axis is taken as the Earth's rotation axis and points to the North Pole, r is the distance from the center O to the measurement point P, θ is the geocentric residual latitude, and λ is the longitude, the above equation is written as

$$\frac{1}{r^2}\frac{\partial}{\partial r}\left(r^2\frac{\partial U}{\partial r}\right)+\frac{1}{r^2\sin\theta}\frac{\partial}{\partial\theta}\left(\sin\theta\frac{\partial U}{\partial\theta}\right)+\frac{1}{r^2\sin^2\theta}\frac{\partial^2 U}{\partial\lambda^2}=0 \tag{4.2}$$

The solution of Eq. (4.2) can be obtained from Gaussian theory as

$$\begin{aligned} U(r,\theta,\lambda) = a\sum_{n=1}\sum_{m=0}\Bigg[&\left(\frac{a}{r}\right)^{n+1}\left(g_n^m\cos m\lambda + h_n^m \sin m\lambda\right) \\ &+\left(\frac{r}{a}\right)^n\left(j_n^m\cos m\lambda + k_n^m\sin m\lambda\right)\Bigg]P_n^m(\cos\theta) \end{aligned} \tag{4.3}$$

Equation (4.3) is called the Gaussian expression for the magnetic potential of the geomagnetic field, and it is the basic formula in geomagnetism. The coefficients $g_n^m, h_n^m, j_n^m, k_n^m$ are called Gaussian coefficients, g_n^m, h_n^m are called internal field coefficients, and j_n^m, k_n^m are called external field coefficients.

Equation (4.3) contains two parts: the first part is the term containing the factor $r^{-(n+1)}$, and the corresponding magnetic potential is denoted by U_i, which represents the magnetic potential generated by the internal field sources of the Earth; the second part is the term containing the factor r^n, and the corresponding magnetic potential is denoted by U_e, which represents the magnetic potential generated by the external field sources of the Earth.

Where, $P_n^m(\cos\theta)$ is defined as the Schmidt orthogonalized connective Lejeune function.

$$P_n^m(\cos\theta)=\left[\frac{2}{\delta_m}\frac{(n-m)!}{(n+m)!}\right]^{1/2}P_n^{(m)}(\cos\theta) \tag{4.4}$$

Among them, $\delta_0 = 2$, $\delta_m = 1 (m > 0)$. And the geomagnetic field strength satisfies

$$B = -\nabla U \tag{4.5}$$

Substituting into Eq. (4.3), we get the expression for the magnetic field strength

$$(X, Y, Z) = \left(\frac{1}{r}\frac{\partial U}{\partial \theta}, -\frac{1}{r\sin\theta}\frac{\partial U}{\partial \lambda}, \frac{\partial U}{\partial r}\right) \tag{4.6}$$

In the geomagnetic field magnetic potential Gaussian series equation, the main magnetic field expression is obtained by the part of internal source field.

$$U(r, \theta, \lambda) = a\sum_{n=1}^{\infty}\sum_{m=0}^{n}[(\frac{a}{r})^{n+1}(g_n^m \cos m\lambda + h_n^m \sin m\lambda)P_n^m(\cos\theta) \tag{4.7}$$

From Eq. (4.7), the triaxial component of the main magnetic field is obtained

$$X = \sum_{n=1}^{\infty}\sum_{m=0}^{n}\left(\frac{a}{r}\right)^{n+2}\left[g_n^m \cos m\lambda + h_n^m \sin m\lambda\right]\frac{dp_n^m(\cos\varphi)}{d\varphi} \tag{4.8}$$

$$Y = \sum_{n=1}^{\infty}\sum_{m=0}^{n}\left(\frac{a}{r}\right)^{n+2} m\left[-g_n^m \sin m\lambda + h_n^m \cos m\lambda\right]\frac{p_n^m(\cos\theta)}{\sin\theta} \tag{4.9}$$

$$Z = -\sum_{n=1}^{\infty}\sum_{m=0}^{n}\left(\frac{a}{r}\right)^{n+2}\left[g_n^m \cos m\lambda + h_n^m \sin m\lambda\right](n+1)p_n^m(\cos\theta) \tag{4.10}$$

Equations (4.7)–(4.10) are the underlying mathematical models of IGRF and WMM.

The above equations provide a complete theoretical formulation of the geomagnetic field distribution, and it can be shown that the terms of $n = 1$ account for about 85% of the whole, and the convergence of the series is quite fast. When n and m are large enough, the weak magnetic anomalies at the Earth's surface can also be calculated by the model. When the Gaussian coefficients are determined, the distribution of the geomagnetic field in near-Earth space can be determined, and the inertial and external fields can be further distinguished. In practice, the Gaussian coefficient can be derived from the geomagnetic data of each discrete point using the least-squares method.

However, the n and m cannot be infinite and need to be replaced by finite terms. The calculated values of the total intensity distribution of the geomagnetic field based on the IGRF2005 and WMM-2005 models are compared with the measured values in the region as shown in Table 4.2. The difference part is the earth anomaly field, so it can be seen that the main earth magnetic field model is difficult to characterize the distribution of the anomaly field, thus for geomagnetic positioning and navigation applications, the earth anomaly field part must be considered.

Table 4.1 Basic characteristics of the main components of the geomagnetic field

		Magnetic fields	Location	Maximum strength	Characteristics	Temporal variation
Internal field	1	Main geomagnetic field	Earth's outer core	50,000–70,000 nT	Dipole field dominated	Annual
	2	Anomaly field	Crust above the curie point	100–10,000 nT	Very irregular, wavelength can be 1m	Geological age
	3	Induction field	Crust, upper mantle and oceans	1/2 of the fields 4, 5, 6 and 7	Global field, but irregular in many places	Same as the fields 4, 5, 6 and 7
External field	4	Regular magnetic storm field	Magnetic layer	150–1000 nT	Nearly uniform outfield	4–10 h, recovery 2–3 days
	5	Irregular magnetic storms and sub-storm fields	Ionosphere and magnetosphere	100–200 nT (aurora bands)	Global field, the strongest in the aurora belt	5–100 min
	6	Daily variation	Ionosphere	50–200 nT (equatorial)	Global field	24h, 12h or 8h
	7	Pulsation	Magnetic layer	0.01–10 nT	Quasi-global field, the strongest in the aurora belt	1–300s

Table 4.2 Comparison of IGRF2005, WMM-2005 and real geomagnetic field values for the same region

	36°45′0″ N	36°47′0″ N	36°49′0″ N	36°51′0″ N
105°45′0″ W	51,715.7	51,734.9	51,754.2	51,773.3
	51,722.0	51,741.3	51,760.5	51,779.8
	51,054.1	51,089.0	51,085.5	51,051.1
105°47′0″ W	51,711.3	51,730.6	51,749.7	51,768.9
	51,717.6	51,736.9	51,756.1	51,775.4
	51,135.7	51,017.5	51,151.4	51,173.7
105°49′0″ W	51,706.9	51,726.2	51,745.3	51,764.5
	51,713.2	51,732.5	51,751.8	51,771.0
	51,020.0	51,146.9	51,071.4	51,077.6
105°51′0″ W	51,702.5	51,721.7	51,740.9	51,760.1
	51,708.8	51,728.1	51,747.4	51,766.6
	50,930.4	50,699.6	51,376.2	51,323.2

4.3 Basic Theory and Modeling Methods of Geomagnetic Anomaly Fields

The interpolation method is a common way to establish the characteristics of a small-scale geomagnetic field such as anomaly field, and its basic principle is the mathematical process of extrapolating the data of unknown point or unknown area according to some mathematical relationship based on a set of a known discrete data or partition data. For the discrete data of geomagnetic field measurement, a finer spatial distribution of magnetic field can be obtained by the interpolation method. The missing observations at a point can be estimated to improve the data density. The data can be griddled to interpolate the non-regularly distributed anomalous field data into the regularly distributed magnetic field data. Interpolation methods for geophysical fields have been studied more, for example, Kriging, Inverse Distance to a Power, Minimum Curvature, Modified Shepard's Method, Nearest Neighbor, Radial Basis Function, Triangulation with Linear Interpolation, Local Polynomial, etc. have been mentioned in much literature in different fields. By analyzing the adaptability of the data distribution and the degree of agreement with the original data after gridding, the Kriging interpolation method is relatively stable in performance, high in accuracy, and more suitable for the implementation of gridding for discrete geomagnetic field data. The relevant performance comparison is summarized in Table 4.3.

With the development of Geo-statistic, of improved Kriging interpolation methods have been proposed and widely used, including ordinary Kriging, indication Kriging, universal Kriging interpolation methods, etc.

Table 4.3 Summary of the performance of various interpolation methods

Interpolation method	Performance	
	Fitting performance	Scope of application
Radial basis function	★★★★	Uniform distribution of raw data
Modified Shepard's method	★	Uniform distribution of raw data
Minimum curvature	★★★★	Widely
Kriging	★★★★	Widely
Triangulation with linear interpolation	★	Uniform distribution of raw data
Inverse distance to a power	★★★	Uniform distribution of raw data
Nearest neighbor	★★	Uniform distribution of raw data
Local polynomial	★★	Generally

4.3.1 Basic Theory of Kriging Interpolation

Theoretically, the Kriging interpolation method considers physical observations of spatial points as random variables. It is an unbiased and optimal interpolation method by using known observations of spatial variables to infer and recognize their patterns and characteristics in space with different positions, especially the values at points outside the observation points. It describes the statistical characteristics of a two-dimensional random field with a spatial variational structure, keeping the overall statistical characteristics of the interpolated data unchanged while describing the local details of the surface. Therefore, it can be used for the interpolation of geomagnetic reference maps.

The Kriging interpolation is implemented based on linear minimum variance estimation as shown in the following equation.

$$E\left\{\left[B(P_i) - B^*(P_u)\right]^2\right\}_{i=1,\cdots,N} \to \min \tag{4.11}$$

where, $B(P)$ is a random function of the geomagnetic intensity with all control points P, and $B(P_i)$ is geomagnetic intensity at P_i. The estimated geomagnetic intensity $B^*(P_u)$ at the unknown point P_u can be shown as

$$B^*(P_u) = \sum_{i=1}^{N} \lambda_i B(P_i) \tag{4.12}$$

where, λ_i is the weighting factor to be determined.

From the estimation theory, it is known that the weighting coefficients can be expressed in terms of the first-order and second-order moments of the random function $B(P_i)$ of the geomagnetic distribution at the control point. Different forms of Kriging interpolation are derived from the different properties of the first and second order moments of the random variable. The properties of the geomagnetic reference map correspond to the following case of the first-order moments of the random function: The mathematical expectation is independent of the position P, and the random function is first-order smooth.

In fact the geomagnetic reference map region conforms to a local continuous field distribution and can be considered to satisfy the second-order smooth assumption and the eigen-assumptions: its mathematical expectation is m (unknown), the covariance function $c(h)$ and the variance function $\gamma(h)$ exist, and h is the separation distance between two points. The Kriging is a linear minimum variance estimate and an unbiased estimate, so

$$E[B(P)] = m \tag{4.13}$$

$$c(h) = E\left[B(P_i)B(P_j)\right] - m^2 \tag{4.14}$$

$$\gamma(h) = \frac{1}{2}E\left\{\left[B(P_i) - B\left(P_j\right)\right]^2\right\} \tag{4.15}$$

The Kriging interpolation actually guarantees the unbiasedness and the optimality of the estimated values.

$$E\left[B^*(P)\right] = E\left[\sum_{i=1}^{N} \lambda_i B(P_i)\right] = \sum_{i=1}^{N} \lambda_i E[B(P_i)] = m \tag{4.16}$$

$$\sigma_E^2 = E\left\{\left[B(P) - B^*(P)\right]^2\right\} = E\left\{\left[B(P) - \sum_{i=1}^{N} \lambda_i B(P_i)\right]^2\right\} \tag{4.17}$$

A condition that the weight coefficient satisfies can be obtained from the following equation

$$\sum_{i=1}^{N} \lambda_i = 1 \tag{4.18}$$

While Eq. (4.17) uses the covariance expression

$$\sigma_E^2 = c(P, P) + \sum_{i=1}^{N} \sum_{j=1}^{N} \lambda_i \lambda_j c\left(P_i, P_j\right) - 2\sum_{i=1}^{N} \lambda_i c(P_i, P) \tag{4.19}$$

To minimize the estimation variance, according to the Lagrange multiplier principle, let

$$F = \sigma_E^2 - 2\mu\left(\sum_{i=1}^{N} \lambda_i - 1\right) \tag{4.20}$$

Find the partial derivatives of F with respect to λ and μ, and let it be zero to obtain the system of Kriging equations

$$\begin{cases} \dfrac{\partial F}{\partial \lambda_i} = 2\displaystyle\sum_{j=1}^{N} \lambda_j c(P_i, P_j) - 2c(P_i, P) - 2\mu = 0 \\ \dfrac{\partial F}{\partial \mu} = -2(\displaystyle\sum_{i=1}^{N} \lambda_i - 1) = 0 \end{cases} \tag{4.21}$$

Simplify the above equation

$$\begin{cases} \sum_{j=1}^{N} \lambda_j c(P_i, P_j) - \mu = c(P_i, P) \\ \sum_{i=1}^{N} \lambda_i = 1 \end{cases} \tag{4.22}$$

However, the above equation still requires that $E[B(P)]$ is known, which contradicts the previous premise. Therefore, using the variational function yields

$$\gamma(h) = c(0) - c(h) \tag{4.23}$$

$$\begin{cases} \sum_{j=1}^{N} \lambda_j \gamma(P_i, P_j) + \mu = \gamma(P_i, P) \\ \sum_{i=1}^{N} \lambda_i = 1 \end{cases} \tag{4.24}$$

$$\sigma_K^2 = \sum_{i=1}^{N} \lambda_i \gamma(P_i, P) - \gamma(P, P) + \mu \tag{4.25}$$

The basis of such an ordinary Kriging interpolation is the determination of the variation function. The variation function is determined in two steps: (1) the expermental variation function is obtained by using the existing reference grid map; (2) the theoretical variation function is obtained by fitting the experimental variation function. After the theoretical variation function is determined, the geomagnetic intensity of the unknown point can be interpolated and calculated.

4.3.2 Problems with Redundancy for Heading Measurement

The experimental variation function is defined as follows: let h be a vector of mode $r = |h|$ and direction a, if there exist N_h pairs of observed data points separated by the vector h, then the experimental variation function $\gamma^*(h)$ corresponding to the vector h in the direction of a can be approximated as

$$\gamma^*(h) = \gamma^*(r, a) = \frac{1}{2N_h} \sum_{i=1}^{N_h} [B(P_i + h) - B(P_i)]^2 \tag{4.26}$$

Problem 1: The computation of the interpolation parameters corresponding to the solution of the experimental variance function of a in all directions is too large.

Problem 2: Reference map data are limited, and when the mode of h (i.e., the distance between data point pairs) is close to the length of the map interval, the number of data pairs N_h is small and the confidence of the resulting experimental variation function becomes low.

Although the geomagnetic measurement is pre-processed by denoising in the process of map formation, there is still a certain amount of interference noise, which is generally desirable as Gaussian distribution.

$$Z(P) = B(P) + e(P) \tag{4.27}$$

$$e(P) \propto N\left(0, \sigma_e^2\right) \tag{4.28}$$

The high-frequency sampling of the magnetometry instrument makes heading redundancy possible, as Fig. 4.1 shown. Redundant data is usually beneficial for the detection of information, but too much redundancy can also increase the oscillation of the interpolation results.

Problem 3: Can the heading redundancy data overcome the effect of lower interference noise and determine a better redundancy so that the interpolation results converge to the true distribution?

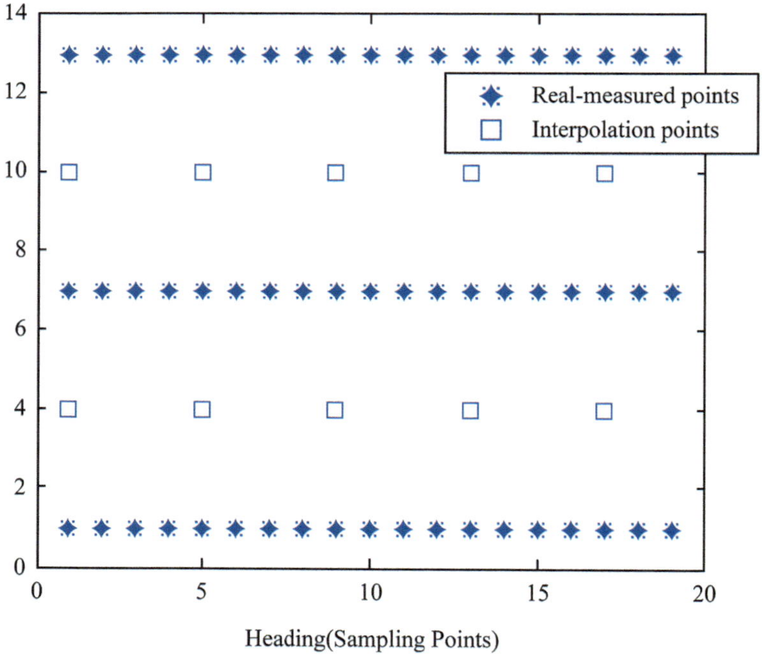

Fig. 4.1 Distribution diagram of heading redundancy measurement data

There are various models of the theoretical variance function, and the more representative ones are the following three.

Spherical Model:

$$\gamma(h) = \begin{cases} 0 & h = 0 \\ c_0 + c(\frac{3h}{2a} - \frac{h^3}{2a^3}) & 0 < h \le a \\ c_0 + c & h > a \end{cases} \tag{4.29}$$

Exponential Model:

$$\gamma(h) = \begin{cases} 0 & h = 0 \\ c_0 + c\left(1 - e^{-\frac{h}{a}}\right) & h > 0 \end{cases} \tag{4.30}$$

Gaussian Model:

$$\gamma(h) = \begin{cases} 0 & h = 0 \\ c_0 + c\left(1 - e^{-\frac{h^2}{a^2}}\right) & h > 0 \end{cases} \tag{4.31}$$

Problem 4: The theoretical model with better fitting effect is selected for lower measurement error and heading redundancy.

4.3.3 Convergent Kriging Interpolation Algorithm

4.3.3.1 Theoretical Basis of the Algorithm

In fact, the convergent Kriging interpolation algorithm takes advantage of the fact that the geomagnetic field is a two-dimensional continuous distributed field, and the magnitude of the magnetic intensity distribution at any point is reflected in the surrounding points. As shown in Fig. 4.2, the interpolated distribution corresponding to the redundant measurements B_r is significantly better than the under-sampled interpolation recovery B_u, where the high sampling rate more effectively 'refracts' the distribution state at the nearby points.

If $s(x, y)$ represents the sampling process, $i_B(x, y)$ is the geomagnetic distribution of the treatment area, and $t(x, y)$ is the reference map.

$$s(x, y) = \sum_{n=1}^{N}\sum_{m=1}^{M}\delta\left(x - n\frac{L_x}{N}, y - m\frac{L_y}{M}\right) \tag{4.32}$$

$$t(x, y) = i_B(x, y) \cdot s(x, y) \tag{4.33}$$

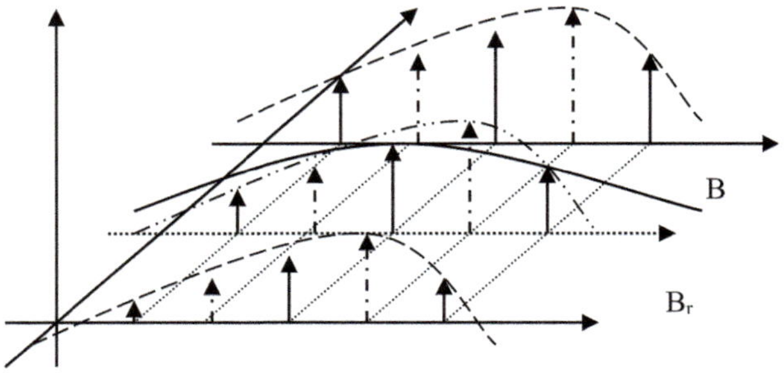

Fig. 4.2 Basis of convergent Kriging implementation for 2D continuous field

Letting $S(u, v)$, $I_B(u, v)$ and $T(u, v)$ be their corresponding Fourier transforms, respectively, then

$$
\begin{aligned}
T(u, v) &= I_B(u, v) * S(u, v) = \sum_{r=1}^{N}\sum_{q=1}^{M} I_B(u, v) * \delta\left(u - r\frac{N}{L_x}, v - q\frac{M}{L_y}\right) \\
&= \sum_{r=1}^{N}\sum_{q=1}^{M} I_B\left(u - r\frac{N}{L_x}, v - q\frac{M}{L_y}\right)
\end{aligned}
\tag{4.34}
$$

where

$$
\begin{cases} s_u = N/L_x \\ s_v = M/L_y \end{cases}
$$

The discrete sampling in the spatial domain corresponds to the period extension in the frequency domain. Since the range of the reference map processing is limited, and thus $I_B(u, v)$ is infinite in the frequency domain. Therefore, the period extension is bound to produce mixing, which is a key factor for the poor accuracy of interpolation as a frequency-to-spatial domain inverse operation.

As shown in Fig. 4.3, increasing s_u and s_v means increasing N and M, which can make the mixing iterations decrease and increase the accuracy of the solution in its inverse solution. It is also the theoretical basis for converging Kriging interpolation algorithm.

4.3.3.2 Algorithm Composition

The specific algorithm is constructed as a sum of solutions for the above four problems based on the redundant additional conditions of the heading.

Firstly, for problem 1, the mean experimental variation function can be defined based on the assumption that the geomagnetic field is isotropic.

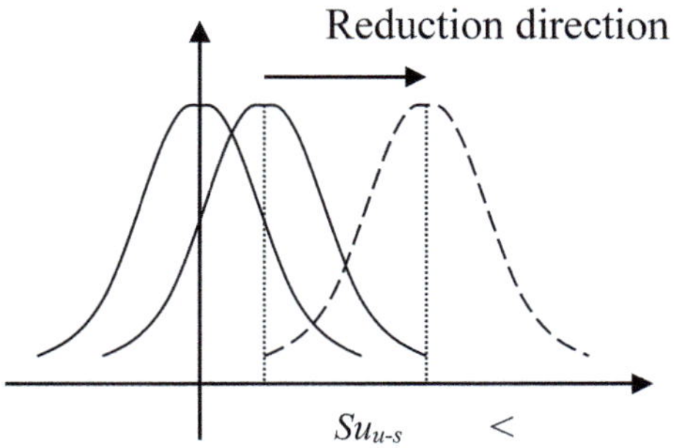

Fig. 4.3 Spectrum aliasing: theoretical basis for convergent Kriging

$$\gamma^*(h) = \left[\sum_{k=1}^{K} N_k(h)\gamma_k^*(h, \alpha_k)\right] \Big/ \left[\sum_{k=1}^{K} N_k(h)\right] \tag{4.35}$$

Secondly, for problem 2, the concept of distance tolerance $\varepsilon(r)$ is introduced to increase the number of interval data pairs. If the distance between two points is $D_{ij} = \|P_i - P_j\|$, and the difference with h is less than $\varepsilon(r)$, then

$$|D_{ij} - h| \leq \varepsilon(r) \tag{4.36}$$

The data pair can be used in the statistics of the pair with the interval h.

Again, for problem 3, the high-frequency sampling characteristics of the geomagnetic measuring instrument are used, and then the Redundancy Average Experimental Variogram Function (RAEVF) is constructed using the redundant data on the measurement line.

$$\gamma^*(h) = \frac{1}{2N_r}\sum_{i=1}^{N_r}\left[B(P_i) - B(P_j)\right]^2 \quad \left|\|P_i - P_j\| - h\right| \leq \varepsilon \tag{4.37}$$

Finally, for problem 4, the representative Gaussian model, spherical model, and exponential model are statistically compared, and the theoretical variational model with a better fitting effect under low-noise conditions is selected.

The mean square value of the error was used as the comparison index for the fitting effect, and this index was also used for the measurement of the interpolation accuracy.

$$E\left[B^2(P_k)\right] \doteq \frac{1}{K}\sum_{k=1}^{K}\left(B^*(P_k) - B(P_k)\right)^2 \tag{4.38}$$

The system of Kriging equations for the redundant model is then solved by

$$K = \begin{bmatrix} \gamma_{11} & \gamma_{12} & \cdots & \gamma_{1N_r} & 1 \\ \gamma_{21} & \gamma_{22} & \cdots & \gamma_{2N_r} & 1 \\ \vdots & \vdots & & \vdots & \vdots \\ \gamma_{N_r1} & \gamma_{N_r2} & \cdots & \gamma_{N_rN_r} & 1 \\ 1 & 1 & \cdots & 1 & 0 \end{bmatrix}, \lambda = \begin{bmatrix} \lambda_1 \\ \lambda_2 \\ \vdots \\ \lambda_{N_r} \\ \mu \end{bmatrix}, D = \begin{bmatrix} \gamma(P_1, P) \\ \gamma(P_2, P) \\ \vdots \\ \gamma(P_n, P) \\ 1 \end{bmatrix} \quad (4.39)$$

$$\lambda = K^{-1}D \quad (4.40)$$

$$\sigma_K^2 = \lambda^T D - \gamma(P, P) \quad (4.41)$$

The convergent Kriging interpolation method with a certain anti-differential effect can be constructed by processing the above-mentioned links.

4.3.4 Numerical Analysis

The reference data are simulated with a 1000*1000 grid distribution. The reference data are sampled at 32 intervals in one direction (1, 33, 65, ...) with Gaussian noise, while the redundant data -1 in heading are sampled at 16 intervals in that direction (1, 17, 33, etc.), and the redundant data –2, –3, –4, –5, etc. The sampling in the vertical direction simulates the survey line spacing with equal 32 interval sampling (1, 33, 65, etc.), so as to form a similar Fig. 4.4 the benchmark data shown.

Statistics of variance function distribution for reference data, reference map, and heading redundancy data –4 respectively, as Fig. 4.4 shown. It can be seen that the reference data has a relatively smooth distribution. The variation function distribution of the reference map is more drastic, and the large grid spacing makes the variation function value missing at low interval distance.

For different Gaussian noise levels and different redundancy levels, the error distribution of the interpolation results fitted with the exponential theory model is shown in Fig. 4.4, while the results of the spherical and Gaussian models are similar. The interpolation accuracy reaches the highest when the redundancy degree is 4, that is, $b = 4$ in the regular function, and the results show that a certain amount of heading redundancy information can effectively improve the interpolation accuracy. At the same time, too much heading redundancy information will lead to the decrease of interpolation accuracy, which is due to the fact that the redundancy information on the geomagnetic distribution trend has been basically reflected at $b = 4$ compared with the relatively flat baseline data, and more redundancy information makes the noise characteristics in it instead This is because the redundant information on the geomagnetic distribution trend is largely reflected in. This actually verifies the original intention of the algorithm in this chapter, that the redundant information of heading can be used to construct the Kriging interpolation algorithm with convergence function within a certain range.

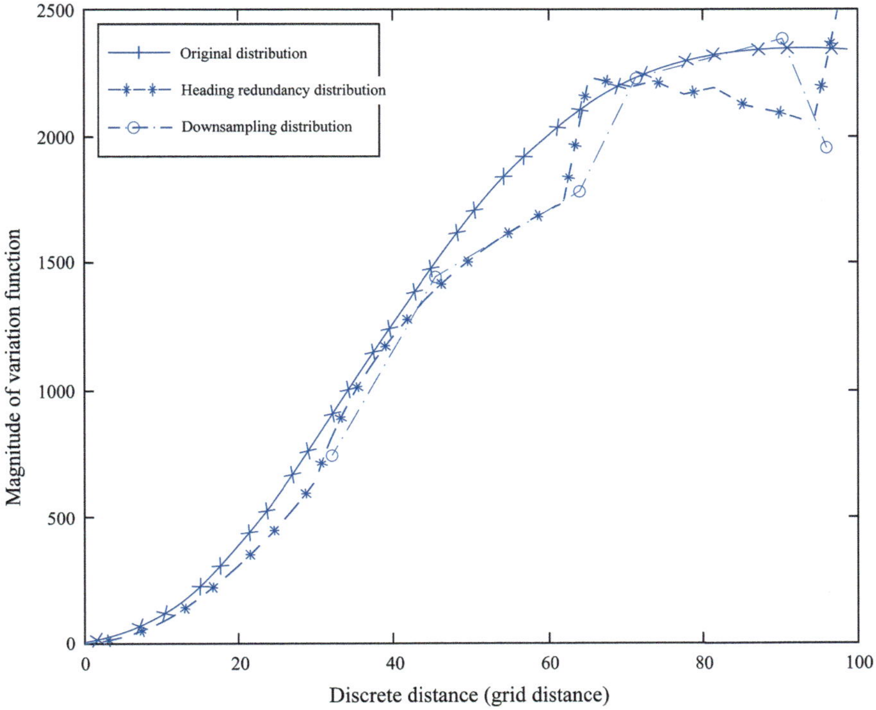

Fig. 4.4 Statistical comparison of variance functions for different data

Taking the lowest interpolation error for the fitting comparison of the theoretical model, the results of the distribution of the error with noise after interpolation are shown in Fig. 4.5, the Gaussian variational model fits the highest degree in the low noise case, while the difference between the theoretical curve obtained from the spherical model and the average variational function curve is the largest. This is also consistent with the statistical distribution of variance in Fig. 4.6, which shows that the variance function of the underlying data approximates the Gaussian distribution pattern.

The above numerical comparison results show that the redundant information not only reduces the influence of noise, but also reflects the information of geomagnetic distribution from several sides with certain convergence function. The algorithm is not only applicable to the optimization of the reference map survey in geomagnetic navigation, but also important for the study of geophysical field distribution laws.

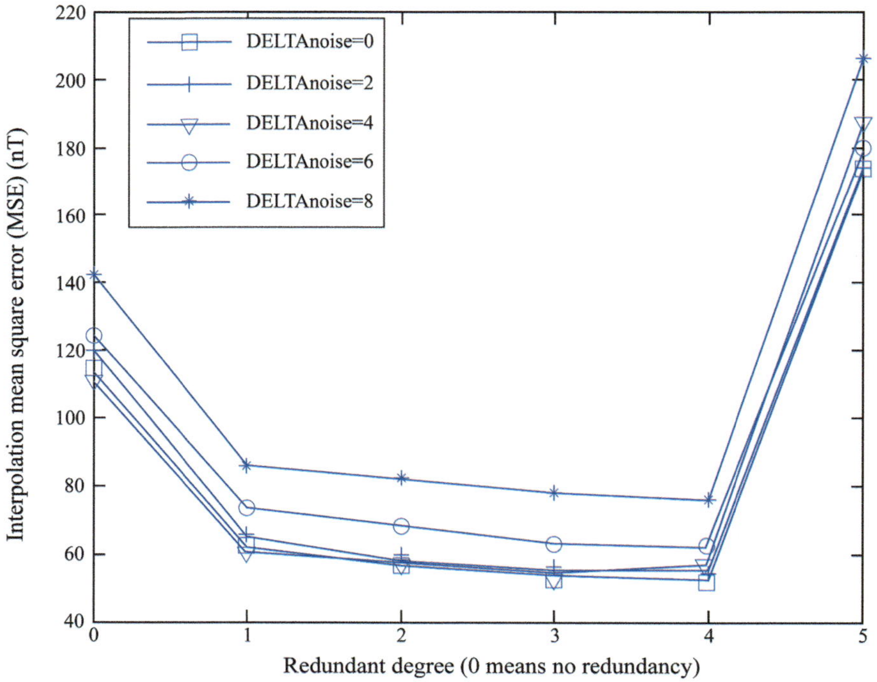

Fig. 4.5 Interpolation error of the redundant grid with the degree of redundancy and noise level

4.4 Theory and Method of Geomagnetic Anomaly Field Continuation

Usually, the geomagnetic anomaly field data obtained from mapping are magnetic measurements at a certain height, while in practical applications, the moving platform may be applied at other heights, which requires the use of magnetic field extension techniques to obtain the geomagnetic anomaly field data at the desired height. According to the physical properties of the geomagnetic anomaly field and the mathematical conditions, the continuation principle of the bit field is applied to extrapolate the data of other spatially distributed surfaces outside the field source from the data of known surfaces [14].

4.4.1 Construction of Local Anomaly Field for Continuation Verification

The current geomagnetic anomaly field continuation verification usually uses the available anomaly distribution data in two planes, but this approach is for special

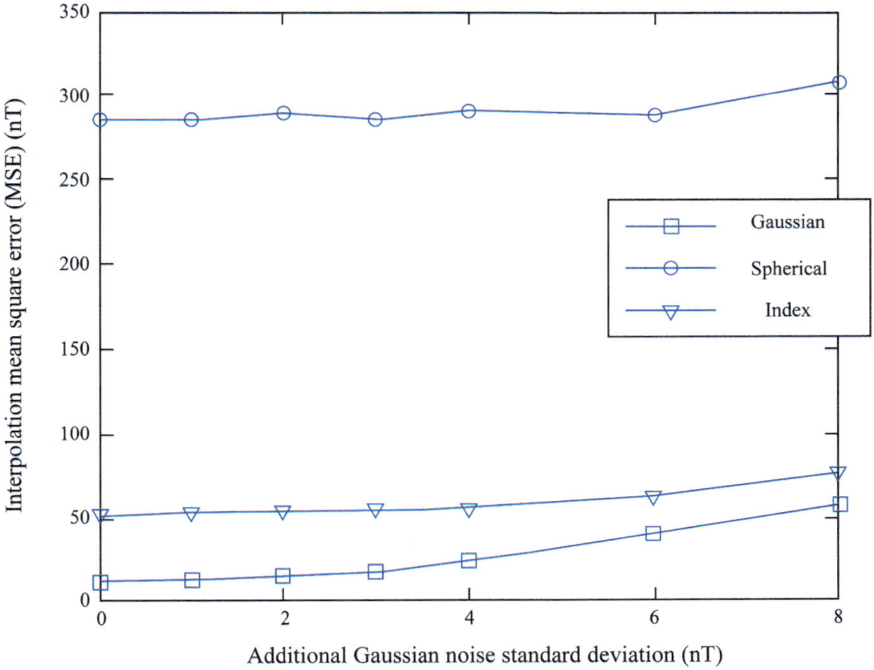

Fig. 4.6 Interpolation error distribution of different theoretical models

cases and does not have the meaning of a universal quantitative comparison. In this chapter, we use a magnetic dipole to generate the most basic anomalous magnetic field and verify the accuracy of upward and downward extension methods, respectively, so that the extension verification can be judged as a unified standard.

The construction process uses magnetic dipoles to generate magnetic field intensity distributions on different distance planes and then verifies the accuracy of the continuation method. The crux of the problem is that the distance from the anomalous intensity distribution plane to the magnetic dipole is unknown during the verification process; that is, the field source depth of the geomagnetic anomaly field is unknown [15].

The total strength of the magnetic field of a magnetic dipole at any point in space is

$$T = \frac{M}{r^3}(1 + 3\sin^2\theta)^{1/2} \tag{4.42}$$

where, M denotes the magnetic moment of the magnetic dipole with the magnitude $M = 2ml$, the direction from $-m$ to m, $-m$ and m denote the two magnetic pole strengths of the magnetic dipole, and $2l$ denotes the distance between them.

The pattern of anomalous magnetic fields generated by a magnetic dipole is shown in Fig. 4.7. Assuming that the height of the two planes of the anomalous magnetic

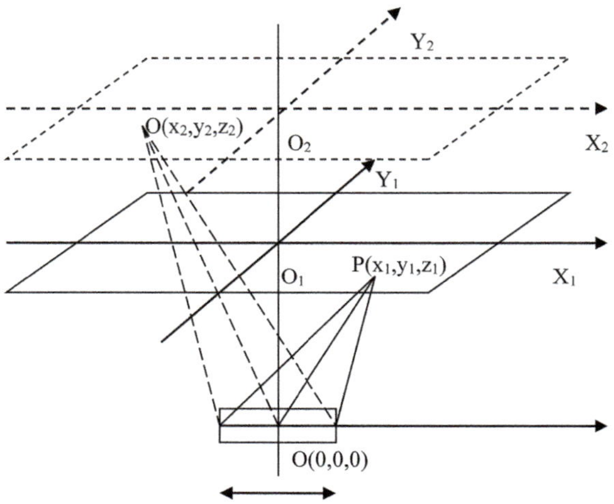

Fig. 4.7 Coordinate generation mode of magnetic dipole anomalous magnetic field

field distribution for verification is z_1 and z_2 respectively, and any point on the two planes is $P(x_1, y_1, z_1)$ and $Q(x_2, y_2, z_2)$. Then the magnetic anomaly field intensity at each point is calculated as follows.

$$T_{aP} = \frac{M}{r_{OP}^3}\left(1 + 3\sin^2\theta_{\angle MOP}\right)^{1/2} = \frac{M}{\left(x_1^2 + y_1^2 + z_1^2\right)^{3/2}}\left(1 + 3\left(\frac{y_1^2 + z_1^2}{x_1^2 + y_1^2 + z_1^2}\right)\right)^{1/2}$$
$$= \frac{M\left(x_1^2 + 4y_1^2 + 4z_1^2\right)^{1/2}}{\left(x_1^2 + y_1^2 + z_1^2\right)^2} \tag{4.43}$$

$$T_{aQ} = \frac{M\left(x_2^2 + 4y_2^2 + 4z_2^2\right)^{1/2}}{\left(x_2^2 + y_2^2 + z_2^2\right)^2} \tag{4.44}$$

Assuming $M = 10^2$, the magnetic field intensity amplitude distribution in the two anomaly planes is as Fig. 4.8 shown.

4.4.2 *Geomagnetic Anomaly Field Continuation Based on Reconciliation Theory*

The geomagnetic anomaly field can be obtained by subtracting the calculated value of the normal field in Sect. 4.1.1 from the geomagnetic survey data. The survey data for geomagnetic navigation are usually targeted to local areas of small extent. As for the selection of the normal field model, since the IGRF model is commonly used for the normal field, and the IGRF model can provide more comprehensive information

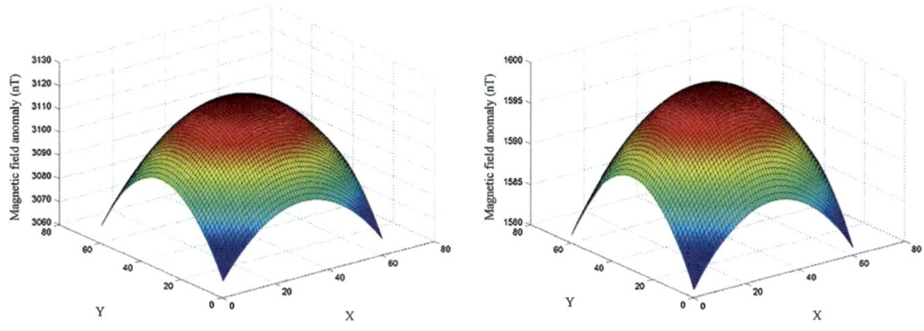

(a) Lower interface (b) Upper interface

Fig. 4.8 Magnetic field intensity distribution in different distance planes of the magnetic dipole anomaly

on the geomagnetic parameters, this model is also used for the calculation of the normal field in this chapter [16].

The geomagnetic anomaly field is a concept to the normal field, and the rest of the field relative to the geocentric dipole field can be considered as a geomagnetic anomaly. If the 12th order geomagnetic field component sum is used as the geomagnetic normal field, the rest of the smaller part can be considered as geomagnetic anomalous field. The geomagnetic anomalous field is the geomagnetic distribution generated by the smaller field sources close to the surface, and the mechanism of magnetic field generation and distribution law of each field source is the same as the magnetic dipole field.

Combined with the current magnetic map of underwater geomagnetic navigation, the problem of geomagnetic anomaly field can be equated to the magnetic field distribution value of a plane in the space outside the known field source. However, the location of the field source and the distance from the plane are unknown, and there is a possibility of multiple field sources at the same time, and then the magnetic field distribution value of the parallel plane is deduced from the magnetic field survey value of the plane, which is the problem of upward or downward extension of the geomagnetic anomaly field.

Upward extension of geomagnetic anomaly field.

The magnetic anomaly extension problem can be reduced to the following mathematical problem: 1. In the space beyond the field source, the magnetic potential and its partial derivatives in all directions are summation functions, which all satisfy the Laplace Eq. (4.2). The distribution of the eigen-components of the geomagnetic reference map for underwater navigation satisfies the first-side-value problem of the Laplace equation, which is the Dirchlet problem.

$$\begin{cases} \Delta T_a|_\Omega = 0 \\ T_a|_R = f(M) \end{cases} \tag{4.45}$$

where, $\Delta T_a|_\Omega = 0$ holds in the space outside the field source Ω, $T_a|_R = f(M)$ holds on the boundary of Ω at R.

Using Green's second formula in (4.32), the integral expression for the solution of Laplace's equation in the (4.45) can be established.

$$\iiint\limits_{\Omega} (T_a \Delta V - V \Delta T_a) dV_\Omega = \iint\limits_{R} \left(T_a \frac{\partial V}{\partial n} - V \frac{\partial T_a}{\partial n} \right) ds_R \tag{4.46}$$

The so-called integral expression of the harmonic function is to use the harmonic function and its normal derivative on the region boundary R along the integral of R to represent the value of the harmonic function at any point in the domain of Ω Let $M_0(x_0, y_0, z_0)$ be a fixed point in the domain of Ω, and introduce another harmonic function.

The integral expression of the harmonic function is used to express the value of the harmonic function at any point in the domain of Ω by using the harmonic function and its normal derivative along R on the regional boundary R. Let $M_0(x_0, y_0, z_0)$ be a fixed point in the Ω domain and introduce another summation function.

$$V = \frac{1}{r} = \frac{1}{\sqrt{(x - x_0)^2 + (y - y_0)^2 + (z - z_0)^2}} \tag{4.47}$$

There is

$$T_a(M_0) = -\frac{1}{4\pi} \iint\limits_{R} \left[T_a(M) \frac{\partial}{\partial n} \left(\frac{1}{r_{MM_0}} \right) - \frac{1}{r_{MM_0}} \frac{\partial T_a(M)}{\partial n} \right] ds \tag{4.48}$$

where,

$$r_{MM_0} = \sqrt{(x - x_0)^2 + (y - y_0)^2 + (z - z_0)^2}$$

But (4.31) only gives the value of the reconciliation function on the boundary, can be considered a special reconciliation function $g(M, M_0)$, it is reconciled in the region Ω about the variable M, and on the boundary of Ω R and the function $-1/4\pi r_{MM_0}$ on the boundary R the same value. At this point if another reconciliation function $V = g(M, M_0)$, then we have.

However, Eq. (4.31) only gives the value of the reconciliation function on the boundary of the region, assuming that the reconciliation function $g(M, M_0)$ is reconciled in the region Ω with respect to the variable M, while on the boundary R of Ω is the same as the value of the function $-1/4\pi r_{MM_0}$ on the boundary R. If there exists another harmonic function $V = g(M, M_0)$, then we have

$$T_a(M_0) = \iint\limits_{R} \left(G \frac{\partial T_a}{\partial n} - T_a \frac{\partial G}{\partial n} \right) ds \tag{4.49}$$

where, G is the Green function.

$$G(M, M_0) = \frac{1}{4\pi r_{MM_0}} + g(M, M_0) \tag{4.50}$$

If there are $g(M, M_0)|_R = \left.\frac{-1}{4\pi r_{MM_0}}\right|_R$, $G(M, M_0)$ is called the Green's function of the Dirchlet problem in the region, then the solution of the Dirchlet problem (4.35) is denoted as

$$T_a(M_0) = -\iint_R \left(T_a \frac{\partial G}{\partial n}\right) ds \tag{4.51}$$

In some specially shaped areas, the Green's function can be determined. The reference area for underwater geomagnetic navigation is usually a small area of the survey map, so the effect of the curvature of the Earth can be ignored. The magnetic source of the geomagnetic anomaly corresponds to a point in the lower half space of the plane, and this problem can be solved by the mirroring method, which yields

$$\left.\frac{\partial G}{\partial n}\right|_{z=0} = \left.\frac{\partial G}{\partial z}\right|_{z=0} = -\frac{1}{2\pi} \frac{z_0}{\left[(x - x_0)^2 + (y - y_0)^2 + (z - z_0)^2\right]^{3/2}} \tag{4.52}$$

Combining Eqs. (4.52) and (4.37) yields

$$T_a(x, y, z) = \frac{z_0}{2\pi} \int_{-\infty}^{+\infty} \int_{-\infty}^{+\infty} \frac{T_a(x, y, 0)}{[(x - x_0)^2 + (y - y_0)^2 + z_0^2]^{3/2}} dxdy \tag{4.53}$$

If the plane in which the geomagnetic reference map is located is considered as the plane $z = 0$, $T_a(\xi, \eta, 0)$ is taken as the measured value of the geomagnetic anomaly, then Eq. (4.39) gives the geomagnetic anomaly value with upward extension.

Considering the plane in which the geomagnetic reference map is located as plane $z = 0$ and $T_a(\xi, \eta, 0)$ as the measured value of the geomagnetic anomaly, Eq. (4.39) gives the value of the geomagnetic anomaly with upward extension.

$$T_a(x, y, z) = \frac{z}{2\pi} \int_{-\infty}^{+\infty} \int_{-\infty}^{+\infty} \frac{T_a(\xi, \eta, 0)}{[(x - \xi)^2 + (y - \eta)^2 + z^2]^{3/2}} d\xi d\eta, z < 0 \tag{4.54}$$

Equation (4.40) denotes the principle for upward extension.

Downward extension of geomagnetic anomaly field.

The downward extension of the geomagnetic anomaly field is the inverse problem of the lower half space of the reference plane containing a field source which is unknown. The bit field below sea level and above the field source in space, it still satisfies the Rasch equation. The lower extension problem can be reduced to solving the following edge value problem.

$$\begin{cases} \dfrac{\partial^2 T_a(x,y,z)}{\partial x^2} + \dfrac{\partial^2 T_a(x,y,z)}{\partial y^2} + \dfrac{\partial^2 T_a(x,y,z)}{\partial z^2} = 0, 0 < z < H \\ T_a(x,y,z)|_{z=0} = T_a(x,y) \end{cases} \tag{4.55}$$

The solution of the above formulation is unstable in the mathematical sense, and can be reduced to a non-deterministic problem in the Hadamard sense. That is, when there is a very small change in the fixed solution data $T_a^0(x, y)$, the resulting solution can vary greatly. The data obtained from field observations always have some approximation and there are some errors in the range of accuracy with the objective reality, and these errors are expressed as large values in the solution, which can mask the real solution.

The mathematical formulation of the above problem is as follows.

Letting $T_a^0(x, y, z)$ be the solution of (4.56), for

$$\begin{cases} \dfrac{\partial^2 T_a(x,y,z)}{\partial x^2} + \dfrac{\partial^2 T_a(x,y,z)}{\partial y^2} + \dfrac{\partial^2 T_a(x,y,z)}{\partial z^2} = 0, 0 < z < H \\ T_a(x,y,z)|_{z=0} = T_a^1(x,y) + \varepsilon \cos Mx \cdot \cos Ny \end{cases} \tag{4.56}$$

Which yields

$$T_a(x,y,z) = T_a^0(x,y,z) + \varepsilon e^{(M^2+N^2)^{1/2} \cdot z} \cos Mx \cdot \cos Ny \tag{4.57}$$

Although ε is small, the difference between $T_a(x, y, z)$ and $T_a^0(x, y, z)$ can be arbitrarily large as long as M and N are sufficiently large.

Therefore, in the extension of the geomagnetic reference map, where accuracy is the first consideration, a new processing method is needed for effective downward extension.

4.4.3 *Fourier Transform-Based Upward and Downward Extension Method*

From the convolution integral formula, the above equation is the two-dimensional convolution of $T_a(x, y, 0)$ and $-z\Big/\left(2\pi\left(x^2+y^2+z^2\right)^{3/2}\right)$ with respect to the variable (x, y). It is also known that the convolution in the spatial domain corresponds to the product in the frequency domain. The Fourier transforms of $T_a(x, y, 0)$ and $-z\Big/\left(2\pi\left(x^2+y^2+z^2\right)^{3/2}\right)$ are solved separately. Let $T_a(x, y, z)$ for the Fourier transform of the variable (x, y).

$$S_{T_a}(u, v, z) = \int_{-\infty}^{\infty}\int_{-\infty}^{\infty} T_a(x, y, z)e^{-2\pi i(ux+vy)}dxdy \tag{4.58}$$

Which yields,

$$S_{T_a}(u, v, 0) = \int_{-\infty}^{\infty}\int_{-\infty}^{\infty} T_a(x, y, 0)e^{-2\pi i(ux+vy)}dxdy \tag{4.59}$$

The Fourier transform of $-z\Big/\left(2\pi\left(x^2+y^2+z^2\right)^{3/2}\right)$, applying the integral transformation table given by Erdelyi yields.

$$\int_{-\infty}^{\infty}\int_{-\infty}^{\infty} \frac{-z}{2\pi\left(x^2+y^2+z^2\right)^{3/2}}e^{-2\pi i(ux+vy)}dxdy = e^{2\pi(u^2+v^2)^{1/2}\cdot z} \tag{4.60}$$

The above equation holds for $z \leq 0$. From (4.59)–(4.61) and the convolution theorem we have.

$$S_{T_a}(u, v, z) = S_{T_a}(u, v, 0)e^{2\pi(u^2+v^2)^{1/2}\cdot z} \tag{4.61}$$

$T_a(x, y, z)$ is the Fourier inverse transform of $S_{T_a}(u, v, z)$, i.e.

$$T_a(x, y, z) = \int_{-\infty}^{\infty}\int_{-\infty}^{\infty} S_{T_a}(u, v, 0)e^{2\pi(u^2+v^2)^{1/2}\cdot z}e^{2\pi(ux+vy)}dudv \tag{4.62}$$

The above equation is the expression of the Fourier transform spectrum for upward extension.

The upward extension method using Fourier transform for the spectrum can be simply formulated as (4.64).

$$U(x, y, h) = F^{-1}\{e^{-\sqrt{k_x^2+k_y^2}h}F[U(x, y, 0)]\} \tag{4.63}$$

where, F and F^{-1} denotes Fourier transform and Fourier inverse transform, respectively. k_x and k_y is the wave number in the direction of x and y, respectively.

A large number of improved algorithms have been proposed for the study of the extension method, such as regularization method, multiscale edge constraint method, Weiner filtering method, singular value decomposition method, derivative and compensation method, etc. However, most of the above methods are proposed for the instability of the downward extension, while the spatial domain methods do not perform the convolution integral directly, but calculates a set of weight coefficients in advance, and calculates the extension value in the form of a weighted sum, which is very unfavorable to the processing process that may change the flight height at any time. The Fourier transform method in the frequency domain is very stable and the algorithm is easy to process and implement numerically, so this chapter adopts the Fourier transform method as the basic method for upward extension of the ocean anomalous magnetic field. The exponent of *e* of the Fourier transform method for downward extension is positive.

4.4.4 Downward Extension Algorithm of Geomagnetic Anomaly Field Based-On the Combination of MAP and Two-Dimensional Wavelet Denoising

1. Theoretical comparison of traditional downward extension methods.

For the ocean magnetic field distribution measured by surface towing, its underwater application requires the anomaly field to be finely extended downward, and it is more applied in the depth range of several hundred meters, which is the research scope of this chapter. However, the downward extension of geomagnetic anomalies is difficult because it is an ill-conditioned problem. In addition to the previously mentioned algorithms, the existing extension methods are mainly the multi-scale edge analysis algorithm proposed by transplanting Mallat on the basis of wavelet multi-scale edge reconstruction and the integral-iterative method from the perspective of digital signal processing, which have received more attention.

The basic idea of the multiscale edge analysis algorithm is that the edges of the two-dimensional signal itself are better maintained in the multiscale analysis, while the edges of noise disappear faster in the multilayer analysis, so the interference of noise can be eliminated by retaining the edges that still exist after the multiscale decomposition for reverse reconstruction. However, the starting point of the Mallat multi-scale edge reconstruction algorithm is image processing with obvious edge features. For geomagnetic anomalous fields with gentle distribution, not only the extraction of edges is affected by the extraction operator and the given threshold, but also the specific implementation of the reverse mapping in the reconstruction process has a greater impact on the reconstruction results. Even for the gently varying anomalous regions, the multi-scale edge analysis method is rendered ineffective due to the inability to determine the effective edges.

Considering that the ship speed is usually slow when measuring the ocean magnetic field, and the sampling frequency of current ocean magnetic measurement instruments, such as G882, is high, there is redundant sampling of ocean magnetic field measurements on the heading. This provides information that can be used to reconstruct a more accurate down-delayed anomaly field distribution. Combined with the idea of super-resolution reconstruction in image restoration processing, this chapter constructs multiple measurement samples of the upper interface corresponding to the lower interface grid distribution by interpolation on the heading combines Maximum a Posteriori Probability Estimation (MAP) and two-dimensional wavelet noise cancellation to construct a new MAP-W algorithm for the downward extension of the ocean magnetic anomaly field. Although the Fourier Transform (FT) extension algorithm usually has a downward extension of no more than 3–5 times the distance between data points. The new algorithm is still based on the Fourier Transform method, considering that the method has a better extension interpretation and algorithm implementation.

2. Mathematical analysis of the downward extension problem.

As described in Sect. 2.1.2, the downward extension of the geomagnetic anomaly field is an inverse problem where the lower half space of the reference plane contains a field source, and the intensity and location of the field source, etc., are unknown. The potential field below sea level, in the space above the field source, still satisfies the Laplace equation. The lower extension problem can be reduced to solving the following edge value problem.

$$\begin{cases} \dfrac{\partial^2 T_a(x, y, z)}{\partial x^2} + \dfrac{\partial^2 T_a(x, y, z)}{\partial y^2} + \dfrac{\partial^2 T_a(x, y, z)}{\partial z^2} = 0, 0 < z < H \\ T_a(x, y, z)|_{z=0} = T_a(x, y) \end{cases} \tag{4.64}$$

Letting $T_a^0(x, y, z)$ be the solution of (4.33), for

$$\begin{cases} \dfrac{\partial^2 T_a(x, y, z)}{\partial x^2} + \dfrac{\partial^2 T_a(x, y, z)}{\partial y^2} + \dfrac{\partial^2 T_a(x, y, z)}{\partial z^2} = 0, 0 < z < H \\ T_a(x, y, z)|_{z=0} = T_a^1(x, y) + \varepsilon \cos Mx \cdot \cos Ny \end{cases} \tag{4.65}$$

Then

$$T_a(x, y, z) = T_a^0(x, y, z) + \varepsilon e^{(M^2+N^2)^{1/2} \cdot z} \cos Mx \cdot \cos Ny \tag{4.66}$$

where, $T_a(x, y, z)$ is the intensity of the magnetic anomaly at (x, y, z), H is the depth of the field source.

The solution of the above formulation is unstable in the mathematical sense and can be attributed to the discomfort problem in the Hadamard sense. That is, the resulting solution can vary greatly when there is a very small change in the fixed solution data $T_a(x, y)$. The data obtained from surface towing measurements always have some approximation, and the corresponding errors show up as large values in the downscaled solutions, thus making the downscaled results deviate from the true distribution.

Mathematically, it is clear from the analysis that ε is an interfering small quantity. An accurate downward extension algorithm needs to effectively overcome the effect of the latter term in Eq. (4.35), i.e., how to characterize the true distribution of ε in redundant measurements.

3. Combined MAP Estimation and Two-Dimensional Wavelet Denoising Downward Extension Algorithm

The downward extended combinatorial algorithm for marine magnetic anomalies uses the maximum a posteriori probability (MAP) estimation as the basic framework for signal reconstruction. In order to increase the utilization of each reconstructed signal, a two-dimensional wavelet denoising method is usually used to remove the interference information from the results of a single iteration.

(1) MAP Estimation

The downward extension framework based on MAP estimation utilizes the recovery idea of super-resolution image reconstruction. The difference is that the sampling points correspond to each other and there is no estimation of the offset, so the three items of migration, blur, and down-sampling do not need to be considered, and only the iterative approximation of the real signal is achieved for the interference of additional noise. The magnetic anomaly distribution of multiple upper interfaces can be obtained by interpolation of redundant measurements, so $y_k, k = 1, \cdots, p$ denotes the number of redundant distributions available for down-sampling at the upper interface, where each point corresponds to each point of the target anomaly distribution at the lower interface.

The FFT downward extension method for constructing multiple redundant lower interface distributions is shown in Eq. (4.36).

$$z(x, y, h) = F^{-1}\{e^{\sqrt{k_x^2+k_y^2}h} F[z(x, y, 0)]\} \tag{4.67}$$

where, F and F^{-1} denotes Fourier transform and Fourier inverse transform, respectively. k_x and k_y are the wave numbers in the direction of x and y, respectively.

According to the knowledge of digital signal processing, the relationship between the planar field distribution after the downscaling of the geomagnetic anomaly field and its true distribution can be characterized as $y = \omega z + n$, where $\omega = \left[I_N^1; I_N^2; \cdots; I_N^p\right]^T$. The vector form is used to characterize as

$$y_{k,m} = \sum_{r=1}^{N} \omega_{k,m,r} z_r + n_{k,m} \tag{4.68}$$

where, y represents a series of two-dimensional downlink anomaly distributions in a one-dimensional sequence; $k = 1, \cdots, p$ represents the number of downlink distributions; $m = 1, \cdots, N$ represents the sampling points of each downlink signal; $N = L_V \times L_H$, L_V and L_H are the number of rows and columns of the downlink anomaly distribution; z represents the true distribution of the geomagnetic anomaly field in one-dimensional form; n is the noise vector.

The noise vector n can be modeled as a Gaussian sequence with zero mean $n \sim N(0, \Lambda_k)$, where $\Lambda_m = \sigma^2 I$, σ^2 is the variance of the noise. Then the probability density function of the noise can be characterized as

$$P(n) = \frac{1}{(2\pi)^{pN/2}\sigma_n^{pN}} \exp\left\{-\frac{1}{2\sigma_n^2} n^T n\right\} = \frac{1}{(2\pi)^{pN/2}\sigma_n^{pN}} \exp\left\{-\frac{1}{2\sigma_n^2} \sum_{m=1}^{pN} n_m^2\right\} \tag{4.69}$$

The estimation idea of MAP is as follows: the posterior probability of z is estimated according to Eq. (4.39) based on the known observation y. The above estimation can be transformed into Eq. (4.40) by Bayes' rule. Since y is known, the denominator can be disregarded and Eq. (4.40) can be transformed into Eq. (4.41), and the form of Eq. (4.42) is used to simplify the process of solving for the result.

$$\hat{z} = \arg\max_{z} P(z|y) \tag{4.70}$$

$$\hat{z} = \arg\max_{z} \frac{P(y|z)P(z)}{P(y)} \tag{4.71}$$

$$\hat{z} = \arg\max_{z} P(y|z)P(z) \tag{4.72}$$

$$\hat{z} = \arg\min_{z}\{-\log[P(y|z)] - \log[P(z)]\} \tag{4.73}$$

The prior probability of the anomaly distribution can be characterized as a Gibbs distribution and according to the Eq. (4.44), then

$$P(z) = \frac{1}{(2\pi)^{N/2}|C_z|^{1/2}}\exp\left\{-\frac{1}{2}z^T C_z^{-1} z\right\} \tag{4.74}$$

$$P(z) = \frac{1}{(2\pi)^{N/2}|C_z|^{1/2}}\exp\left\{-\frac{1}{2\lambda}\sum_{i=1}^{N}\left(\sum_{j=1}^{N} d_{i,j}z_j\right)^2\right\} \tag{4.75}$$

where

$$d_{i,j} = \begin{cases} 1 & if\,(i=j) \\ -\frac{1}{4} & if\,(j=i-1)or(j=i+1)or(j=i-L_H)or(j=i+L_H) \end{cases}$$

C_z is the covariance array of z.

MAP estimation means solving for the minimal value of the objective function of the following equation.

$$L(z) = \frac{1}{2\sigma_n^2}(y-\omega z)^T(y-\omega z) + \frac{1}{2}z^T C_z^{-1} z \tag{4.76}$$

z can be obtained by $\nabla_z L(z) = 0$.

$$\hat{z}^{n+1} = \left[\omega^T\omega + \sigma_n^2 C_z^{-1}\right]^{-1}\omega^T y \tag{4.77}$$

The equation is usually solved by the gradient decreasing method, and the objective function can be characterized as

$$L(z) = \frac{1}{2\sigma_n^2}\sum_{m=1}^{pN}\left(y_m - \sum_{r=1}^{N}\omega_{m,r}z_r\right)^2 + \frac{1}{2\lambda}\sum_{i=1}^{N}\left(\sum_{j=1}^{N}d_{i,j}z_j\right)^2 \tag{4.78}$$

Solved iteratively by Eq. (4.48).

$$\hat{z}_k^{n+1} = \hat{z}_k^n - \varepsilon^n g_k\left(\hat{z}_k^n\right) \tag{4.79}$$

The local gradient is

$$g_k(z) = \frac{1}{\sigma_n^2}\sum_{m=1}^{pN}\omega_{m,k}\left(\sum_{r=1}^{N}\omega_{m,r}z_r - y_m\right) + \frac{1}{\lambda}\sum_{i=1}^{N}d_{i,k}\left(\sum_{j=1}^{N}d_{i,j}z_j\right) \tag{4.80}$$

The decreasing factor is

$$\varepsilon^n = \frac{\frac{1}{\sigma_n^2}\sum\limits_{m=1}^{pN}\gamma_m\left(\sum\limits_{r=1}^{N}\omega_{m,r}\hat{z}_r^n - y_m\right) + \frac{1}{\lambda}\sum\limits_{i=1}^{N}\overline{g}_i\left(\sum\limits_{j=1}^{N}d_{i,j}z_j\right)}{\frac{1}{\sigma_n^2}\sum\limits_{m=1}^{pN}\gamma_m^2 + \frac{1}{\lambda}\sum\limits_{i=1}^{N}\overline{g}_i^2} \tag{4.81}$$

where, $\gamma_m = \sum\limits_{r=1}^{N}\omega_{m,r}g_r(\hat{z}^n)$, $\overline{g}_i = \sum\limits_{j=1}^{N}d_{i,j}g_j(\hat{z}^n)$.

The solution of Eq. (4.48) that stabilizes is the final anomaly field distribution.

(2) Two-dimensional Wavelet Denoising

In order to increase the efficiency of utilizing the reconstructed signal for each iteration in the iterative process, a two-dimensional wavelet denoising method is used to weaken as much as possible the interference information in the single iteration results. For the noise cancellation of two-dimensional wavelets, the thresholding process is required for each layer of its horizontal, vertical and oblique decomposition signals separately. As shown in Fig. 4.9, after the two-dimensional wavelet decomposes the signal in multiple layers, the cA_{j+1}, $\underbrace{cD_{j+1}^{(h)}}_{horizontal}$, $\underbrace{cD_{j+1}^{(v)}}_{vertical}$, and $\underbrace{cD_{j+1}^{(d)}}_{diagonal}$ coefficients are thresholded and the signal is reconstructed.

The two-dimensional discrete wavelet decomposition can be expressed as

$$\begin{aligned} f(x,y) = &\sum_{k,l\in Z}a_{J,k,l}\phi_{J,k,l}(x,y) + \sum_{j\geq J}\sum_{k,l\in Z}b_{j,k,l}^h\psi_{j,k,l}^h(x,y) \\ &+ \sum_{j\geq J}\sum_{k,l\in Z}b_{j,k,l}^v\psi_{j,k,l}^v(x,y) + \sum_{j\geq J}\sum_{k,l\in Z}b_{j,k,l}^d\psi_{j,k,l}^d(x,y) \end{aligned} \tag{4.82}$$

where, $\phi(x, y)$ and $\psi(x, y)$ are canonical orthogonal basis functions, respectively.

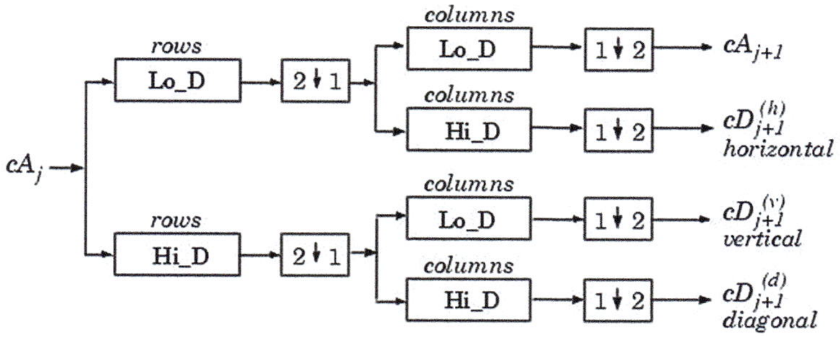

Fig. 4.9 Decomposition and reconstruction process of two-dimensional wavelet

Different thresholds are taken for the three directions on each scale j, λ_j, then

$$\lambda_j = \sigma\sqrt{2\log(F)}\Big/\log(j+1) \tag{4.83}$$

where, σ^2 is the variance of the noise; F and N are the lengths of the corresponding discrete sampled signals; j is the decomposition scale.

In practical applications, because of the convergence smoothing effect of MAP estimation, the process of denoising each iteration can usually be simplified by taking the approximate coefficient cA_{j+1} of its two-dimensional decomposition to reconstruct the signal as the initial value for the next iteration.

(3) Flow of the Combinatorial Algorithm

n = 0, the mean value of FT downward extension distributions $\hat{z}^0 = \bar{y}_i$ is selected as the initial estimate.

A two-dimensional wavelet denoising method was used to obtain $\hat{z}'$ by $\hat{z}$.

Calculating the local gradient $g_k\left(\hat{z}'^n\right)$ according to Eq. (4.49), where $k = 1, 2, \cdots, N$.

Calculating the optimal step size ε^n according to Eq. (4.50).

$$\hat{z}_k^{n+1} = \hat{z}_k^{\prime n\prime n} - \varepsilon^n g_k\left(\hat{z}'^n\right),$$

where $k = 1, 2, \cdots, N$.

If $\left\|\hat{z}_k^{n+1} - \hat{z}_k'^n\right\| / \left\|\hat{z}_k'^n\right\| < \alpha$ or the number of iterations reaches the given upper bound, the loop stops.

Otherwise, $n = n + 1$ and go back to step (2) and re-execute.

After the above steps, the distribution of the oceanic magnetic field anomaly at the lower interface is obtained.

3. Numerical Analysis

The detection data are obtained using a simulation generation method. The magnetic anomaly field is generated by three magnetic dipole sources with a distribution of

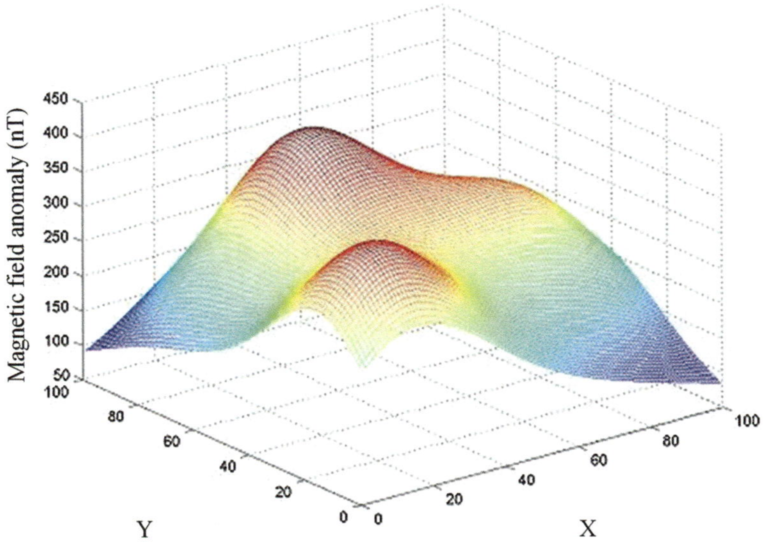

Fig. 4.10 Baseline magnetic field simulation data

1000*1000. As shown in Fig. 4.10, the equally spaced sampling distribution of magnetic anomalies on the plane 100*100, which is taken as the lower interface. During the simulation test, the magnetic anomaly distribution above its corresponding 2 times point distance is considered as the upper interface. The Gaussian-like white noise is attached to this upper interface, and the standard deviation of the noise is increased sequentially, and the magnitude of the lower delay difference is compared respectively.

In this chapter, the Fast FT (FFT) transform is used as the basis for the implementation of the basic deferral processing, and the ML estimation method is used along with the MAP estimation to form five downward methods for comparison. These five methods are FT, FFT + ML, FFT + MAP, FFT + ML and two-dimensional wavelet denoising, FFT + MAP and two-dimensional wavelet denoising. Where the threshold value α is taken as 10^{-7}, the iterative process is terminated when this condition is satisfied.

The convergence of the proposed method in this chapter is given in Fig. 4.7, where the horizontal coordinate is the number of iterations and the vertical coordinate corresponds to the logarithm of the ratio value $\left\| \hat{z}_k^{n+1} - \hat{z}{\prime}_k^n \right\| / \left\| \hat{z}{\prime}_k^n \right\|$ in the threshold judgment. After adding two-dimensional wavelet denoising during each iteration, the MAP estimation downscaling is close to the true distribution than the MAP estimation alone with fewer iterations, and the estimation results can be achieved close to the true distribution with fewer iterations, which greatly improves the execution efficiency of the algorithm.

Considering that when the threshold value cannot be determined, the difference of multiple consecutive iterations is usually used to determine whether convergence

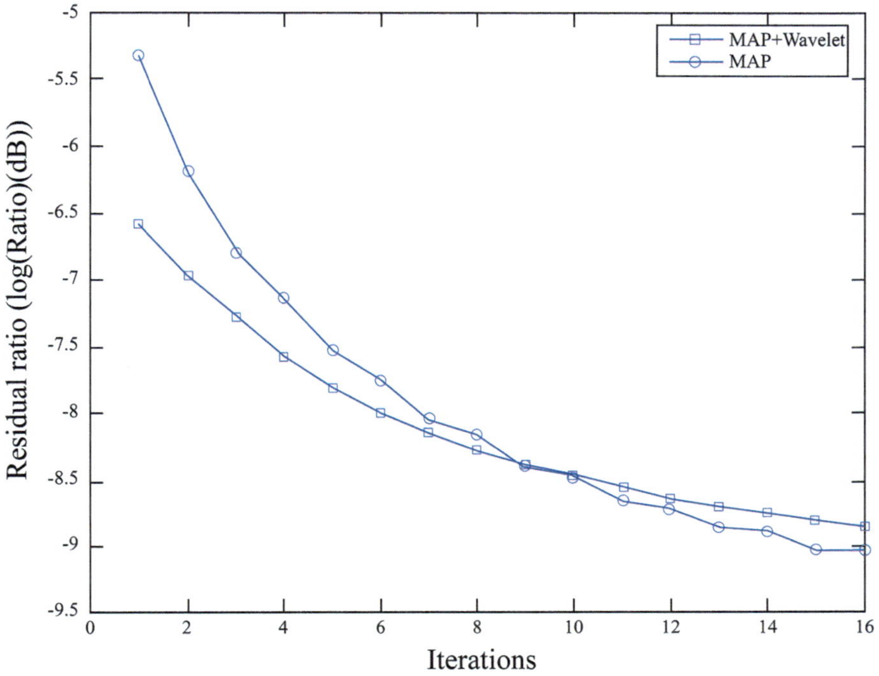

Fig. 4.11 Convergence of MAP estimation extended downward algorithm

is less than a small value, such as the iterative difference $\Delta = 0.2$(dB), the iterative approximation process of MAP-W is also smoother and easier to judge convergence compared to simple MAP estimation, which is evident in the second half of Fig. 4.11 ll these factors increase the applicability of MAP-W algorithm.

A comparison of the five downscaling methods is given in Fig. 4.8, where the horizontal coordinate is the standard deviation of the upper interface data with additional Gaussian white noise, and the vertical coordinate corresponds to the single-point average error of the downscaled distribution and the true distribution. From Fig. 4.8, it can be seen that the MAP-W downscaling method with the combination of MAP estimation and two-dimensional wavelet denoising has the highest accuracy among the five downscaling methods. The average single-point error of the downscaling results is below 1nT when the Gaussian noise with 5nT standard deviation is added to the upper interface, which is sufficient for most applications within a few hundred meters below the sea surface, so the combined method has greater practicality (Fig. 4.12).

Drawing on the idea of image super-resolution reconstruction, this chapter addresses the difficult problem of downward extension of anomalous fields in sea surface magnetic field measurements from the perspective of two-dimensional digital signal processing, taking advantage of the actual situation of high sampling rate of

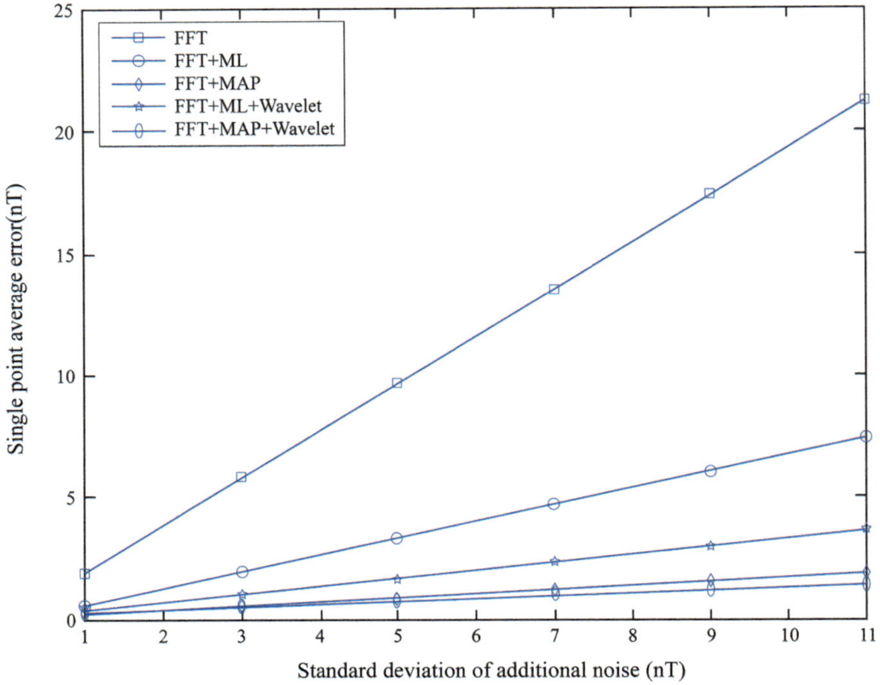

Fig. 4.12 Accuracy comparison of different extended downward algorithms

current marine magnetic field measurement instruments and slow speed of measurement vessels. A new MAP estimation and two-dimensional wavelet denoising algorithm is proposed for the downward extension of magnetic anomalies in the space of several hundred meters below the sea surface, which uses a combination of MAP estimation and two-dimensional wavelet denoising to approximate the real distribution of the lower interface and fully exploits the implicit information of the heading redundancy measurement data. When a greater depth of extension is required, the basic method of the lower extension processing can be used iteratively. Numerical analysis shows that the accuracy of the down-delayed magnetic field generated by this method basically meets the requirements of practical applications, which opens up a new way for the down-delaying of magnetic anomalies at the sea surface.

4.5 Mapping and Generation of Geomagnetic Reference Maps

The basic composition of the geomagnetic field and the analysis of the basic components and characteristics of the geomagnetic field that can be used for positioning navigation are given in the first four sections. This section discusses the process and methods for producing geomagnetic reference maps suitable for geomagnetic navigation based on the measured data [17–19].

4.5.1 Preliminary Processing of Measurement Data

Due to random interference and other errors, there are inevitably obvious measurement wild values in the measurement data that need to be removed, as shown in Fig. 7.3. The contour map of the total geomagnetic field intensity after removing the measurement wild values is shown in Fig. 4.13 [20].

4.5.2 Position Correction

There is a fixed relative relationship between the position of the magnetometer sensor and the position of the measurement carrier positioning system, so the position of each measurement line needs to be corrected according to the carrier position information.

4.5.3 Correction of Daily-Varying Data

The selection of the base value of the daily geomagnetic variation is an important part of the geomagnetic daily variation correction. The existing methods for determining the base value of the daily variation include the average method, the base moment method, the base time method, and the time-zone-weighted average method [21].

4.5.4 Data Smoothing

In order to reduce the measurement error, 7-point smoothing is performed on the measured line data after each daily variation correction, and the smoothing formula is as follows.

$$X_i = \frac{1}{7}(X_{i-3} + X_{i-2} + X_{i-1} + X_i + X_{i+1} + X_{i+2} + X_{i+3}) \tag{4.84}$$

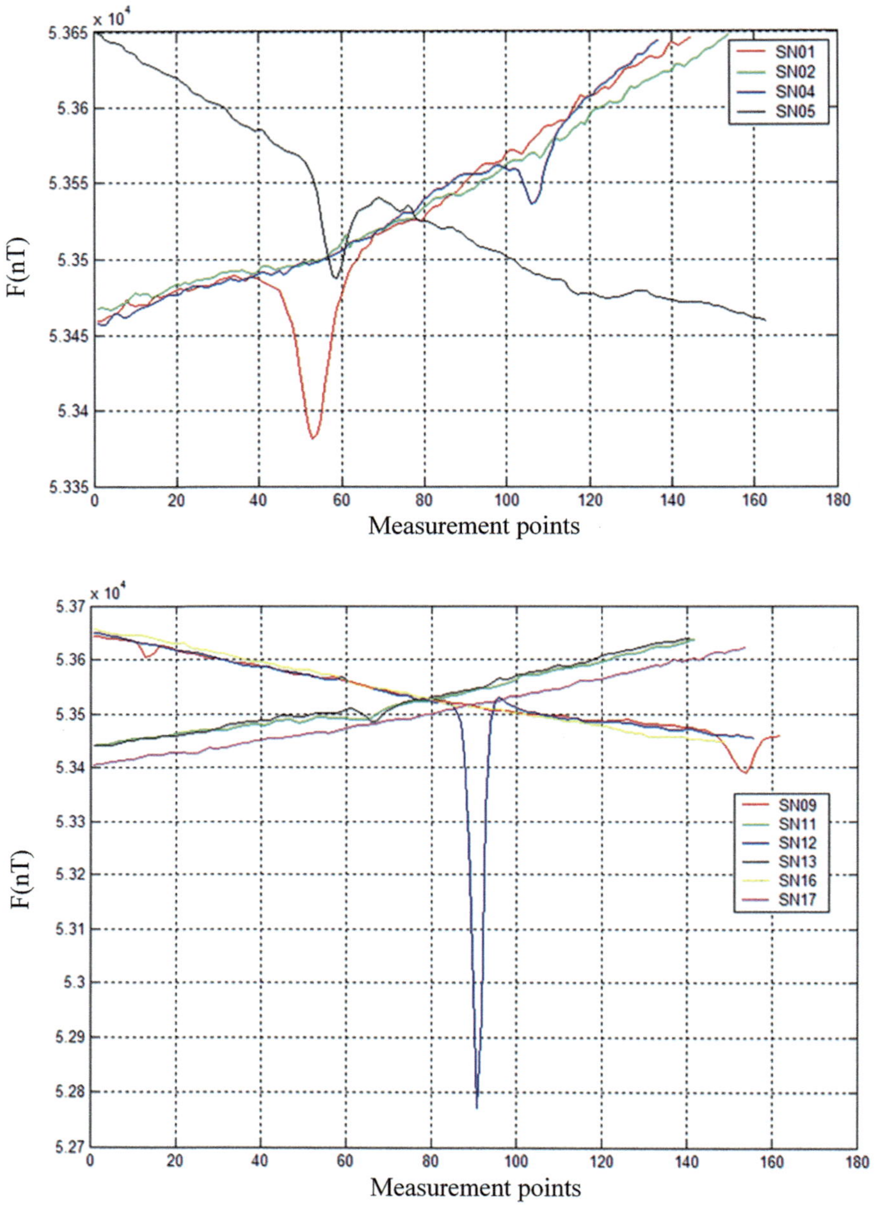

Fig. 4.13 Possible measurement anomaly measurement line geomagnetic field total intensity curve

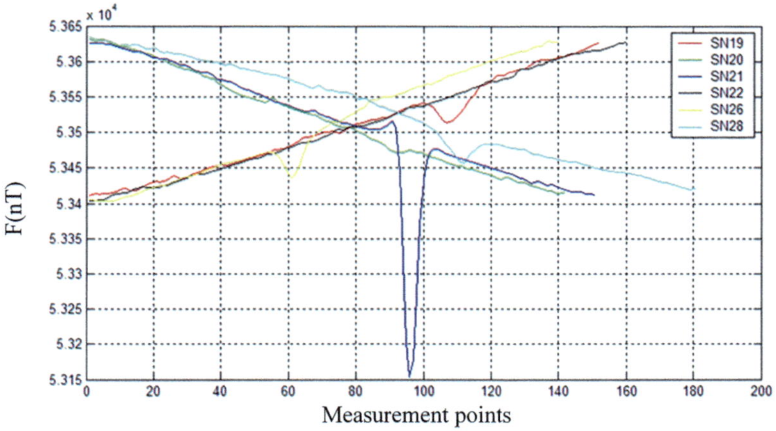

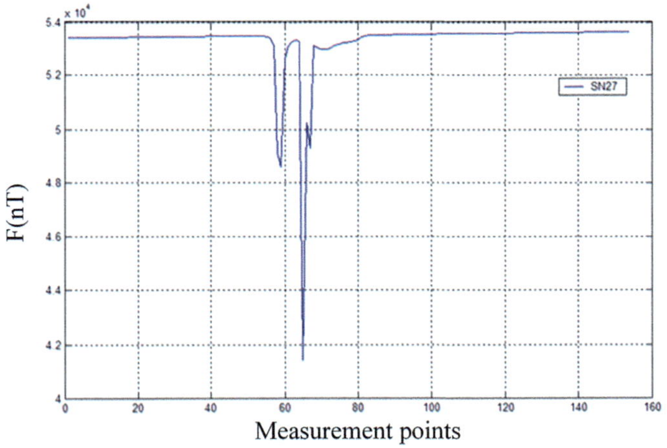

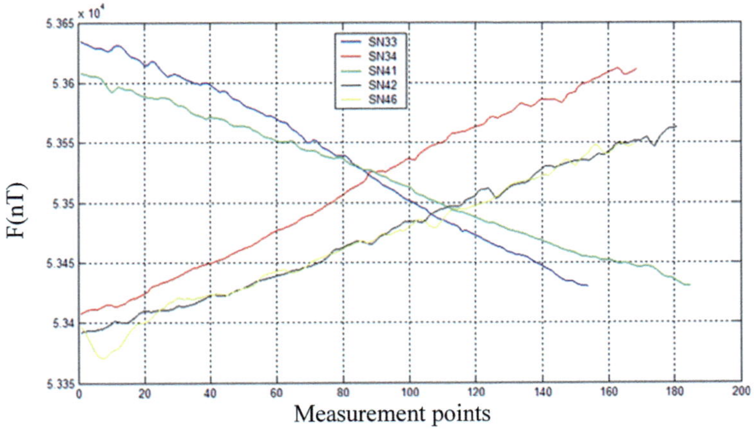

Fig. 4.13 (continued)

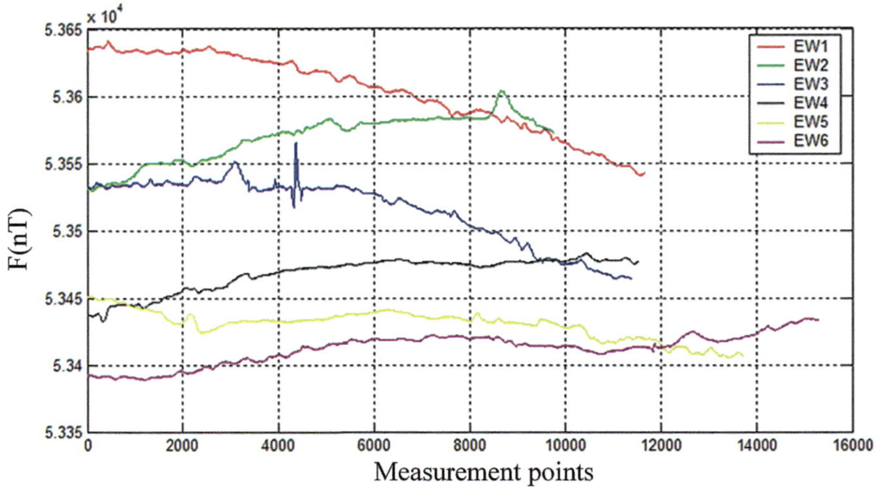

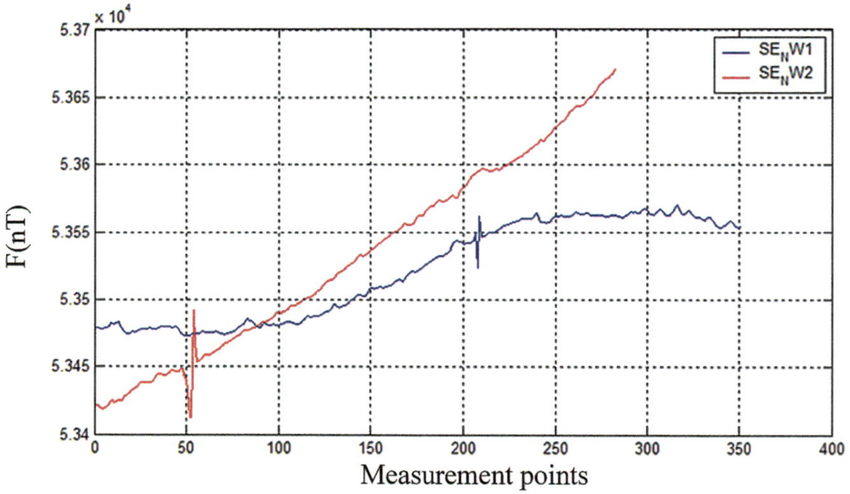

Fig. 4.13 (continued)

4.5.5 Data Parity Processing

In the geomagnetic measurement, the survey line is usually laid out as the main survey line and the sub-survey line intersecting directly to form a survey line network, the intersection of the main survey line and the sub-survey line is called the intersection point, and the sub-survey line is also called the cut line, which is mainly used to check and process the measurement data of the main survey line.

At present, the methods of using the intersection measurement difference to level the measurement data are mainly based on the least squares principle and the adjustment treatment methods based on the uniform rank deficit network adjustment treatment model. These methods are mainly based on finding the level difference value on each measurement line and subtracting the systematic error value from the corresponding measurement line to obtain the level difference processed data. When the measurement area is small, the measurement line is relatively short and the actual measurement trajectory deviates from the planned measurement line, these adjustment treatment methods can better eliminate the influence of systematic error, and the adjustment treatment accuracy is higher. However, when the measurement area is large, the measurement line is long and the actual measurement trajectory deviates from the planned measurement line, the adjustment treatment process by the above methods results in lower adjustment treatment accuracy and cannot meet the actual measurement requirements. This book proposed a adjustment treatment method based on polynomial fitting model of different orders for the characteristics of the measured marine magnetic data, and the actual data calculation and processing show that the method can better eliminate the influence of systematic errors, and the adjustment treatment accuracy is high and the computer implementation is relatively simple.

Measurement adjustment treatment is the adjustment of measurement data, which is based on some optimization criteria to determine the best valuation and accuracy of the unknown quantity based on a series of measurement data with observed errors. The common methods of measurement adjustment treatment include condition adjust method, indirect adjustment method, condition adjust method with unknown parameters, and indirect adjustment method with constraints, etc. Any practical adjustment treatment problem must first choose to establish some kind of function model.

Let there be n data points in the measurement area $(x_i, y_i, z_i), i = 1, 2, \cdots, n$, the independent variable (x_i, y_i) is the plane coordinate of the point, which is a non-random quantity. The dependent variable is the observed quantity z_i at the point, which is regarded as a random variable. Although there is no definite functional relationship between the independent variable (x_i, y_i) and the dependent variable z_i, the measured data can be expressed in terms of its trend variation $f(x, y)$ and random error ε, which yields

$$z = f(x, y) + \varepsilon \tag{4.85}$$

When the $f(x, y)$ is taken as a polynomial in the independent variable x, y, Eq. (7.2) is called a polynomial fit model. The general form of the polynomial fit model is

$$\begin{aligned} z = f(x, y) &= b_0 + b_1 x + b_2 y + b_3 x^2 + b_4 xy + b_5 y^2 \\ &\quad + b_6 x^3 + b_7 x^2 y + b_8 xy^2 + b_9 y^3 + \cdots \end{aligned} \tag{4.86}$$

where, $b_0, b_1, b_2 \cdots$ is the polynomial coefficient.

When the highest order of the independent variable is 1, it is called a first-order polynomial fit model, and when it is 2, it is called a second-order polynomial fit model.

The first-order polynomial fit model is

$$z_i = f(x_i, y_i) = b_0 + b_1 x_i + b_2 y_i + \varepsilon_i \tag{4.87}$$

The second-order polynomial fit model is

$$z_i = f(x_i, y_i) = b_0 + b_1 x_i + b_2 y_i + b_3 x_i^2 + b_4 x_i y_i + b_5 y_i^2 + \varepsilon_i \tag{4.88}$$

where, z_i is the total geomagnetic field intensity difference; $b_{0,} \cdot \cdot\cdot, b_5$ is the fitting coefficient; x_i, y_i is the plane right angle coordinates of the point; and ε_i is the residual.

The difference of the total intensity of the geomagnetic field from more than the number of cross-measurement points is obtained by the least squares method ($[\varepsilon_i]^2 = \min$) to find the fitting coefficient $b_{0,} \cdot \cdot\cdot, b_5$, and the plane coordinates of the measurement points are substituted into Eqs. (4.88) and (4.89) to obtain the adjustment treatment value of each measurement point, and then the total intensity of the magnetic field of the measurement points is subtracted from the systematic error value to obtain the leveled data.

From Eq. (4.89), the first-order polynomial fit model is geometrically represented as a plane, and therefore, it is more suitable for the case when the linear map of the total local magnetic field intensity difference is planar.

For Eq. (4.89), the first-order partial derivative should be found firstly

$$\begin{cases} \dfrac{\partial z_{i'}}{\partial x_{i'}} = b_1 + 2b_3 x_i + b_4 y_i \\ \dfrac{\partial z_{i'}}{\partial y_{i'}} = b_2 + b_4 x_i + 2b_5 y_i \end{cases} \tag{4.89}$$

Then the second-order partial derivative can be denoted.

$$\begin{cases} \dfrac{\partial z_i''}{\partial x_{i'}} = 2b_3 \\ \dfrac{\partial z_i''}{\partial y_{i'}} = 2b_5 \end{cases} \tag{4.90}$$

When Eq. (4.90) is determined, $b_{0,} \cdot \cdot\cdot, b_5$ are constants, the second-order polynomial fit model with a tangent point to all planes, which indicates that there is only one convex or concave surface of the second-order polynomial fit model. It is important to understand the geometric characteristics of the second-order polynomial fit model, which cannot be represented in mathematical form by a single second-order polynomial function if there are multiple concave and convex surfaces in the geomagnetic intensity difference isopleth maps. This requires dividing the

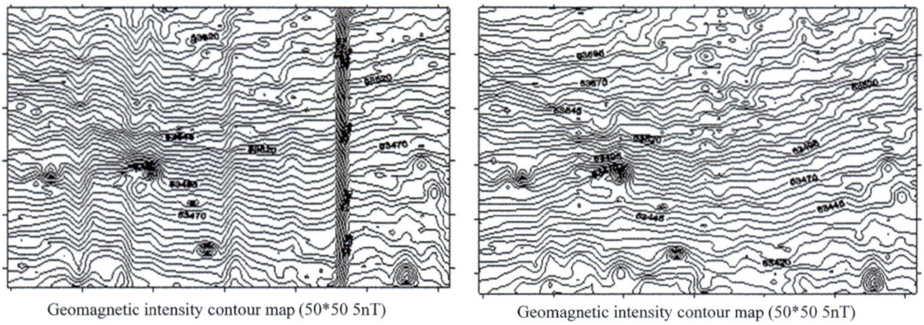

(a) Without adjustment treatment (b) With adjustment treatment

Fig. 4.14 Contour plot of magnetic field intensity before and after adjustment treatment

geomagnetic intensity difference contour map into several regions according to its characteristics and defining a second-order polynomial function for each region separately for adjustment treatment. At the same time, when finding the fitting coefficients of the polynomial fitting model, the selected intersection points should be distributed as evenly as possible in the measurement area, and the number should not be smaller than the number of fitting coefficients.

Figure 4.14a is the contour map of magnetic field intensity generated from the north–south line data without adjustment treatment. Figure 4.14b is the contour map of magnetic field intensity generated from the north–south line data after adjustment treatment, where the magnetic field data grid interval is 50 m*50 m and the contour interval is 5nT.

As can be seen from Fig. 4.14a, since the geomagnetic data of the north–south line are not acquired on the same day, the geomagnetic data of different days appear to be different due to the influence of the measurement conditions and the measurement environment, and the magnetic field intensity contour map reflects the sudden upward and downward movement of the contour. After the leveling process, as shown in Fig. 4.14b, the generated contour map of magnetic field intensity is relatively smooth, without sudden jumps, and basically conforms to the distribution trend of geomagnetic field in the measurement area.

4.6 Summary

This chapter introduces the analytical representation of the Earth's magnetic field and the mechanism of using the Earth's magnetic field for positioning and navigation, focusing on the detailed description of the process of creating geomagnetic reference maps for geomagnetic positioning and navigation, which lays a theoretical foundation for the implementation of geomagnetic positioning and navigation.

References

1. Zhang, M-L. (2019). *Introduction to geophysics* (pp. 23–25). Petroleum Industry Press.
2. (Japan) Kamide, Y., & (Brazil) Chian, A. C. L. (2010). *Handbook of the solar-terrestrial environment. Jin-Long Jian* (pp. 91–96). Science Press.
3. Liu, K. (2019). *Research on long-distance geomagnetic navigation based on magnetotactic search* (pp. 45–50). Northwestern Polytechnical University.
4. Chen, K., Liang, W.-C., Zeng, C.-Z., & Guan, R. (2021). Multi-geomagnetic-component assisted localization algorithm for hypersonic vehicles. *Journal of Zhejiang University-Science A (Applied Physics & Engineering), 22*(5): 357–368.
5. Tian, F.M., Bian, L.-X., & Wang, H.-B. (2021). Autonomously navigation simulation system based on global geomagnetic anomaly grid. *Journal of Projectiles, Rockets, Missiles and Guidance, 41*(02), 10–14.
6. Wang, G.-G., Shen, Z.-F., & Li, S.-M. (2019). A geomagnetic localization method based on road magnetic field characteristics. Beijing: CN107621263B (pp. 12–27).
7. Li, X., Cheng, D.-F., & Zhou, Z.-J. (2017). A matching algorithm based on the gradient of the total geomagnetic. *Chinese Journal of Sensors and Actuators, 30*(12), 1869–1875.
8. Zhu, Z.-S., Yuan, C.-Z., & Zhou, P. (2011). Situation and development of passive navigation localization technology. *Progress in Geophysics, 26*(04), 1473-1477.
9. Xu, W. (2009). *Physics of electromagnetic phenomena of the earth* (Vol. 01, pp. 78–82). Press of University of Science and Technology of China.
10. Xu, W. (2003). *Geomagnetism* (pp. 67–69). Seismological Press.
11. Xu, W. (2014). *Introduction to geomagnetic activity* (pp. 112–120). Science Press.
12. (US) Menke, W. (2019). *Geophysical data analysis discrete inverse theory.* Zou, Z.-H., Zhang, J.-Z. (pp. 204–210). Science Press.
13. Chang, Y.-F. (2015). *Research on satellite geomagnetic data process and geomagnetic model recovery theory and method* (pp. 34–45). PLA Information Engineering University.
14. Shao, G.-Z., Ma, J.-Q., & Li, Q.-C. (2021). *Inverse theory in geophysics.* China Communication Press.
15. Yao, C.-L., Guan, Z.-N., & Wu, Q.-B., et al. (2004). An analysis of EULER deconvolution and its improvement. *Geophysical and Geochemical Exploration, 2004*(02), 150–155.
16. Xu, S.-Z., & Yu, H.-L. (2007). The interpolation-iteration method for potential field continuation from undulating surface to plane. *Chinese Journal of Geophysics, 2007*(06), 1811–1815.
17. Wang, G.-G., Li, S.-M., & Shen, Z.-F. (2019). *An indoor positioning terminal incorporating a visual odometer and a magnetic sensor*. Beijing:CN107576325B, 2019–10–11.
18. Huang, H., Zhao, Y., & Wang, C.-L. et al. (2017). Design of the acquisition system of indoor positioning reference map based on magnetic field data. *Bulletin of Surveying and Mapping, 2017*(02), 54–59.
19. Wang, M.-J., & Yu, Y.-Z. (2012). Application of SLAM in submarine's geomagnetism navigation and positioning based on EKF. *Ship Electronic Engineering, 32*(08), 52–54.
20. Xu, Z.-Y., Wei, D., & Yan, L. (2012). On-line identification and correction of the outlier for oceanic geomagnetic parameter measurement. *Progress in Geophysics, 27*(03), 1034–1039.
21. Huang, C.-Y., Deng, C.-T., & Zhao, H. (2013). The affection of magnetic diurnal variation in geomagnetic matching, *22*(04), 17–21.

Chapter 5
Geomagnetic Navigation and Positioning Methods

Abstract Chapter 4 introduces the composition of the geomagnetic field and the basic principles of the geomagnetic field that can be used for navigation and positioning, as well as the method of creating geomagnetic reference maps for navigation and positioning. This chapter focuses on the process and method of implementing navigation and positioning on the basis of geomagnetic reference maps. As one of the important technologies in the multi-source fusion navigation and positioning system, geomagnetic navigation and positioning have the advantages of a wide range of applications, strong anti-electromagnetic interference capability, and the ability to work around the clock, providing a passive and autonomous navigation and positioning technology path for motion carriers in underground, underwater, and other scenarios where satellite signal reception is limited (Fisher and Raquet in Air & Space Power Journal, 24 –33, 2011; Goldenberg et al. in Location, And Navigation Symposium, 684–694, 2006; Wilson et al. in Proceedings of the 2006 National Technical Meeting of The Institute of Navigation, Monterey, CA, January 2006, pp. 770–779; Li et al. in A geomagnetic positioning method suitable for indoor free motion carriers, Beijing, 2018). Geomagnetic navigation and positioning use the characteristic that the geomagnetic field intensity has a different distribution with a change in geospatial location to achieve the positioning of the motion carrier. The geomagnetic navigation and positioning system obtains the geomagnetic field characteristic data on the motion trajectory through the magnetometer installed on the motion carrier and matches it with the pre-stored geomagnetic field model or geomagnetic reference map to determine the real-time position and correct the inertial guidance error. All seven geomagnetic elements of the geomagnetic field described in Chap. 4 can be used as geomagnetic feature data or a combination of them. The selection of geomagnetic positioning feature data mainly requires the following considerations: (1) significant spatial variation; (2) slow change with time; (3) accurate measurement by measuring instruments; and (4) low influence by external interference. According to the basic principle of geomagnetic navigation and positioning, which is similar to the gravity aided navigation and positioning method. Two working systems are required: a pre-processing system and a real-time positioning system. The pre-processing system is mainly based on the process introduced in Chap. 4 to complete the preparation of geomagnetic reference data. This chapter introduces the implementation process

L. Yan et al., *Mathematical and Physical Fundamentals of Navigation and Positioning with Earth's Natural Fields*, Navigation: Science and Technology 13,
https://doi.org/10.1007/978-981-97-5524-0_5

of the real-time positioning system, which mainly includes real-time measurement of geomagnetic fields and positioning calculation. The implementation of geomagnetic navigation and positioning focuses on three aspects: (1) magnetometer sensors; (2) real-time measurement methods; and (3) navigation and positioning algorithms. According to the different carriers and scenarios of geomagnetic positioning applications, there are certain differences in the process of geomagnetic navigation and positioning implementation, which are mainly reflected in the above three aspects. In this chapter, different application scenarios of geomagnetic positioning are categorized, and the specific implementation process for different application scenarios is given (Jun et al. in Journal of Astronautics 05:1467–1472, 2008). For the convenience of description, "geomagnetic navigation and positioning" is referred to as "geomagnetic positioning" in this chapter.

5.1 Magnetometer Sensors for Geomagnetic Navigation and Positioning

The mainstream magnetometer sensors currently on the market are shown in Table 5.1 and Fig. 5.1.

The scalar measurement of the geomagnetic field mainly uses sensors such as proton precession magnetometers and optical-pumping magnetometers, which are notable for their high measurement resolution, accuracy, and stability, but the corresponding devices are larger and have higher power consumption. Therefore, scalar measurement is suitable for vehicle carriers with a large form factor. The vector measurement of the geomagnetic field mainly uses fluxgate magnetometers, magnetoresistive magnetometers, etc., which can measure the three components of the geomagnetic field vector simultaneously. The vector measurement can obtain the three-component information of the geomagnetic field, and the total intensity of the magnetic field can also be indirectly obtained by vector modulo calculation to achieve an indirect scalar measurement. The measurement accuracy and resolution level of a vector magnetometer are relatively low (0.1–5nT), but the volume is relatively small, and the energy consumption and cost are low, so it is suitable for small vehicle carriers and the construction of multi-sensor measurement systems oriented to data fusion.

In accordance with Table 5.1, a suitable magnetometer is selected according to the specific application scenario and platform, which will be expanded below.

Table 5.1 Mainstream magnetometer sensors

Index	Fluxgate magnetometer	Optical-pumping magnetometer	Proton precession magnetometer	Magnetoresistive magnetometer
Principle	Faraday electromagnetic induction	Zeeman split	Larmor precession	Magnetoresistive effect
Physical quantities	Vector	Scalar	Scalar	Vector
Resolution (nT)	0.01 nT	0.001 nT	0.01 nT	300 nT
Frequency (Hz)	0–1000	0–10	0–1	$0\text{–}0^6$
Power consumption	<1w	~5w	~5w	~1mw
Volume	32 × 32 × 150 mm	φ60 mm × 150 mm	φ75 m × 175 mm	5 mm*5 mm*1 mm
Weight	400 g	600 g	1000 g	1 g
Stability	Large zero drift with time	Good	Fine	Large zero drift with time and temperature
Impact resistance	Fine	General	General	Very good
Dynamic response Performance	Better	General	Poor	Fine
Applicable scene	Vehicle, aircraft, ship, space	Aircraft, ground	Static observation	Pedestrian, robot

5.2 Aeromagnetic Navigation and Positioning

Aeromagnetic navigation and positioning technology refers to the application of geomagnetic navigation and positioning technology within the vehicle carrier, where the vehicle refers to the manned or unmanned vehicle that can fly off the ground in space within the atmosphere, and can control its flight attitude and flight altitude. Geomagnetic navigation and positioning can be used as a complement to GNSS, inertial navigation, and other navigation and positioning systems. When the satellite navigation system is interfered or unavailable, the combination of geomagnetic navigation and positioning system and inertial navigation system can provide position coordinate information for the vehicle and improve the reliability and stability of the whole navigation and positioning system.

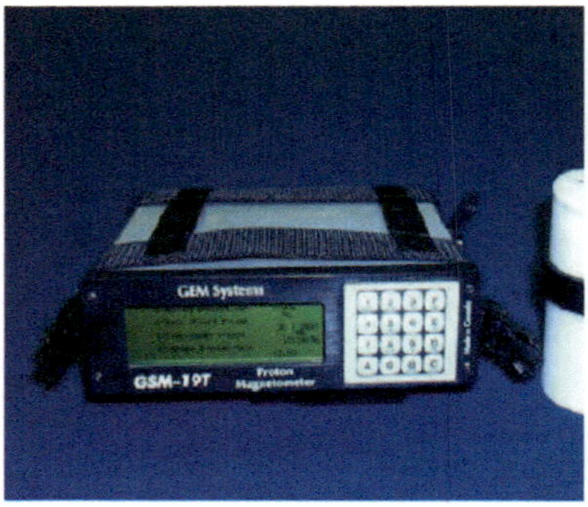

(a) Proton precession magnetometer

(b) Optical-pumping magnetometer

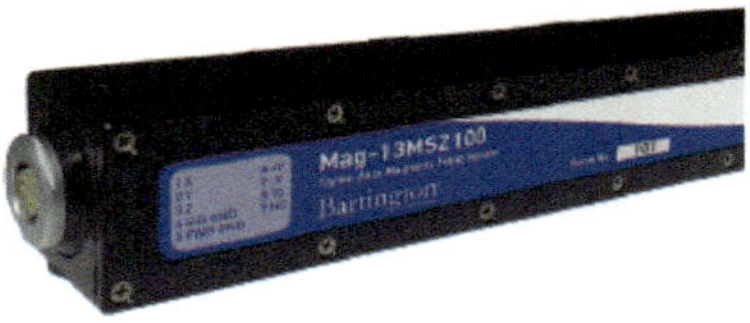

(c) Fluxgate magnetometer

(d) Magnetoresistive magnetometer

Fig. 5.1 Mainstream magnetometer sensors

5.2.1 Selection of Characteristic Quantities for Aeromagnetic Positioning

The features of geomagnetic positioning reference can be divided into direct features and indirect features. Direct features refer to the seven elements of the geomagnetic field, such as the total intensity of the geomagnetic field, the magnetic declination angle, and the northward component of the geomagnetic field; indirect features refer to features other than the seven elements of the geomagnetic field, such as the spatial gradient and spectral features of the geomagnetic field. The selection criteria of feature quantities can be summarized as follows: (1) the long-term changes of feature quantities are relatively stable; (2) the short-term changes of feature quantities have less impact; (3) the real-time measurement of feature quantities does not require high equipment performance; and (4) the acquisition of benchmark maps is relatively easy.

According to the characteristics of the flight platform for aeromagnetic positioning, the total field strength of the geomagnetic field or the gradient of the total field strength is the characteristic quantity more suitable for aeromagnetic positioning. The sensors are selected as light pump magnetometers or fluxgate magnetometers.

5.2.2 *Implementation of Aeromagnetic Measurement*

Real-time accurate measurement of geomagnetic field under carrier magnetic interference is the key factor to achieve accurate positioning. All kinds of magnetic bodies and metal conductors in the carrier components generate various complex carrier interference magnetic fields in the space around them that change with the carrier motion state. Especially when the carrier space is small and the installation position is limited, the interference magnetic field has a certain intensity, which affects the accurate measurement of the geomagnetic field, so the interference magnetic field generated by the carrier motion must be removed in the real-time measurement process [6].

The mechanical structure of the carrier, the electric control device, the engine, and the follower components all have strong local magnetic fields. In addition, the change of carrier attitude causes a change of magnetic flux inside the carrier and generates an eddy current electric field, which in turn induces an eddy current magnetic field. These constitute a complex interference with the geomagnetic field measurement.

Inherent magnetic disturbance of the carrier mainly includes residual magnetic disturbance generated by the hard magnetic material in the carrier structure, and magnetic disturbance generated by the electric control device. The carrier's intrinsic magnetic interference is not related to the motion state, but the angle between the intrinsic magnetic field vector and the geomagnetic field vector changes continuously when the carrier is in motion, which causes the change of the synthetic magnetic field, and then generates the interference to the measurement results. According to the carrier structure characteristics, the interference caused by the intrinsic magnetic interference is the most serious background interference, and its interference intensity is proportional to the intrinsic magnetic value.

The dynamic magnetic interference of the carrier changes with the change of the carrier motion state, mainly including the inducing magnetic field generated by the magnetization of the carrier structure by the geomagnetic field, and the eddy current magnetic field generated by the movement of the shell conductive material in the geomagnetic field, etc. The distribution characteristics are relatively complex, which significantly interferes with the real-time geomagnetic field measurement results.

The Optical-pumping magnetometer is used as the direct measurement instrument of total geomagnetic intensity, while the three-axis vector sensor and others are used to provide the attitude information of the carrier in the background field, and the background interference magnetic field is compensated with the background interference magnetic field compensation model to eliminate the influence of the carrier remanent magnetism, induced magnetic field, and eddy current magnetic

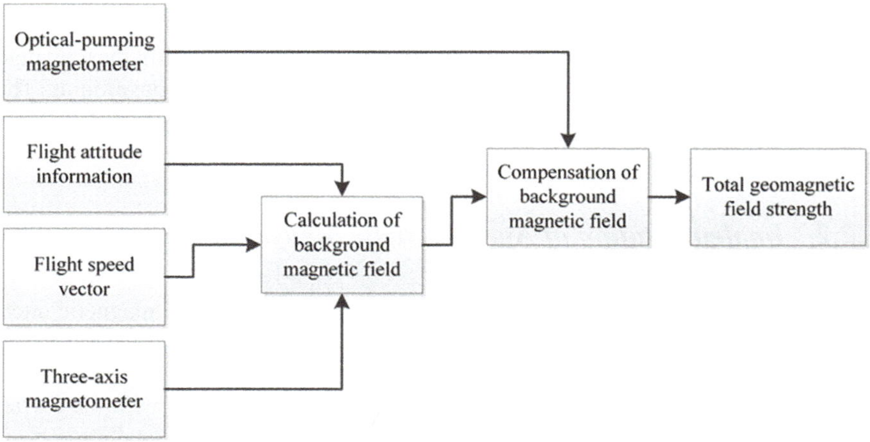

Fig. 5.2 Measuring system

field on the measurement results and achieve the accurate measurement of magnetic fields. The measurement system is shown in the Fig. 5.2.

Firstly, the coordinate system of aeromagnetic measurement is established, with the magnetometer as the coordinate origin, the X-axis is parallel to the longitudinal axis of the carrier, forward is positive; the Y-axis points to the horizontal transverse of the carrier, right is positive; the Z-axis is the vertical axis, down is positive.

The directional angle formed by the geomagnetic field vector H_d and the coordinates X, Y, and Z axis is $\theta_{i0} = \left(\theta_1, \theta_2, \theta_3\right)$, and its directional cosine is $(\cos\theta_1, \cos\theta_2, \cos\theta_3)$.

(1) Intrinsic magnetic field

The inherent magnetic field is the residual magnetic field, which can be expressed as

$$\overline{H_p} = \left\{p_1\ p_2\ p_3\right\} \tag{5.1}$$

The projection at the measurement point along the direction of the geomagnetic vector is

$$H_{pd} = \left(p_1\ p_2\ p_3\right)\begin{pmatrix}\cos\theta_1\\ \cos\theta_2\\ \cos\theta_3\end{pmatrix} = \sum_{i=1}^{3} p_i u_i \tag{5.2}$$

where, $u_i = \cos\theta_i$ is the directional cosine of the geomagnetic vector.

(2) Induced magnetic field

The vector of the induced magnetic field at the measurement point is

$$H_i = \begin{pmatrix} i_{11} & i_{12} & i_{13} \\ i_{21} & i_{22} & i_{23} \\ i_{31} & i_{32} & i_{33} \end{pmatrix} \begin{pmatrix} B_1 \\ B_2 \\ B_3 \end{pmatrix} = \begin{pmatrix} i_{11} & i_{12} & i_{13} \\ i_{21} & i_{22} & i_{23} \\ i_{31} & i_{32} & i_{33} \end{pmatrix} \begin{pmatrix} \cos\theta_1 \\ \cos\theta_2 \\ \cos\theta_3 \end{pmatrix} B \tag{5.3}$$

where, B_i is the vector triplet component of the Earth's magnetic field; B is the intensity of the Earth's magnetic field; i_{ij} is the coefficient of the induced magnetic field model.

The projection of the induced magnetic field along the direction of the geomagnetic vector is

$$\begin{aligned} H_{id} &= \left(\cos\theta_1 \ \cos\theta_2 \ \cos\theta_3\right) \begin{pmatrix} i_{11} & i_{12} & i_{13} \\ i_{21} & i_{22} & i_{23} \\ i_{31} & i_{32} & i_{33} \end{pmatrix} \begin{pmatrix} \cos\theta_1 \\ \cos\theta_2 \\ \cos\theta_3 \end{pmatrix} B \\ &= \sum_{i=1}^{3}\sum_{j=1}^{3} \left(i_{ij}B\right) u_i u_j = \left(\sum_{i=1}^{3}\sum_{j=1}^{3} a_{ij} u_i u_j - a_{33} u_3^2\right) + a_{33} \end{aligned} \tag{5.4}$$

When $a_{ij} = a_{ji} = \frac{1}{2}\left(i_{ij} + i_{ji}\right)B, \ i \neq j$, then

$$a_{11} = (i_{11} - i_{33})B \tag{5.5}$$

$$a_{22} = (i_{22} - i_{33})B \tag{5.6}$$

The above equation does not include the $a_{33}u_3^2$ term in the parentheses, and a_{33} outside the parentheses is a constant, independent of the heading, which can be calculated using a filter to remove the static signal. The a_{33} term in the modeling has no effect on the subsequent processing, Therefore, when $a_{33} = 0$, $\sum_{i=1}^{3}\sum_{j=1}^{3} a_{ij}u_i u_j$ contains only 5 coefficients.

(3) Eddy current magnetic field

From the study of the mechanism of vortex magnetic field generation, it is known that the vortex magnetic field is linearly related to the rate of change of the Earth's magnetic field, which is expressed vectorially as

$$H_e = \left(B_1' \ B_2' \ B_3'\right) \begin{pmatrix} e_{11} & e_{12} & e_{13} \\ e_{21} & e_{22} & e_{23} \\ e_{31} & e_{32} & e_{33} \end{pmatrix} = \left(u_1' \ u_2' \ u_3'\right) \begin{pmatrix} e_{11} & e_{12} & e_{13} \\ e_{21} & e_{22} & e_{23} \\ e_{31} & e_{32} & e_{33} \end{pmatrix} B \tag{5.7}$$

By the nature of the direction cosine, there is $\sum_{i=1}^{3} u_i^2 = 1, u_1u_1' + u_2u_2' + u_3u_3' = 0$. Letting

$$\begin{cases} b_{11} = (e_{11} - e_{33})B \\ b_{22} = (e_{22} - e_{33})B \\ b_{33} = 0 \\ b_{ij} = e_{ij}B, \ i \neq j \end{cases} \tag{5.8}$$

The projection of the eddy current magnetic field along the direction of the geomagnetic vector can be expressed as

$$H_{ed} = \begin{pmatrix} u_1' & u_2' & u_3' \end{pmatrix} \begin{pmatrix} e_{11} & e_{12} & e_{13} \\ e_{21} & e_{22} & e_{23} \\ e_{31} & e_{32} & e_{33} \end{pmatrix} \begin{pmatrix} u_1 \\ u_2 \\ u_3 \end{pmatrix} B = \sum_{i=1}^{3} \sum_{j=1}^{3} b_{ij} u_j u_i' \tag{5.9}$$

The total disturbance magnetic field is

$$\Delta B = \sum_{i=1}^{3} p_i u_i + \sum_{i=1}^{3} \sum_{j=1}^{3} a_{ij} u_i u_j + \sum_{i=1}^{3} \sum_{j=1}^{3} b_{ij} u_j u_i'. \tag{5.10}$$

5.2.3 *Aeromagnetic Positioning Methods*

In systems that use geophysical fields for navigation, batch correlation processing algorithms are often used. Underwater geomagnetic navigation systems using batch processing techniques are often referred to as UGMCOM. The basic idea of the batch correlation processing algorithm is that when the underwater carrier runs for a period of time, the position estimation sequence output from the inertial navigation system is input into the pre-stored geomagnetic field reference map to get the corresponding magnetic field intensity sequence (inferred intensity sequence), and it is correlated with the geomagnetic intensity sequence under the real track measured according to the magnetic sensor (measured intensity sequence), and the position corresponding to the correlation pole point obtained is the matching position, which is then used to make corrections to the inertial guidance system (Fig. 5.3).

Suppose the characteristic function of the inferred intensity sequence is $T(x, y, z)$, the characteristic function of the measured intensity sequence is $D(x, y, z)$, and the geometric measure of both sequences is the same as V. Then the commonly used correlation extremum function is defined as

Cross Correlation Algorithm.

$$COR(T, D) = \frac{1}{V} \oint_V T(x, y, z) \cdot D(x, y, z) dv \tag{5.11}$$

Correlation Coefficient algorithm.

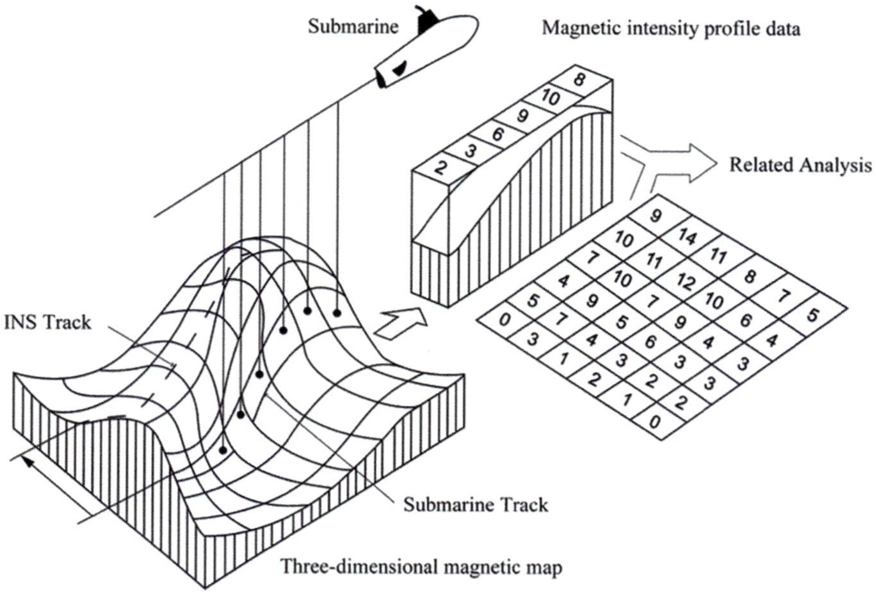

Fig. 5.3 The basic principle of correlation matching algorithm based on batch processing technology

$$CC(T, D) = \frac{\oint_V \{[T(x, y, z) - \overline{T}] \cdot [(D(x, y, z) - \overline{D})]\}dv}{\sqrt{\oint_V [T(x, y, z) - \overline{T}]^2 dv}\sqrt{\oint_V [D(x, y, z) - \overline{D}]^2 dv}} \tag{5.12}$$

Among them, $\overline{T} = \frac{1}{V}\oint_V T(x, y, z)dv, \overline{D} = \frac{1}{V}\oint_V D(x, y, z)dv$.
Mean Absolute Difference algorithm.

$$MAD(T, D) = \frac{1}{V}\oint_V |T(x, y, z) - D(x, y, z)|dv \tag{5.13}$$

Mean Square Difference algorithm.

$$MSD(T, D) = \frac{1}{V}\oint_V [T(x, y, z) - D(x, y, z)]^2 dv \tag{5.14}$$

The batch correlation matching algorithm has the following advantages.

(1) Low requirement for initial position error.
(2) It is not necessary to make any linearization assumptions about the magnetic field, and it works as long as the magnetic field variation is well characterized.
(3) The global optimal solution is obtained.

The batch correlation matching algorithm has the following areas for further improvement.

(1) Searching across the entire underwater geomagnetic digital reference map to determine if the inferred intensity sequence is generated globally.
(2) Sensitivity to velocity errors and heading errors during matching data acquisition by the navigation system, and therefore limited length of data acquisition.
(3) The existence of quantization errors in the search interval of the baseline magnetogram, which can be reduced by using a small search interval but will increase the computational effort.
(4) It is difficult to guarantee real-time performance because of the large amount of data processing and the high requirement for digital computers. Both the search and transformation of the reference magnetogram and the related calculation and comparison have a large amount of data to be processed, so the real-time is poor.
(5) Since the measured intensity sequence is always based on the data collected by the magnetic sensor in the correlation processing algorithm, the error is not taken into account, so the robustness of the algorithm is poor.

Usually, the geomagnetic reference map is only the existing data in a certain plane, and it is usually measured by sea surface survey or even aerial survey. For this reason, when exploring the problem of matching localization, a situation that cannot be ignored is that the reference map is not uniform in height with the measurement sequence. To address this situation, the algorithm for geomagnetic batch correlation matching is explored while introducing the treatment of the special case that the magnetogram is higher than the measurement sequence to increase the extended processing of geomagnetic data.

For the situation that the downward extension error of geomagnetic anomaly is large, this chapter investigates the batch correlation matching algorithm (MSUC-UGMCOM, UGMCOM based on Measurement Sequence Upward Continuation) based on the fact that the upward extension of geomagnetic anomaly is relatively stable by combining the traditional matching localization algorithm.

1. **MSUC-UGMCOM Algorithm**

The MSUC-UGMCOM algorithm consists of two segments: upward extension of the two-dimensional magnetic anomaly and batch correlation matching. The measured magnetic field intensity sequence is first decomposed to obtain its normal and anomalous field parts separately. The normal field is calculated using the negligible degree of magnetic field intensity difference over a finite distance, and then the IGRF model is used to calculate the normal field value at the height of the magnetogram, while the upward extension operation is performed for the anomalous field part. The calculated normal field values at the height of the magnetogram are summed with the abnormal field values to regain the geomagnetic distribution sequence at that location. Then the UGMCOM algorithm is used to match the location. The main components of the algorithm are as follows.

(1) Real-time geomagnetic measurement sequences are accumulated to form $T_S^A = \{T_1^A, T_2^A, \ldots, T_n^A\}$.
(2) Decompose the geomagnetic real-time measurement sequence into a normal field component and an anomalous field component; $T_{nS}^A = \{T_{n1}^A, T_{n2}^A, \ldots, T_{nn}^A\}$ and $T_{aS}^A = \{T_{a1}^A, T_{a2}^A, \ldots, T_{an}^A\}$.

$$T_S^A = T_{nS}^A + T_{aS}^A \tag{5.15}$$

where, T_{nS}^A is calculated from the inertial guidance indication position according to the IGRF model.
(3) The geomagnetic normal field component is recalculated by IGRF at the height of the geomagnetic reference map $T_{nS}^B = \{T_{n1}^B, T_{n2}^B, \ldots, T_{nn}^B\}$.
(4) The geomagnetic anomaly distribution $T_{aS}^B = \{T_{a1}^B, T_{a2}^B, \ldots, T_{an}^B\}$ at the new height is calculated from Eq. (5.22).

$$T_{aS}^B = F^{-1}\left[e^{-|k_x|h} \cdot F\left(T_{aS}^A\right)\right] \tag{5.16}$$

(5) An upward continuation sequence of geomagnetic reference map heights summing the normal and anomalous fields obtained in the above two steps.

$$T_S^B = T_{nS}^B + T_{aS}^B \tag{5.17}$$

(6) Within the search window determined by the inertial guidance estimation error, assuming that the running direction of the underwater carrier is the same as the direction indicated by the inertial guidance, the MAD criterion is used to determine the closeness of each parallel track to the upper delay sequence in turn, and the track corresponding to the very small value is selected to determine the position of the underwater carrier.

$$MAD(T_S^B, T_S^M) = \frac{1}{V}\int_V \left|T_S^B(x, y) - T_S^M(x, y, \tau)\right| d\tau. \tag{5.18}$$

2. Feasibility Analysis of the Algorithm

The drift of the inertial navigation system is usually in the range of kilometers or so, so an auxiliary position correction is required, and according to this principle the search window of the matching algorithm can be determined. In the limited distance range, there are differences in the magnitude of the geomagnetic intensity calculated by IGRF, as shown in Tables 5.2 and 5.3. In the case where only the inertial navigation position is available and the true position is unknown, it is not possible to accurately decompose the geomagnetic normal and abnormal fields. The extraction of the normal field in the geomagnetic real-time measurement sequence becomes the key problem of geomagnetic composition decomposition.

Table 5.2 Comparison of IGRF calculated magnetic field strengths for smaller intervals (100 m below geoid)

Normal field (nT)	125°21′0″	125°22′0″	125°23′0″	125°24′0″
40°31′0″	52,879.7	52,875.7	52,871.8	52,867.8
40°32′0″	52,888.7	52,884.7	52,880.7	52,876.7
40°33′0″	52,897.6	52,893.6	52,889.6	52,885.6
40°34′0″	52,906.5	52,902.5	52,898.5	52,894.5

Table 5.3 Comparison of IGRF calculated magnetic field strengths for smaller intervals (geoid)

Normal field (nT)	125°21′0″	125°22′0″	125°23′0″	125°24′0″
40°31′0″	52,876.9	52,872.9	52,869.0	52,865.0
40°32′0″	52,885.9	52,881.9	52,877.9	52,873.9
40°33′0″	52,894.8	52,890.8	52,886.8	52,882.8
40°34′0″	52,903.7	52,899.7	52,895.7	52,891.8

However, comparing Tables 5.2 with 5.3, it can be seen that the differen values of the corresponding points of the geomagnetic normal field are consistent in the region of $3' \times 3'$, as shown in Table 5.4. The normal field and the abnormal field are a relative concept, so the normal field value at the location of the inertial navigation indication can be selected as the normal field component of the whole reference map, and then the upward operation of the normal field can be obtained by adding the difference shown in Table 5.4, and the rest of the amplitude components can be processed as the upward extension of the abnormal field.

Within a small area, the degree of difference in the amplitude of the geomagnetic normal field intensity at different levels remains basically the same, which provides a solution to the inertial navigation indication position deviating from the real position cannot accurately provide the accurate geomagnetic field intensity, which also provides a theoretical basis for the feasibility of the MSUC-UGMCOM algorithm.

3. Experiment and Analysis

The geomagnetic map of a local area is selected and up-scaled as the reference map. A sequence of total geomagnetic intensity amplitudes is extracted from the known

Table 5.4 Difference in intensity of different levels of IGRF calculated magnetic field for smaller intervals

Difference of normal field (nT)	125°21′0″	125°22′0″	125°23′0″	125°24′0″
40°31′0″	2.8	2.8	2.8	2.8
40°32′0″	2.8	2.8	2.8	2.8
40°33′0″	2.8	2.7	2.8	2.8
40°34′0″	2.8	2.8	2.8	2.7

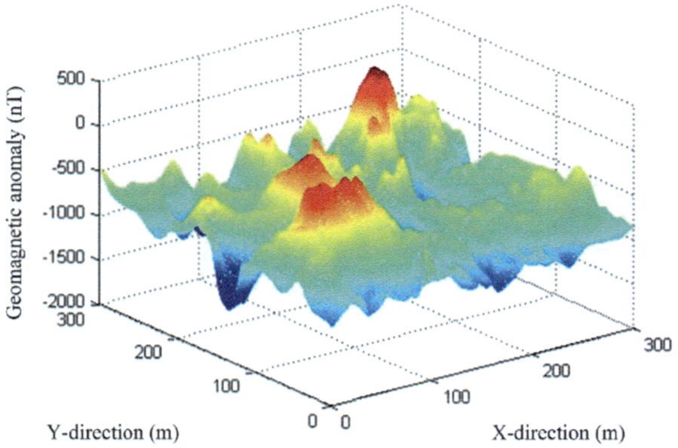

Fig. 5.4 Magnetic map of a local area in the actual plane of measurement

geomagnetic map according to the inertial navigation indication trajectory, and the measured Gaussian white noise is added as the true measurement sequence. Then the MSUC-UGMCOM algorithm is used to match the localization (Fig. 5.4).

The relevant parameters of the used regional magnetic field are: 105.7920°–105.6208° of west, 37.2128°–37.3282° of north, sampling spacing of 50 m, and height of 150 m, as shown in Fig. 5.5. The upper extension height for constructing the baseline magnetogram is 100 m. The average value of the normal field calculated using the IGRF model is 52009.1 nT, and the difference of the normal field part of the upper extension of 100 m is 2.6 nT.

The additional error of the geomagnetic measurement is Gaussian white noise of amplitude 4 nT. The inertial navigation offset error is 200 m. The cumulative length of the sampling sequence is 30 sampling points, and the localization results using the MSUC-UGMCOM algorithm are shown in Fig. 5.6. Since the breakpoints at both ends of the measurement sequence have an impact on the results after the up-delay, while the up-delay makes the distribution features weaker, further reflection on the geomagnetic matching localization algorithm is needed.

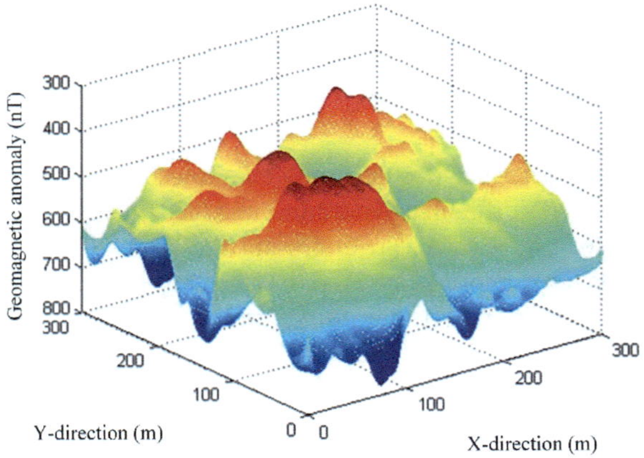

Fig. 5.5 Epistatic distribution of the baseline magnetogram

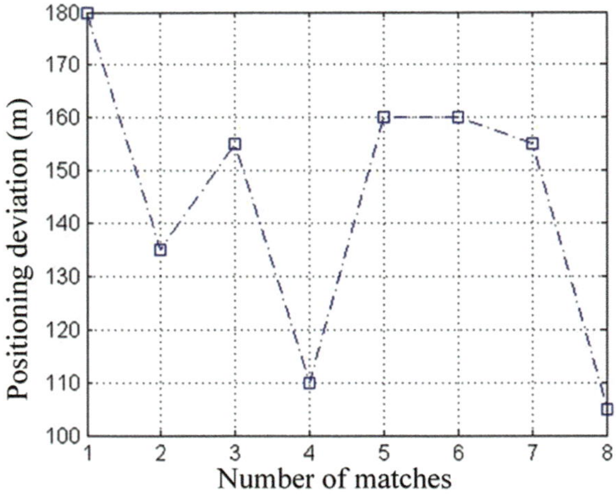

Fig. 5.6 Localization results of MSUC-UGMCOM algorithm

5.3 Underwater Geomagnetic Navigation and Positioning

Underwater geomagnetic navigation and positioning technology refers to the application of geomagnetic navigation for positioning in surface or underwater vehicles, which are navigable bodies that can navigate on the surface or underwater, or both, including manned and unmanned vehicles. In the increasingly important marine development nowadays, navigators are able to accomplish tasks such as transportation, exploration, detection and even military offense and defense, playing an important role in both civil and military applications. Navigation and positioning technology plays an important role as one of the key technologies for the successful execution of the mission of the navigators. Geomagnetic navigation and positioning technology, as a means of navigation and positioning, provides an additional option for navigation and positioning of navigators [7–9].

5.3.1 Selection of Characteristic Quantities for Underwater Geomagnetic Positioning

Underwater carriers generally take towed measurement, the attitude of the magnetometer is difficult to determine, and the attitude of the magnetometer and the attitude of the underwater carrier are not the same. Therefore, the attitude of the magnetometer cannot be obtained through the attitude measurement system of the underwater carrier, and the attitude of the magnetic sensor in the geodetic coordinate system cannot be obtained. So, according to the motion of the underwater platform and the characteristics of the magnetic measurement, the total field strength of the geomagnetic field should be preferred as the characteristic quantity when the geomagnetic navigation and positioning technology is applied underwater, and the magnetometer sensor can choose the optical-pumping magnetometer or the fluxgate magnetometer. The light pump magnetometer can directly measure the total geomagnetic field strength, and the fluxgate magnetometer can calculate the total geomagnetic field strength by the three-component vector summation.

5.3.2 Implementation of Underwater Geomagnetic Measurements

In order to reduce the influence of the ship's hull on the measurement, the underwater geomagnetic measurement tows as far as possible. The towing distance is related to the interference size of the hull, the larger the magnetic field of the hull, the farther the towing distance, in practical application, 50–100 m is generally used.

For each measurement line, the measurement direction is determined according to the starting position. For north–south measurement line, from south to north, the cable length between the towing and the DGPS antenna is subtracted from the Y direction; conversely, the cable length between the towing and the DGPS antenna is added. For the east–west direction, the measurement data should be corrected in the X-direction, from west to east, the cable length between the towing and the DGPS antenna should be subtracted from the X-direction; conversely, the cable length between the towing and the DGPS antenna should be added. For the oblique direction, the X and Y direction correction values are first calculated according to the slope, and then corrected according to the X and Y directions respectively [10, 11] (Fig. 5.7).

Fig. 5.7 Marine geomagnetic survey vessel and main instrumentation

5.3.3 *Underwater Geomagnetic Positioning Methods*

Due to the inherent perturbation of the geomagnetic field and the existence of some oscillations in the magnetometer drag, there are still non-measurement error disturbances even if a higher-order approximation is applied to the nonlinear function. In addition, the geomagnetic field distribution is not guaranteed to satisfy the Gaussian distribution in the finite interval of the measurement sequence. In this section, a particle filtering algorithm based on Monte Carlo random sampling [12, 13], which is applicable to a large interference environment, is used to solve the filtering problem of non-Gaussian distribution of process states of nonlinear dynamic systems.

5.3.3.1 Fundamentals of Particle Filtering

The application of Monte Carlo methods in filtering can be traced back to 1969–70. Particle filtering is a nonlinear filtering method that can be used for nonlinear and non-Gaussian models. Particle filtering is an algorithm that provides an iterative Monte Carlo approximation for a given sequence of state variable distributions and can be used to approximate the desired probability density function (PDF) by collecting a large number of points in the state space, which are called particles. Each particle is assigned a probability proportional to the weight, and these particles are sampled randomly selected according to the desired PDF. Thus, as the number of particles increases, they effectively provide a good approximation to the desired PDF. Particle filtering is applied not only for heading estimation and terrain navigation, but also for target tracking, and some improved algorithms are currently proposed for particle filtered navigation [14].

1. Dynamic State Space Model

The general state space model can be divided into state transfer and state measurement models.

$$x_{k+1} = f_k(x_k, w_k) \tag{5.19}$$

$$z_k = h(x_k, v_k) \tag{5.20}$$

The states follow a first-order Markov process, while the state given time observations are not correlated. The model considered is a nonlinear and non-Gaussian regression process with the expected distribution at $k = 0$ is given by $p(x_0)$. The posterior density $p(x_{0:k}|z_{1:k})$ contains the complete solution of the sequential estimation problem.

The particle filter algorithm is generally executed in three steps.

a. Sampling element: A stochastic model of the system is used to generate the possible states for each sample, and the particles are obtained from the importance density function that plays a key role in the particle filter design.
b. Weight reassignment step: this step is related to the observation conditions, each sample is given a weight by the probability of the observed quantity being sampled to characterize the state, so that the sample with the better-predicted observation has a higher weight.
c. Resampling element: To generate a weight posterior with mean 1, a set of sampling points with weight mean 1 is resampled from the distribution characterized by the weight sampling. In resampling, the probability of a new sample is proportional to its weight, so a high-weight sample can be replaced by several samples and a low-weight sample will disappear.

The first two steps form the particle filter of the serial importance sampling filter, and all three links performed are called the importance resampling filter.

2. Sequential Importance Sampling

The Sequential Importance Sampling (SIS) algorithm is a Monte Carlo sampling method. The key idea is to represent the desired posterior probability density based on a set of random samples with corresponding weights, and to compute estimates based on these samples and weights. Because the number of these samples is very large, the Monte Carlo description becomes an equivalent representation of the posterior probability density function described by an ordinary function, and thus the SIS filter approximates the optimal Bayes filter.

If the probability density function is $p(x) \propto \pi(x)$, and $\pi(x)$ is also a probability density function, so it's difficult to sample, but it is possible to compute. Let $\widehat{x}^{(i)} \propto q(x)$ $i = 1, 2, \ldots, N$, $\pi(x)$ be the samples generated by sampling a proposed easily sampled probability density function $q(\cdot)$, called the sampled particles, and $q(\cdot)$ is called the importance density function. From this, the weighted approximation to the probability density function $p(\cdot)$ can be expressed as

$$p(x) \approx \sum_{i=1}^{N} \lambda^{(i)} \delta(x - \widehat{x}^{(i)}) \tag{5.21}$$

where, $\lambda^{(i)} \propto \frac{\pi(\widehat{x}^{(i)})}{q(\widehat{x}^{(i)})}$, $i = 1, 2, \ldots, N$ is the canonical weight of ith particle, which satisfies $\sum_{i=1}^{N} \lambda^{(i)} = 1$.

For the filtering problem of nonlinear dynamic system, let $X^k = \{x_1, x_2, \ldots, x_k\}$, $Z^k = \{z_1, z_2, \ldots, z_k\}$ denote the set of vectors consisting of all states and all measurements at the moment k, respectively, and

$$\begin{cases} \left\{x_{0:k}^{(i)}\right\}_{i=1}^{N} \overset{def}{=} \left\{x_0^{(i)}, x_1^{(i)}, \ldots, x_k^{(i)}\right\} \\ \left\{\lambda_k^{(i)}\right\}_{i=1}^{N} \overset{def}{=} \left\{\lambda_0^{(i)}, \lambda_1^{(i)}, \ldots, \lambda_k^{(i)}\right\}^T \end{cases} \tag{5.22}$$

The above equation represents the samples of capacity N sampled for all states at time k, and their corresponding weights, respectively. The above is a random quantity, then the posterior probability density function at moment k can be approximated as

$$p(X^k|Z^k) \approx \sum_{i=1}^{N} \lambda_k^{(i)} \delta(X^k - X_{(i)}^k) \tag{5.23}$$

The above equation is a discrete weighted approximation of the true posterior probability density function $p(X^k|Z^k)$.

If the samples $\left\{\widehat{x}_{0:k}^{(i)}\right\}_{i=1}^{N}$ are obtained from the extraction of the importance density function $q(X^k|Z^k)$, while $q(x_k|X^{k-1}, Z^k) = q(x_k|x_{k-1}, z_k)$. The modified weights can be simplified as

$$\lambda_k^{(i)} \propto \lambda_{k-1}^{(i)} \frac{p(z_k|\widehat{x}_k^{(i)})p(\widehat{x}_k^{(i)}|\widehat{x}_{k-1}^{(i)})}{q(\widehat{x}_k^{(i)}|\widehat{x}_{k-1}^{(i)}, z_k)} \tag{5.24}$$

The posterior probability density function can be approximated as

$$p(x_k|Z^k) \approx \sum_{i=1}^{N} \lambda_k^{(i)} \delta(x_k - \widehat{x}_k^{(i)}) \tag{5.25}$$

3. Resampling

The variance of the importance weights in SIS grows randomly with time, and the importance weights are proportional to the importance ratio, whose variance also grows with time. In fact, after several iterations, one of the normalized importance weights will approach 1, while the rest will approach 0, so a large number of sampled points become invalid and can be ignored. To avoid this particle degradation phenomenon, particles need to be resampled to reduce the number of particles with lower importance weights and double the number of particles with higher importance weights. In this step, sample points $(\hat{x}_{0:k}^{(i)}, P_{0:k}^{(i)})$ with higher/lower importance weights $\tilde{w}_k^{(i)}$ are increased/compressed respectively to obtain N random samples. The resampling method is as follows.

a. Keep $s_i = [N\tilde{w}_k^{(i)}]$ copies of $x_k^{(i)}$, $i = 1, \ldots, n$, where $\tilde{w}_k^{(i)}$ is the normalized weight of $w_k^{(i)}$, and $N_r = N - \sum_{i=1}^{n} s_i$.

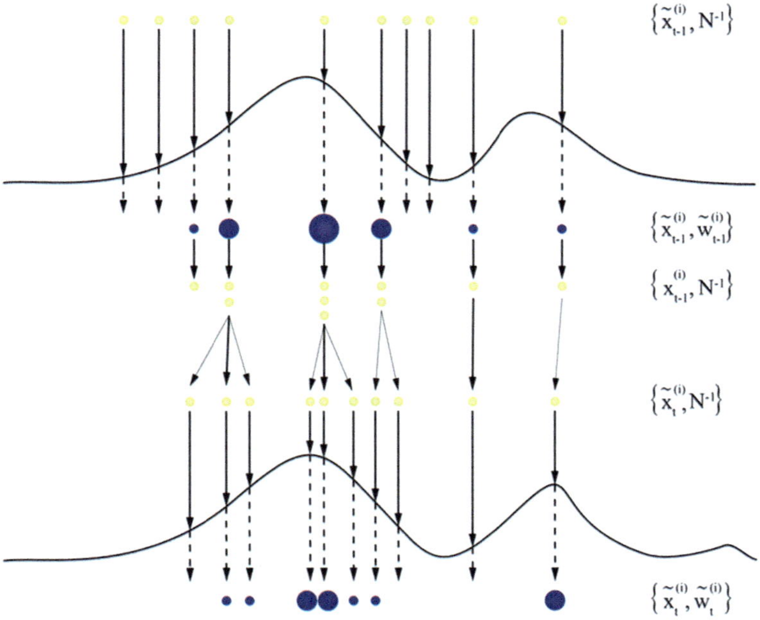

Fig. 5.8 Schematic diagram of particle filter resampling

b. Draw from $x_{1:k}^{(i)}$ and retain the independent and uniformly distributed N_r with probability proportional to $N\tilde{w}_k^{(i)} - s_i, i = 1, \ldots, N$.
c. $\tilde{w}^{(i)} = 1/N, i = 1, \ldots, N$ (Fig. 5.8).

4. Result Output

In Monte Carlo simulation, a weighted set of particles drawn from the posterior distribution is used to map the integral operation to the discrete summation. More precisely, the posterior can be approximated by the following empirical estimate

$$p(x_{0:k}|z_{1:k}) \approx \widehat{p}(x_{0:k}|z_{1:k}) = \frac{1}{N}\sum_{i=1}^{N} \delta_{x_{0:k}^{(i)}}(dx_{0:k}) \tag{5.26}$$

where, the random sample $\left\{x_{0:k}^{(i)}, i = 1 : N\right\}$ is drawn from the posterior distribution; $\delta(d\cdot)$ denotes the Dirac-Delta function.

5.3.3.2 Geomagnetic Navigation Based on Particle Filtering

1. State Equation

The PF filter is processed using a simplified model of the system state error equation, and therefore the state equation of this model is linear.

$$X_{t+1} = F_t \cdot X_t + w_t \tag{5.27}$$

2. Measurement Equation

After discretization, the total geomagnetic intensity amplitude measured by the magnetometer in real time is a nonlinear function of the underwater carrier position, expressed in the geographic coordinate system as

$$z = h_t(\varphi_T, \lambda_T) + v_t \tag{5.28}$$

The measurement equation is the nonlinear function of the state error component $(\delta\varphi, \delta\lambda)$.

According to the inertial guidance system indication position $(\varphi_{INS}, \lambda_{INS})$, and the error component $(\delta\varphi, \delta\lambda)$ of the state error equation, the measurement equation can be expressed as

$$z = h_t(\varphi_{INS} + \delta\varphi, \lambda_{INS} + \delta\lambda) + v_t \tag{5.29}$$

(φ_T, λ_T) is the true position of the underwater carrier in the geodetic coordinate system. v_t is the real-time measurement error.

Unlike UKF, the error vectors w_t and v_t in the state error equation and the measurement equation can't satisfy the Gaussian distribution, which increases the complexity of the PF algorithm processing.

3. Importance Density Function

The transition probability density function is usually chosen as the importance density function.

$$q(x_t|X_{t-1}, Z_{t-1}) = p(x_t|X_{t-1}, Z_{t-1}) \tag{5.30}$$

where $\lambda_t^{(i)} \propto \lambda_{t-1}^{(i)} p\left(z_t|\widehat{x}_{t-1}^{(i)}\right)$, which satisfies the optimal importance density function.

4. Rao-Blackwellized Algorithm

The Rao-Blackwellized algorithm divides the system into a nonlinear part and an approximately linear part when the system has a high dimension. For the nonlinear part of it, the PF is used for filtering; while the remaining linear part is filtered using Kalman filter.

The state vector is divided into the nonlinear and linear components of the system

$$
\begin{aligned}
X_t &= \left(x_t^p + x_t^k\right)^T \\
&= \left(\left(\delta\varphi\ \delta\lambda\ 0\ 0\ 0\ 0\ 0\right)^p + \left(0\ 0\ \delta V_E\ \delta V_N\ \phi_E\ \phi_N\ \phi_U\right)^k\right)^T
\end{aligned}
\tag{5.31}
$$

State error equation

$$x_{t+1}^k = F_t \cdot x_t^k + w_t^k \tag{5.32}$$

$$x_{t+1}^p = F_t \cdot x_t^p + w_t^p \tag{5.33}$$

Measurement equation

$$z_t = h(x_t^p) + v_t \tag{5.34}$$

Importance density function

$$q\left(x_t^p | X_{t-1}^p, Z_{t-1}\right) = p\left(x_t^p | X_{t-1}^p, Z_{t-1}\right) \tag{5.35}$$

The above state error equation and measurement equation can be modified as follows

$$X_t^p = \left[\delta\widehat{\varphi}\ \delta\widehat{\lambda}\right]^T \tag{5.36}$$

$$z_t = h(\left[\varphi^{INS}\ \lambda^{INS}\right]^T + X_t^p) + v_t \tag{5.37}$$

Since the established strapdown inertial navigation model is only a theoretical discussion, it is impossible to obtain its real and accurate error sources, so the inertial guidance velocity and acceleration error correction based on simplified modelling is actually not too meaningful for engineering practice. Therefore, this section discusses the role of particle filtering in position correction with position correction as the main discussion point and the Rao-Blackwellized algorithm as the theoretical basis.

5.3.3.3 Simulation Test and Numerical Analysis

The simulation test uses the measured ocean local magnetic field data, which is complemented by another airborne magnetic measurement data to compensate for the diversity of the validation samples. Both of them are in the form of a regular grid. The grid spacing of the aeromagnetic data is 50 m, with 480 × 480 points. The grid spacing of the marine magnetometer data is 50 m, with 50 × 50 points. The particle sampling number is 50,000. The results of the simulation are shown in Figs. 5.9, 5.10, 5.11, 5.12, 5.13 and 5.14. The simulation results are shown as follows.

For the aeromagnetic measurement data, the positioning errors in latitude and longitude directions are shown in Figs. 5.10 and 5.11, respectively. The dashed line is the particle filter iterative localization error in one direction, while the solid line is the deviation between the INS indicated position and the true track, which is the same as below. From the figures, both directions converge to the true position after the initial oscillation, while the rate of convergence is slower because of the larger number of particles.

The same piece of real geomagnetic measurement data is selected as the simulation to verify the PF algorithm, and the iterative positioning results are shown in Fig. 5.12 The iterative positioning results are shown in Fig. 5.12. Among them, the

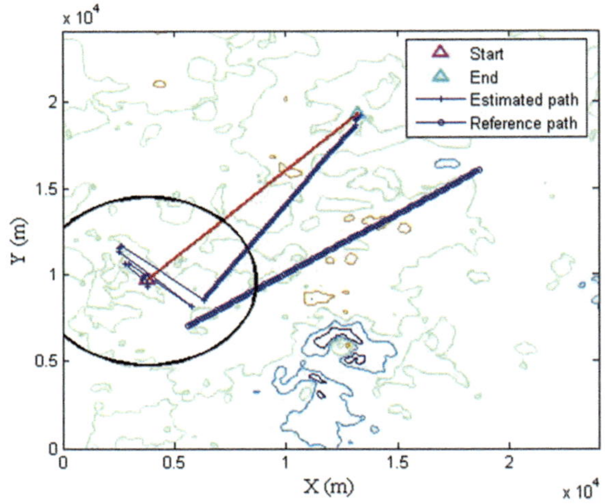

Fig. 5.9 Simulation test results of aeromagnetic measurement data

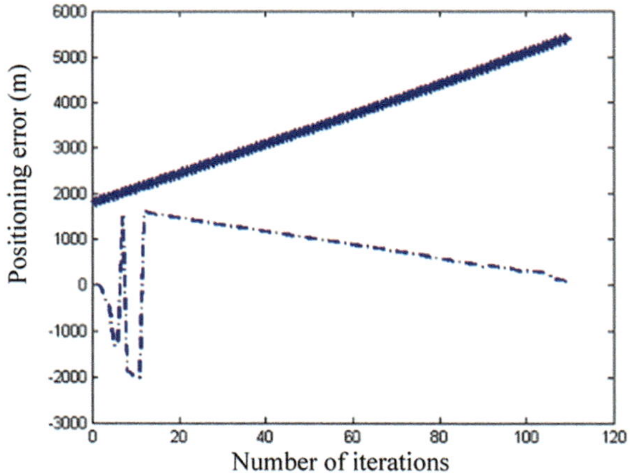

Fig. 5.10 Northward positioning error of aeromagnetic

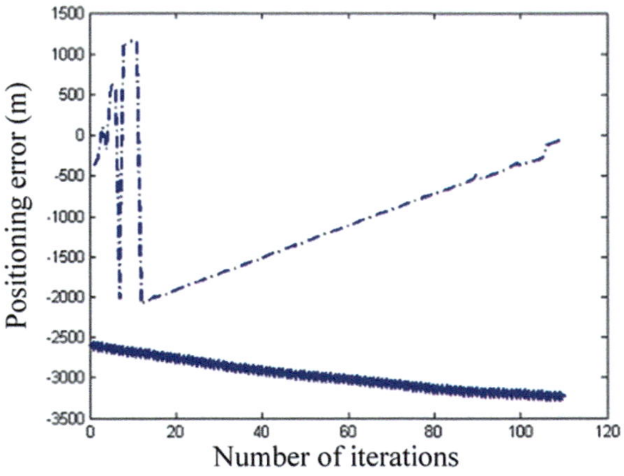

Fig. 5.11 Eastward positioning error of aeromagnetic

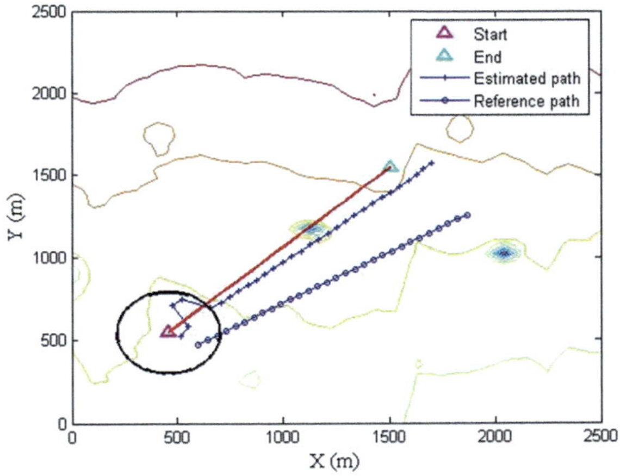

Fig. 5.12 Simulation test results of marine magnetic measurement data

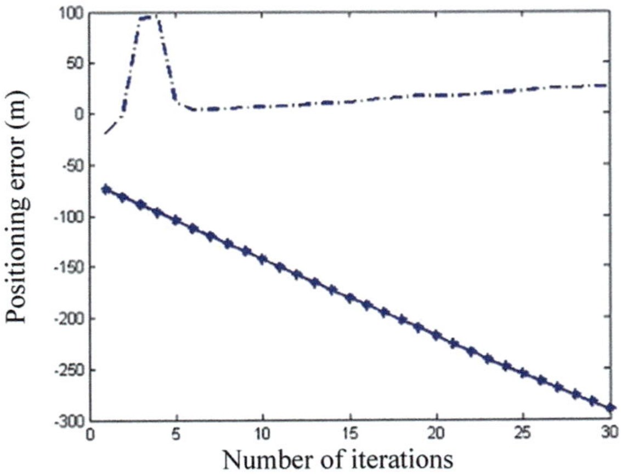

Fig. 5.13 Northward positioning error of marine magnetic

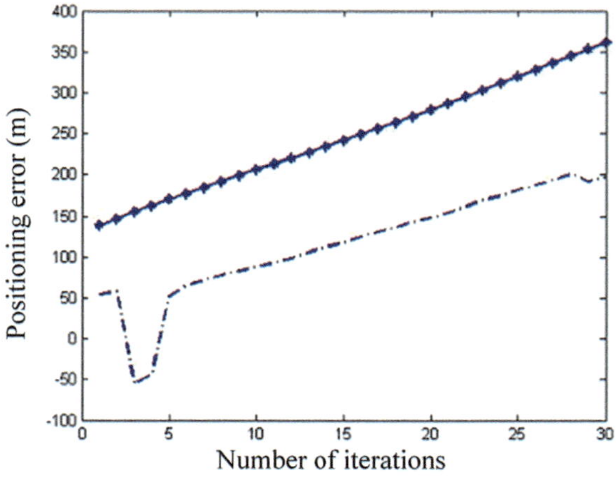

Fig. 5.14 Eastward positioning error of marine magnetic

eastward positioning deviation increases with the increase of positioning times, and the positioning result deviates from the real trajectory by nearly 200 m. Although it is still within the target range, the operation effect is not very satisfactory, and the ability of the algorithm to approximate the real position needs to be further improved.

The particle filtering algorithm has the possibility of deviating from the true position of the carrier during the initial sampling estimation, but as the sampling estimation is recursively pushed, the filtering effect increases in deviation due to problems such as particle collapse. The resampling method is used to restore the localization results to the vicinity of the real position of the underwater carrier. Meanwhile, although the low speed of underwater carrier navigation does not require high real-time performance of the algorithm, the computational speed of PF algorithm still needs to be improved.

There are many improved algorithms for particle filtering, among which one of the approximations using prior knowledge to give a possible distribution is one of the effective improvements to the method, and these methods offer the possibility of improving the accuracy and running speed of particle filtering localization results. In fact, trying to make the measurement sequence present a Gaussian state is also an effective measure for improvement, but considering that the filtering pattern does not store information about the previous state, the filtering denoising of the sequence cannot be performed. Combined with the existence of redundant information due to high-frequency sampling before the iteration point and the previous point, the offline identification method is used to remove the real-time measurement wild values that lead to obvious non-Gaussianity, reduce the number of particles, and improve the localization accuracy and operation efficiency.

5.4 Ground Vehicle Geomagnetic Navigation and Positioning

Driverless vehicles and robots and other unmanned ground carriers bring more and more convenience to people's work and life. With the development of 5G, high-resolution maps, sensor technology, artificial intelligence technology, etc., driverless vehicles have received more and more attention and become a research hotspot in the field of new-generation information technology [15, 16]. Real-time accurate positioning of moving vehicles is very important for intelligent vehicle technologies such as in-vehicle navigation systems, vehicle networking and driverless cars. Independent positioning solutions using a single positioning source all have different types of drawbacks: GNSS is unable to output at high frequencies and is seriously interfered by obstacles; inertial navigation system positioning errors accumulate over time and there are large deviations in positioning results after a longer period of time. The reliability of vehicle positioning methods using a single positioning source cannot be guaranteed, and the use of multi-sensor fusion for vehicle positioning has received increasing attention in recent years, especially geomagnetic navigation and positioning technology, which is an effective supplement to existing navigation and positioning technologies [17].

5.4.1 Selection of Characteristic Quantities for Geomagnetic Positioning of Ground Vehicles

According to the movement characteristics of the ground vehicle, the magnetometer can be fixedly installed in the car, so the vehicles driving on the road can choose the total field strength of the geomagnetic field in addition to the three components as the characteristic quantity of geomagnetic positioning. The force measuring instrument generally chooses the optical-pumping magnetometer or fluxgate magnetometer.

5.4.2 Implementation of Ground Vehicle Geomagnetic Measurements

The theoretical model of the relationship between the three inherent parameters of fluxgate magnetometer zero bias, sensitivity and quadrature error angle and output magnetic field strength is established, and the nine inherent parameters of the sensor are solved by the scalar matching method using the least squares method, and then the nine inherent parameters of the fluxgate magnetometer are corrected.

Define the vector non-orthogonal output as F, and the high precision scalar magnetometer output as B_{SC}.

Assign initial values to intrinsic parameters: zero bias $b_1 = b_2 = b_3 = 0$; sensitivity $s_1 = s_2 = s_3 = 1$; orthogonal error angle $u_1 = u_2 = u_3 = 0$. And write them in vector form: $m = [b_1, b_2, b_3, s_1, s_2, s_3, u_1, u_2, u_3]^T$.

From the intrinsic parameters and the vector non-orthogonal output, the scalar output in the corrected orthogonal coordinate system is calculated as B_{FG}, it yields

$$B_{FG} = \sqrt{(F - b)^T \cdot S^{-1} \cdot \left(P^{-1}\right)^T \cdot P^{-1} \cdot S^{-1} \cdot (F - b)} \tag{5.38}$$

Calculating the deviation B_{FG} from B_{SC}, denoted as

$$d = B_{SC} - B_{FG} \tag{5.39}$$

Calculate the partial derivative matrix G of the deviation d with respect to the intrinsic parameters

$$G = \frac{\partial d}{\partial m} \tag{5.40}$$

The incremental values of the intrinsic parameters Δm are calculated from the partial derivative matrix G, Huber matrix w and deviation d.

$$\Delta m = \left[G^T \cdot w \cdot G\right]^{-1} \cdot \left[G^T \cdot w \cdot G\right] \tag{5.41}$$

The iterative value of $m = m + \Delta m$ under the intrinsic parameters until the cycle is completed, the fluxgate magnetometer output deviation relative to the minimum scalar magnetometer, the intrinsic parameters at this time is the best value, in the use of the parameters and can achieve the fluxgate magnetometer output correction (Fig. 5.15).

The maximum deviation before calibration reached over 1800 nT with a standard deviation of 436 nT; the maximum deviation after calibration was 19.4 nT with a standard deviation of 5.4 nT.

Comparison of the solved and measured values of zero bias and sensitivity (Table 5.5).

From the above comparison, the difference between the results obtained by the two methods is very small, which also further proves the effectiveness of the three-axis vector magnetometer consistency error correction algorithm.

5.4.3 *Ground Vehicle Geomagnetic Positioning Methods*

The matching algorithm is more effective in localizing with low measurement noise. Due to the presence of perturbed conditions in the geomagnetic field, this places a requirement on the noise reduction of the measured sequence. Firstly, the wild

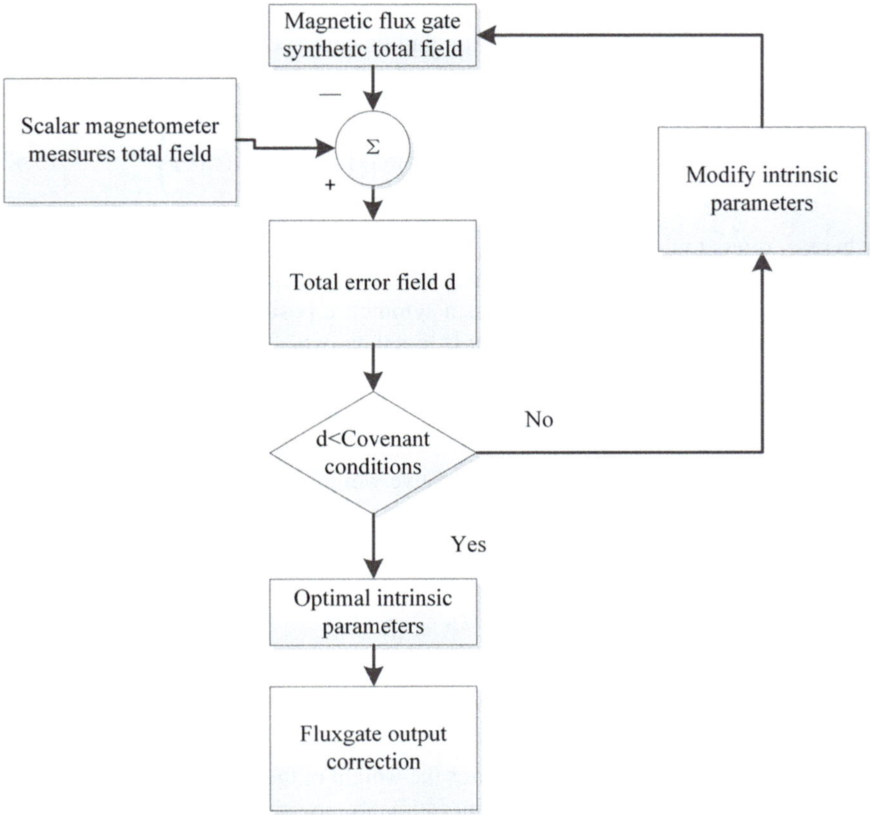

Fig. 5.15 Fluxgate sensor calibration process

Table 5.5 Calculated solved and measured values

Zero bias		Sensitivity	
Solution	Measured	Solution value	Measured
-35.8	−36	0.96732	0.9671
46.4	46	0.96844	0.96798
-52.7	−55	0.97042	0.97031

values triggered by magnetic storms, etc. need to be removed, and secondly, the geomagnetic field micro-perturbations need to be attenuated. Although the noise reduction method proposed in Chap. 2 has a good effect on the whole, it still cannot avoid the divergence of the results of the localization determination function due to the large offset of a single point. Therefore, in this chapter, the same localization decision rule is considered, and the Huber-type M-estimation method is used to improve the batch correlation function proposed in the previous paper.

M-estimation is a common form of maximum likelihood estimation, where the parameter P can be obtained by maximizing its corresponding likelihood equation F. The procedure for solving P^* is shown below.

$$P^* = \arg\max(-\ln F) = \arg\max\left(-\sum_i \ln f(r_i|P)\right) \quad (5.42)$$

where, r_i refers to the residual at the point i.

As shown in Eq. (5.40), the M-estimator can be characterized as a solution process of a more general form, where $\rho(\cdot)$ is a symmetric positive definite function with unique and zero extrema and the order is less than two.

$$P^* = \arg\max\left(M = \sum \rho(r_i, P)\right) \quad (5.43)$$

The problem is often solved as an iterative weighted least squares problem. The parameter vector $P = (p_1, p_2, \ldots, p_n)^T$ can be estimated by solving the following two equations, $j = 1, 2, \ldots, n$.

$$\sum \psi(r_i, P)\frac{\partial r_i}{\partial j} = 0 \quad (5.44)$$

$$\psi(r) = \frac{d\rho(r)}{dr} \quad (5.45)$$

The influence function $\psi(\cdot)$ measures the weight of the data points on the parameter estimates. For a good error-tolerant estimator, any single point of data does not introduce an obvious error, which makes it insensitive to wild values. The specific M-estimator should be bounded in P and be a convex set. And when $\partial\rho^2(\cdot)/\partial P^2 = 0$, $\partial\rho(\cdot)/\partial P \neq 0$.

The Huber function is an efficient implementation of the M-estimator. It is a parabola around the value of 0 that increases linearly over a given range of $|r| > k$, which limits the effect of wild values. With this estimator, the asymptotic efficiency of the standard normal distribution close to 95% can be obtained by tuning the constant $k = 1.345\sigma$, where σ is the estimated standard deviation of the error. For the same values as the tuning constant k, it is equally valid for many normal distributions. The equations $\rho(\cdot)$ and the influence function $\psi(\cdot)$ that form the M-estimator are shown in Eqs. (5.43) and (5.44), and $k = 4$ is shown in Fig. 5.16.

$$\rho(r) = \begin{cases} r^2/2 & |r| < k \\ k(|r| - k/2) & |r| \geq k \end{cases} \quad (5.46)$$

$$\psi(r) = \begin{cases} r & |r| < k \\ k \cdot sign(r) & |r| \geq k \end{cases} \quad (5.47)$$

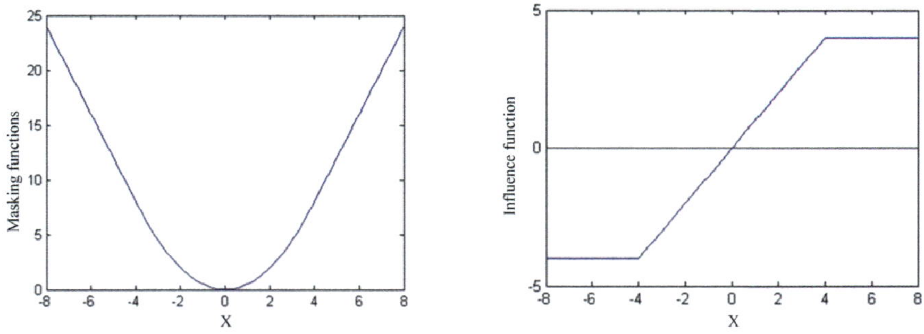

(a) Masking functions (b) Influence function

Fig. 5.16 M-estimator of Huber

Huber's M-estimation Correlation (HMC) gives a measure of similarity between the measured sequence and the comparison sequence in the benchmark magnetogram, and its value can be calculated using Huber statistics. The correlation mask function can compress the effect of wild values. The standard deviation σ of the correlation is a key parameter that can be calculated by the adaptive finding process of the residuals of the measured and compared sequences.

It is assumed that s_i and s_i represent the data point values of the measurement sequence and the matching sequence, respectively, while m and M represent the influence functions, respectively.

Step 1: Using the influence function of Eq. (5.44), the mask functions $m(i)$ and $M(i)$ are calculated as follows.

$$m(i) = \begin{cases} r_i & |r_i| < k_1 \\ k_1 \cdot sign(r_i) & |r_i| \geq k_1 \end{cases} \tag{5.48}$$

$$M(i) = \begin{cases} r_i' & |r_i'| < k_2 \\ k_2 \cdot sign(r_i') & |r_i'| \geq k_2 \end{cases} \tag{5.49}$$

where, $r_i = s_i - \overline{s}, r_i' = S_i - \overline{S}, k_1 = 1.345\sigma_1,$ and $k_2 = 1.345\sigma_2$ are residuals; σ_1 and σ_2 are the standard deviations of the residuals of the measured and matched sequences, respectively.

Step 2: The HMC value between the measured sequence and the matched sequence is calculated iteratively according to Eq. (5.47).

$$HMC = \frac{\sum_{i=1}^{n} M(i)(r_i')m(i)(r_i)}{\sqrt{\sum_{i=1}^{n} m(i)(r_i)^2}\sqrt{\sum_{i=1}^{n} M(i)(r_i')^2}} \tag{5.50}$$

Step 3: The position corresponding to the maximum value of the HMC is calculated, which can be considered as the position of the underwater carrier at the current moment.

The same real geomagnetic grid data from Sect. 4.3.3 is used as the baseline magnetogram for simulation validation. The measurement sequence is obtained by appending Gaussian white noise to any of the sequences in the map and adding a position offset. The standard deviation of the Gaussian noise is 2% of the geomagnetic amplitude variation range, while two times the amplitude range is used for randomly increasing wild values.

Next, HMC is executed and compared with the CC determination rule. To check the noise diversity, each localization algorithm was executed 10 times and the results are shown in Fig. 5.17. From the results, we can see that the localization results of the HMC algorithm are generally more accurate and stable than the traditional CC algorithm. In fact, from the perspective of one-dimensional signal processing, this localization rule overcomes the disadvantages of large dispersion and sample loss in noise cancellation. The applicability is extended to non-Gaussian noise processing, which also makes the algorithm more suitable for engineering applications.

The Huber-type M-estimator, the objective function of the robust localization algorithm, has a mask function that reduces the effects of high noise and missing samples. It has advantages over traditional localization algorithms that solve for correlation. Combining the localization orientation algorithm focusing on feature extraction with the Huber-type M-estimator rule focusing on target confirmation can provide better localization results.

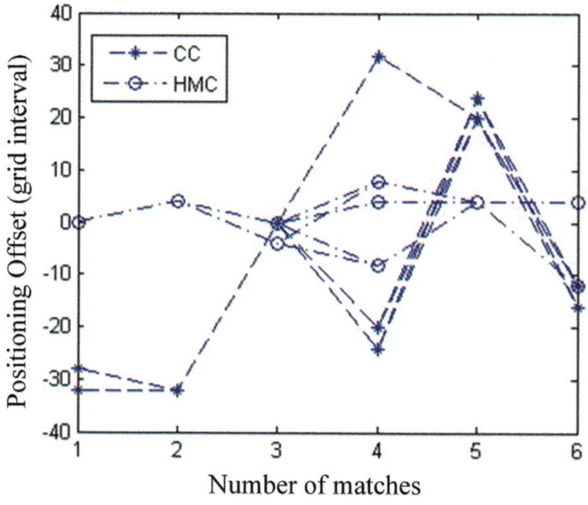

Fig. 5.17 Comparison of positioning results using HMC and conventional CC

5.5 Geomagnetic Navigation and Positioning for Pedestrians

With the maturity and application of GNSS technologies (including GPS, GLONASS, Galileo and Beidou), pedestrian-oriented navigation and positioning, especially pedestrian positioning in indoor/underground and other enclosed spaces, have become a hot research topic. Pedestrian navigation and positioning can be used in indoor scenes of consumers in shopping centers, stations, airports, venues, etc. to help consumers quickly find people, stores, cars and parking spaces to enhance user travel, consumption and entertainment experience; in some special industries, such as chemical plants, mines, power plants, the positioning tracking of personnel is required, and timely alarm is needed when personnel mistakenly enter a dangerous area, stay in an area overtime, or remain stationary for a long time (such as fainting, unconscious); visualization shows the walking path of inspection personnel, early detection of equipment failure, combined with the location of the fault, improve the efficiency of inspection and maintenance, and enhance the intelligent level of enterprise safety management; in the application fields of emergency rescue search and rescue, anti-terrorism, fire control, military, law enforcement, criminal tracking, etc., we can grasp the location and status of task performers through positioning to ensure safe, reliable and efficient execution of tasks [18–20].

Location services are significant for pedestrians, but 70% of human activities are indoors, and GNSS is suitable for outdoor unobstructed areas, so pedestrian navigation and positioning cannot rely entirely on GNSS, and geomagnetic navigation and positioning technology provides a technical way to provide location services for pedestrians in enclosed spaces such as indoors or underground.

5.5.1 Selection of Characteristic Quantities for Geomagnetic Positioning of Pedestrians

The process of magnetic field localization is as follows. The magnetic sensor chip inside the smartphone or other customized terminal continuously collects magnetic field data along the walking/driving route as the person walks or the car moves, and the actual magnetic field data measured by the magnetic sensor is searched and matched with the pre-collected geomagnetic map. The location within the geomagnetic map where the magnetic field data most similar to the measured magnetic field data is found is considered to be the location of the smartphone, i.e. the location of the pedestrian. Therefore, applying geomagnetic localization on pedestrians, the total intensity of the geomagnetic field can only be selected as the feature quantity because the magnetic sensor cannot be fixedly connected to pedestrians [21].

5.5.2 Implementation of Pedestrian Geomagnetic Measurements

Pedestrian positioning places special demands on the cost, size and easy to use. With the development of chip technology, magnetoresistive sensors can now meet the demand for personnel positioning, so magnetoresistive sensors are chosen as magnetic field measurement instruments for personnel positioning, and positioning is achieved by synthesizing the total field strength through a three-component magnetic field.

5.5.3 Geomagnetic Positioning Method for Pedestrians

The geomagnetic localization scheme for personnel is shown in Fig. 5.18. The construction of the geomagnetic database is completed first. In the localization stage, the attitude information (mainly heading, since most smart devices provide mature filtering algorithms for obtaining the device attitude), acceleration, magnetic and signal strength obtained from Bluetooth scanning are extracted from the smartphone, firstly, the pedestrian gait detection and step length estimation are performed based on the acceleration data, and the pedestrian heading estimation is combined with the heading information. Then the magnetic field data and PDR results are corresponded (10 steps) to obtain a certain length of magnetic field data containing relative position, and on the basis of Bluetooth positioning results, if the current trajectory meets the discriminative conditions, magnetic field matching is performed on the basis of Bluetooth positioning to output magnetic field positioning results, and if the current trajectory does not meet the discriminative conditions, Bluetooth positioning results are output directly.

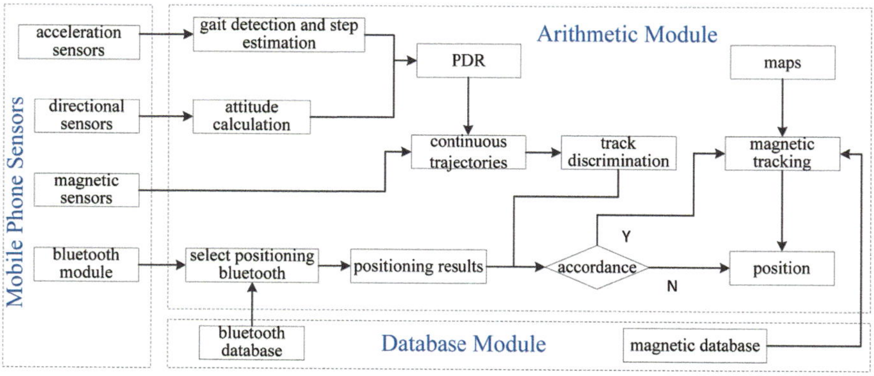

Fig. 5.18 Indoor positioning solution with multi-source information fusion

The magnetic data are collected using special equipment to obtain both indoor floor plans as well as highly accurate location and magnetic field information. The final database architecture is shown in Eq. (5.51).

$$DB_{mag} = \left\{ v_{i,j} \middle| i = 1, \ldots M, j = 1, \ldots N \right\} \tag{5.51}$$

where, $M = row/2$, $N = col/2$; row and col denote the number of rows and columns of the corresponding pixel point matrix of the indoor floor plan (20 pixel points represent the actual 1 m); $v_{i,j} = \begin{cases} 0, & obstruction \\ mag, & other \end{cases}$, where mag indicates the total magnetic field strength at that location, and the actual plane position can be obtained by converting the position of the point in the database.

The result of pedestrian position reckoning projection is calculated by Eq. (5.52).

$$\begin{cases} Pos\_X_i = Pos\_X_{i-1} + Len\_step \times \cos(\psi) \\ Pos\_Y_i = Pos\_Y_{i-1} + Len\_step \times \sin(\psi) \end{cases} \tag{5.52}$$

where, Pos_X_i and Pos_Y_i indicate the current relative position; Len_step indicates the current step length; and ψ indicates the heading information output by the smart device.

Pedestrian gait detection is obtained from Eq. (5.52).

$$\begin{cases} Var(Acc) > throld1\_acc \\ Acc\_\max - Acc\_\min > throld2\_acc \end{cases} \tag{5.53}$$

where, Acc is the resultant acceleration; $throld1\_acc$ is the threshold of Acc; Var is the variance; $Acc\_\max$ is the maximum resultant acceleration in the set period; $Acc\_\min$ is the minimum resultant acceleration in the set period; and $throld2\_acc$ is the peak-to-peak threshold of the resultant acceleration in the set period.

The pedestrian step estimate is obtained from Eq. (5.54).

$$Len\_step = \frac{0.15 \times fs}{fs\_step} + 0.3 \tag{5.54}$$

where, fs is the sampling frequency of inertial sensors; fs_step is the duration of the current step.

In the fusion localization scheme shown in Fig. 5.18, the trajectory discrimination is mainly based on the heading information of the currently extracted trajectory, and the subsequent fusion localization is performed when the heading change is less than a certain threshold value. The magnetic field localization process is shown in Fig. 5.19.

After obtaining the Bluetooth positioning result, the Bluetooth positioning result is used as the center to generate the samples to be positioned in a certain range. In this chapter, the Bluetooth positioning result is used as the center, and the samples to

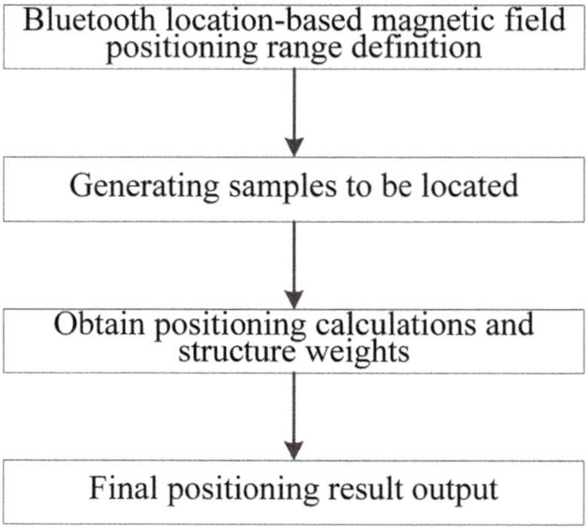

Fig. 5.19 Magnetic field positioning process

be positioned are generated in a circle of 5 m length radius with a 0.5 m interval grid, and further generation is carried out on a grid with the current heading as the center and a 5° heading change, assuming that the number of samples to be positioned generated is *R*.

In the localization calculation, for a sample to be localized, the current trajectory is projected onto the plane map according to its position and heading information, the corresponding magnetic field data in the magnetic field database is extracted, and the distance is calculated based on the magnetic field data collected in real time and the extracted magnetic field data in the database, and the calculation formula is shown in (5.55).

$$dis_i = \begin{cases} 100000000, & blocking \\ \frac{1}{n}\sum_{j=1}^{n} \left|M_{now,j} - M_{db,i}\right|, & other \end{cases} \tag{5.55}$$

where, dis_i denotes the distance calculated from the sample i; $M_{now,j}$ denotes the magnetic field value corresponding to the jth point of the trajectory; $M_{db,i}$ denotes the jth magnetic field value in the magnetic field database extracted based on the current sample to be located; and n denotes the number of magnetic field data contained in the trajectory.

After obtaining the distance information based on Eq. (5.55), the initial weights of each sample to be located are calculated according to Eq. (5.56).

$$w_i = \frac{1}{dis_i} \tag{5.56}$$

The weights are then normalized based on Eq. (5.57).

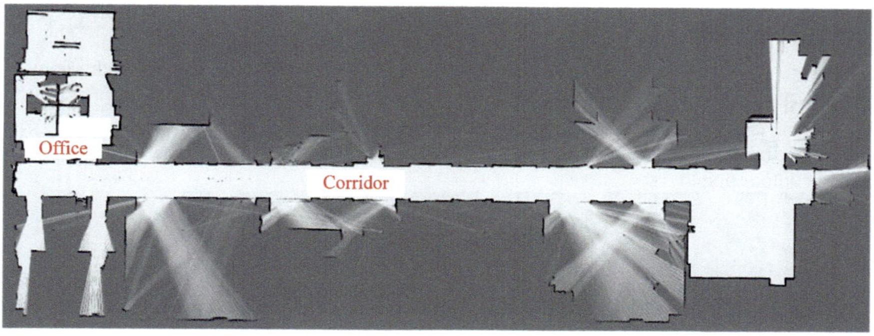

Fig. 5.20 Plan view of the test scene

$$w_i = \frac{w_i}{\sum_{j=1}^{R} w_j} \tag{5.57}$$

Finally, the final positioning result is obtained by Eq. (5.58).

$$Pos = \sum_{j=1}^{R} w_j(x_j, y_j) \tag{5.58}$$

where, (x_j, y_j) denotes the position of the plane where the jth sample to be located is located.

Experimental validation of the multi-source information fusion localization method described in this chapter. The validation site is selected as an office environment and contains a long straight corridor. The plan view of the scene scanned by the scanning device is shown in Fig. 5.20, with a length of 50 m and a width of 20 m. The Bluetooth base station installation location is shown in Fig. 5.21, with a total of 14 Bluetooth base stations installed. The generated magnetic field plane distribution is shown in Fig. 5.22.

The walking trajectory during the positioning verification is shown in Fig. 5.23, the real-time positioning results are shown in Fig. 5.24, and the cumulative distribution of error probability is shown in Fig. 5.25. The probability of error is <1.4 m at 60% probability.

5.6 Summary

This chapter provides a detailed description of the implementation process of geomagnetic positioning under different application scenarios and application platforms. Magnetic field information naturally exists and changes abundantly, so

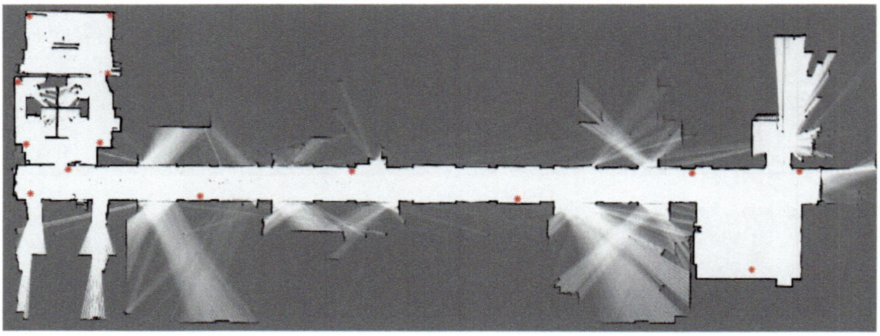

Fig. 5.21 Bluetooth base station installation location

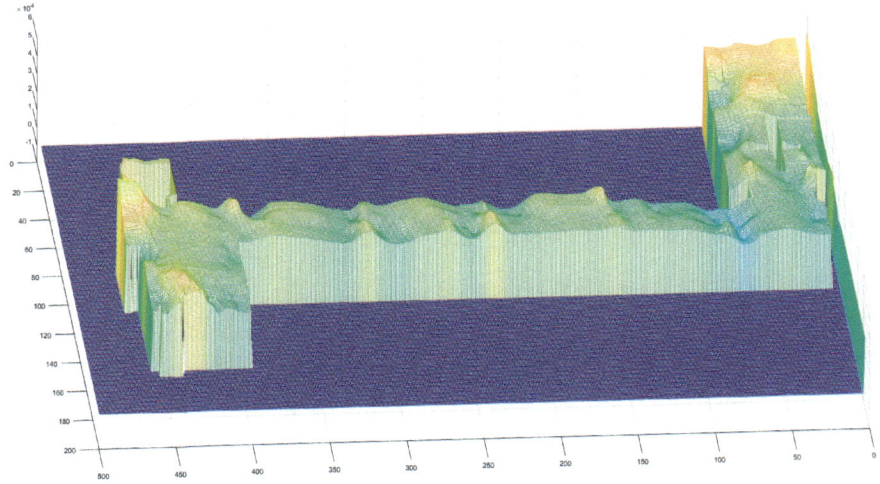

Fig. 5.22 Magnetic field distribution map of the validation scenario

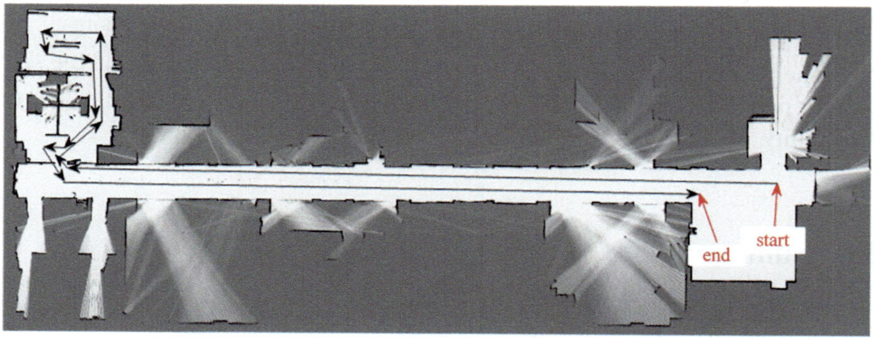

Fig. 5.23 Walking trajectory during positioning verification

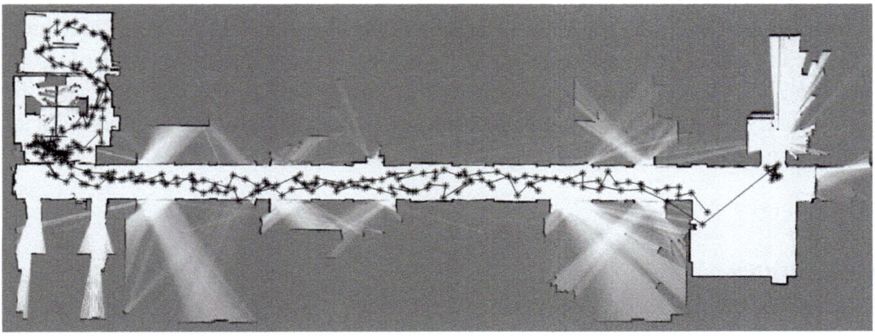

Fig. 5.24 Positioning results

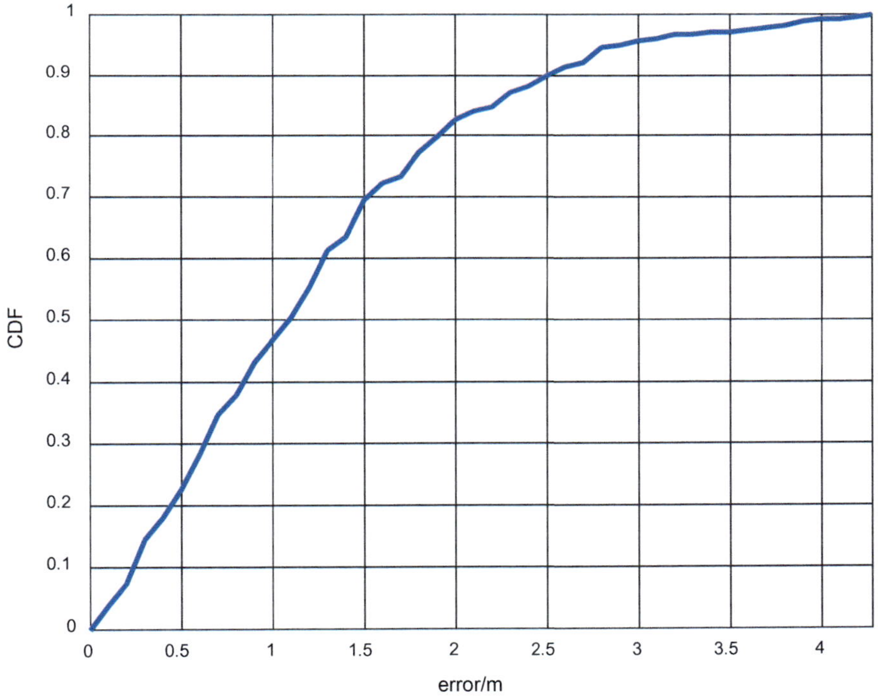

Fig. 5.25 Cumulative distribution of error probabilities

magnetic field positioning is passive positioning, without the help of external communication signals, without radiating energy to the outside, and with high concealment. It is difficult to artificially generate a wide range of inherent magnetic interference that affects the characteristics of the environmental magnetic field distribution. In addition, the random interference magnetic field generated by the current in the environment, electrical equipment switching on and off, radio, etc. has obvious differences in its spectrum and the spectrum of the inherent magnetic field of the environment, which can be identified and processed by professional magnetic signal processing methods. Therefore, magnetic field positioning has strong anti-interference capability and is highly adaptable, providing an important technical pathway for navigation and positioning in special scenes.

References

1. Fisher, K., & Raquet, J. (2011). Precision position, navigation, and timing without the global positioning system. *Air & Space Power Journal*, 24–33.
2. Goldenberg, F. (2006). Geomagnetic navigation beyond the magnetic compass. In: 2006 IEEE/ION Position, Location, And Navigation Symposium (pp. 684–694).
3. Wilson, J. M., Kline-Schoder, R. J., Kenton, M. A., et al. (2006). Passive navigation using local magnetic field variations. In: *Proceedings of the 2006 National Technical Meeting of The Institute of Navigation* (pp. 770–779), Monterey, CA, January 2006.
4. Li, S.-M., Wang, G.-G., & Shen, Z.-F. (2019, November 8). A geomagnetic positioning method suitable for indoor free motion carriers, Beijing, CN107543541B.
5. Zhou, J., Ge, Z.-L., Shi, G.-G., et al. (2008). Key technique and development for geomagnetic navigation. *Journal of Astronautics*, (05), 1467–1472.
6. Xiong, S.-Q. (2009). The present situation and development of airborne gravity and magnetic survey techniques in China. *Progress in Geophysics, 24*(01), 113–117.
7. Naomi, K., & Shigetomi, T. (2009). Underwater navigation for long-range autonomous underwater vehicles using geomagnetic and bathymetric information. *Advanced Robotics, 23*, 787–803.
8. Cai, Z.-Y., Wei, H.-P., & Ren, Z.-X. (2007). A review of underwater geomagnetic navigation technology research. *National Defense Technology*, (03), 28–29.
9. Zhang, W.-Y., Qiu, D.-F., & Hu, X.-T. (2006). Development of underwater robotics, military applications and implications. *China Shiprepair*, (06), 37–39.
10. Yang, Y.-T. (2019). Research on geomagnetic measurement error compensation in navigation and localization. *Ship Science and Technology, 41*(11), 125–128.
11. Lin, Y., Yan, L., & Zeng, Q.-X. (2008). Outlier identification in offline mode for dynamic measurements in underwater geomagnetism navigation. *Journal of Wuhan University of Technology*, (09), 112–115.
12. Ma, M.-Z. (2019). Research on underwater geomagnetic assisted navigation matching algorithm. *Southeast University*, 78–85.
13. Mu, H., Ren, Z.-X., Hu, X.-P., et al. (2007). Information fusion strategy and performance for marine Inertial/Geomagnetic navigation system. *Journal of Chinese Inertial Technology*, (03), 322–326.
14. Hao, Y.-L., Zhao, Y.-F., & Hu, J.-F. (2008). Preliminary analysis on the application of geomagnetic field matching in underwater vehicle navigation. *Progress in Geophysics*, (02), 594–598.
15. Rahok, S. A., Shikanai, Y., & Ozaki, K. (2010). Trajectory tracking using environmental magnetic field for outdoor autonomous mobile robots. *IEEE/RSJ International Conference on Intelligent Robots and Systems, 2010*, 1402–1407.
16. Rahok, S. A., & Ozaki, K. (2011). Play-back navigation for outdoor mobile robot using trajectory tracking based on environmental magnetic field. *IEEE International Conference on Robotics and Automation, 2011*, 625–630.
17. Li, S.-M., & Zhang, W.-Q. (2004). Study on the application of geomagnetic field contour in matching guidance. *Guidance & Fuze*, (03), 21–23.

18. Wang, G.-G., Li, S.-M., & Jiang, H.-L., et al. (2020). Multi-source information fusion for smartphone indoor positioning technology. *Selected Papers on Satellite Navigation and Positioning Techniques*, 260–266.
19. Liu, Y.-C., Cai, C.-L., Wei, Z.-C., et al. (2020). Floor positioning method based on fusion of PDR and geomagnetic matching. *Chinese Journal of Sensors and Actuators, 33*(04), 557–563.
20. Li, S.-M., Wang, G.-G., & Shen, Z.-F. (2019, October 11). A passive combination localization method based on magnetic field characteristics of spatial environment, Beijing. CN107421546B.
21. Lu, Y.-L., Wei, J.-L., Liu, C.-W., et al. (2020). An indoor localization method based on adaptive particle filter fusion of geomagnetism, WiFi and PDR. *Bulletin of Surveying and Mapping*, (06), 1–6.

Chapter 6
Topographic Elevation Navigation and Positioning Fundamentals and Theoretical Models

Abstract Topographic elevation navigation positioning is a navigation method that uses the specificity and uniqueness of elevation differences at each point in space for positioning. Because it is a passive navigation method that does not depend on external signals, it is not easily affected by objective factors such as electromagnetic interference and weather and is valued in existing combined navigation methods. The basic principle of topographic elevation navigation and positioning is that the geographic coordinates of any location on the earth's surface can be determined singularly based on the contours or landforms of its surrounding territory. Firstly, the terrain elevation model (map) along the route is stored in the motion carrier, and when the motion carrier flies over an area with known terrain elevation, the current terrain elevation is measured by the altimetric (depth) sensor on the motion carrier, and the measured data is correlated and analyzed with the pre-stored terrain elevation model (map) to determine the corresponding grid position of the motion carrier in the terrain elevation model (map) because the terrain The latitude and longitude corresponding to the grid position in the elevation model (map) is known in advance so that the position of the motion carrier can be determined. Although the combination of inertial navigation and positioning system (INS) and satellite navigation and positioning system (GNSS) navigation can meet the requirements of high precision positioning, this combination of navigation autonomy is insufficient and easy to expose the target. The combination of INS and terrain elevation navigation and positioning meets various navigation requirements and does not depend on an artificial satellite navigation system. In addition to the navigation and positioning of motion carriers, terrain elevation positioning navigation also has many functions such as terrain tracking/terrain avoidance, rational routing, threat avoidance, obstacle alert, ground alert, target interception, marine exploitation, marine wreck salvage, and underwater launch. Terrain elevation navigation and positioning are most widely used in air vehicles and underwater vehicles because it is easier to obtain airborne terrain elevation maps, so the application of terrain elevation navigation and positioning is more mature in air vehicles; underwater artifacts are the least, and satellite positioning is limited, so it is the easiest background condition for terrain elevation navigation and positioning to work, so the academic research is focused on underwater terrain. Regardless of underwater and aviation, the basic theory of terrain

L. Yan et al., *Mathematical and Physical Fundamentals of Navigation and Positioning with Earth's Natural Fields*, Navigation: Science and Technology 13,
https://doi.org/10.1007/978-981-97-5524-0_6

elevation positioning navigation and positioning navigation methods are common, only in the means of terrain elevation map acquisition, altimetry (depth) sensors exist differently: this chapter will be developed to describe respectively.

6.1 Digital Terrain Elevation Model

Digital Elevation Model (DEM) is a new generation of a topographic map, which is one of the main contents of spatial databases and the basic data for various terrain analyses. As a digital topographic map, DEM contains a large variety of topographic structures and feature information, which is the basic data for the quantitative description of the geomorphological structure, hydrological process, biological distribution, and other spatial The DEM is the basic data for quantitative description of spatial changes of landscape structure, hydrological processes, and biological distribution.

The DEM provides information on the relief of the Earth's surface and is mainly used for mission planning, terrain simulation, and 3D display and contour mapping of terrain. The DEM consists of a terrain height matrix. Topographic points are positioned by a latitude and longitude grid, and the grid spacing represents the resolution of the DEM. With a certain resolution to achieve a discrete representation of continuously changing terrain surface, the resolution brings two aspects: the terrain representation scale and the elevation range represented by the grid points, which directly impacts the navigation and positioning accuracy. In the terrain elevation navigation and positioning, two discretization standards are generally used; the first discretization standard is the great circle of 3 s longitude on the equator with arc length (about 100 m); the second standard is the great circle of 1-s longitude on the equator with arc length (about 30 m).

DEM is an approximation of the actual terrain in an optimal sense, a simulation of the terrain surface. Therefore, the degree of approximation of terrain by DEM depends on the quality and distribution of original terrain sampling points, terrain roughness, grid resolution, and other factors. Therefore, it inevitably contains errors. Data interpolation technique is necessary for DEM, but the actual application process should focus on the control of data source and input quality rather than learning complex interpolation methods; as long as the original data quality is good, simple interpolation methods can also achieve satisfactory results; on the contrary, if the original data quality is not high, then the best interpolation methods are difficult to get higher accuracy. However, different interpolation methods significantly impact specific navigation and positioning applications.

DEM is generally organized in the following three ways, regular grid structure, irregular triangular grid structure, and contour structure. All the above three structures can be applied to terrain elevation navigation and positioning. However the advantages of high automation of regular grid data acquisition, convenient storage, management, calculation, and easy integration with remote sensing raster images make the specification grid structure terrain elevation navigation and positioning dominant. At present, the data provided by many countries are in the form of a data matrix of a regular grid. Nevertheless, the data structure of digital elevation models is not static, and a large number of algorithms are available for interconversion between different structures. In the following discussion of this book, the regular grid structure is used unless otherwise specified.

The data acquisition methods of terrestrial DEM and underwater DEM are different due to the characteristics of air and seawater, so there are differences in data sources and processing methods, which are described separately in this section.

6.1.1 Terrain Elevation Model

The terrain elevation model is an altitude matrix consisting of the elevation of the land surface, with data consisting of (X, Y, and H). (X, Y) is the geodetic coordinates, and H is the elevation at that coordinate position. The process of obtaining the geodetic terrain elevation model is the process of mapping the elevation of the land surface.

At present, the more commonly used methods of DEM data acquisition are mainly ground survey (point group-based method), digitization of existing topographic maps (map-based method), and photogrammetry (image-based method, high definition aerial film, satellite remote sensing telemetry).

Large-scale topographic elevation mapping is mainly obtained through the total station, satellite differential positioning (GNSS-RTK) mapping technology, and other methods. But with the rapid development of mapping technology, many new technologies are gradually applied to the production of large-scale topographic maps, such as three-dimensional laser scanning technology, low-altitude satellite remote sensing technology, etc. For a long time, China has been using traditional instruments for the production of large-scale topographic maps, such as level, total station, satellite receiver, and other measuring instruments to collect feature point coordinate data in the field, combined with the internal computer compilation into topographic maps. The operation method is greatly affected by the environmental factors in the field and has the disadvantages of a tedious process, slow speed, and high operation costs. When using satellite differential technology for field mapping, in addition to full-field measurement, data processing is also required, which includes control measurement, regional coordinate conversion parameter solution, and other work, and the operation is vulnerable to the constraints of satellite signals. Topographic maps based on satellite remote sensing photography technology are susceptible to meteorological conditions and geomorphological conditions, and currently cannot meet the requirements of 1:1000 and 1:500 large-scale terrain elevation mapping; although topographic map production based on 3D laser technology can improve the production speed, it is difficult to ensure good mapping results due to the large survey area and small terrain fluctuations and complex feature characteristics, and its cost is high [1].

Tilt photogrammetry is a new measurement technology developed in the surveying and mapping industry in recent years, which collects ground image data from multiple angles and conducts data processing to generate regional realistic 3D model results, solving the problem that traditional aerial photogrammetry technology can only obtain orthophoto results, but not complete topographic features and landforms and other information. The UAV flying platform with a five-lens camera can obtain complete and accurate ground information from five angles: vertical, front view, back view, left view, and right view, and the realistic 3D model established by this technology has high accuracy, good model detail expression and fast modeling speed, which is suitable for topographic map mapping in areas with complex terrain conditions.

In summary, the current terrain elevation navigation and positioning technology on the land surface using the required terrain elevation model technology is mature and well-secured, laying a better foundation for the application of terrain elevation navigation and positioning technology.

6.1.2 Underwater Topography Elevation Model

Unlike the geodetic topographic elevation model, the underwater topographic elevation model is a depth matrix consisting of the depth of the seafloor relative to the sea surface. The underwater topographic elevation model data also consists of (X, Y, H), where (X, Y) is the position coordinate and H is the depth of the seafloor at that coordinate position. The process of obtaining the underwater topographic elevation model is the process of mapping the depth of the seafloor.

Traditional seafloor topography measurement needs to be obtained by depth, tide level, draft, surge, and other measurements, modern underwater and coastal zone topography measurement has undergone profound changes due to the progress of related technology, showing a three-dimensional, efficient, and high-precision measurement posture. Bathymetry is an important part of underwater topographic survey, and at present, single-beam bathymetry, multibeam bathymetric system (MBS) and airborne laser (LiDAR) full-coverage bathymetry and other technologies are mainly used. Single-beam measurement can only obtain one bathymetric point at a time, which is suitable for small and medium scale or small area large scale underwater topography measurement; MBS can obtain hundreds to several hundred points in the navigational orthogonal sector at a time to achieve full coverage sweep of the seafloor.

GNSS integrated ship-based bathymetry technology is the representative of modern ship-based bathymetry, integrating single-beam and multi-beam bathymetry technology, GNSS-RTK and acoustic velocity underway measurement technology, which achieves integrated acquisition and fusion of multi-source information underway, minimizes the influence of each error, and improves the accuracy and efficiency of seafloor topography measurement.

AUV (Antonio-mous underwater vehicle), ROV (remotely operated vehicle), the deep towing system as the platform, carrying multibeam bathymetry system, pressure sensor, an underwater acoustic positioning system (USBL, ultra-short baseline), and other equipment of the submarine-based seafloor the submarine-based topographic and geomorphological survey system has been introduced and applied in some key survey waters and projects in China. The submarine-based seafloor topography measurement system provides absolute plane and vertical coordinates for the platform with the help of a compass, attitude sensor, and pressure sensor acquires the seafloor topography with multibeam, and transmits the information to the shipboard storage and processing unit through the cable to obtain the seafloor topography by comprehensive calculation. The submarine-based measurement technology applies to deep-sea topographic measurement. With the deepening of China's deep-sea investigation activities, its application will certainly become more and more extensive.

In addition to the above measurement platforms, obtaining seafloor topography by inversion techniques is also a means, but with lower accuracy. Satellite remote sensing inversion of bathymetry is to achieve large-area bathymetry inversion with the help of spectral changes of visible light after propagation and reflection in the water, combined with bathymetry inversion model, and then combined with remote sensing imaging moment water level back calculation to obtain the topography of the seafloor. The currently available satellites are IRS, Ikonos, QuickBird, AVIRIS, Sentinel-2, Landsat, TM, SPOT, and so on.

In addition to satellite remote sensing inversion of bathymetry, gravity anomalies and seafloor topography are highly correlated in a certain waveband, according to which large-scale seafloor topography can be inverted. Gravity inversion topography has undergone the development from one-dimensional to two-dimensional linear filtering, the core of which is inversion model construction.

In summary, due to the seawater blockage of visible light, radio, and other signals, underwater topographic elevation measurement is difficult and costly compared with geodetic topographic elevation measurement technology, and the application of underwater topographic elevation navigation and positioning technology is limited to a certain extent due to the lack of data security.

6.1.3 Common Methods of Topographic Elevation Model Resampling

Topographic elevation model resampling is to convert the measured 3D relative coordinates of irregularly distributed terrain points into the topographic elevation of each grid center of the regular distribution, to be used for topographic elevation navigation and positioning purposes. There are usually three methods for resampling random terrain points: the first is the interpolation method used in remote sensing digital photogrammetry systems, that is, the randomly distributed discrete points are interpolated to become the elevation of each grid center point in a regular arrangement; the second is the block averaging method usually used in the early terrain telemetry system, the elevation of each grid center is expressed by the average elevation of each point in the grid area; the third is random linearization technique of terrain the method used in modern terrain matching system (e.g., Sandia inertial terrain navigation and positioning system). The interpolation method requires high accuracy of node data, and the resampling accuracy is low when there is a large observation error; the selection criteria and search algorithm of interpolation nodes are complicated, and for a fixed number of nodes, the nodes are required to meet the interpolation conditions and at the same time, the nearest node is required to be selected, which makes the node search algorithm complex and the real-time performance is poor. Therefore, the nearest neighbor pixel method is a common alternative method when real-time performance is required, but its resampling accuracy and noise immunity are lower than interpolation. The block averaging method divides the terrain into a square grid, and the elevation of each sampling point falling into the grid is averaged as the elevation of the center of the grid. When each sampling point is close to the center of the grid, it has certain estimation accuracy, and the block averaging method has certain noise immunity for random noise with 0 mean distribution due to the noise averaging effect; the advantage of this algorithm is that the algorithm complexity is low and it is easy to process in real-time. The essence of random linearization of the terrain is to linearize the nonlinear measurement equation of the terrain surface based on the least squares error criterion. The algorithm has the best estimate in the least-squares sense, and thus is more resistant to random noise than the previous schemes, and has moderate algorithmic complexity if an appropriate implementation method is used [2–4].

1. Weighted average interpolation

Weighted average interpolation is relatively simple, the method is to first weigh the known points and then find the weighted average interpolation of the key issue is to use what weighting rules, the existing weighting methods are distance weighting method, and Tyson polygon, weighting method distance weighting method. Distance weighting is a function of the distance from a known point to the point to be found, such as inverse distance weighting, Gaussian distance weighting, and harmonic distance weighting. Tyson polygon weighting belongs to the study of natural proximity interpolation, and the influence area ratio of the Tyson polygon is used as the weight. For simplicity, the nearest neighbor method and the arithmetic average method are sometimes used in the construction of interpolation. The nearest neighbor method searches for the nearest known point from the distance or the influence area of the Tyson polygon and assigns its elevation to the point to be interpolated, and the weight is polarized at this time, that is, the weight of the nearest known point is, and the weight of other points is. The arithmetic average method of interpolation uses the arithmetic average of the known points as the elevation of the points to be interpolated, and the weight of the known points is equalized to 1/n, n being the number of known points.

2. Multifaceted superposition interpolation

The principle followed by the multi-surface superposition interpolation method is that any regular or irregular continuous surface can be viewed as several simple surfaces to Shengga pass close, its original form from the U.S. Iowa proposed multi-surface function method specific interpolation model is to create a surface at each known point, and then in the vertical direction of each surface in a certain proportion of the benefit of the addition, the formation of a whole continuous surface, the surface strictly The surface is passed through each data point. The core of the multi-layer surface interpolation method is the design of simple surfaces, also known as kernel function congregation, kernel function is the main factor to determine the weight, in recent years, domestic and foreign scholars have designed a variety of kernel functions, such as cone, hyperboloid, cubic surface, Gaussian surface to Gaussian curve as the bus of the rotation surface. The multi-surface superposition interpolation includes a large part of interpolation models, such as the commonly used sample interpolation, radial basis function interpolation, kriging interpolation, multi-surface function method interpolation, etc. The only difference between these interpolation models is the physical exhaustion of the kernel function in the interpolation process. The advantage of multifaceted superposition interpolation is the flexibility and controllability of the kernel function design, the user can add the required information in the kernel function according to their specific requirements, for example, if you want to consider the ground slope information in the interpolation process, you can design the kernel function with slope characteristics.

6.2 Aerial Terrain Elevation Measurement Method

The aeronautical terrain elevation measurement method refers to the measurement method to obtain the terrain elevation underneath the current aeronautical vehicle in real-time in the aeronautical terrain elevation navigation and positioning system. Generally, the sensors used are an altitude sensor and a barometric altimeter, using the absolute altitude h_a measured by the barometric altimeter and the relative ground height h_c measured by the altitude sensor to subtract the actual elevation h_r of the terrain, and then correlate with the terrain elevation h_j calculated according to the position information of inertial guidance system and digital terrain elevation database (DEM) according to a certain algorithm, and the related extreme value point corresponding to the position corresponding to the relevant extreme value point is the flight position of the vehicle, and then corrects the navigation parameters of the main navigation system, as shown in Figs. 6.1 and 6.2 [5].

The altitude sensor is the key sensor for terrain elevation navigation and positioning, generally commonly used are FM continuous wave radio altimeter, pulse radar altimeter, and laser altimeter, no matter what system or type of altitude sensor is chosen, the basic principle is ranging, i.e., measuring the distance from the vehicle to directly below, the following is an introduction to the measurement principle and main measurement errors of three commonly used altitude sensors [6].

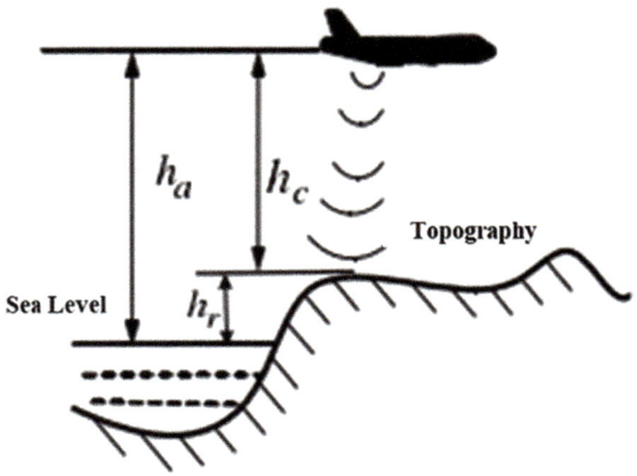

Fig. 6.1 Principle of aerial terrain elevation measurement

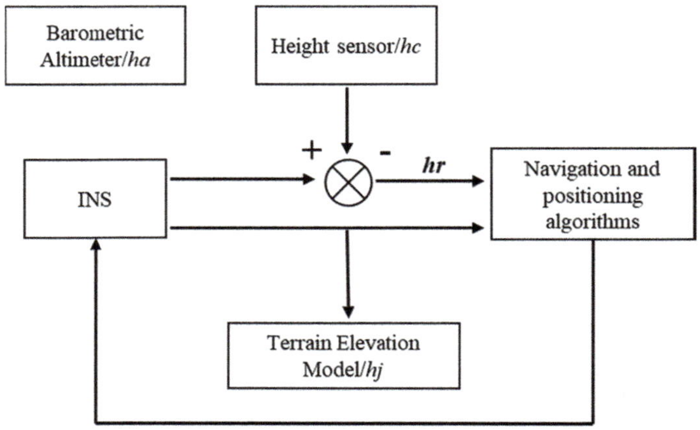

Fig. 6.2 Principle of aerial terrain elevation navigation and positioning

6.2.1 FM Continuous-Wave Radio Altimeter

The FM continuous wave radio altimeter uses the frequency difference between the transmit signal and the return signal to determine the height to the measured surface, i.e., the frequency coherence detection method is used. The single value of the measurement is guaranteed by the choice of modulation frequency, and the accuracy is determined by the size of the FM bandwidth. In the aspect of height measurement the FM continuous wave system implementation technology is relatively simple, with good high-low compatible height measurement capability and better measurement accuracy is widely used [7].

FM continuous wave altimeter can be divided into the open-loop system with differential beat counting and the closed-loop system with the constant differential beat. The open-loop system has the characteristics of simple implementation technology, low altimetric range, and accuracy, and is less used in modern altimetry technology, while the constant differential beat the closed-loop system is the most widely used, with relatively complex implementation technology and high altimetric range and accuracy. The following is a description of the working principle of altimetry with a constant differential beat FM continuous wave radio altimeter modulated by a sawtooth wave.

As in Fig. 6.3, a sawtooth wave signal with constant amplitude and controlled period is generated by the sawtooth wave generator, and this signal causes the altimeter transmitting and receiving unit to generate a continuous wave signal with center frequency F_0 and linear FM width ΔF. The signal is transmitted to the ground by the altimeter transmitting antenna, reflected by the ground, and received by the altimeter receiving antenna, from transmitting to receiving after a time τ, the received reflected signal is then Mixing with the transmitted signal to produce a difference frequency for f_b low-frequency signal, because the FM characteristics of linear, then there is:

$$\frac{f_b}{\tau} = \frac{\Delta F}{T} \tag{6.1}$$

Since, $\tau = \frac{2h}{c}$, the flight altitude is:

$$H = \frac{c \times f_b}{2 \times \Delta F} T \tag{6.2}$$

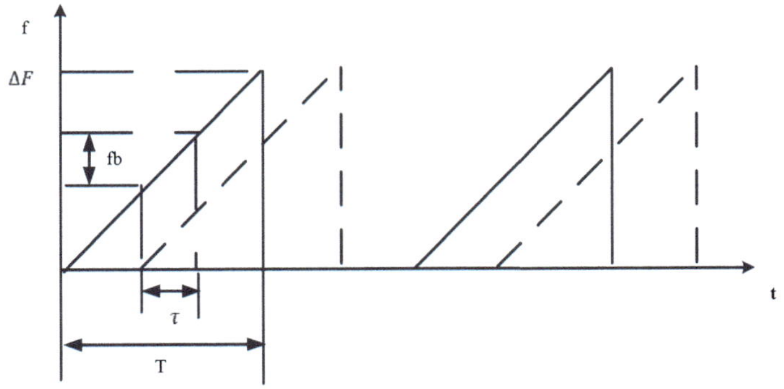

Fig. 6.3 Transmit and receive signal frequency profile

6.2.2 *Pulsed Radar Altimeter*

The pulsed radar altimeter irradiates an electromagnetic wave pulse signal onto the earth's surface and measures the delay time of the return signal relative to the transmitted pulse so that the altitude is determined, i.e., by the envelope coherence detection method. The singularity of the measurement is determined by the choice of the pulse repetition period and the accuracy of the measurement is determined by the pulse width. In the aspect of altimetry, the FM continuous wave system to achieve a simpler technology has a good high-low altitude compatible altimetry capability and better measurement accuracy is widely used. The pulse radar altimeter is limited by the pulse width, and its low altitude measurement accuracy is lower than that of the radio high altitude meter, but it is more often used in the field of high altitude or space measurement.

The radar altimeter transmits narrow pulses of high-frequency electromagnetic waves from the air to the ground, and the reflected echoes are formed after the electromagnetic waves are transmitted to the ground, and the relative delay time between the received echo pulses and the transmitted pulses is shown in Fig. 6.4 by using the isokinetic nature of electromagnetic wave propagation in free space ($c = 3 \times 10^8$ m/s) [7].

$$t_H = \frac{2H}{c} \tag{6.3}$$

Therefore, by accurately measuring the time delay we can obtain the height of the vehicle relative to the ground as:

$$H = \frac{1}{2} c \times t_H \tag{6.4}$$

The radar altimeter emits electromagnetic waves downward. According to the working principle of the radar altimeter, the radar altimeter receives the electromagnetic waves emitted back from the nearest point terrain of the vehicle instead of the reflected electromagnetic waves directly below, as shown in Fig. 6.5. Therefore it measures the height of the nearest point instead of the height directly below, when the attitude of the vehicle is more demanding, therefore, under the condition

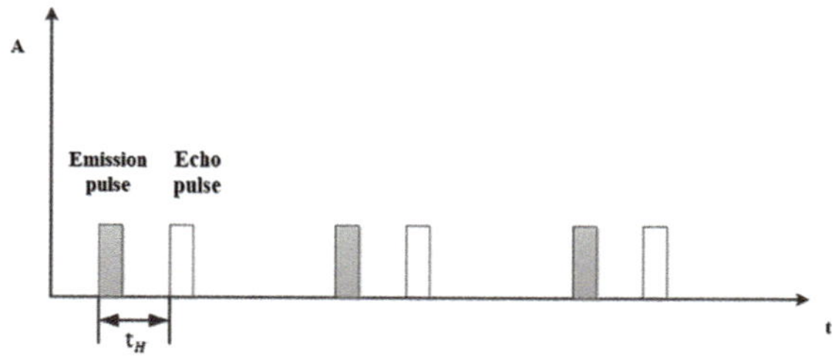

Fig. 6.4 Radar altimeter pulse schematic

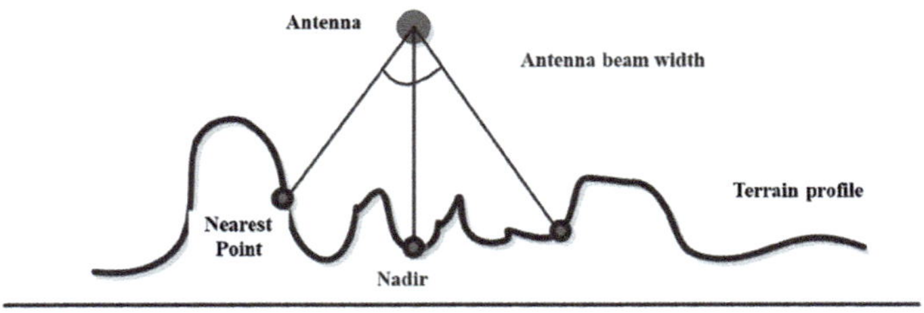

Fig. 6.5 Altimeter measurement error diagram

that the platform is stable and the pointing is unchanged, the narrow-beam radar has higher measurement accuracy compared with the wide-beam radar, but due to the unstable platform of the vehicle, under such conditions, the narrow-beam radar altimeter cannot guarantee that the beam can be accurately pointed to the sky bottom point, thus in the vehicle Generally use wide-beam radar altimeter, in the face of wide-beam radar altimeter, commonly used half-power method and average height and other methods to improve the accuracy of wide-beam radar measurement. Half-power method refers to the half-power point position of the radar-reflected echo front relative to the positive radar sky bottom point. The basic idea of the average height radar altimetry algorithm is that the echo power at a certain moment can be represented by the superposition of the echo power of all scattering elements within a certain spatial distance at that moment, and the average height can be obtained by calculating the average distance of the radar and these scattering elements [8].

6.2.3 Laser Altimeter

Radio and radar altimeter is more commonly used on aircraft to detect the height of the terrain device, are less affected by the weather and environment, work stability, measurement accuracy is relatively low, and are vulnerable to electromagnetic interference. The altimetric error increases with the height of the aircraft from the ground, when the height from the ground is below 100 m, the error is within 0.6 m; above 100 m height from the ground, the error is about 3% of the measured height. Moreover, the pulse radar altimeter has a wide transmitting beam, irradiating a large footprint to the ground, and the undulating terrain will produce serious clutter interference. To make up for the lack of radio altimeter and detection performance. Laser altimeter measurement accuracy is higher and resistant to electromagnetic interference, but it is affected by weather and battlefield environments, and the measurement accuracy will be greatly reduced in fog, haze, rain, snow, smoke, and other environments [9].

A laser altimeter uses a pulse method to measure height, and the altimeter emits one or one very narrow light pulse to the ground target and measures the time for the light pulse to reach the ground and return from the ground to the receiver, from which the height of the vehicle from the ground can be calculated. The laser altimeter is relatively accurate due to its narrow beam, which is less affected by ground undulations. The main factor affecting the accuracy of laser altimeter measurement is the flight attitude of the vehicle. Let the real height of the vehicle from the ground be H. Since the flight transient attitude of the vehicle is not parallel to the ground, it will cause a certain angle between the emission path of the laser altimeter and the vertical line between the vehicle and the ground, and the size is related to both the cross-roll angle φ and the pitch angle θ in the attitude angle of the vehicle, which can be expressed as $\gamma = f(\theta, \varphi)$. The presence of γ leads to the measurement distance $L = h/\cos\gamma$, resulting in a measurement error $\Delta L = L - h = h(\sec\gamma - 1)$. Assuming that the beam angle of the emitted electromagnetic wave is β, when $\gamma > \beta$, its measurement error ΔL is not negligible. For the laser altimeter, its beam angle is small, so the impact of the vehicle attitude motion on its measurement error is even greater. Therefore, when the laser altimeter is used to measure the terrain height, the attitude angle of the vehicle needs to be measured in real-time by the inertial guidance system on the vehicle, and the measurement results of the altimeter need to be corrected according to the observed attitude values.

6.3 Underwater Topographic Elevation Measurement Methods

The underwater topographic elevation measurement method refers to the measurement method to obtain the current water depth of the underwater vehicle (hereinafter referred to as "vehicle") in real-time in the underwater topographic elevation navigation and positioning system, as shown in Fig. 6.6. The sensor generally used is a bathymetry dive meter. Depth sounder has two functions: one is to measure the distance between the position of the vehicle and the water surface, generally using hydrostatic pressure sensors; the second is to measure the height of the vehicle from the bottom of the water, generally using acoustic measurement methods. The sum of the two indicates the depth value of the position where the carrier is located. Due to the presence of tides, the water depth value measured by the hydrostatic pressure sensor includes the tidal depth value above the average sea level and the depth value between the carrier and the sea level. It is customary to call the distance of the carrier from the seafloor also the depth value, and if not specifically stated, the depth measured by the bathymetry mentioned later refers to the depth of the carrier from the seafloor [6, 9–12].

Most of the testers use ultrasonic pulse as a detection signal, vertical to the water bottom to launch sound waves, through the water medium to the water bottom propagation. A part of the sound energy from the bottom of the water or other solid media has reflected the receiver, through the technical emission and reception time, you can get the current carrier location at the water depth value, to achieve the purpose of measuring the water depth.

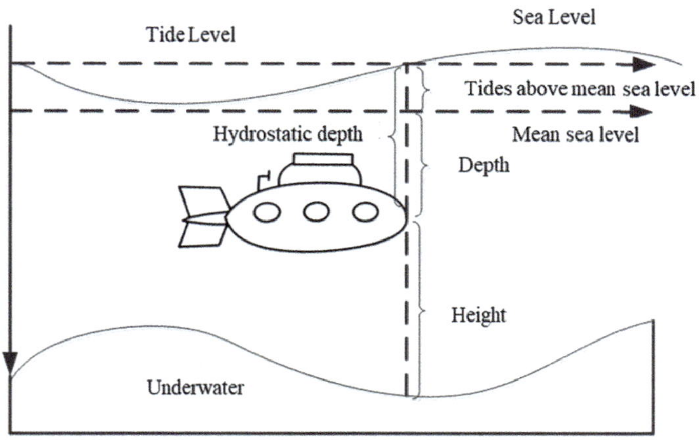

Fig. 6.6 Schematic diagram of underwater topographic elevation measurement method

Carrier distance from the bottom of the acoustic distance measurement method is divided into single-beam measurement and multi-beam measurement. Single-beam bathymetry process is to use the transducer's vertical downward emission of a short pulse acoustic wave, when the pulse acoustic wave meets the seafloor reflection, the reflected echo back to the sonar and received by the transducer. Its water depth value is determined by the two-way travel time of the acoustic wave between the seafloor and the average speed of sound in the water medium.

$$D_v = \frac{1}{2} v_m t \tag{6.5}$$

where D_v is the distance between the transducer and the seafloor, v_m is the average sound speed of the water body, and t is the two-way travel time of the acoustic wave. The above water depth value D_v plus the transducer draft depth D_d and the tidal correction value ΔD_t gives the actual water depth value D_s.

$$D_s = D_v + D_D + \Delta D_t \tag{6.6}$$

The single-beam bathymetry process takes a single-point continuous measurement method, and the bathymetry data are densely distributed along the course. Multi-beam bathymetry is a kind of bathymetry equipment that can realize large range and high precision bathymetry, and it can obtain the bathymetry value of multiple measurement points perpendicular to the course direction at the same time. Multi-beam bathymetry based on the coherence principle can obtain hundreds or even thousands of bathymetry points at the same time, and it detects the incident angle of the beam by comparing the phase difference between two given receiving units of the transducer, and the phase difference increases with the increase of the incident angle, so that the detection accuracy is also greatly improved. The schematic diagram of

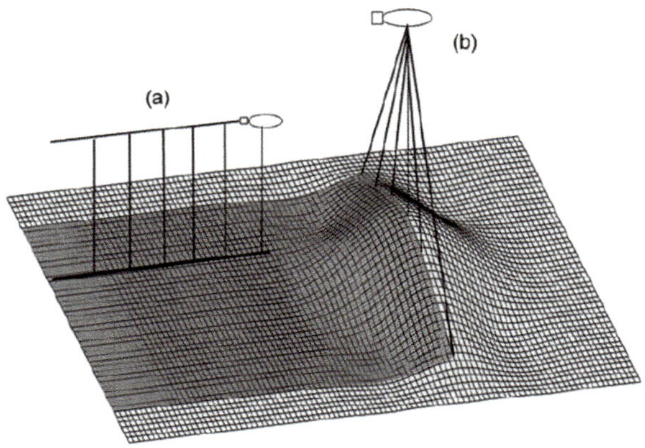

Fig. 6.7 Schematic diagram of single-beam and multi-beam seafloor topography survey

single-beam and multi-beam seafloor topography measurement is shown in Fig. 6.7. Multi-beam measurement bathymetry principle and data processing methods will be introduced in detail in the next section. Due to more topographic elevation data obtained, the performance of underwater topographic elevation navigation and positioning can be further improved by two-dimensional topographic elevation navigation and positioning.

The instruments that can provide bathymetric values also include a Doppler range finder, interferometric sonar, laser bathymetry, etc. This book focuses on single-beam as well as multi-beam acoustic bathymetry as examples.

6.3.1 Underwater Multi-Beam Bathymetry Principle

In Fig. 6.8, multi-beam means is used to transmit and receive multiple beams at a time from a wide angle at each measurement point on the carrier route, to measure the depth (height) data of the terrain area near the carrier route, and to obtain the position of the carrier in the map after two-dimensional matching with the map. Multi-beam bathymetry system is a complex system composed of multiple sensors, which is developed based on a single-beam echosounder. It can give the bathymetric value of dozens or even hundreds of measured points on the seabed in the vertical plane perpendicular to the heading at one time, or the full-coverage bathymetric strip of a certain width, with a wider sweep and higher density of measured points; on the other hand, the adoption of the narrower beam, advanced detection technology, and precise acoustic correction method also ensures the accuracy of homing calculation of hull coordinates of measured points, thus multibeam bathymetry has full coverage, high accuracy, The characteristics of multibeam bathymetry are full coverage, high accuracy, high density and high efficiency [13–15].

There are 11 common indicators to describe the working performance of multi-beam system, which are as follows: (1) beam incidence angle, the angle between

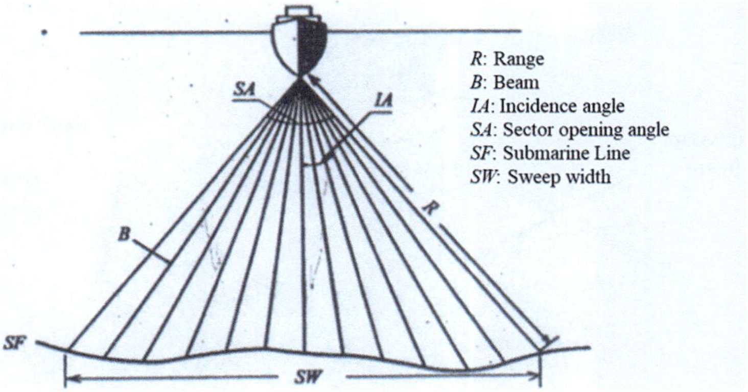

Fig. 6.8 Schematic diagram of the inter-beam geometric relationship

the symmetry axis of beam stereo angle and the vertical line; (2) beam angle or beamwidth, the stereo angle of a beam in space; (3) beam (3) beam grazing angle, the angle between the symmetry axis of the beam stereo angle and its orthographic projection on the projection interface; (4) footprint, the intersection tangent between the beam stereo angle and the seafloor surface; (5) inter-beam angle or beam spacing, the angle between two adjacent beam stereo angles and the symmetry axis; (6) range, the range of the beam (6) range, the actual travel distance between the beam from the transducer to the projection point on the seabed; (7) sector opening angle or fan width, a complete scan formed by the outermost edge of the beam on both sides of the sector opening angle; (8) sector scan, the completion of a complete sector scan of the pulse transmission and reception process; (9) update (9) update rate, the number of times to complete the sector scan per unit of time; (10) bathymetric cross-section, a series of beam measurement points formed by a sector scan under the vertical heading of the transducer composed of the bathymetric profile of the seafloor; (11) coverage rate, the ratio of the width of the multibeam bathymetric profile to the water depth.

The transmitting base array and receiving a base array of the multibeam strip bathymetry system adopt the "T" shape structure which is perpendicular to each other. The transmitting base array is installed parallel to the bow and stern line of the ship, while the receiving base array is installed perpendicular to the bow and stern line of the ship. The transmitting and receiving signals form a small directional angle in one direction and a wide directional angle in the other direction, which are perpendicular to each other.

As shown in Fig. 6.9, the basic working principle of a multibeam sonar strip bathymetry system is: in the direction perpendicular to the mother ship's heading, several beams are formed in several preformed beam angle directions by multibeam formation technology, and then the depth of the corresponding point with each beam is measured separately. Several measurement periods are combined to form a strip bathymetric map with the mother ship's track as the centerline, so it is called a strip.

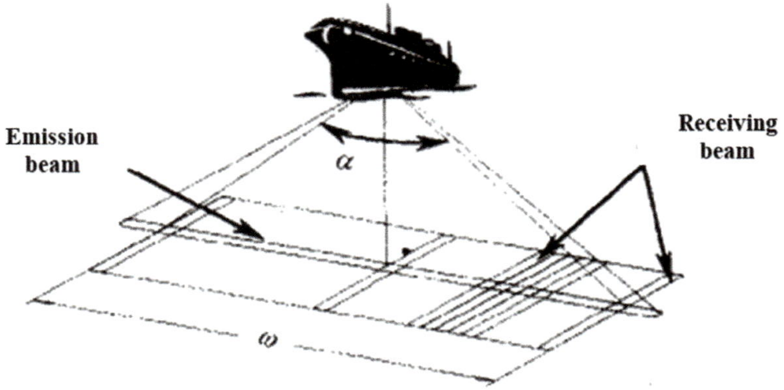

Fig. 6.9 Schematic diagram of the working principle of multi-beam strip bathymetry system

The most important index of a strip sounder is the swept width W. As can be seen in Fig. 6.4, W depends on the actual depth $H$of the most marginal beam and the angle α at which the most marginal beam is tilted:

$$W = 2Htg\left(\frac{\alpha}{2}\right) \tag{6.7}$$

6.3.2 *Position Measurement Principles and Methods for Multi-Beam Systems*

The multibeam sonar strip bathymetry system based on the principle of seafloor acoustic scattering is an active multibeam bathymetric sonar system, as shown in Fig. 6.10 [16].

Figure 6.10 shows a kind of submarine acoustic scattering bathymetry model, assuming that the transmitting beam is incident to the seafloor at an angle of concerning the earth's gravity line, the incident acoustic signal will be scattered along all directions, among which there must be some acoustic waves reflected along the incident direction, if we can receive and measure the time delay Δt of this part of the backward reflected echo signal, that is, we can calculate the right angle ΔOPR in Fig. 6.10. The length of the beveled edge L (sound range).

$$L = c \times \frac{\Delta t}{2} \tag{6.8}$$

In the above equation, c denotes the speed of sound, Δt and denotes the time delay from the time of signal emission to the time of arrival of the echo from the active sonar. Thus, from the geometric relationship in Fig. 6.10, we can calculate:

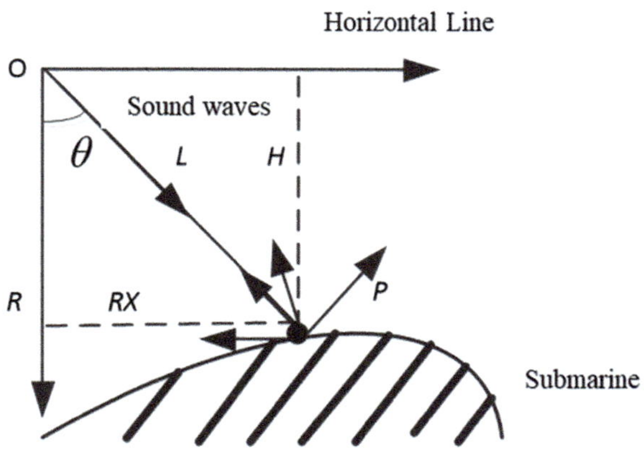

Fig. 6.10 Seafloor scattering bathymetry model

$$\begin{cases} H = L \times \cos\theta \\ RX = L \times \sin\theta \end{cases} \quad (6.9)$$

where H denotes the depth at the measured point P and RX denotes the horizontal relative position of the measured point P. The top view of Fig. 6.10 is plotted as shown in Fig. 6.11, and the strip scanning direction is perpendicular to the heading direction. Let the ship's heading angle be φ, and define the sign of RX as negative (−) when it is to the left of the heading direction and positive (+) when it is to the right of the sailing direction, then the coordinates of the measured point $P(x, y)$ are:

$$\begin{cases} x = RX \times \cos\varphi + x_0 = L\sin\theta\cos\varphi + x_0 \\ y = -RX \times \sin\varphi + y_0 = -L\sin\theta\sin\varphi + y_0 \end{cases} \quad (6.10)$$

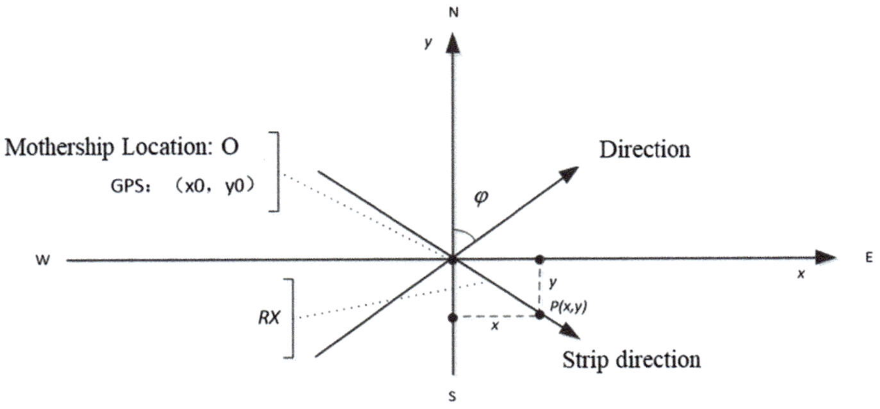

Fig. 6.11 Schematic diagram of the positioning principle of the strip measurement system

Where are the absolute Earth INS coordinates of the mother ship's position, which can be obtained by the inertial guidance system mounted on the ship, and the heading angle φ can be measured by the auxiliary equipment rosette system. From Eqs. (6.8) to (6.10), we can get

$$\begin{cases} x_\theta = c \times \frac{t}{2} \times \sin\theta\cos\varphi + x_0 + \tilde{x}(P_\varepsilon, \Phi_\varepsilon, C_\varepsilon, h, p, r) \\ y_\theta = -c \times \frac{t}{2} \times \sin\theta\cos\varphi + y_0 + \tilde{y}(P_\varepsilon, \Phi_\varepsilon, C_\varepsilon, h, p, r) \\ H_\theta = c \times \frac{t}{2} \times \cos\theta + \tilde{h}(P_\varepsilon, \Phi_\varepsilon, C_\varepsilon, h, p, r) \end{cases} \tag{6.11}$$

The subscript θ indicates that each set $(x_\theta, y_\theta, H_\theta)$ of values corresponds to a specific beam of the multibeam system. This is the principle formula of depth and position measurement of a multi-beam strip bathymetry system. Because of the sound velocity bending of the sound line C_ε, the INS positioning error of the mother ship's position P_ε, the measurement error of the mother ship's heading Φ_ε, the attitude of the mother ship, such as transverse rocking r, longitudinal rocking p and lifting, and sinking h, etc. will affect the measurement results, as a practical application of the strip measurement system must also consider the correct and effective correction and compensation of these parameters. The above equation $\widetilde{x}$, $\widetilde{y}$ and $\widetilde{h}$ is the comprehensive correction function of these parameters, and its complex functional form is related to many specific structures and installation parameters of the measurement system.

6.3.3 Multi-Line Correction for Ocean Bathymetry

Due to the influence of instrument error, external environmental influence and human error and other factors, the measurement of bathymetric diving instruments inevitably has measurement errors. The measurement error is mainly caused by the sound velocity error caused by the influence of external environmental factors, thus translating into the estimation of the accuracy of the sound velocity. Generally speaking, to make the accuracy of depth reach 1% (national marine measurement specification), it is required that the error due to the inaccuracy of sound velocity does not exceed 0.25%, so that the accuracy of the measured average sound v_m velocity must be ensured at about 4 m/s (1500 × 0.25%). To meet this standard, related parameters such as the average temperature of seawater measurement need to be accurate to 1%, and the salinity measurement needs to be accurate to 1‰. In practical engineering measurements, the determination of sound velocity in seawater can also be calculated by adopting some empirical formulas [17].

In addition to the sound velocity error, compared with the air topographic elevation measurement, the ocean bathymetry has many interference factors, which makes the bathymetry in a very complex environment, all these factors have an impact on the bathymetry, thus causing various effects, these effects include: beam angle effect, geopotential effect, wind and wave effect, delay effect, offset effect, water quality effect, etc.

Beam angle effect: Due to physical reasons, the beam angle of the sounder always exists at a certain width, usually the transducer launches a conical beam to the seafloor, and the beam has reflected the transducer after reaching the seafloor, and the water depth value can be determined according to the sending and receiving time. However, the measured bathymetry value is generally not the vertical distance from the transducer to the seafloor, but the shortest distance from the seafloor to the transducer in the area irradiated by the bathymetric signal. This situation causes the deviation and distortion of seafloor topography measurement, which is called the beam angle effect of seafloor topography measurement.

Topographic effect: It is due to the different relative positions between the undulation of the terrain and the direction of the survey line, which has an impact on the bathymetry.

Wind and wave effect: the carrier is in a bumpy and undulating marine environment, the impact of waves, and the blowing of the sea wind will cause the change of carrier attitude, thus affecting the attitude of the transducer installed in the bottom of the ship, causing the change of bathymetry, called the wind and wave effect.

Delay effect and offset effect: As positioning and bathymetry still belong to two independent technical systems, the non-synchronization of the two systems in time during bathymetry causes the bathymetric image offset, which is called delay effect; the offset of the two systems in space causes the bathymetric image offset, which is called the offset effect of positioning center and bathymetric center.

Water quality effect: In ocean bathymetry, sound waves are propagated in seawater, and the sound speed changes with the salinity, temperature, and depth of seawater as well as other factors, resulting in sound waves not propagating along a straight line. Also cause changes in bathymetry, known as the water quality effect.

Many factors affecting ocean bathymetry exist, and the first three effects are dealt with by transforming the coordinate system to include one case by transforming the line of sight coordinate system and the different relative positions of the terrain to the line of sight coordinate system in a unified way. The multi-survey line correction technique is proposed, and the effects of beam angle and terrain undulation as well as carrier attitude change on bathymetry are corrected comprehensively by the above means to ensure measurement accuracy.

Analysis of beam angle effect and geopotential effect

(1) Topographic coordinate system: The coordinate system is established to describe the tilt direction of the seafloor topography, as shown in Fig. 6.12. The x'-axis points horizontally to the direction of the largest upward steepness of the seafloor slope, the y'-axis points vertically to the seafloor, the z'-axis is horizontal, and P is the origin, forming a right-handed right-angle coordinate system, which is a dynamic coordinate system with the origin P moving along the line of measurement and α is the tilt angle of the seafloor (Fig. 6.13).
(2) The line of sight coordinate system: It is a coordinate system established to describe the direction of the line of sight advance. x-axis coincides with the

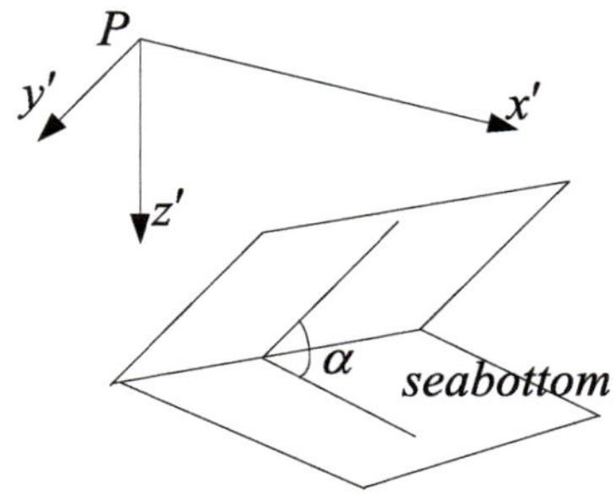

Fig. 6.12 Topographic coordinate system

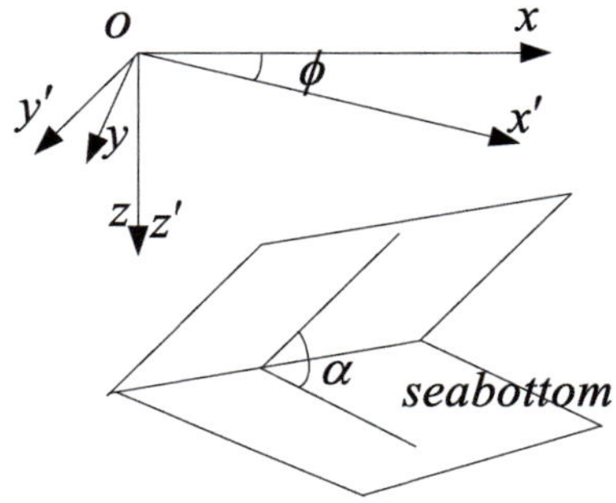

Fig. 6.13 Relationship between the topographic coordinate system and the survey line coordinate system

line of sight and points to the direction of the line of sight advance, the z-axis points vertically to the seafloor, the y-axis is horizontal, and the origin is o, forming a right-handed right-angle coordinate system, which is a static coordinate system, see Fig. 6.12, due to the different inclination of the seafloor and the line of sight direction to form a different angle of rotation ϕ ($0° < \phi < 360°$). There is a rotation between the two coordinate systems, then the transformation relationship between the two coordinate systems is:

$$\begin{bmatrix} x \\ y \\ z \end{bmatrix} = \begin{bmatrix} \cos\phi & -\sin\phi & 0 \\ \sin\phi & \cos\phi & 0 \\ 0 & 0 & 1 \end{bmatrix} \cdot \begin{bmatrix} x' \\ y' \\ z' \end{bmatrix} \tag{6.12}$$

As shown in Fig. 6.14, the geopotential coordinate system is a dynamic coordinate system, the origin P is moving on the measuring line, the inclined seafloor is in the geopotential coordinate system, the transducer is located at point P, θ is the half-beam angle of the sounder, α is the inclination angle of the seafloor the origin O is the starting point of the measuring line.

The measured water depth is $r(x, y) = PS$, the real water depth is $z(x, y) = PS_1$, the translation on the x'-axis $\Delta x'$. When $|\alpha| > \theta$, the relationship between the measured water depth $r(x, 0)$ and the true water depth $z(x, 0)$ is:

$$z'(0, 0) = \sec\alpha \cdot \cos(-\alpha - \theta) \cdot r(x, 0) \tag{6.13}$$

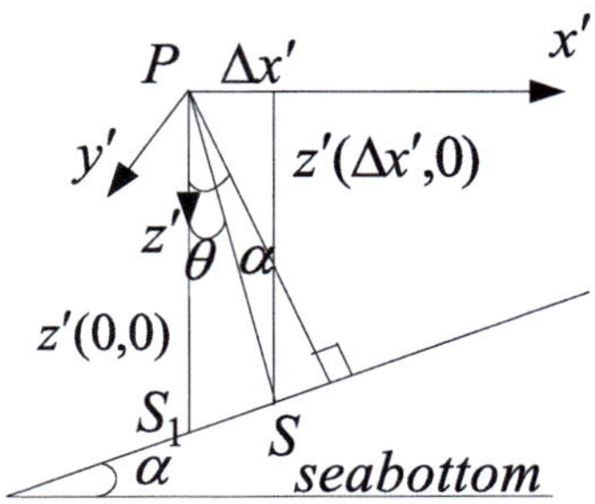

Fig. 6.14 Translational effect diagram of linear transition section with large tilt angle

$$r(x,0) = \cos\alpha \cdot \sec(-\alpha - \theta) \cdot z'(0,0) \tag{6.14}$$

the error between the two is:

$$\Delta z'(0,0) = z'(0,0) - r(x,0) = [\sec\alpha \cdot \cos(-\alpha - \theta) - 1] \cdot r(x,0) \tag{6.15}$$

and the time shift effect produced in the middle is:

$$\Delta x' = \sin\theta \cdot r(x,0) \tag{6.16}$$

$$z'(\Delta x', 0) = \cos\theta \cdot r(x,0) \tag{6.17}$$

$$r(x,0) = \sec\theta \cdot z'(\Delta x', 0) \tag{6.18}$$

Obviously, $z(x,0) = z'(0,0)$, where $(x, 0)$ is the coordinate of point P in the line of sight coordinate system, due to the influence of terrain in the terrain coordinate system on the x'-axis to produce a translation $\Delta x'$, in the line of sight coordinate system on the x-axis to produce translation effect:

$$\Delta x = \Delta x' \cdot \cos\phi \tag{6.19}$$

On the y-axis to produce translation effect:

$$\Delta y = \Delta x' \cdot \sin\phi \tag{6.20}$$

When $|\alpha| < \theta$, as shown in Fig. 6.15, the same can be obtained from the formula:

$$z'(0,0) = \sec(-\alpha) \cdot r(x,0) \tag{6.21}$$

$$r(x,0) = \cos(-\alpha) \cdot z'(0,0) \tag{6.22}$$

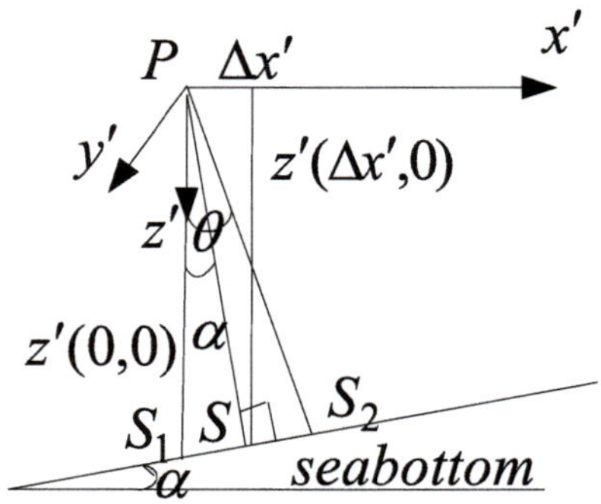

Fig. 6.15 Translational effect diagram of linear transition section with small tilt angle

$$\Delta z'(0,0) = z'(0,0) - r(x,0) = [\sec(-\alpha) - 1] \cdot r(x,0) \tag{6.23}$$

$$\Delta x' = \sin(-\alpha) \cdot r(x,0) \tag{6.24}$$

$$z'(\Delta x', 0) = \cos(-\alpha) \cdot r(x,0) \tag{6.25}$$

$$r(x,0) = \sec(-\alpha) \cdot z'(\Delta x', 0) \tag{6.26}$$

The same will also produce a translation effect $\Delta x = \Delta x' \cdot \cos\varphi$, $\Delta y = \Delta x' \cdot \sin\varphi$.

See Fig. 6.16. At $(x_0, 0)$, the vertical distance from the transducer to the top of the convex seafloor s is $z(x_0, 0)$.

Referring to Fig. 6.16 $(x_0, 0)$, the vertical distance from the transducer to the top of the convex seafloor s is $z(x_0, 0)$. At that time, no deformation occurs. Then, when the ship transducer deviates from the point $(x_0, 0)$ and lies on a segment of the chord where the line of measurement intersects the circular region V, a deformation will occur. The recorded image is a hyperbolic shape with the equation:

$$r^2(x,0) = z^2(x_0,0) + (x - x_0)^2 + y^2 \tag{6.27}$$

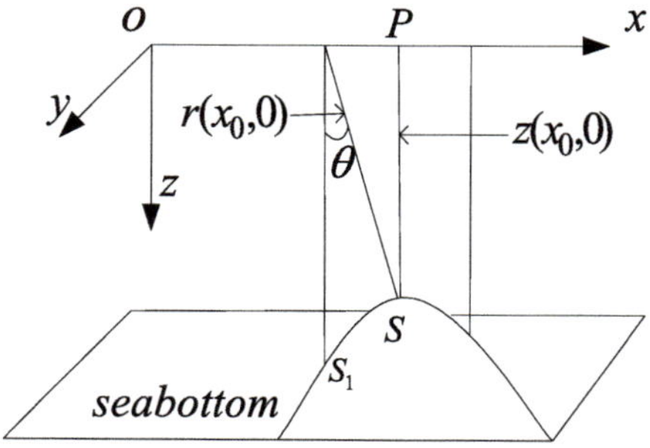

Fig. 6.16 Graph of the hyperbolic hypothesis-increasing effect of a convex point

where V is a circular region with $(x_0, 0)$ as the center and $\Delta x_0 = r(x_0, 0)tg\theta$ as the radius.

This means that the $r(x, 0)$ recorded by the ship in the circular region comes from the reflected signal at the tip S only. In this paper, we call this effect of beam angle the hyperbolic falsification effect This hyperboloid falsification effect does not lose the seafloor shape information, but adds pseudo seafloor shape information to the extent that the survey line lies within a segment of the chord in the circular region.

In Fig. 6.16, when the transducer is at point P on the measurement line, the real water depth is:

$$z(x, y) = PS_1 \tag{6.28}$$

and the recorded water depth is:

$$r(x, y) = \min(PS_2, PS, PS_1, PS_3) = PS \tag{6.29}$$

where S is the point on the edge of the concave bottom surface, S_1 is the point where the center of the beam is projected vertically downward onto the bottom surface, and S_2 、 S_3 are the contacts between the edges of the beam and the bottom surface on both sides, respectively. As long as the transducer is located directly above the concave bottom surface, the recorded water depth is not the real water depth. In this case, the real seafloor information within a section of chord AB where the measurement line intersects the concave seafloor edge will be lost in the bathymetric image $r(x, y)$, and this effect is called the loss effect. The range is the whole concave area of the seafloor, and the true water depth within this range will not be known from the recorded bathymetric image $r(x, y)$. The width of this range is AB (Fig. 6.17).

2. Analysis of wind and wave effects

As the carrier is in the seawater, it is always subject to the blowing of the sea wind and the impact of the waves, causing changes in the attitude of the carrier, and then causing changes in the attitude of the bathymetric system carried by the carrier,

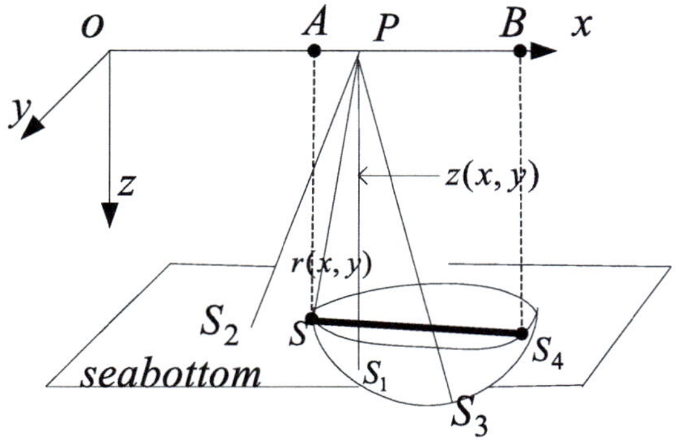

Fig. 6.17 Depth loss effect plot for concave points

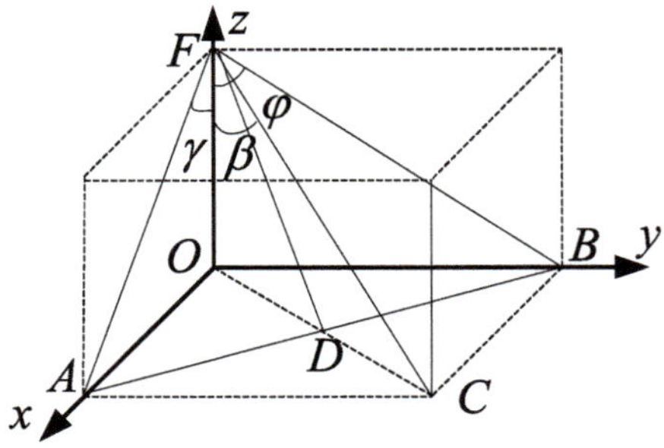

Fig. 6.18 Pitch angle, transverse winding angle, and beam declination relationship

which is called the wind and wave effect of bathymetry. Take a single-beam echo sounder as an example, and set its yaw angle ψ, pitch angle φ, and cross-winding angle γ. The yaw angle only causes the rotation of the transducer, which does not affect the bathymetry, while the pitch and cross-winding of the carrier will cause the change of the carrier beam projection to the position of the seafloor measurement point relative to the seafloor, which in turn affects the change of the bathymetry position, i.e., produces the translation effect. The changes caused by this pitching and cross-winding are analyzed below.

Disregarding the yaw, only the effect of carrier pitch and transverse winding changes on bathymetry is analyzed, as shown in Fig. 6.14.

In Fig. 6.18, let the transducer be at the point F, $\angle OFB = \phi$, $\angle OFA = \gamma$, $\angle OFC = \beta$, β is the angle of deviation of the transducer beam from the vertical direction due to the change in the pitch angle φ, the angle of transverse winding γ, that is, the angle of incidence, in the above rectangular block, it is easy to deduce the relationship between β and γ, φ.

According to $OA = OF \cdot tg\gamma$, $OB = OF \cdot tg\,\varphi$, then we have:

$$OC^2 = OA^2 + OB^2 = OF^2 \cdot (tg^2\gamma + tg^2\varphi) \tag{6.30}$$

$$OD = \frac{1}{2}OC = \frac{1}{2}OF \cdot \sqrt{tg^2\gamma + tg^2\varphi} \tag{6.31}$$

and thus,

$$tg\angle OFD = \frac{OD}{OF} = \frac{1}{2} \cdot \sqrt{tg^2\gamma + tg^2\varphi} \tag{6.32}$$

then

$$\angle OFD = arctg\left(\frac{1}{2} \cdot \sqrt{tg^2\gamma + tg^2\varphi}\right) \tag{6.33}$$

From the point of view of vector rotation, the new vector produced by two consecutive rotations of a vector is the sum of the vectors produced by the separate single rotations.

Also available,

$$\begin{aligned} FC^2 &= OF^2 + OC^2 = OF^2 + OA^2 + OB^2 \\ &= OF^2 \cdot (1 + tg^2\gamma + tg^2\varphi) \end{aligned} \quad (6.34)$$

$$\cos \angle OFC = \frac{OF}{FC} = \frac{1}{\sqrt{1 + tg^2\gamma + tg^2\varphi}} \quad (6.35)$$

There

$$\angle OFC = \arccos \frac{1}{\sqrt{1 + tg^2\gamma + tg^2\varphi}} \quad (6.36)$$

Or

$$tg\angle OFC = \frac{OC}{OF} = \sqrt{tg^2\gamma + tg^2\varphi} \quad (6.37)$$

$$\beta = \angle OFC = arctg(\sqrt{tg^2\gamma + tg^2\varphi}) \quad (6.38)$$

As seen from Eq. 3.32, the beam deflection angle increases with the increase of carrier pitch angle and transverse winding angle, and the two have the same effect on the deflection angle. Therefore, in the case of carrier attitude change, bathymetric deformation will occur even for flat seafloor.

Comparing the topographic effect of bathymetry, it is then further found that the change in the relative topographic position of the beam projected onto the seafloor measurement point caused by the carrier attitude change is essentially the same as the change in the location of the measurement point caused by the topographic change, both of which cause the change in the location of the measurement point relative to the seafloor topography.

3. Line correction for multi-beam measurement systems

As mentioned earlier, since the topography of the seafloor measured by each beam angle is undulating, each measurement line is not a straight line, but has concave and convex variations. The lateral lines will protrude outward where the topography is relatively depressed, and the lateral lines will depress inward where the topography is relatively prominent on the contrary. The multi-beam system measures the seafloor topography, and what we get is only a regular matrix of the corresponding beam angle depth values, which does not represent the real topography of the measurement.

Let the multibeam system consist of m beams and the scanning length is n, then the multibeam system scans the seafloor topography data as an m*n matrix, whose grid spacing is very different from the original seafloor DEM after interpolation, and the subsequent 2D topographic elevation navigation and positioning calculation also needs the multibeam scanned seafloor DEM data or images with the same sampling interval as the original seafloor reference DEM.

Based on the above two points, the line correction of the multibeam system must be carried out and resampled into the same sampling interval as the datum seafloor DEM data.

The algorithm for line correction and resampling imaging of a multibeam system is as follows.

(1) The horizontal distance of each beam footprint from the center point is found by Equation $Y_i = r_i(x_n, y_m) \cdot \sin(i - \frac{1}{2})\theta$.
(2) Calculate the actual grid spacing grid of the base map of the baseline DEM data after interpolation, and find the grid number $M_i = Y_i/grid$ of the measured horizontal distance of each beam from the center.
(3) The maximum grid number of the measured footprint of the first beam $R_1max = max(M_1)$ and the maximum grid number of the measured footprint of the m-*th* beam are obtained, to determine the size of the matrix used to store the multi-lateral line correction and resampled multibeam scanned images as $(R_1max + R_mmax) * n$.
(4) The measured values of m/2 beam footprints on each port and starboard side of the hull are stored in the newly generated matrix according to the grid number M_1 of the calculated horizontal distance from the center, and the remaining positions of the matrix are assigned with 0.
(5) The one-dimensional line difference $P(i+k,j) = (Z(i+1,j) - Z(i,j))/(M(i+1,j) - M(i,j)) * k + M(i,j)$ is performed between two different beam footprints at the same moment to supplement the elevation value of the grid between the two beam footprints, and finally the multi-metric corrected and resampled into a grid with the original seabed reference, the final image or scanned seafloor DEM data from the multi-beam system with the same sampling interval as the original seafloor reference DEM.

6.4 Topographic Elevation Information Characteristics

Terrain feature analysis is the basis for establishing a causal relationship between terrain features and matching accuracy by studying the trend of terrain feature parameters closely related to terrain-aided navigation to judge the characteristics of terrain from a quantitative perspective, and is the basis for track planning, terrain matchable area selection, and terrain navigation algorithm feasibility study in terrain-aided navigation systems.

Theoretically, once the elevation value of the underwater terrain is given, information about the terrain features is available. To judge the navigability of a specific sea area is to analyze the terrain parameters of the sea area and judge the uniqueness of the terrain based on the value of the terrain parameters. The stronger the uniqueness, the easier it is to find out the differences between the terrain of the area and other areas, which is helpful for terrain matching.

In the existing related literature, there are dozens of parameters for describing terrain features, and the methods and conclusions for selecting the parameters of underwater terrain features are not entirely consistent, and the perspective of description often starts from the terrain undulation degree, terrain fragmentation degree, terrain self-similarity, and based on general DEM model analysis, the fractal Brownian motion model FBM, trend surface regression model, To model and quantify the underwater terrain from various aspects, theoretical and mathematical physical models such as information entropy are introduced.

In this chapter, we first analyze the statistical parameters of many terrain features, discuss their meanings and interconnections in the expression of terrain features, summarize the laws of underwater terrain features, and provide the theoretical basis for extracting the set of terrain features suitable for underwater terrain-aided navigation. Then, the feature parameters are calculated for the existing 7 underwater terrains, and based on the calculation results, a set of feature parameters suitable for describing the terrain is proposed to create conditions for judging the navigability of the terrain. However, the conclusions obtained here are for the seven terrains analyzed, and the number of samples analyzed is relatively small. In practical applications, a more adequate experimental study can be conducted with the method proposed here to propose a terrain feature parameter set that better meets the application requirements. Finally, the simulation experiments and the results are analyzed to achieve the following two purposes: (1) comparing the matching effects in different terrains to judge the nature of terrain features from the perspective of the matching algorithm; (2) correlating the terrain feature parameter values and the matching error values to determine which terrain feature parameters can accurately describe the terrain features [8].

6.4.1 Topographic Feature Parameter Algorithm and Its Significance

First, we start from the relevant parameters describing the terrain information, through which we can understand the terrain structure and describe the terrain features. To establish the Grid-DEM model, take the terrain size as $M \times N$ (elevation matrix of M rows and N columns, with rows along the latitude direction and columns along the longitude direction), and the elevation value at (i, j) (the i-th row and j-th column position elements, i and j are counted from 0) as $H(i, j)$, which mainly includes the following information features.

1. Topographical elevation parameters

(1) Extreme value

The maximum and minimum values of the elevation of the selected sea area, denoted as $H_{\max}$ and $H_{\min}$, respectively, reflect the range of regional topographic relief. At the same time, since the research object is oriented to underwater navigation, the actual diving depth of the underwater carrier should be taken into account.

(2) Maximum elevation difference

It is the difference between the maximum and minimum values of elevation in the selected sea area, reflecting the maximum value of regional topographic relief. In statistics, it is also called the full distance of the sample data. The calculation formula is

$$H = H_{\max} - H_{\min} \tag{6.39}$$

(3) Mean elevation

$$\overline{H} = \frac{1}{MN}\sum_{i=0}^{M-1}\sum_{j=0}^{N-1} H(i,j) \tag{6.40}$$

2. Standard deviation

The standard deviation of elevation reflects the dispersion of water depth values and the undulation of submerged topography in the whole sub-region. The larger the standard deviation, the greater the degree of topographic relief; the smaller the standard deviation, the smaller the degree of topographic relief.

$$\sigma_T = \sqrt{\frac{1}{MN-1}\sum_{i=0}^{M-1}\sum_{j=0}^{N-1}\left(H(i,j) - \overline{H}\right)^2} \tag{6.41}$$

3. Coefficient of skewness

The coefficient of skewness reflects the degree of asymmetry or skewness of the distribution of topographic elements.

$$C_s = \frac{1}{MN}\sum_{i=0}^{M-1}\sum_{j=0}^{N-1}\frac{\left(H(i,j) - \overline{H}\right)^3}{\sigma_T^3} \tag{6.42}$$

In describing the frequency distribution, skewness refers not only to the magnitude of skewness but also to the direction of skewness. When the distribution is symmetric with $C_s = 0$, the mean, median and plurality are the same; when the distribution is positively skewed with $C_s > 0$, the plurality is smaller than the median; when the distribution is negatively skewed with $C_s < 0$, the median is smaller than the plurality.

4. Coefficient of kurtosis

The coefficient of kurtosis reflects the degree of concentration of the topographic data, if the kurtosis coefficient is large, some data are more concentrated, on the contrary, the data distribution is more uniform. From the perspective of mathematical statistics, the coefficient of kurtosis refers to the shape of the peak of the distribution curve.

$$C_e = \frac{1}{MN} \sum_{i=0}^{M-1} \sum_{j=0}^{N-1} \frac{\left(H(i,j) - \overline{H}\right)^4}{\sigma_T^4} \tag{6.43}$$

When normally distributed $C_e = 3$, $C_e > 3$ for high narrow peaks and low broad peaks.

5. The degree of slope

The degree of slope is a vector that represents the degree of slope of the ground. The mode of the slope vector at a point on the ground is equal to the magnitude of the angle between the tangent plane of the surface function at that point and the horizontal plane (or the tangent of the angle). It can be shown that the slope of any slope is equal to the vector sum of its slope components in two mutually perpendicular directions on that slope.

At present, the methods for calculating slope based on grid DEM broadly include the following categories: four-block method, space vector analysis method, fitted plane method, fitted surface method and direct solution method. Among them, the first three methods are designed for solving the average slope of the ground; the last two methods are designed for solving the maximum slope of the ground. In this study, considering the computational efficiency and the extraction of regional topographic information, the fitted plane method is chosen to calculate the slope of the selected sea topography.

First, the four vertices of each rectangular grid are used as a calculation unit, so that the fitted plane can be formed by (i, j), $(i + 1, j)$, $(i, j + 1)$ and $(i + 1, j + 1)$ as:

$$Z = a_0 + a_1 X + a_2 Y \tag{6.44}$$

The elevation values of the four vertices are substituted into the above equation to solve for:

$$\begin{cases} a_0 = \frac{1}{4}(3H(i,j) + H(i,j+1) + H(i+1,j) - H(i+1,j+1)) \\ a_1 = \frac{1}{2\Delta X}(H(i+1,j) + H(i+1,j+1) - H(i,j) - H(i,j+1)) \\ a_2 = \frac{1}{2\Delta Y}(H(i,j+1) + H(i+1,j+1) - H(i,j) - H(i+1,j)) \end{cases} \tag{6.45}$$

By taking the partial derivatives of both sides of Eq. (6.45) concerning X, we can get $\frac{\partial Z}{\partial X} = a_1$, which is the component of the slope of the point in the X direction.

Similarly, the component of the slope in the Y-direction $\frac{\partial Z}{\partial Y} = a_2$ can be obtained, and the average slope value in this rectangular area can be obtained as:

$$\alpha = arctg\sqrt{a_1^2 + a_2^2} \tag{6.46}$$

The slope direction (aspect) is the direction of the slope vector, which is equal to the direction of the projection of the vector along the maximum inclination direction in the horizontal direction on this tangent plane. Taking the positive south direction as the starting direction and the counterclockwise as the positive direction, the main value of the slope direction is calculated as:

$$\theta = arctg\frac{a_2}{a_1} \tag{6.47}$$

Due to the nature of the function, the value of θ is in the range of $\left[-\frac{\pi}{2}, \frac{\pi}{2}\right]$, so the slope direction should be assigned to the actual situation in the test process.

6. The degree of roughness

The degree of roughness is a measure of the roughness (or average smoothness) of the terrain, which describes the finer undulations of the terrain; the greater the degree of roughness, the richer the terrain features. Numerically, the degree of roughness is the ratio of the surface area (S_1) on the DEM corresponding to a grid point cell to its horizontal projection area (S_2).

$$Q = \frac{S_1}{S_2} \tag{6.48}$$

When $Q = 1$, the roughness is minimized, the actual surface of the grid cell is horizontal at this time.

The table area is calculated as follows: first, each grid of the grid DEM is decomposed into two triangles (as shown in Fig. 6.19).

Calculate the surface area of each triangle using Heron's formula:

$$S = \sqrt{P(P - D_1)(P - D_2)(P - D_3)} \tag{6.49}$$

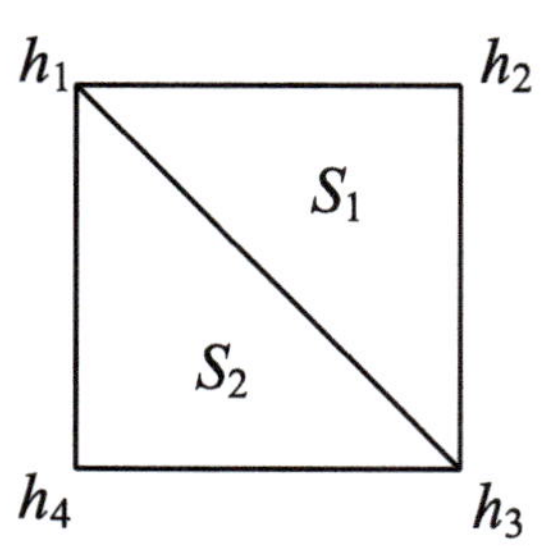

Fig. 6.19 Triangulation of individual grids

$$P = \frac{1}{2}(D_1 + D_2 + D_3) \tag{6.50}$$

where, $D_i = \sqrt{\Delta X^2 + \Delta Y^2 + \Delta Z^2} \quad (i = 1, 2, 3)$

7. Undulating abundance

As mentioned earlier, the standard deviation σ_T and the roughness Q reflect the undulation and the average smoothness of the underwater terrain in the whole sub-region, respectively. However, it is still not sufficient to characterize the underwater terrain by standard deviation σ_T or roughness Q alone. When σ_T is large, the topography is undulating but not necessarily rough ($Q$is large), and vice versa. Therefore, a new variable Q/σ_T, the topographic relief feature abundance, is defined to describe underwater topographic relief. Also, the literature states that when the topographic relief abundance of an area is greater than 0.2, then the area can be used as a matching area.

8. Local difference entropy

Let the number of grids in an electronic chart be $N \times N$ the water depth value of grid the point (i, j). In the $N \times N$ region, a local sea area of window size $(M < N)$ is selected as the matching sea area, and the topography in this window is called local topography. The average water depth of the window can be expressed as:

$$\overline{H} = \frac{\sum_{i=1,j=1}^{M,M} H(i,j)}{M \times M} \tag{6.51}$$

The value of the difference in water depth at any measurement point within the window is:

$$D_H(i,j) = \frac{\left|H(i,j) - \overline{H}\right|}{\overline{H}} \tag{6.52}$$

The local variance probability can be expressed as follows:

$$P_{i,j}(H) = \frac{D_H(i,j)}{\sum\limits_{i=1,j=1}^{M,M} D_H(i,j)} \tag{6.53}$$

According to the definition of SHANNON information entropy, the local variance entropy is calculated as:

$$H_E = -\sum_{i=1,j=1}^{M,M} P_{i,j}(H) \log P_{i,j}(H) \tag{6.54}$$

From the definition of terrain information difference entropy, it is clear that terrain information difference entropy can be used to describe the complexity of terrain undulation, which has the following characteristics.

(1) The entropy of the terrain reflects the size of the information contained in that terrain; if the value of that local elevation changes more sharply, the more unique the terrain is and the richer the information is, the smaller the entropy calculated at that time, the more favorable to the matching operation.
(2) The entropy is insensitive to a certain degree of baseline error, so the matching method established with the entropy of terrain information difference has good resistance to baseline error.

9. Fractal characteristics

FBM is a generalization of general Brownian motion, which is a continuous real-valued function defined as follows:

Letting $x \in \mathrm{E}^n$ (E^n is an n-dimensional Euclidean space) and $f(x)$ is a real random function about the point x. If there exists a constant $H (0 < H < 1)$ such that the formula:

$$F(x) = P\left\{\frac{f(x+\Delta x) - f(x)}{\|\Delta x\|^H} < y\right\} \tag{6.55}$$

Equation (6.55) is a distribution function independent of x and Δx, then $f(x)$ is considered to be consistent with the FBM model. In general, assuming $F(x) \sim N(0, \sigma^2)$, FBM can be more intuitively expressed as:

$$\langle|\Delta f|\rangle \propto |\Delta x|^H \tag{6.56}$$

The FBM satisfies self-similarity in a statistical sense, called the self-similarity parameter with fractional dimension: A is the Euclidean space dimension

$$D_f = E + 1 - H \text{ (A is the Euclidean space dimension)} \tag{6.57}$$

The FBM is a continuous but non-differentiable non-smooth stochastic process, and its incremental process is smooth and isotropic. Equation (6.56) can be further expressed as:

$$\mathrm{E}\left(|f(x+\Delta x) - f(x)|^2\right) \propto \|\Delta x\|^{2H} \tag{6.58}$$

Letting $f(0) = 0$, the past increment be $f(0) - f(-x)$ and the future increment be $f(x) - f(0)$. Then the covariance between two adjacent increments is:

$$\begin{aligned} \mathrm{cov} &= E\{[f(0) - f(-x)][f(x) - f(0)]\} \\ &= E\left\{\frac{1}{2}[f(x) - f(-x)]^2 - \frac{1}{2}[f(0) - f(-x)]^2 - \frac{1}{2}[f(x) - f(0)]^2\right\} \end{aligned}$$

$$= \frac{1}{2}\left(\|2x\|^{2H} - 2\|x\|^{2H}\right) \tag{6.59}$$

The correlation coefficient between the increments is:

$$\rho = \text{cov}/\|x\|^{2H-1} - 1 \tag{6.60}$$

Thus, the smaller the H, the more irregular the change of $f(x)$, the more complex the shape, and the relatively larger the amount of information; conversely, the larger the H, the smoother the change of $f(x)$, the simpler the shape, and the relatively smaller the amount of information. Therefore, H can be used as a quantitative indicator of surface complexity or roughness. According to the literature, the characteristics of the terrain in the region can be determined depending on the interval in which the value of H. The characteristics are:

$0.7 < H \le 1.0$, Information-poor areas;

$0.3 < H \le 0.7$, matchable areas;

$0 < H \le 0.3$, terrain hazard areas.

Of course, this judgment has yet to be further verified.

In addition, the larger σ is, the higher the slope of the topographic shape undulation; conversely, the smaller σ is, the lower the slope of the topographic shape undulation. Therefore, it can be used as the general shape and undulation characteristics of the regional terrain.

Through the above analysis, the fractal characteristic value can be used as the quantitative index of the complexity and undulation of the terrain surface respectively. The characteristic value H reflects the complexity of the terrain micro-undulation or the degree of surface fragmentation, which is an abstraction and generalization of the natural terrain complexity. The standard deviation feature σ reflects the degree of undulation of the terrain and represents the macroscopic shape characteristics of the terrain. The combination of H and σ can reflect the global characteristics of regional terrain and realize the classification of terrain. H and σ are calculated as follows:

Take the log of both sides of this formula:

$$E\left[|f(x+1) - f(x)|\right] = C\|\Delta x\|^{H} \tag{6.61}$$

We get:

$$\log \mathrm{E}\left[|f(x+1) - f(x)|\right] = H \log\|\Delta x\| + \log C \tag{6.62}$$

In a rectangular plane coordinate system, if $\log \mathrm{E}\left[|f(x+1) - f(x)|\right]$ is the y-axis coordinate and $\log\|\Delta x\|$ is the x-axis coordinate, since H and σ are both constants, the result is a line with a slope equal to H and an intercept on the y-axis $\log C$. And then we can derive from the previous formula.

Since H and σ can be used as quantitative indexes of terrain surface complexity and relief degree respectively, the combination of H and σ can comprehensively reflect the global features of regional terrain. The experiment shows that the mountainous terrain generally has a large topographic standard deviation, and the surface is relatively broken, which is reflected in the numerical value, H is small, while σ is large. The topographic standard deviation of plain terrain is small, and the surface is smooth, which is reflected in the numerical value, H is larger and σ is smaller. And hilly and in between. Therefore, terrain classification can be realized by using H and σ.

10. Relevant length

Topographic correlation length L represents the degree of similarity between topographic data, which is not only a key parameter to characterize topographic features, but also the main basis for selecting a topographic matching area and measuring belt length. In addition, in reference, the author proposed an underwater terrain regression model based on correlation length, which describes the global features of the terrain through L.

It is generally believed that the autocorrelation coefficient $\rho_{h,v}$ of the terrain is approximately exponential and separable, i.e., the formula:

$$\rho_{h,v} = e^{-|h|/L_h} e^{-|v|/L_v} \tag{6.63}$$

where, h and v respectively represent lateral and longitudinal displacement increments. When h and v are less than or equal to e^{-1}, h and v are L1 and L2, namely the lateral and longitudinal topographic correlation lengths. The correlation length of the whole base plot is half the sum of the horizontal and vertical correlation lengths $(L_h + L_v)$. If the carrier is assumed to enter the matching area horizontally, the longitudinal correlation of the data is very weak and can be ignored, and only the transverse correlation length of the data can be calculated. The following is the relevant length calculation for this case, the calculation steps are as follows:

(1) Calculate the covariance matrix

$$C_T = \begin{bmatrix} C_T(i_1, i_1) & C_T(i_1, i_2) & \dots & C_T(i_1, i_M) \\ C_T(i_2, i_1) & C_T(i_2, i_2) & \dots & C_T(i_2, i_M) \\ \vdots & \vdots & \dots & \vdots \\ C_T(i_M, i_1) & C_T(i_M, i_2) & \dots & C_T(i_M, i_M) \end{bmatrix}$$

The formula for calculating covariance is as follows:

$$C_T(i_k, i_l) \approx \frac{1}{M-1} \sum_{j=0}^{M-1} [H(j, i_k) - m_T(i_k)][H(j, i_l) - m_T(i_l)] \tag{6.64}$$

where the mean is:

$$m_T(i_k) \approx \frac{1}{M}\sum_{j=0}^{M-1} H(j, i_k) \tag{6.65}$$

Thus, the normalized covariance matrix can be obtained:

(2) Calculate the correlation length L

The matrix R_T is symmetric. Starting from the diagonal element $R_T(i_k, i_k)$ on each row, compare the element $R_T(i_k, i_k + L)$ in this row from left to right to find the first element in this row that is less than or equal to e^{-1}. So if we have:

$$R_T(i_k, i_k + \tau_k) = e^{-1} \quad k = 1, 2, \ldots, \mathrm{k}_1 \tag{6.66}$$

then the correlation length is:

$$L = \frac{1}{k_1}\sum_{k=1}^{k_1} \tau_k \tag{6.67}$$

6.4.2 Classification of Feature Parameters

(1) Undulating features

The elevation is the basic data, and the continuous change of elevation in the region leads to the undulation of the terrain. According to the degree of undulation, it is usually divided into mountains, hills, plains and basins. Therefore, the undulating feature of the terrain is one of the most basic characteristics of underwater terrain.

Among the above 16 topographic parameters, the minimum water depth, maximum water depth, maximum elevation difference and average water depth express the information of "elevation difference" by absolute value, and focus on reflecting the size of topographic undulation; while the topographic standard deviation, topographic kurtosis coefficient and topographic skewness coefficient can be combined with the topographic elevation histogram, through statistical The standard deviation of terrain, crest factor and skew factor of terrain can be combined with the histogram of terrain elevation to reflect the normal distribution of regional elevation values through statistical methods, which also expresses the undulating characteristics of terrain from one side. In addition, the FBM model parameters can also reflect the degree of regional topography undulation, and it is also a macroscopic indicator to reflect the undulation characteristics of the whole region.

(2) Fragmentation characteristics (or smoothness degree)

In the three-dimensional geographic space, the undulating characteristics of underwater terrain are expressed as the terrain characteristics in the vertical direction, while in the horizontal direction, the terrain exists to describe the degree of fragmentation.

The degree of fragmentation reflects the finer undulating situation of the terrain, and the greater the degree of fragmentation, the richer the terrain information. Roughness and FBM model parameters reflect the fragmentation characteristics of the terrain, but the difference between them is that the greater the roughness, the richer and more complex the terrain information is, and the smaller it is. The slope and slope direction analysis of terrain can get two parameters, which not only reflect the roughness of the surface, but also reflect the degree and direction of terrain slope. In addition, the local difference entropy of terrain information is the application of entropy theory to terrain description, and the spatial distribution of difference entropy also reflects the complexity of underwater terrain from another perspective.

Since the undulation feature and fragmentation degree of underwater terrain are two different terrain features, it is impossible to describe the underwater terrain information completely by only one of them. To solve this problem when selecting terrain feature parameters, either one from each of these two types of parameters is chosen to jointly portray the underwater terrain features, or a comprehensive index is used which can reflect both the terrain undulation features and describe the smoothness of the terrain, such as the abundance of terrain undulation features.

(3) Autocorrelation (similarity) feature

According to fractal theory, there is a fundamental property of matter widely existing in nature: self-similar characteristics of local and local, local and whole in reflecting the topography of selected sea areas. The morphological, temporal and spatial aspects have a statistical sense of similarity. This self-similarity is also an important feature of underwater topography. Using self-similarity can build an effective terrain regression model, and more importantly, in terrain matching navigation, the magnitude of terrain correlation directly affects the matching probability and matching accuracy of terrain. If the correlation degree of underwater terrain is large, i.e., there are similar terrains in the terrain matching area, then it is easy to cause terrain mismatching when searching for matching locations. Similarity analysis includes point, line, strip and block correlation analysis.

The "correlation length" among the 16 terrain features listed above reflects the self-similarity of the underwater terrain and is the main basis for selecting the terrain matching area and the length of the measurement zone. Combined with the autocorrelation curve, it can effectively express similar characteristics of the navigation area (Table 6.1).

6.5 Summary

This chapter introduces the representation method of the terrain elevation model, and the mapping method of the elevation model, focusing on the detailed description of aerial and underwater terrain elevation measurement methods, which is an important methodological basis for the implementation of terrain elevation navigation and positioning applications.

Table 6.1 Classification of terrain feature parameters

No	Parameter	Means	Category
1	Minimum depth of water	The highest undulating position of the terrain	Characteristic of fluctuation
2	Maximum water depth	The lowest undulating position of the terrain	
3	Maximum height difference	The absolute elevation difference of topographic relief	
4	Mean water depth	Average elevation value	
5	Standard deviation	Reflect the dispersion and fluctuation degree of regional elevation value	
6	Coefficient of skew	Reflects the normal distribution of regional elevation values	
7	Peak state coefficient	Reflects the peak state of regional elevation distribution	
8	Parameter of fractal	Reflect the overall topographic relief degree of the region	
9	Roughness of surface	Reflects the roughness of the terrain	Characteristics of breakage
10	Slope	Reflects the slope of the terrain (topography)	
11	Aspect	Reflects the slope of the terrain (topography)	
12	Parameter of fractal	Reflect the overall terrain complexity of the region	
13	Local differential entropy	Reflects the complexity of terrain information	
14	Length of correlation	Reflect the zone correlation of regional topography	Self-similar feature
15	Abundance of fluctuation	Comprehensively reflect the relief characteristics and the degree of fragmentation of the terrain	
16	Spacing of grid	Reflects the spatial scale and is related to the matching accuracy	

References

1. Wang, H. (2020). *Surveying and mapping of large-scale topographic map and quality evaluation of UAV oblique photography*. Anhui University of Science and Technology, MA thesis:1–2.
2. Liu, Z. (2002). Creating a DEM based on terrain mapping with laser radar. *Remote Sensing Technology and Application, 17*(5), 266.

3. Lu, H. X. (2008). *Research on DEM error model.* Nanjing Normal University, Ph.D. dissertation:32–34.
4. Zhou, L. (2018). On seabed terrain aided navigation technology for autonomous underwater vehicles. *Southeast University, 23*, 34.
5. Li, X. W., Liu, J. Y., & Kang, G. H. (2006). Development and application of TERCOM elevation-aided navigation system. *Journal of Chinese Inertial Technology, 14*(1), 35.
6. Zhao, Y. Z. (2008). *Digital terrain analysis based on DEM*. Wuhan University of Technology, MA thesis:7–12.
7. Lin, L. (2012). *Multiple beam height velocity radar theory and algorithm research.* University of Electronic Science and Technology of China, MA thesis:6–7.
8. Xiao, L., & Zhou, X. Z. (2018). Height-finding algorithm of wide beam radar. *Shipboard Electronic Countermeasure, 41*(3), 105.
9. Bi, K. B., Yao, S. M., & Xie, C. S. (2016). Research into composite altimetry technology based on radio/laser altimeter. *Shipboard Electronic Countermeasure, 39*(3), 20.
10. Chen, X. L. (2012). *A study on underwater terrain matching aided navigation technology of AUV*. Harbin Engineering University, Ph.D. dissertation:13–15.
11. Deng, W. (2006). *Combination of underwater TERCOM and STIAN algorithm and simulation study*. Peking University, MA thesis.
12. Liu, C. X. (2003). *A study on terrain matching assistant positioning technology for underwater vehicle*. Harbin Engineering University, Ph.D. dissertation.
13. Liu, X. P., Zhang, L. H., Jia, S. D., & Shi, Y. (2018). Selection analysis of nautical chart used in underwater terrain matching location. *Bulletin of Surveying and Mapping, 08*, 74–77.
14. Tian, Y., Li, G. Q., & Song, X. (2018). A novel 3D terrain feature detecting and matching method. *Journal of Astronautics, 39*(06), 690–696.
15. Wang, J., Tan, Y. G., & Chen, C. X. (2020). Terrain matching localization for a hybrid underwater vehicle in the challenger deep of the Mariana Trench. *Frontiers of Information Technology and Electronic Engineering, 21*(5), 749–760.
16. Wang, K. D. (2005). *Research-assisted navigation for underwater contour matching algorithms*. Beijing: Beijing University Postdoctoral Research Report.
17. Wang, R. P., Li, Y., & Ma, T. (2018). Terrain reference navigation fast convergence filtering of AUV. *Journal of Huazhong University of Science and Technology (Natural Science Edition), 46*(07), 94–97.

Chapter 7
Topographic Elevation Field Navigation and Positioning Technology Methods

The terrain elevation field navigation and positioning can overcome the shortcomings of inertial navigation accuracy which decreases with time accumulation and meet the needs of long-time and high-precision navigation of motion carriers by comprehensive correction of inertial navigation (INS) errors with terrain elevation field information. According to the principle of terrain elevation field navigation and positioning, terrain elevation field navigation and positioning mainly includes three key technologies, such as the establishment of terrain elevation field model, the research and development of altimetric sensors and the technical methods of terrain elevation field navigation and positioning. Chap. 6 focuses on the foundation and theoretical model of terrain elevation field navigation and positioning, mainly including digital terrain elevation field model and terrain elevation field measurement method, which is the basis of terrain elevation field navigation and positioning. This chapter focuses on the technical methods of terrain elevation field navigation and positioning, which is the core of terrain elevation field navigation and positioning implementation and plays a key role in the performance of terrain elevation field navigation and positioning.

The current topographic elevation field navigation and positioning technology methods are divided into two main categories according to the positioning algorithms: batch correlation matching algorithms and iterative recursive filtering matching algorithms.

The basic idea of the batch correlation matching algorithm is: during a period of carrier operation, the altitude sensor acquires a section of terrain elevation field information, takes the position indicated by INS as reference, correlates the measured elevation sequence with the elevation sequence of the topographic map in a certain range corresponding to the digital elevation topographic map, and considers that when the two meet certain conditions or indicators, it is considered that the topographic map sequence corresponding to The latitude and longitude coordinates of the topographic map sequence are the real trajectory passed by the carrier, and the trajectory is used to replace the INS indication trajectory to complete the task of aiding

L. Yan et al., *Mathematical and Physical Fundamentals of Navigation and Positioning with Earth's Natural Fields*, Navigation: Science and Technology 13,
https://doi.org/10.1007/978-981-97-5524-0_7

navigation. According to the practical implementation of batch processing related matching algorithm, it includes terrain elevation field contour matching algorithm TERCOM [1], terrain elevation field contour matching algorithm ICCP (Iterated Closest Contour Point), etc.

Because the batch matching algorithm must complete a period of track operation before it can match once, there is a certain time lag. The whole matching process is done in one sampling cycle. Therefore, the advantage of iterative recursive matching algorithm is good real-time performance. However, due to the use of single-point matching, it is easily affected by terrain similarity, which leads to mis-matching. At present, iterative recursive matching algorithms mainly include SITAN (Sandia Inertial Terrain-Aided Navigation) algorithm, VITERBI maximum a posteriori estimation algorithm and BITAN (Baihang Inertial Terrain-Aided Navigation) algorithm, etc. The latter two algorithms are improvements and developments of the SITAN algorithm. The authors optimized and improved the above methods based on them, which are described in detail below.

7.1 Topographic Elevation Field Contour Matching Method

Terrain Elevation Field Contour Matching TERCOM (Terrain Contour Matching) method (hereinafter referred to as Terrain Matching) is a terrain contour matching technique related to the submarine topography, which was developed by E Systems, Inc. in the 1970s. The working principle of Terrain Contour Matching is as follows: after a period of navigation, the terrain elevation field sequence under the real track of the motion carrier is measured at $H_r = [h_r(1), h_r(2), h_r(3), \cdots, h_r(n)]$, and the measured terrain elevation field sequence data is correlated with the terrain elevation field model at H_j. The point with the correlation peak is determined as the estimated position of the motion carrier, and the error correction is made to the inertial navigation system accordingly.

7.1.1 One-Dimensional Correlation Matching Method for Topographic Elevation Field

The main correlation matching algorithms commonly available for TERCOM systems are the Cross Correlation (CC-Cross Correlation), Mean Square Difference (MSD-Mean Square Difference) and Mean Absolute Difference (MAD-Mean Absolute Difference) algorithms, which are defined as.

$$J_{CC}(x, y) = \frac{1}{n}\sum_{i=1}^{n} h_j(x + i\Delta x, y + i\Delta y) \cdot h_r(i) \tag{7.1}$$

$$J_{MAD}(x, y) = \frac{1}{n}\sum_{i=1}^{n} |h_j(x + i\Delta x, y + i\Delta y) - h_r(i)| \tag{7.2}$$

$$J_{MSD}(x, y) = \frac{1}{n}\sum_{i=1}^{n} [h_j(x + i\Delta x, y + i\Delta y) - h_r(i)]^2 \tag{7.3}$$

where, $h_r(i)$ is the actual measured topographic elevation field of the motion carrier, $h_j(x+i\Delta x, y+i\Delta y)$ is the topographic elevation field value stored in the topographic elevation field model, (x, y) is any search position, and n is the number of topographic elevation field sequences. In the cross-correlation algorithm, the points satisfying $\max_j J_{CC}(x, y)$ are the matching points; in the MAD and MSD algorithms, the points satisfying $\min_j J_{MAD}(x, y)$ and $\min_j J_{MSD}(x, y)$ are the matching points.

The characteristics of the one-dimensional TERCOM system are mainly determined by its algorithm. Since the best matching position is performed by searching each grid position in the position uncertainty region without any omission after measuring a sequence of topographic elevation fields, the one-dimensional TERCOM correlation algorithm does a posteriori correlation analysis on a sequence of topographic elevation fields, and there is a process of collecting data before each matching solution, and the distance passed by the carrier in this process is the sampling distance. The sampling distance is equal to the product of the sampling step size and the number of samples.

The sampling distance should be chosen long enough to obtain an accurate and unique polar value, and short enough to keep the INS error between each localization to a minimum. Too long a sampling distance will distort the path generated by the INS, and in addition the selection of the sampling distance is also affected by terrain features and measurement noise. In particular, the carrier's motion is slow, and the INS error of the carrier is much larger than the INS error of the vehicle for the same distance sampling, which becomes the biggest difficulty for the TERCOM algorithm to be applied to navigation.

The above contradiction can be solved by analyzing the navigability of the terrain in advance and selecting only those regions with large terrain undulations and richer features to apply the TERCOM algorithm, where only short sampling distances are required to achieve satisfactory matching results, and we call these regions as adaptation zones. However, this requirement of navigating only in the adaptation zone obviously reduces the adaptability of the algorithm to the terrain.

Since the multi-beam system can be scanned laterally to obtain sufficient terrain information, the requirement for sampling distance is relaxed, making the sampling distance and sampling time more freely selectable. Accordingly, it can be suitable for more terrain features.

In summary, 1D terrain matching naturally has few feature points and many similar terrain elevation field values between different terrains, so the possibility of false matching caused by using 2D terrain matching is greatly reduced. From the terrain analysis, 2D matching can obtain the terrain features of the area around the carrier route, covering the surrounding terrain that is not available for 1D matching, thus enriching the matching terrain elevation field feature point information, improving the matching accuracy, reducing the mismatch, and correspondingly improving the terrain adaptability of terrain elevation field navigation and positioning. Later in this section, we focus on the 2-D matching method of terrain elevation field.

7.1.2 Two-Dimensional Correlation Matching of Topographic Elevation Fields Method

The most significant advantage of two-dimensional matching is that there are few terrain features or even close to a blank slate in the route area, but there are abundant terrain features around the route, so that the terrain features on both sides of the route can be fully used in matching, as Fig. 7.1 shown.

To compare the difference between one-dimensional matching and two-dimensional matching, let the maps F1, and F2, where the lines are L1 and L2, respectively, as 图 7.2 and shown in Fig. 7.3.

If:, $h_{11} = h_{21}$, $h_{12} = h_{22}$, $h_{13} = h_{23}$, L1 and L2 are identical. Obviously map F1 is not the same as map F2, but the elevation values of the three points on their midline are exactly the same, and it is obvious that the two maps cannot be distinguished if only the points on the midline are examined.

In order to increase the sampling information and also to reduce the mis-matching phenomenon, the terrain information on both sides of the route is also introduced into the matching, so that a two-dimensional TERCOM is formed, and the corresponding matching accuracy and stability are also improved.

The main components of the 2D matching method for topographic elevation fields contain the following four aspects.

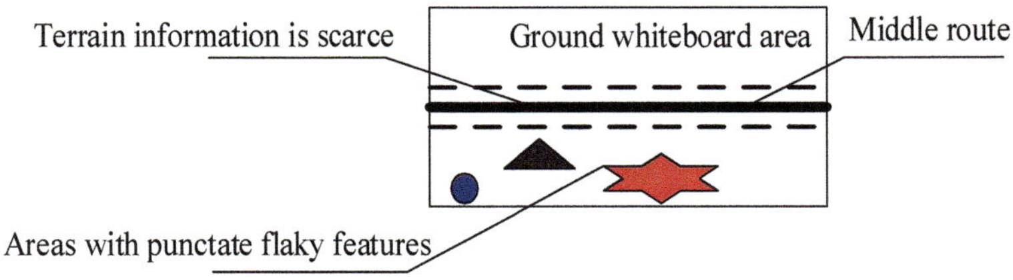

Fig. 7.1 Relationship between terrain features and route

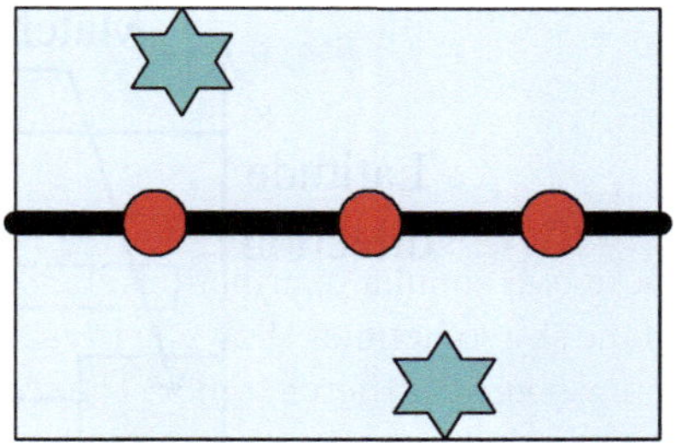

Fig. 7.2 Map F1

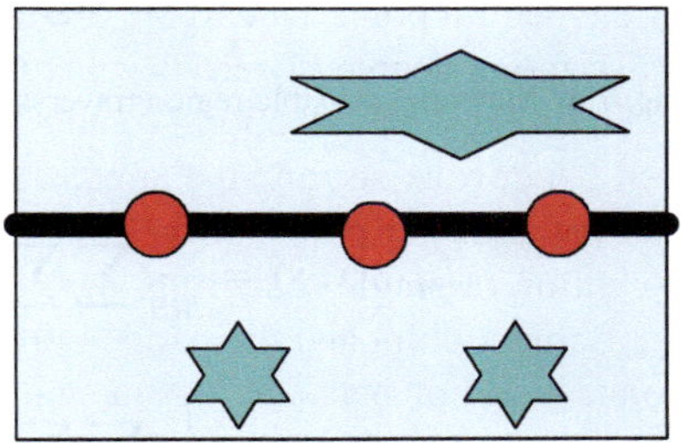

Fig. 7.3 Map F2

1. Determine the Search Area

A rectangular area, centered on the inertial guidance output position (x_{ins}, y_{ins}), can be used as the search area even if the topographic map of the scanned area is completely included in the case of maximum inertial guidance error.

2. Determine the Matching Template

Since the real-time topographic elevation field image imaged by the line correction and resampling of the multibeam system is not a regular rectangle, the rectangular area image is taken as the matching template with the narrowest position length of the beam scanning range at the same moment as the side length.

3. Topographic Elevation Field Filtering

Since the scanned seafloor topographic data of the multibeam system with errors have an impact on the matching accuracy, and the systematic errors and chance errors are generally high-frequency information in the measured image data, a 3*3 Gaussian template is used for low-pass filtering to reduce the impact of errors on matching.

4. Determine the Matching Criteria

In two-dimensional terrain matching, the new correlation algorithm is

$$J_{CC}(x, y) = \frac{1}{mn} \sum_{j=1}^{m} \sum_{i=1}^{n} h_j(x + j\Delta x, y + i\Delta y) \cdot h_r(j, i) \tag{7.4}$$

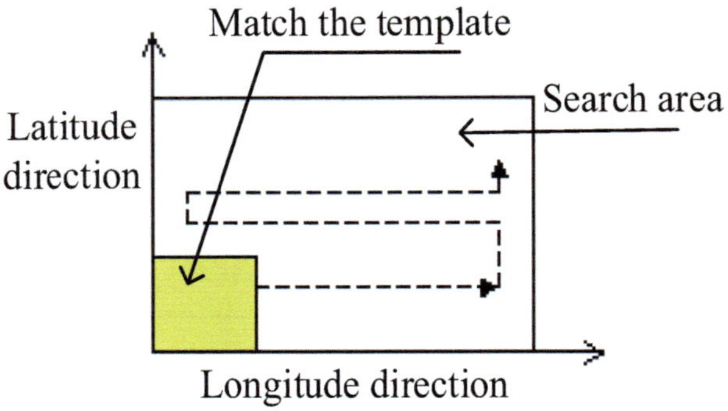

Fig. 7.4 Matching possible region traversal search graph

$$J_{MAD}(x, y) = \frac{1}{mn}\sum_{j=1}^{m}\sum_{i=1}^{n}|h_j(x + j\Delta x, y + i\Delta y) - h_r(j, i)| \quad (7.5)$$

$$J_{MSD}(x, y) = \frac{1}{mn}\sum_{j=1}^{m}\sum_{i=1}^{n}[h_j(x + j\Delta x, y + i\Delta y) - h_r(j, i)]^2 \quad (7.6)$$

where m is the number of beams in the multibeam system and n is the number of route samples.

5. Design Search Strategy

The matching sub-blocks in the search area are searched in the order of first along the latitude line and then along the longitude line, as Fig. 7.4 shown.

Topographic two-dimensional matching has a large amount of data due to the processing between maps, which seriously affects the speed of matching. Modern navigation requires a high positioning speed with a high accuracy, and in order to meet the need of real-time navigation, it is important to improve the speed of matching, and the sequential judgment algorithm of hierarchical search is a common fast search algorithm.

This hierarchical search algorithm is based on people's habit of finding things coarsely first and then finely. For example, when looking for the location of Beijing on a world map, one can first find out the vast territory of China, which is called coarse correlation. The location of Beijing is carefully determined within this territory, which is called fine correlation. So the hierarchical search algorithm formed by this idea has a fairly high processing speed, and the algorithm consists of two steps.

(1) Map Layering

First, the image being matched is preprocessed in layers. The method is to average the neighborhoods of the Figure 2 × 2 dimensions grid by grid, thus obtaining an image with lower resolution and smaller dimension. Then, this image is processed again in the same way to obtain an image with even lower resolution and smaller dimension, and so on. If a total of K layering processes are performed, then K processed images can be obtained. Together with the original image, a sequence of images with high to low resolution and large to small dimensionality can be formed. This technique is called hierarchical preprocessing. If the above layering technique is applied to the baseline image and the real-time image, two such image sequences can be obtained: a baseline image sequence and a real-time image sequence. In practice, due to the measurement error, if the map is layered too much, it will inevitably lead to distortion, and generally two to three layers are reasonable.

(2) The process related to the first coarse and then fine

As mentioned before, the first correlation starts from H_{rk} and H_{jk}, and in order to find possible matching positions, correlation should be performed on all search positions of H_{rk} and H_{jk}, , and coarse matching positions (u^k, v^k) should be identified. Since H_{rk} and H_{jk} have the smallest number of dimensions at this point, the search process is fast and may yield several coarse match positions. The second correlation is done between the higher resolution images $H_{r(k-1)}$ and $H_{j(k-1)}$. So $H_{r(k-1)}$ only needs to perform a correlation search near one or several coarse matches with $H_{j(k-1)}$ to find one or a few more likely matches (u^{k-1}, v^{k-1}). In the above correlation process, several additional search positions should be added near the coarse match position (u^k, v^k) in order not to lose the match point. Continue in this order until the highest resolution real-time map H_{r0} finds a match on the base map. H_{j0}.

Sometimes, the map is searched on each sublevel to further improve the computational efficiency without necessarily panning to every point. For example, a combined coarse–fine search can be used, search for the next match every m $(m > 1)$ points, and then find the maximum match by matching each reference point position in the local range around the great match value.

7.1.3 Fourier Transform Based Terrain 2-D Matching Algorithm

Since Fourier proposed the Fourier transform in mathematics, it has been rapidly applied to the field of signal processing, thus broadening the processing of signals in the air domain to the frequency domain and greatly enriching the means of signal processing. A signal undergoes Fourier transform, the spectrum of that signal can be analyzed, and the spectrum responds to which frequencies the signal is composed of [2].

1. Fourier Transform Theory

A one-dimensional signal $f(x)$ To implement the Fourier Transform FT (Fourier Transform), the Dirichlet condition must be satisfied;

(1) with finite number of interruption points.
(2) with finite number of extremes.

(3) Absolute productability.

Then its Fourier transform pair (Fourier transform with its inverse transform) must exist. In practical applications, these conditions can always be satisfied in general.

In terrain matching, since two-dimensional matching is used, a two-dimensional Fourier transform is required. If the two-dimensional function $f(x, y)$ satisfies the Dirichlet condition, its two-dimensional Fourier transform pair is

$$\Im[f(x, y)] = F(u, v) = \int_{-\infty}^{+\infty}\int_{-\infty}^{+\infty} f(x, y) \cdot e^{-j2\pi(ux+vy)}dxdy \tag{7.7}$$

$$\Im^{-1}[F(u, v)] = f(x, y) = \int_{-\infty}^{+\infty}\int_{-\infty}^{+\infty} F(u, v) \cdot e^{j2\pi(ux+vy)}dudv \tag{7.8}$$

To apply the Fourier transform in a digital topographic image, the Discrete Fourier Transform (DFT) is required. Consider a digital image with M rows and N columns and the DFT is defined as

$$\Im[f(x, y)] = F(u, v) = \sum_{x=0}^{M-1}\sum_{y=0}^{N-1} f(x, y) \cdot e^{-j2\pi(\frac{ux}{M}+\frac{vy}{N})} \tag{7.9}$$

$$\Im^{-1}[f(x, y)] = f(x, y) = \frac{1}{MN}\sum_{u=0}^{M-1}\sum_{v=0}^{N-1} F(u, v) \cdot e^{j2\pi(\frac{ux}{M}+\frac{vy}{N})} \tag{7.10}$$

The coefficient $\frac{1}{MN}$ can appear in either the forward or inverse transformation, or can be multiplied by the coefficient $\frac{1}{\sqrt{MN}}$ in front of the forward and inverse transformation, as long as the product of the two coefficients is equal to. $\frac{1}{MN}$

The Fourier spectrum, phase spectrum, and energy spectrum of the two-dimensional discrete function are

$$|F(u, v)| = \sqrt{R^2(u, v) + I^2(u, v)} \tag{7.11}$$

$$\phi(u, v) = \arctan\frac{I(u, v)}{R(u, v)} \tag{7.12}$$

$$E(u, v) = R^2(u, v) + I^2(u, v) \tag{7.13}$$

where $R(u, v)$ and $I(u, v)$ are the real and imaginary parts of $F(u, v)$, respectively.

2. Image Spectrum Matching

In two-dimensional image alignment, it is often necessary to obtain the three parameters of scale S (Scale), rotation R (Rotation) and translation T (Translation) between two images. The core idea of image matching using FFT is to use the phase correlation of the spectrum to obtain the detection of the translation parameters of the two images. At the same time, the rotation and scale factors in the two images are transformed into the translation components of the log-polar coordinate plane by combining the log-polar transformation to obtain the detection, and then image matching based on translation, rotation and scale invariance is achieved.

(1) Realization of phase transformation for image translation detection.

Let f_s be the input image, f_r be the reference image, and there is a translation between the two at $(\Delta x, \Delta y)$,

$$f_s(x, y) = f_r(x - \Delta x, y - \Delta y) \tag{7.14}$$

According to the phase shift theorem of the Fourier transform, the Fourier spectrum between the two images corresponding to F_s, F_r satisfies.

$$F_s(u, v) = e^{-j2\pi(u\cdot\Delta x + v\cdot\Delta y)} F_r(u, v) \tag{7.15}$$

After the Fourier transform, the relative advection information is transferred to the phase factor. In order to solve for $(\Delta x, \Delta y)$, the normalized correlation power spectrum R of F_s and F_r is defined as follows.

$$R = \frac{F_s(u, v)\cdot F_r^*(u, v)}{|F_s(u, v)\cdot F_r(u, v)|} = e^{j2\pi(u\cdot\Delta x + v\cdot\Delta y)} \tag{7.16}$$

where F^* is the complex conjugate of F. By taking the Fourier inverse transform of R, an impulse function in the XOY plane can be obtained, by detecting the peak position of the impulse function, the relative translations of the two images can be obtained.

(2) Parameter detection when scale, rotation and translation variables exist simultaneously.

When the input image f_s is the reference image f_r transformed by the scale scaling factor k, the rotation factor $\Delta\theta$ and the translation factor $(\Delta x, \Delta y)$,

$$f_s(x, y) = f_r[k\cdot(x\cdot\cos\Delta\theta - y\cdot\sin\Delta\theta) - \Delta x, [k\cdot(x\cdot\sin\Delta\theta + y\cdot\cos\Delta\theta) - \Delta y] \tag{7.17}$$

Then the Fourier spectra of the input image f_s and the reference image $f_r F_s$ and F_r are related as follows

$$F_s(u, v) = \frac{1}{k^2} e^{-j2\pi(\frac{u\cdot\Delta x}{k} + \frac{v\cdot\Delta y}{k})}\cdot F_r\left(\frac{u\cdot\cos\Delta\theta - v\cdot\sin\Delta\theta}{k}, \frac{u\cdot\sin\Delta\theta - v\cdot\cos\Delta\theta}{k}\right) \tag{7.18}$$

Taking the magnitude spectrum for both sides of the above equation, we have

$$|F_s(u, v)| = \frac{1}{k^2} \cdot \left| F_r\left(\frac{u \cdot \cos \Delta\theta - v \cdot \sin \Delta\theta}{k}, \frac{u \cdot \sin \Delta\theta - v \cdot \cos \Delta\theta}{k} \right) \right| \tag{7.19}$$

To simplify the notation, let G_s and G_r be the corresponding amplitude spectra of F_s and F_r, respectively, and we have

$$G_s = \frac{1}{k^2} \cdot G_r\left(\frac{u \cdot \cos \Delta\theta - v \cdot \sin \Delta\theta}{k}, \frac{u \cdot \sin \Delta\theta - v \cdot \cos \Delta\theta}{k} \right) \tag{7.20}$$

Let H_s and H_r be the results of converting the amplitude spectra G_s and G_r from the Cartesian to polar coordinate system, respectively,

$$H_s(\rho, \theta) = \frac{1}{k^2} \cdot H_r(\rho/k, \theta + \Delta\theta) \tag{7.21}$$

Among them.

$$\begin{cases} \rho = (u^2 + v^2)^{1/2} \\ \theta = \tan^{-1}(u/v) \end{cases} \tag{7.22}$$

Taking a log-polar transformation of ρ, we have

$$\begin{cases} r = \log_a \rho \\ \phi = \theta \end{cases} \tag{7.23}$$

Let Q_s and Q_r be the result of the transformation of H_s and H_r in the log-polar coordinate system, then we have

$$Q_s(r, \phi) = Q_r(r - \log_a k, \phi + \Delta\theta) \tag{7.24}$$

Both the scale factor and the rotation factor are reduced to additive operations, so that $\log_a k$ and $\Delta\theta$ can be obtained using the phase detection method of 2.1. Thus, k and can be obtained. $\Delta\theta$.

This method of performing the Fourier transform in the log-polar coordinate plane is equivalent to performing the Fourier-Mellin transform on the function. $f(x, y)$

$$F_M(k_1, k_2) = \int_{-\infty}^{+\infty} \int_{0}^{2\pi} f(e^r \cos \phi, e^r \sin \varphi) \cdot e^{j(k_1 r + k_2 \varphi)} dr d\varphi \tag{7.25}$$

The schematic diagram of FFT-based translation, rotation and scale invariance image alignment is shown in Fig. 7.5. After the input image f_s and the reference image f_r are FFT transformed, their amplitude spectra (abs) are taken, filtered, and then transformed from the Cartesian coordinate system to the log-polar coordinate system

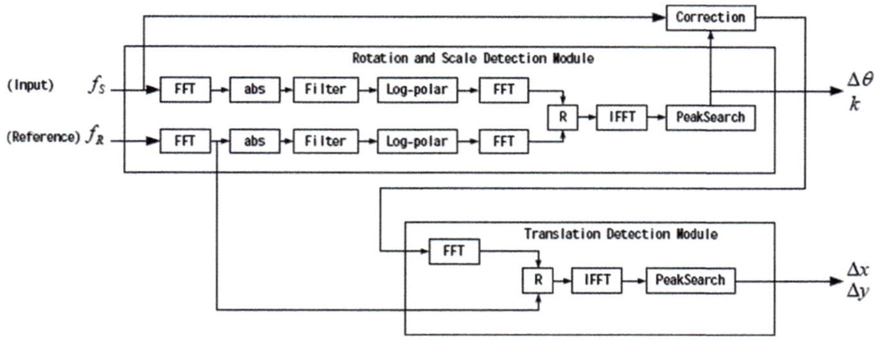

Fig. 7.5 Schematic diagram of image alignment based on translation, rotation and scale invariance of FFT

(Log-Polar). After the FFT transform in the log-polar plane, the normalized phase correlation function (R) is obtained, and after the IFFT operation, the correlation peak is obtained, and the location of the correlation peak is searched (PeakSearch) to obtain the rotation factor $\Delta\theta$ and the scale factor k. These two parameters are used to correct the input image so that only the translation component exists between the corrected image and the reference image. Then, the same phase detection operation is performed to obtain the translation component. At this point, the relative translation, rotation and scale change factors are all obtained, thus allowing the topographic image to be aligned.

(3) Matching criteria

$$\min\{abs[MOD(MODEL) - MOD(searchblock)]\} \tag{7.26}$$

where $MOD(MODEL)$ denotes the mode of the Fourier transform of the template and $MOD(search\ block)$ denotes the mode of the Fourier transform of the corresponding block in the search region.

7.1.4 Matching Algorithm Based on the Entropy of Topographic Elevation Field Information

The matching algorithm based on the information entropy of terrain elevation field (terrain entropy for short) is obtained by combining the information entropy theory with the terrain elevation field navigation and positioning theory. The algorithm fully utilizes the advantages of information entropy by calculating the entropy value from the elevation information value obtained after the carrier runs for a period of time using the definition of terrain entropy, and this entropy value is compared with the entropy value of matching elevation value to get the correct position of the carrier.

Topographic entropy has certain resistance to elevation measurement error, which can smooth out the measurement error of the altimetric sensor and the measurement error caused by transient changes of the environment.

1. The Basic Principle of Information Entropy

In statistical physics, thermal entropy is a measure of the clutter (disorder) of a physical system. Shannon introduced the concept of entropy in physics to information theory and introduced the concept of information entropy. The information entropy of a source is considered in terms of the statistical properties of the whole source. It characterizes the overall information measure of a source in an average sense. For a particular source (with a given probability space), its information entropy is a definite value. Different sources have different entropies due to their statistical properties.

In information theory, entropy is defined as the entropy of an information source $\chi = (S, P)$ as the statistical average of the self-information,

$$H(S) = -\sum_{i=1}^{n} p_i \log p_i \tag{7.27}$$

where the self-information of the event $x_i (x_i \in S)$ is $-\log p(x_i)$.

It expresses the amount of information brought by a series of events x_i. The smaller the probability of x_i occurring, the greater the amount of information it brings. The more even the probability of occurrence of each event, the greater the entropy value.

Information entropy has three physical meanings.

(1) Information entropy is an indication of the average amount of information provided by each message after the source output.
(2) Information entropy is the average uncertainty of the source before the output of the source is expressed.
(3) The information entropy is used to characterize the randomness of the variable X.

The entropy characterizes the statistical properties of the total source, the overall average uncertainty. This also shows that the defined information entropy has its limitations, it cannot describe the specific implication and subjective value of the event itself. In order to overcome this drawback of information entropy, the introduction of the concept of information entropy into the fields of signal processing, image processing and navigation needs to be combined with specific knowledge of these fields to define specific definitions of entropy in these fields, thus overcoming the drawbacks of entropy.

2. Mathematical Model of Terrain Entropy Matching Algorithm

The basic idea of this algorithm is to intercept the current measurement points and historical measurement points to form an elevation sequence, calculate the topographic entropy value, compare it with the topographic entropy of each matching curve in the topographic elevation field model within the search range, get the grid point with the smallest absolute difference or the smallest variance, and determine that point as the best matching point.

To apply the concept of terrain entropy to the terrain entropy matching algorithm, it needs to be applied according to the characteristics of the terrain. The definition of terrain entropy determines that the prerequisite for its application must be that the source conforms to a normal distribution. When the submarine terrain is relatively flat, the discrete source's information point elevation values conform to a normal distribution, and the definition of terrain entropy can be directly applied,

$$H = -\sum_{i=1}^{n} P_i \log P_i \tag{7.28}$$

Among them.

$$P_i = \frac{h_i}{\sum_{i=1}^{n} h_i} \tag{7.29}$$

It represents the "probability" of the elevation value in the sequence, where the topographic entropy can be one-dimensional, or it can be transformed into two-dimensional topographic entropy accordingly if a multibeam measurement system is used.

When the seafloor topography is relatively variable, the elevation values of the information points of the discrete sources do not conform to a normal distribution, and the concept of entropy of local topographic differences needs to be defined.

The average elevation of the measured elevation sequence is

$$\overline{h}_{ms} = \frac{\sum_{k=1}^{n} h_k(z+m)}{n} \tag{7.30}$$

The elevation difference values for each measurement point are

$$C_k = \frac{\left| h_k(z+m) - \overline{h}_{ms} \right|}{\overline{h}_{ms}} \tag{7.31}$$

The elevation difference probabilities are.

$$P_k(z+m) = \frac{C_k}{\sum_{k=1}^{n} C_k} \tag{7.32}$$

The entropy of local topographic differences where the elevation sequence is located is

$$H_{ms} = -\sum_{k=1}^{n} P_k \log P_k \tag{7.33}$$

The improved algorithm meets the applicable conditions of the terrain entropy definition. The algorithm needs to set the terrain feature threshold when applying these two operators, and if the terrain feature value is greater than a certain threshold, it means the seafloor terrain is more undulating, and the algorithm selects the local terrain difference entropy operator. If the terrain feature value is less than or equal to a certain threshold, the seafloor terrain is flat and the matching algorithm selects the terrain entropy operator.

3. Characteristics of Topographic Entropy

The topographic entropy value is independent of the order of the topographic elevation field sequence and the absolute size of the topographic elevation field, it is related to the relative relationship of the topographic elevation field sequence and can reflect the degree of complexity of the relationship between the points in the sequence. Therefore, topographic entropy can be used to describe the complexity of topographic relief, which has the following characteristics.

(1) The terrain entropy reflects the size of the information contained in that terrain, and thus the undulating characteristics of the local terrain. The more dramatic the local elevation change, the more complex the terrain undulation and the richer the information, the larger the calculated entropy is and the more favorable the matching.
(2) The topographic entropy has the effect of smoothing the singular elevation measurement points.
(3) Topographic entropy has the ability to resist elevation measurement error, mainly because it can smooth out the measurement error of the altimetric sensor itself and the elevation measurement error caused by transient changes of the environment.

7.2 Topographic Elevation Field Contour Matching Method

Iterated Closest Contour Point (ICCP) is also a batch correlation matching method, which is often used in geomagnetic fields to describe the geomagnetic properties of a region. ICP is a method to match the measured data with an existing reference by iterating the points on the nearest contour. The algorithm first finds the coordinates of the nearest point on the geomagnetic field contour. Then it performs a rigid transformation using the "Euclidean distance squared minimum" as the evaluation function. By repeated iterations, the trajectory closest to the real track is obtained and outputted to correct the accumulated errors of the inertial guidance system [3].

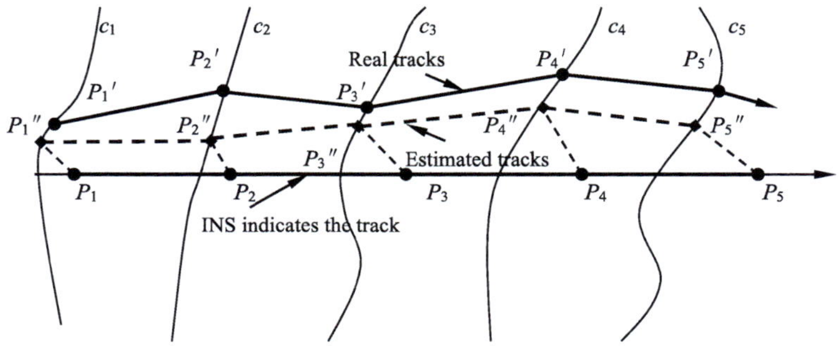

Fig. 7.6 Schematic diagram of the contour algorithm

7.2.1 Fundamentals

As Fig. 7.6 shown is the basic schematic of ICCP algorithm. The motion carrier has an actual running track (shown as the thick solid line in the figure), called the "real track", consisting of points such as P_i' $(i = 1, 2, \cdots, M)$, where M is the estimated track length (number of points); meanwhile, the INS gives a measured track (shown as the thin solid line in the figure), called the "INS track ", consisting of points such as P_i; in addition, the elevation measurement sensor measures the actual local elevation values c_i, each of which corresponds to an elevation contour line in the elevation map.

Due to the measurement error of INS, there is inevitably some deviation between P_i and P_i'. The idea of the contour algorithm is that P_i' must lie on the contour of c_i, then P_i can be brought close to c_i according to certain principles to find the optimal estimation point P_i'' and the trajectory to achieve the error compensation of the INS. This is actually an optimization process, for the curve segment composed of M points, the only thing that can be adjusted is that these points are located on their respective contours accordingly, and can be slid on their respective contours to finally minimize the objective function value. In addition, the following assumptions can be made: although the INS output has a certain amount of cumulative error, but the indicated position and the real position of the carrier is not far away. Thus the objective function can be expressed as follows.

$$E = d(\mathbf{x}_1, \mathbf{a}_1) + \sum_{i=2}^{M} d(\mathbf{x}_i - \mathbf{x}_{i-1}, \mathbf{a}_i - \mathbf{a}_{i-1}) + K \sum_{i=1}^{M} d(\mathbf{x}_i, \mathbf{y}_i) \tag{7.34}$$

where: E is the objective function, the total error; $d(\mathbf{p}, \mathbf{q})$ is the distance between $\mathbf{p}$ and $\mathbf{q}$; $\mathbf{x}_i$ is the position of the estimated point P_i'', $\mathbf{y}_i$ is the projection of $\mathbf{x}_i$ on the contour or the closest point to the contour; $\mathbf{a}_i$ is the INS indicator position; and K is the stiffness factor. The first and second terms of the objective function shown in Eq. (7.34) are to restrict the estimated point to the vicinity of the INS indicator

Fig. 7.7 Effect of carrier velocity and orientation errors on position errors

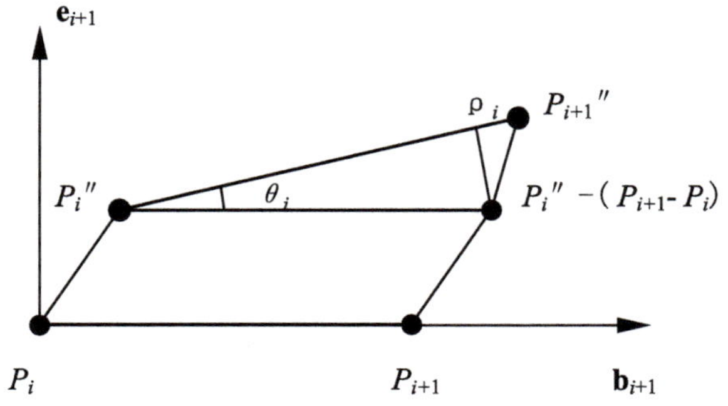

position, and the third term is to bring the estimated point closer to the contour of the measured value.

Therefore, this algorithm considers that the true position is in the vicinity of the INS indicated position and projects it onto the corresponding contour according to the nearest path principle to achieve the initial estimation of the INS position; then, according to the optimization principle, the objective function of Eq. (7.34) of the current INS sequence, the estimation of the P_i'' sequence is completed. Optimization methods can use various well-established algorithms, and in this project the tuned simplex method is used.

As shown in Fig. 7.7, the effect of velocity and direction error of the carrier on one-step navigation is shown in ρ_i for the carrier velocity error, θ_i for the carrier heading error, $\mathbf{b}_{i+1}$ for the unit vector along the navigation direction, and $\mathbf{e}_{i+1}$ for the unit vector in the direction perpendicular to $\mathbf{b}_{i+1}$. From Fig. 7.7, the following equation can be obtained.

$$\begin{aligned} \mathbf{x}_i &= \mathbf{x}_{i-1} + (\|\mathbf{a}_i - \mathbf{a}_{i-1}\| + \rho_i)(\cos\theta_i \mathbf{b}_i + \sin\theta_i \mathbf{e}_i),\ i > 1; \\ \mathbf{x}_1 &= \mathbf{a}_1 + \rho_1(\cos\theta_1 \mathbf{b}_1 + \sin\theta_1 \mathbf{e}_1),\ i = 1\,. \end{aligned} \tag{7.35}$$

If both ρ_i and θ_i are small quantities, then Eq. (7.35) can be approximated as

$$\begin{aligned} \mathbf{x}_i &\doteq \mathbf{x}_{i-1} + \mathbf{a}_i - \mathbf{a}_{i-1} + \rho_i \mathbf{b}_i + \zeta_i \mathbf{e}_i,\ i > 1; \\ \mathbf{x}_1 &\doteq \mathbf{a}_1 + \rho_1 \mathbf{b}_1 + \zeta_1 \mathbf{e}_1,\ i = 1. \end{aligned} \tag{7.36}$$

where: $\zeta_i = \|\mathbf{a}_i - \mathbf{a}_{i-1}\|\theta_i$, $\zeta_1 = \rho_1\theta_1$. Substituting Eq. (7.36) into Eq. (7.34) yields

$$E = \sum_{i=1}^{M} (\rho_i^2 + \zeta_i^2) + K \sum_{i=1}^{M} \|\mathbf{x}_i - \mathbf{y}_i\|^2 \tag{7.37}$$

For $\mathbf{x}_i$ ($i = 1, 2, \cdots, M$), it can be calculated based on Eq. (7.36) if $\mathbf{a}_i$, ρ_i and θ_i are known, while $\mathbf{a}_i$ is the INS indication position, which is known, so $\mathbf{x}_i$ is a function of ρ_i and θ_i. In addition, is derived from, so $\mathbf{y}_i\mathbf{x}_i\mathbf{y}_i$ is also an implicit function of and. $\rho_i\theta_i$

The optimization process is to optimize Eq. (7.37). Since $\mathbf{x}_i$ and $\mathbf{y}_i$ are functions of ρ_i and θ_i, the optimization variables are ρ_i and θ_i, and the objective function is E. In this project, the optimization of Eq. (7.37) is performed by the tuned simplex method. The advantage of the simplex method is that the gradient of the objective function does not need to be solved, but the objective function value is calculated and compared directly.

In the process of optimization, there is also the problem of choosing the length of the optimization sequence M. If M is too large, the computation time is too long and the real-time performance is not good; if M is too small, the optimization adjustment range is too small, and only the local optimal point may be found instead of the overall optimal point, which is not conducive to improving the matching accuracy. The determination of the suitable matching length will be studied in detail in Chap. 6 of this paper.

$K \propto K$ In Eq. (7.37), the stiffness coefficient K determines the flexibility of the estimated trajectory, the larger K is, the greater the flexibility of the estimated trajectory, the greater the range of adjustment of the points on the estimated trajectory in their respective contours. The smaller the flexibility of the estimated trajectory, that is, the smaller the adjustment range of each point on the estimated trajectory, if K tends to 0, then any small adjustment of the estimated trajectory (varying the values of ρ_i and θ_i), then it will lead to a substantial change of the target error, which drowns out the error generated by the mis-match, and therefore, it can no longer be optimized according to Eq. (7.34). At K tends to 0, since any small change in the trajectory may lead to a large change in the objective function value, it is only possible not to make any adjustment to the trajectory, but to make the objective function converge to the optimal value by rigid transformations (rotation and translation), at which point the objective function is expressed as follows.

$$E = \sum_{i=1}^{M} w_i \|\mathbf{y}_i - T_i \mathbf{x}_i\|^2 \tag{7.38}$$

where: w_i is the scale factor and T_i is the rotation and translation matrix. The optimal trajectory is finally found by optimizing Eq. (7.38). This method is suitable for two-dimensional map matching or one-dimensional contour matching, it requires a relatively accurate measurement of the current geographic feature by the geographic feature sensor with a small error.

7.2.2 Method for Determining the Nearest Contour Point

An equivalence line is generally composed of segments, as shown in Fig. 7.2, 7.3, where $\mathbf{f}_i$ $(i = 1, 2, \cdots, n)$ is the endpoint of the segment that makes up the equivalence line. For a given point outside the equivalence line $\mathbf{x}$, the closest point to the equivalence line can be determined as follows (Fig. 7.8).

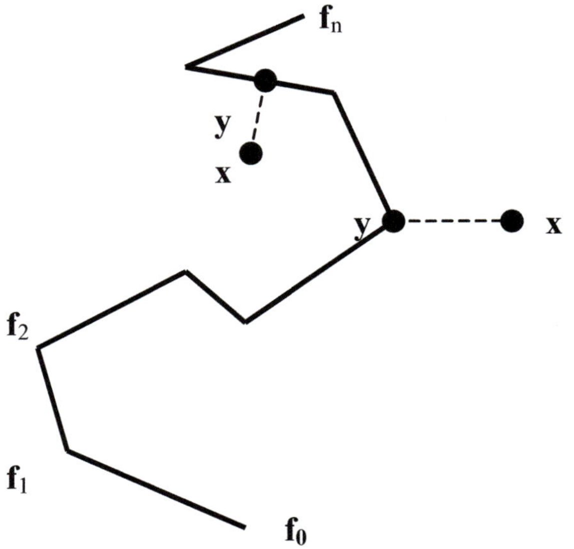

Fig. 7.8 Schematic diagram of the determination of the nearest contour point

First, find the nearest point from $\mathbf{x}$ to each segment of the line $\mathbf{y}_i$ $(i = 1, 2, \cdots, n-1)$; then, find the nearest point from the set of these nearest points $\{\mathbf{y}_i\}$ $(i = 1, 2, \cdots, n-1)$ to the point $\mathbf{x}$, which is the nearest point from $\mathbf{x}$ to the equivalence line.

For a given line segment F_i (consisting of two points from $(\mathbf{f}_i, \mathbf{f}_{i+1})$) on the equivalence line, the projection point of the point $\mathbf{x}$ on that line segment or its extension is

$$\mathbf{x}_i' = (1-\alpha)\mathbf{f}_i + \alpha\mathbf{f}_{i+1} \tag{7.39}$$

$$\alpha = \frac{(\mathbf{x}-\mathbf{f}_i)^{\mathrm{T}}(\mathbf{f}_{i+1}-\mathbf{f}_i)}{l_i^2} \tag{7.40}$$

where: $l_i = \|\mathbf{f}_{i+1} - \mathbf{f}_i\|$. When $0 \le \alpha \le 1$, the projection point $\mathbf{x}_i'$ is on the line segment F_i. Therefore, from Eqs. (2–6) and (2–7), the point $\mathbf{x}$ is nearest to the line segment F_i as

$$\mathbf{y}_i = \begin{cases} \mathbf{f}_i, \ \alpha \le 0; \\ \mathbf{x}_i', \ 0 < \alpha < 1; \\ \mathbf{f}_{i+1}, \ \alpha \ge 1. \end{cases} \tag{7.41}$$

If there are several points in the set of points $\{\mathbf{y}_i\}$ $(i = 1, 2, \cdots, n-1)$ on an equivalence line that have equal and smallest distances to the point $\mathbf{x}$, then the point

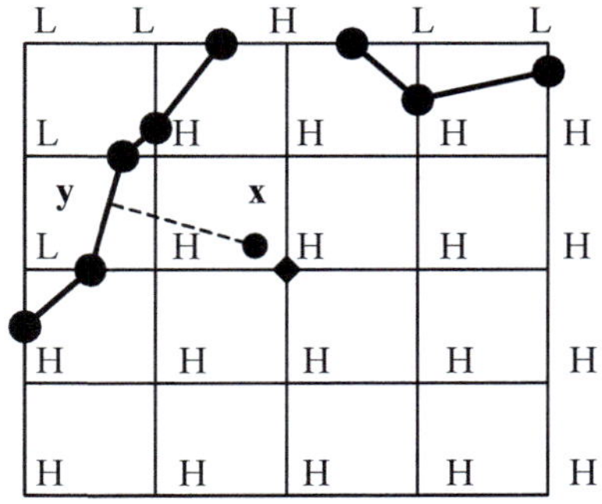

Fig. 7.9 Schematic diagram of the nearest contour point calculation for the point **x**

with the smallest ordinal number is used as the nearest point of the point **x** on that equivalence line.

7.2.3 Extraction of Contours

For a given point **x** (with error), the geographic feature measurement for that location is c and the contour extraction process is as follows (Fig. 7.9).

(1) Find the nearest map grid point to **x** (i, j).
(2) Find the contours in the 3×3 domain at the map centered on the points (i, j) and proceed to step (4) if there are contours, otherwise proceed to step (3).
(3) expand the search range ($5 \times 5, 7 \times 7, \ldots, n \times n$) until there are contours, if the search range is expanded beyond the map range and still cannot find contours, the point is considered problematic for geographic feature measurement and terminate the calculation, or further search in a larger map range.
(4) Calculate the nearest point position from the point **x** to the contour **y** in the same way as before.
(5) In the range of $(n + 2) \times (n + 2)$ further search for the equivalence point, calculate the nearest point position from the point **x** to the equivalence line $\mathbf{y}'$, if the distance from the point **x** to $\mathbf{y}'$ is less than the distance to **y**, then the point $\mathbf{y}'$ is used as the nearest point position from the point **x** to the equivalence line. The judgment formula is as follows.

$$d\left(\mathbf{x}, \mathbf{y}\right) < \min\left(\tfrac{n+1}{2} - |\mathbf{x}_x - i|, \tfrac{n+1}{2} - \left|\mathbf{x}_y - i\right|\right) \tag{7.42}$$

Otherwise, **y** is used as the closest point from the point **x** to the contour.

The specific process is shown in Fig. 7.2–7.4, where the grid in Fig. 7.2–7.4 denotes the map grid, and H and L denote the geographic feature values of the grid point greater than and less than c, respectively, where the thick black lines denote the contours within the range of the grid.

7.2.4 Further Improvement of the Algorithm

From the matching criterion shown in Eq. (2–1), it is clear that the ICCP algorithm is based on the following two assumptions.

(1) The true position of the carrier is in the vicinity of the INS indicated position.
(2) The true position of the carrier is near the contour of the measured elevation value.

For the assumption (2), there is generally no problem because the matching is based on the condition that the elevation measurement is accurate, then the true position is on the contour of that elevation value. However, for assumption (1), when the initial INS error before matching is relatively large, it may cause the algorithm to fail. That is, the actual situation is that the INS indicated position is far from the true position, and the contour matching algorithm still matches according to assumption (1) and searches near the INS indicated position, which will lead to a slowing down or even an increase in the trend of decreasing the objective function value, the result is scattered. Therefore, from the principle, the contour matching algorithm is more suitable for fine matching with small errors.

Therefore, a matching method is given below when the initial error is relatively large, TERCOM is fused with ICCP.

In the case of large INS initial error, a coarse search can be performed first to make the searched tracks closer to the real position, and then, the searched tracks can be used as the input tracks for contour matching (alternative INS tracks) for further matching to improve the matching accuracy.

In the initial search phase, the mean square difference matching criterion MSD (Mean Square Difference) and mean absolute difference matching criterion MAD (Mean Absolute Difference) are used for search matching, referring to Equations. (7.2) and (7.3). Calculate the magnitude of their MAD and MSD, respectively. Among all the MADs and MSDs, the elevation sequences corresponding to the smallest MAD and MSD are found out respectively. If the elevation sequences corresponding to the two are at the same position, the trajectory is used instead of the INS trajectory for subsequent contour matching; if the trajectories corresponding to the two are different, the absolute difference between the elevation sequence values of the two and the measured elevation sequence value is compared again, and the one with the smaller difference is used as the coarse matching output trace. Then, further matching is performed according to the previous ICCP algorithm. The improved algorithm gives full play to the respective advantages of the original algorithm and can avoid the traditional ICCP algorithm which is easy to diverge and fail to match properly under the large INS initial error; in addition, the estimated trajectory is made closer to the real one by further tuning in the fine matching stage. Therefore, the improved contour algorithm not only expands the application range, but also improves the matching accuracy.

7.3 Iterative Recursive Filtering Matching Method

The iterative recursive filter matching method (SITAN algorithm) uses the extended KALMAN filter algorithm to recursively process the underwater measurement data to correct the INS drift error in real time. Using the INS error model, the state equation of the system is established; meanwhile, the elevation value corresponding to the INS output position is obtained on the reference terrain elevation field map (model), and the measurement equation of the system is established based on the actual measured elevation value. Based on the established system state equation and system measurement equation, KALMAN optimal estimation is performed to correct the INS error in real time. The principles of the underwater and aerial algorithms are similar, so for the convenience of description, this chapter introduces and describes the underwater application as the background.

7.3.1 Extended Kalman Filtering Principle

KALMAN filter estimation is a recursive linear minimum variance estimation method developed by KALMAN R.E. in the early sixties and is an important part of modern control theory. It uses state space and recursive computational methods to estimate smooth and non-smooth stochastic processes. Therefore, linear minimum variance estimation of various information (states) of stochastic dynamic systems can be obtained with the KALMAN filter. If combined with optimal control, optimal control of stochastic dynamic systems can be achieved.

Let the equation of state and the equation of measurement of the discrete system be

$$\begin{aligned} \mathbf{X}_k &= \mathbf{\Phi}_{k,k-1}\mathbf{X}_{k-1} + \mathbf{\Gamma}_{k-1}\mathbf{W}_{k-1} \\ \mathbf{Z}_k &= \mathbf{H}_k\mathbf{X}_k + \mathbf{V}_k \end{aligned} \tag{7.43}$$

where is the $\mathbf{X}_k n$-dimensional state vector at the moment of k, which is the estimated quantity; is the $\mathbf{Z}_k m$-dimensional measurement vector at the moment of; $k\mathbf{\Phi}_{k,k-1}$ is the one-step transfer matrix at the moment of to $k - 1kn \times n$ (order); $\mathbf{W}_{k-1}$ is the system noise vector at the moment of $k - 1$ (r dimension); $\mathbf{\Gamma}_{k-1}$ is the system noise matrix ($n \times r$ order), which represents the degree to which each noise from $k - 1$ to k affects each state separately; $\mathbf{H}_k$ is the measurement matrix at the moment of k ($m \times n$ order); and $\mathbf{V}_k$ is the m-dimensional measurement noise vector at the moment of k.

Solving for the Eq. (7.43), it is required that $\{\mathbf{W}_k\}$ and $\{\mathbf{V}_k\}$ are mutually uncorrelated zero-mean white noise sequences with.

$$\begin{aligned} E\{\mathbf{W}_k\} &= 0 \; E \; \{\mathbf{W}_k\mathbf{W}_j^T\} = \mathbf{Q}_k\delta_{kj} \\ E\{\mathbf{V}_k\} &= 0 \; E \; \{\mathbf{V}_k\mathbf{V}_j^T\} = \mathbf{R}_k\delta_{kj} \end{aligned} \tag{7.44}$$

$\mathbf{Q}_k$ is called the system noise variance matrix and $\mathbf{R}_k$ is called the measurement noise variance matrix, and both are required to be known in the KALMAN filtering process, and $\mathbf{Q}_k$ is a non-negative fixed array and $\mathbf{R}_k$ is a positive fixed array. δ_{kj}*is Kronecker* δ function.

$$\delta_{kj} = \begin{cases} 0, \; k \neq j \\ 1, \; k = j \end{cases} \tag{7.45}$$

Also, KALMAN filtering requires that the statistical characteristics of the initial state are known quantities, $E[\mathbf{X}_0] = \boldsymbol{\mu}_0$, $Var[\mathbf{X}_0] = \mathbf{C}_{\mathbf{X}_0} = \mathbf{P}_0$, and requires that $\mathbf{X}_0$ be uncorrelated with $\{\mathbf{W}_k\}$ and $\{\mathbf{V}_k\}$,

$$\text{cov}(\mathbf{W}_k\mathbf{X}_0) = 0, \; E\{\mathbf{W}_k\mathbf{W}_j^T\} = \mathbf{Q}_k\delta_{kj} \tag{7.46}$$

For the above system and measurements, and the statistical characteristics about them, a set of KALMAN filter equations according to Eqs. (1–3) to (3–5) can be used to find the estimated value $\hat{X}_k$ using the measured value $\mathbf{Z}_k$ and the estimated value $\hat{\mathbf{X}}_{k-1}$ derived from the previous moment.

State one-step prediction equation is

$$\hat{\mathbf{X}}_{k,k-1} = \mathbf{\Phi}_{k,k-1}\hat{\mathbf{X}}_{k-1} \tag{7.47}$$

State estimation equation is

$$\hat{\mathbf{X}}_k = \hat{\mathbf{X}}_{k,k-1} + \mathbf{K}_k\boldsymbol{\varepsilon}_k \tag{7.48}$$

Filter gain equation is

$$\mathbf{K}_k = \mathbf{P}_{k/k-1}\mathbf{H}_k^T(\mathbf{H}_k\mathbf{P}_{k,k-1}\mathbf{H}_k^T + \mathbf{R}_k)^{-1} \tag{7.49}$$

One-step prediction mean square error equation is

$$\mathbf{P}_{k,k-1} = \mathbf{\Phi}_{k,k-1}\mathbf{P}_{k-1}\mathbf{\Phi}_{k,k-1}^T + \mathbf{\Gamma}_{k-1}\mathbf{Q}_{k-1}\mathbf{\Gamma}_{k-1}^T \tag{7.50}$$

Estimated mean square error equation.

$$\mathbf{P}_k = (\mathbf{I} - \mathbf{K}_k\mathbf{H}_k)\mathbf{P}_{k,k-1} \tag{7.51}$$

Or

$$\mathbf{P}_k = (\mathbf{I} - \mathbf{K}_k\mathbf{H}_k)\mathbf{P}_{k,k-1}(\mathbf{I} - \mathbf{K}_k\mathbf{H}_k)^T + \mathbf{K}_k\mathbf{R}_k\mathbf{K}_k^T \tag{7.52}$$

where, $\boldsymbol{\varepsilon}_k$ is the new interest measured for the kth

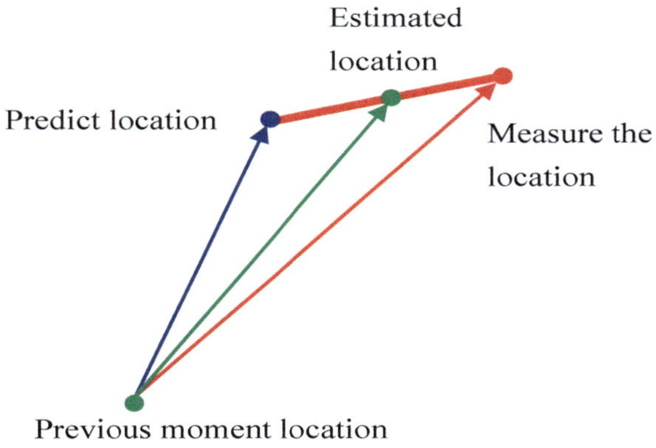

Fig. 7.10 Schematic diagram of KALMAN filtering principle

$$\boldsymbol{\varepsilon}_k = \mathbf{Z}_k - \mathbf{H}_k \hat{\mathbf{X}}_{k,k-1} \tag{7.53}$$

$\boldsymbol{\varepsilon}_k$ indicates $\mathbf{Z}_k$ the "new" information remaining after subtracting the "old" information about $\mathbf{Z}_k$ predicted in the previous k observation from the k observation.

The KALMAN one-step filtering process can be represented by Fig. 7.10, in the case of the previous moment position is known, the filtering algorithm first predicts a current moment position, at the same time, the system measurement unit also provides a measurement position, and the filtering algorithm balances the information of both to get the estimated position of the current moment as the final output quantity. When the system model and the error model are accurate and the previous setting conditions are met, then the estimated position is optimal, it is the closest to the current true position.

In underwater terrain elevation field navigation and positioning engineering applications, it is impossible to construct a simple linear system due to the complex motion model of underwater motion carriers and the complex and variable seafloor topography. To solve this problem, a technique called extended Kalman filtering is used, which linearizes the nonlinear system and then performs the solution of the Kalman filter equation.

The following models are available for nonlinear discrete systems.

$$\begin{aligned} \mathbf{X}_k &= f(\mathbf{X}_{k-1}, k-1) + \boldsymbol{\Gamma}_{k-1}\mathbf{W}_{k-1} \\ \mathbf{Z}_k &= h(\mathbf{X}_k, k) + \mathbf{V}_k \end{aligned} \tag{7.54}$$

where $f(*)$ is an n-dimensional nonlinear vector function; $h(*)$ is an m-dimensional nonlinear vector function.

The extended Kalman filter is the linearization of the nonlinear functions $f(*)$, $h(*)$ around the values $\hat{X}_k$. The usual linearization method used is to expand the nonlinear function Taylor series and omit the second order and higher terms.

Expanding the nonlinear function $f(*)$ in the state equation of Eq. 3–13 around the filtered value $\hat{X}_k$ into a Taylor series and omitting terms above the second order, we get

$$\mathbf{X}_k \approx f(\hat{\mathbf{X}}_{k-1}, k-1) + \frac{\partial f}{\partial \hat{\mathbf{X}}_{k-1}}(\mathbf{X}_{k-1} - \hat{\mathbf{X}}_{k-1}) + \mathbf{\Gamma}_{k-1}\mathbf{W}_{k-1} \tag{7.55}$$

Let $\frac{\partial f}{\partial \hat{\mathbf{X}}_{k-1}} = \left.\frac{\partial f(\hat{\mathbf{X}}_{k-1}, k-1)}{\partial \mathbf{X}_{k-1}}\right|_{\mathbf{X}_{k-1}=\hat{\mathbf{X}}_{k-1}} = \mathbf{\Phi}_{k,k-1}$, the $f(\hat{\mathbf{X}}_{k-1}, k-1) - \left.\frac{\partial f(\hat{\mathbf{X}}_{k-1}, k-1)}{\partial \mathbf{X}_{k-1}}\right|_{\mathbf{X}_{k-1}=\hat{\mathbf{X}}_{k-1}} * \hat{\mathbf{X}}_{k-1} = \phi_{k-1}$.

Then the equation of state is

$$\mathbf{X}_k = \mathbf{\Phi}_{k,k-1}\mathbf{X}_{k-1} + \mathbf{\Gamma}_{k-1}\mathbf{W}_{k-1} + \phi_{k-1} \tag{7.56}$$

At the same time will be the type

$$\begin{aligned} \mathrm{X}_k &= f(\mathrm{X}_{k-1}, k-1) + \Gamma_{k-1}\mathrm{W}_{k-1} \\ \mathrm{Z}_k &= h(\mathrm{X}_k, k) + \mathrm{V}_k \end{aligned} \tag{7.57}$$

The nonlinear function in the observation equation $h(*)$ expands into a Taylor series around the filtered values $\hat{\mathbf{X}}_{k,k-1}$ and omits terms above the second order, giving

$$\mathbf{Z}_k = h(\hat{\mathbf{X}}_{k,k-1}, k) + \frac{\partial h}{\partial \hat{\mathbf{X}}_{k,k-1}}(\mathbf{X}_k - \hat{\mathbf{X}}_{k,k-1}) + \mathbf{V}_k \tag{7.58}$$

make

$$\frac{\partial h}{\partial \hat{\mathbf{X}}_{k,k-1}} = \mathbf{H}_k,\ y_k = h(\hat{\mathbf{X}}_{k,k-1}, k) - \frac{\partial h}{\partial \hat{\mathbf{X}}_{k,k-1}} * \hat{\mathbf{X}}_{k,k-1}$$

Then the observation equation is

$$\mathbf{Z}_k = \mathbf{H}_k\mathbf{X}_k + y_k + \mathbf{V}_k \tag{7.59}$$

After linearizing the state and observation equations, the filtering solution can be performed by applying the Kalman fundamental filtering equation. This method gives a better solution when the nonlinearity of the system is not high, but if the nonlinearity is high, the filtering tends to diverge, so an appropriate way needs to be found to improve this situation.

7.3.2 Extended Kalman Filtering Algorithm Applications

The equation of state of the terrain elevation field navigation and positioning system is a simplification of the equation of state of the inertial guidance system, and the indirect method is used here to estimate the error of the inertial guidance system.

1. The Establishment of the System Equation of State

The inertial guidance system uses the inertial guidance of the north pointing platform and the northeast sky coordinate system. The 3D position errors δx, δy and δz and 2D velocity errors δv_x, δv_y are taken as states and noted as

$$X = \left[\delta x \ \delta y \ \delta v_x \ \delta v_y\right]^T \tag{7.60}$$

Using the difference method to establish the equation of state of the system, in order to meet the requirements of the Kalman filter implementation calculation, the equation is simplified and the higher order minima are omitted, we can get

$$\dot{X} = FX + W \tag{7.61}$$

where, F is the system array, W is the noise array, and

$$F = \begin{bmatrix} 0\,0\,0\,1\,0 \\ 0\,0\,0\,0\,1 \\ 0\,0\,0\,0\,0 \\ 0\,0\,0\,0\,0 \\ 0\,0\,0\,0\,0 \end{bmatrix}; W = \begin{bmatrix} w_x \\ w_y \\ w_z \\ w_{v_x} \\ w_{v_y} \end{bmatrix} \tag{7.62}$$

The system is discretized with filtering period T. The equation of state of the system is

$$X_{k+1} = \Phi_{k+1,k} X_k + \omega_k \tag{7.63}$$

Among them.

$$\Phi_{k,k-1} = \begin{bmatrix} 1 & 0 & T & 0 \\ 0 & 1 & 0 & T \\ 0 & 0 & 1 & 0 \\ 0 & 0 & 0 & 1 \end{bmatrix}$$

2. Figure 7.11 is a Schematic Diagram of the Parameters of the Carrier During Navigation.

Where h_t is the depth of the seafloor at the location of the submarine as a function of the submarine's position (x, y), h_q is the dive depth of the submarine, and h_s is the distance of the submarine from the seafloor.

Let the 3D position information indicated by the inertial guide be $(x_{ins}, y_{ins}, h_{ins})$, the predicted elevation can be read from the digital chart $\hat{h}_t(x_{ins}, y_{ins})$, and

$$\hat{h}_t(x_{ins}, y_{ins}) = h_t(x_{ins}, y_{ins}) + \gamma_{map} \tag{7.64}$$

Among other things, γ_{map} includes measurements and quantization of noise during digital map production.

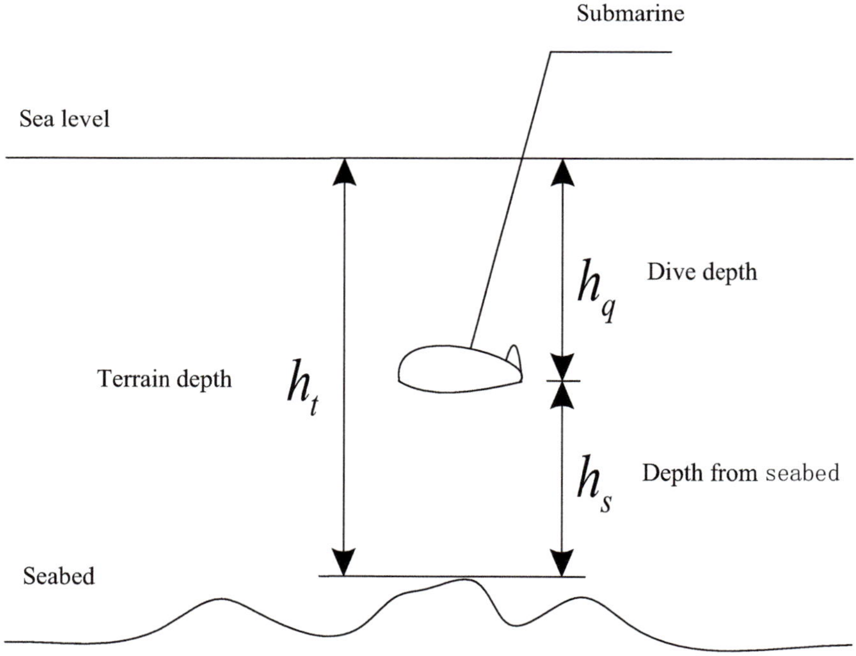

Fig. 7.11 Schematic diagram of the parameters of the submarine while underway

The depth of the submarine from the seafloor is estimated using the position given by the inertial guidance system and the value of the dive sounder, noted as $\hat{h}_s$, and

$$\hat{h}_s = \hat{h}_t(x_{ins}, y_{ins}) - \hat{h}_q(x, y) \tag{7.65}$$

In addition, the depth of the submarine from the bottom can be obtained from the bathymetry $\tilde{h}_s(x, y)$, and

$$\tilde{h}_s(x, y) = h_s(x, y) + \gamma_s \tag{7.66}$$

Observing the error of both, we have

$$\begin{aligned} Z = \hat{h}_s - \tilde{h}_s &= \hat{h}_t(x_{ins}, y_{ins}) - \hat{h}_q(x, y) - h_s(x, y) - \gamma_s \\ &= h_t(x_{ins}, y_{ins}) - h_q(x, y) - h_s(x, y) + \gamma_{map} - \gamma_q - \gamma_s \end{aligned} \tag{7.67}$$

A first-order linearized expansion of $h_t(x_{ins}, y_{ins})$ in Eq. yields

$$h_t(x_{ins}, y_{ins}) = h_t(x, y) + \frac{\partial h_t(x, y)}{\partial x}\delta x + \frac{\partial h_t(x, y)}{\partial y}\delta y + \gamma_l \tag{7.68}$$

where, γ_l is the linearization error.

Also from $h_t(x, y) = h_q(x, y) + h_s(x, y)$, and let $h_x = \frac{\partial h_t(x,y)}{\partial x}$, $h_y = \frac{\partial h_t(x,y)}{\partial y}$, $\gamma = \gamma_{map} - \gamma_s - \gamma_q + \gamma_l$, then we have

$$Z = HX + \gamma \tag{7.69}$$

Among them.

$$H = \begin{bmatrix} h_x & h_y & 0 & 0 \end{bmatrix} \tag{7.70}$$

For discrete systems, the

$$Z_k = H_k X_k + \gamma_k \tag{7.71}$$

Equation (7.56) is the measurement equation of the system.

3. Topographic Elevation Field Linearization Method

The terrain linearization technique applied in the nonlinear Kalman filter is the key to the success of terrain elevation field navigation and positioning systems. Terrain linearization is achieved by replacing surface equations with plane equations within a region whose size is proportional to the immediate position uncertainty. The terrain linearization accomplishes two tasks: (1) obtaining the predicted values of the resulting elevation measurements; and (2) deriving the slope for calculating the Kalman gain at h_x, h_y.

Fig. 7.12 Topographic linearization nine-point fitting method

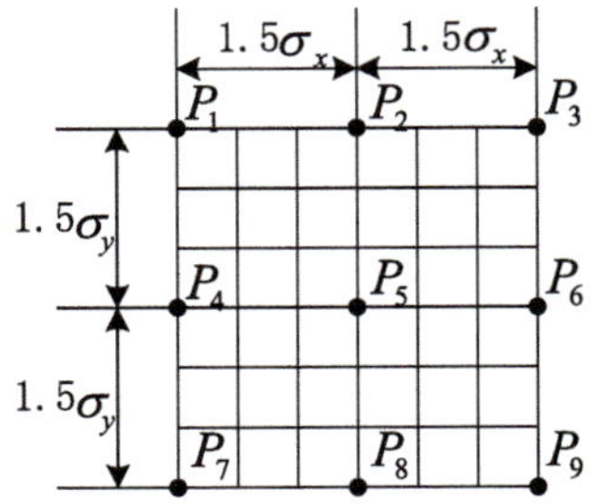

The common methods of random linearization of terrain are first-order Taylor expansion, nine-point fitting method, full-plane fitting method and so on. When the uncertainty range is large, we cannot simply use the first-order Taylor expansion, and the full-plane fitting method is too large, so we choose the "nine-point fitting method", as shown in Fig. 7.4 (Fig. 7.12).

Let σ_x and σ_y be the position errors obtained in the Kalman filtering process plus the root mean square value, and take the prediction point P_5 as the center $\pm 1.5\sigma$ range as the fitting area, and take the depth values of the center point and the surrounding eight points for calculation, we can get the corresponding values of h_x and h_y as

$$h_x = \frac{(h_3 + h_6 + h_9) - (h_1 + h_4 + h_7)}{6Md} \tag{7.72}$$

$$h_y = \frac{(h_7 + h_8 + h_9) - (h_1 + h_2 + h_3)}{6Nd} \tag{7.73}$$

where d is the grid interval of the map and M and N are integers, determined by the area taken for the fit.

7.3.3 *Extended Kalman Filtering Algorithm Working Mode*

The extended Kalman filter algorithm has three modes of operation, namely search mode, tracking mode and loss mode. The control logic determines the transition between the three modes: the algorithm enters the search mode when the system has a large initial error, the algorithm enters the tracking mode when the position error is in a smaller range, and the algorithm enters the loss mode if the algorithm control logic concludes during the search phase that the carrier is not in a certain search area.

1. Search mode.

The search mode is used to reliably and efficiently locate a carrier within a large uncertainty region. When the initial uncertainty error reaches hundreds or even thousands of meters, the navigation distance required for a single Kalman filter to reach the steady state is too large, and the excessive terrain linearization error will lead to filter scattering, the common method is to introduce a parallel Kalman filter array over the entire search area.

2. Tracking mode.

The purpose of the tracking mode is to estimate the position of the carrier continuously as accurately as possible. After the algorithm enters the tracking mode, the mathematical model introduced in 2.3.2 is used to perform the filtering solution. During the solving process, the main factors affecting the performance of the algorithm are the degree of nonlinearity of the terrain and the size of the linearization fitting error. In this thesis, the linearization region is determined based on the present covariance estimate Pk and the state error then linearized using the nine-point fitting method. Refer to Sect. 2.3.2 for other technical details.

3. Lost mode.

When the algorithm's judgment logic concludes that the true location of the carrier is not within the search area, the algorithm enters the lost mode. When in lost mode, the algorithm does not provide any valuation.

4. Mode Control Logic

Design of search, tracking and lost modes As mentioned earlier, the mode control logic is used to determine which mode the algorithm should work in, with the algorithm working in search mode when the error is large and in tracking mode when the error is small. Keeping the false locus rate low during the mode transition is the main guiding principle for designing the pattern control logic. To achieve this goal, the control logic considers the following two aspects.

A. When switching from search mode to tracking mode, it should be ensured that the correct filter is found.
B. In the tracking mode, if there is filtering divergence, the algorithm should switch to other modes in time.

Following this principle, the algorithm designs a detailed transformation logic between the three modes.

1. Search mode into tracking mode conditions.
2. Tracking mode to search mode conditions.

In tracking mode, if the following conditions occur:

A. *SWRS* is greater than a certain threshold for several consecutive times.
B. If the filtering does not converge after a period of time, the tracking filtering is considered to have failed, and the tracking mode is exited, and the search mode is re-entered.

The smoothed weighted residual squared (*SWRS*) is used here as a metric to examine the degree of matching of the filter. *SWRS* The smaller the sub-filter is, the better the match is considered roughly. It is calculated as follows.

Residuals is

$$e_t = Z_t - X_t \tag{7.74}$$

where Z_t and X_t are the observed and state values at the time of t, respectively.

Normalized residuals squared.

$$WRS_t = \frac{e_t^2}{P_t + R} \tag{7.75}$$

where P_t and R are the state variance array and measurement noise variance array of the system, respectively.

Smoothed weighted residuals squared.

$$SWRS_t = \alpha WRS_t + (1 - \alpha) SWRS_{t-1} \tag{7.76}$$

where α is the weighting factor.

It can be seen that $SWRS$ examines the overall situation of the filter state at all past moments and adds heavier weighting factors to the covariance values for moments closer to the current moment.

7.3.4 Improvement of SITAN Algorithm

The mathematical model state equations established in topographic elevation field navigation and positioning applications are linear and the measurement equations are nonlinear, thus requiring adjustments to the mathematical model.

The state vector is dimensioned and the observation vector Z is added to the state vector X,

$$X = \left[\delta x \ \delta y \ \delta h \ \delta v_x \ \delta v_y \ z \right]^T \tag{7.77}$$

When the state of each set member is predicted, the transfer equation of the first five quantities remains unchanged and the transfer equation of the sixth quantity is

$$z = h_{x_e, y_e} - h_{x_{ins}, y_{ins}} = h'_{x_e, y_e} - h'_{x_{ins} y_{ins}} + \omega_z \tag{7.78}$$

$$\begin{aligned} x_e &= x_{ins} + \delta\varphi \\ y_e &= y_{ins} + \delta\lambda \end{aligned} \tag{7.79}$$

h_{x_e, y_e} is the true elevation at (x_e, y_e), $h_{x_{ins}, y_{ins}}$ is the true elevation at (x_{ins}, y_{ins}), h'_{x_e, y_e} is the elevation of the map at (x_e, y_e), $h'_{x_{ins}, y_{ins}} h_{x_{ins}, y_{ins}}$ is the elevation of the map at (x_{ins}, y_{ins}), and ω_z is the interpolation error plus the mapping error.

Measurement equation is

$$
\begin{aligned}
Z_k &= h_{measure} - h'_{x_{ins},y_{ins}} = h_{true} - h_{x_{ins},y_{ins}} + \gamma_{measure} - \gamma_{map} \\
&= HX_k + V_k
\end{aligned}
\tag{7.80}
$$

$$H = [0, 0, 0, 0, 0, 1] \tag{7.81}$$

$$V_k = \gamma_{measure} - \gamma_{map} \tag{7.82}$$

where $h_{measure}$ is the elevation measurement, $h'_{x_{ins},y_{ins}}$ is the elevation at (x_{ins}, y_{ins}) in the map, $h_{x_{ins},y_{ins}}$ is the true elevation at (x_{ins}, y_{ins}), $\gamma_{measure}$ is the measurement noise, and γ_{map} is the mapping error.

The newly established mathematical model state equation is nonlinear and the measurement equation is linear. The improved algorithm has no linearization error effect and works well when the uncertainty region is large, and the topographic linearization error is small when the uncertainty region is small. Therefore, this improved algorithm is used to work instead of the extended Kalman filtering algorithm in the search phase, and then enter the tracking phase when the filtering is stable.

The smoothed weighted residual squared ($SWRS$) is used as a metric to examine the degree of matching of the Kalman filter, which is calculated as

$$\overline{A_k^a} = A_k^a 1_N = (X_k^a, X_k^a, \ldots, X_k^a) \in R^{n\times N} \tag{7.83}$$

$$e_k = X_k^a[6] - Z_k \tag{7.84}$$

$$WRS_k = \frac{e_k^2}{P_{e,k}^a + R_e} \tag{7.85}$$

$$SWRS_k = \alpha WRS_k + (1-\alpha)SWRS_{k-1} \tag{7.86}$$

where $P_{e,k}^a$ is the approximate state variance array of the Kth filter, R_e is the Kth measurement noise approximate variance, Z_k is the Kth observation, and α is the weighting factor; the control logic for the search phase into the tracking phase is: N consecutive $SWRS < 1.4$.

7.4 Summary

This chapter introduces the positioning methods commonly used in the navigation and positioning of terrain elevation fields, explains the basic principles of each positioning method, analyzes the technical characteristics and applicable conditions of each positioning method, and proposes corresponding solutions. In actual use, one or more different positioning and navigation methods should be combined with the actual situation and selected to solve specific problems with the idea of fusion to achieve satisfactory positioning results.

References

1. Golden, J. P. (1980). Terrain Contour Matching (TERCOM): A cruise missile guidance aid. *SPIE*.
2. Yu Jiacheng et al. (2007). Technique of two searches on SITAN algorithm for seabed terrain-aided navigation. *Acta Electronica Sinica*, (03): 474-477.
3. Xu Xiaosu et al. (2014). ICCP algorithm for underwater terrain matching navigation based on affine correction. *Journal of Chinese Inertial Technology*, *22*(03): 362–367

Chapter 8
Formation Mechanism of Earth-Sky Polarization Field and Basis for Navigation and Positioning

When solar radiation enters the earth, it will be affected by the refraction, scattering and reflection of atmospheric particles and the ground surface, resulting in the polarization of sunlight. The academic journals of Science and Nature have sporadically reported related discoveries, such as the visual polarized light of bats for precise navigation [1], the polarized light mystery for ancient navigators to identify the direction in the Atlantic Ocean, the earth's vegetation traces selected by the ESA from the weakly polarized reflected light of the moon, and the solar blaze spectrum first time obtained by polarization detection. Thus, the scientific question is that: (1) whether the solar polarization phenomenon can be understood and described by the concept of "field"; (2) whether it can be proved that it has the detection ability of "weak light strengthening or strong light weakening" to accurately identify the atmosphere, ocean transpiration and surface interaction of energy in the ecosphere for the innovation and development of remote sensing technology; (3) whether the polarized field can be combined with the earth's gravitational field and geomagnetic field to form the "three field theories" for deepening earth science research?

After summarizing the commonality of different fields, differential expression and energy transfer methods, our team has pioneeringly created the study of sky polarization fields in the world. The main contents can be generally described as:

(1) The force line of the polarized field and the law of the polarization pattern in the sky were explored. Based on the long-term observations in different times and latitudes, a sun-centred sky polarization pattern is formed through model analysis, the concept of polarized force lines is defined by connecting different intensity values, the space-based/ground-based land-sea surface areas and clean/dust atmospheric polarization images are obtained, and a three-dimensional tomography method for non-contact measurement of atmospheric particles is formed. We have published the benchmark article on polarization biomimetic navigation pattern map with the highest citation rate in China [2], which promoted the development and application in the field of navigation polarized light in our

L. Yan et al., *Mathematical and Physical Fundamentals of Navigation and Positioning with Earth's Natural Fields*, Navigation: Science and Technology 13,
https://doi.org/10.1007/978-981-97-5524-0_8

country, proved the objective stability and model availability of force lines in the polarized field.

(2) To develop the theoretical "atmospheric window" of the polarization field source point to form the "new atmospheric window" of remote sensing, i.e., the field "axis". This field axis is crucial for using remote sensing data to recognize the characteristic spectrum of ground objects and is traditionally defined as the frequency region with the least attenuation of different ground objects. Taking the sun as the centre, the area where the atmospheric polarization effect is zero in the centre of the sun is called the polarization neutral point and is defined as the source point of the polarization field. The "New Atmospheric Window" theory is generally applicable to different ground objects; through a series of aerial earth observations to verify and transform the relevant auxiliary mirrors on the National Astronomical Observatory's Sun Tracking Telescope, it is found that the observation information obtained based on the "New Atmospheric Window" has increased by over 70% compared with the traditional method, and then propose a design plan for the observation of solar synchronous polarized satellites.

(3) The basic theory of sky polarization field combined with field source and force line is formed. Which can be simply summarized as:

   (a) To clarify the nature of the polarized field. It transmits energy to objects in the form of solar electromagnetic waves through the polarization of the earth's geosphere.
   (b) To clarify the nature of polarized light waves. The scalar propagation equation of light has three elements: intensity, phase, and frequency, and is the physical basis for the geometric-spectral-radiometric-temporal resolution of remote sensing images. Based on the optical theory of light or vector transverse wave, light vibration is perpendicular to the direction of light propagation; polarization is a measure of the asymmetry of light transverse vibration under the action of light and the surface medium. It is the vector analysis of light intensity and constitutes the fourth element of light waves.
   (c) To clarify the nature of polarization remote sensing. It is to perfect the scalar equation of light intensity, which is traditionally based on remote sensing, into a vector equation containing all optical and physical elements; thus, it can "cut" the sunlight to measure the microscopic particle effect of the remote sensing medium, so that the three elements of remote sensing ground objects, atmosphere and instruments can be measured. The quantitative physical limit can be approached to the light particle scale. In order to promote the application of remote sensing in the fields of geography, atmosphere, ocean and even the earth, and greatly improve the accuracy of remote sensing inversion, new elements are added. A. Stewart, an academician of the United States and the United Kingdom, commented that the work "has added a new dimension to the understanding of the earth's environmental radiation field, and is an outstanding work in the field of earth observation".

The basic principle of polarized light field navigation is that when sunlight or moonlight enters the atmosphere and is scattered by atmospheric molecules, a polarization pattern map symmetrical to the line connecting the zenith and the sun is formed in the sky. Animals are based on their own axes and polarization forces. The included angle of the line is used to detect its own position, that is, the direction information required for navigation is obtained from the polarized vision of the animal; at the same time, the navigation function is completed based on a certain reference point such as a nest.

What is the polarization state of sunlight after being scattered by atmospheric molecules? How does the atmospheric environment affect the scattering of sunlight? How to obtain the state information of the polarized light field vector after being scattered by atmospheric molecules by means of measurement and realize navigation? This is the main content that will be discussed in this chapter.

8.1 Mathematical and Physical Basis of Polarized Light Fields

Light waves are transverse waves, and the vibration direction of the light vector is perpendicular to the propagation direction of the light wave, as shown in Fig. 8.1. In the plane perpendicular to the propagation direction, the electric field strength vector may also have various vibration directions, showing different polarization states. Light waves with different polarization states have different properties, and we call the property that the vibration direction of light is asymmetric relative to the direction of light propagation as the polarization property of light. The most common polarization states of light in practice are roughly divided into five types: unpolarized, linearly polarized, partially polarized, circularly polarized and elliptically polarized.

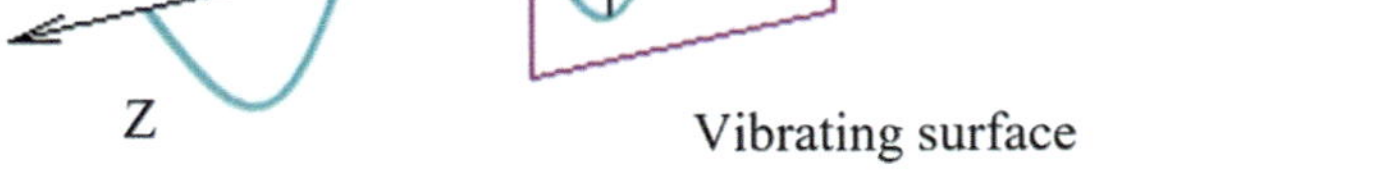

Fig. 8.1 The schematic diagram of light transverse

8.1.1 Polarization States of Light Waves

According to whether the vibration direction of the light vector has symmetry with respect to the propagation direction in the plane perpendicular to the propagation direction, light waves can be divided into unpolarized light and polarized light, and polarized light with asymmetry can be divided into fully polarized light and partially polarized light. Polarized light [3]. Only a brief introduction to the polarization state of light waves is given here. Interested scholars can refer to the author's other monograph "Polarization Remote Sensing Physics" [4].

1. Unpolarized light (natural light)

The light waves emitted by ordinary light sources are not a single plan-polarized light, but the sum of many light waves: they have all possible vibration directions, the amplitudes in each vibration direction have the same average value in the observation time, and the initial phases are not related to each other. Light with this characteristic is called unpolarized light or natural light. Natural light can be represented by two mutually perpendicular light vectors. The amplitudes of these two light vectors are the same, but the phase relationship is uncertain and changes rapidly. These two light vectors must not be synthesized into a stable or regularly changing complete polarized light. Figure 8.2 is usually used to represent natural light, where (a) represents the distribution of natural light in a plane perpendicular to the propagation direction, (b) represents that natural light can be represented by two mutually perpendicular light vectors, and (c) represents that natural light can be equivalent to Superposition of two plane-polarized lights with uncertain phase relationship and perpendicular vibration directions.

2. Fully polarized light

The vibration direction of the light vector does not change, but the magnitude changes with the phase. This kind of light is called fully polarized light. If the light vector of each point in the direction of light propagation is in a certain plane, this light is called plane-polarized light. plane-polarized light is often represented in Fig. 8.3, where (a) represents plane-polarized light with electric vector perpendicular to the surface of the drawing, (b) represents plane-polarized light with electric vector parallel to the surface of the drawing.

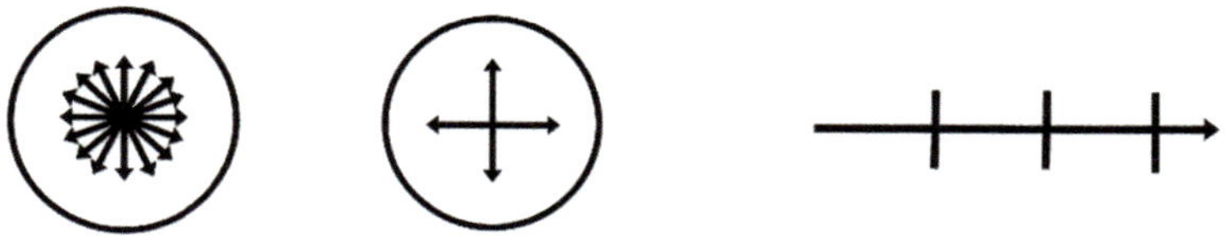

(a) No dominant direction (b) Decomposition of natural light (c) Superposition of two polarized light

Fig. 8.2 The graphical method of natural light

(a) The electrical vector is perpendicular to the plane

(b) The electrical vector is parallel to the plane

Fig. 8.3 The schematic diagram of plane-polarized light

According to the different trajectories of the vector end of the optical field E at any point in space at different times, fully polarized light can be divided into elliptically polarized light (Fig. 8.4), linearly polarized light and circularly polarized light. Linear polarization (sometimes also called plane polarization) means that the amplitude component of the polarized electric field is always in a plane perpendicular to the propagation direction of the electromagnetic wave, and the normal direction of the plane does not change with time, so it is in a plane perpendicular to the propagation direction. The electric field components are projected as a straight line. In addition, the superposition (or combination) of two plane-polarized lights with the same propagation direction, perpendicular vibration directions and constant phase difference can synthesize circularly polarized light and elliptically polarized light whose light vector changes regularly. Circular polarization means that the electromagnetic wave has a constant electric field amplitude in the direction perpendicular to the propagation direction, but the electric field direction rotates with time, so the projection of the electric field component in the plane perpendicular to the propagation direction is a circle. Electromagnetic waves whose amplitudes are perpendicular to each other and have a fixed phase difference have the same effect. Ellipsometry is when the projection of an electric field component in a plane perpendicular to the direction of propagation not only rotates with time, but also changes its amplitude with time (Fig. 8.5). Obviously linear polarization and circular polarization are special cases of elliptical polarization.

3. Partially polarized light

If the vibration in one vibration direction of natural light is dominant over other directions due to the action of the outside world, it becomes partially polarized light. Partially polarized light can be viewed as a mixture of fully polarized light and natural light. Thus, partially polarized light can be represented by two mutually perpendicular light vectors, whose amplitudes are not equal and whose phase relationship is also uncertain. Figure 8.6 is usually used to represent partially polarized light, (a) represents partially polarized light dominant in a certain direction, (b) represents partially polarized light can be represented by two mutually perpendicular light vectors with unequal amplitudes, (c) represents parallel Partially polarized light with a strong surface electric vector.

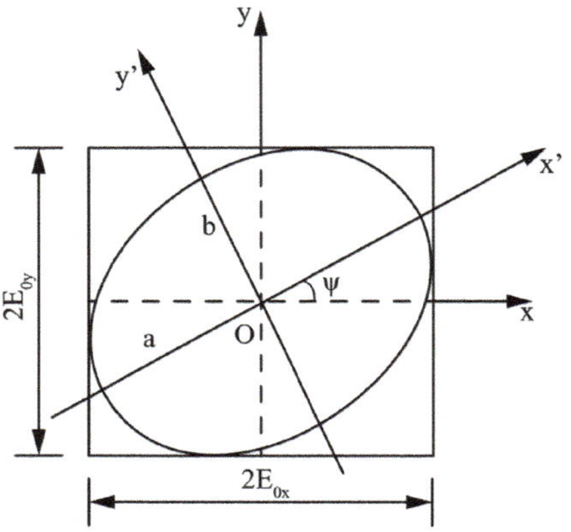

Fig. 8.4 Parameters of elliptically polarized light

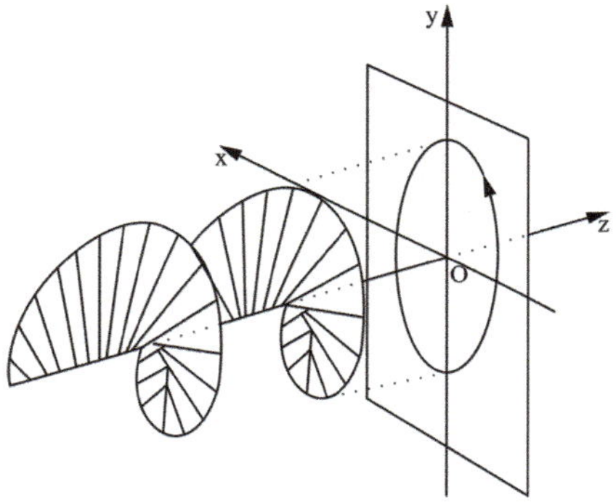

Fig. 8.5 The vector distribution of elliptically polarized light

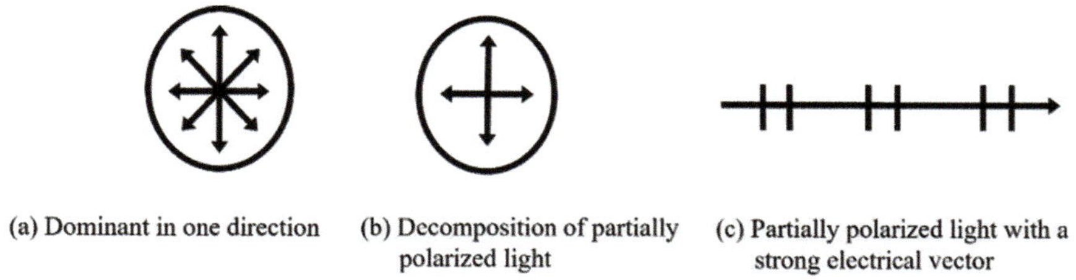

Fig. 8.6 The schematic diagram of partially polarized light

8.1.2 *Mathematical Description and Scattering Properties of Polarized Light*

The mathematical description of polarized light generally adopts four representation methods: electric vector component, Jones vector, Stokes vector and Boukar sphere diagram [5]. The process is simplified. When analyzing and calculating the transmission of polarized light in a single optical device, the electric vector component method is most used. The first three representation methods are briefly introduced here.

1. Electric vector component method

The electric vector component method is a method based on the classical wave theory. In the Cartesian coordinate system, suppose a monochromatic plane-polarized wave propagates along the axis, and its three electric vector components are:

$$\begin{aligned} E_x &= A\cos(\omega t - kz + \delta_x) \\ E_y &= A\cos(\omega t - kz + \delta_y) \\ E_z &= 0 \end{aligned} \tag{8.1}$$

where A, B is the amplitude, δ_x, δ_y is the initial phase of the two components E_x, E_y, k is the wave vector size, also known as the space circular frequency, and ω is the time circular frequency.

In general, the trajectory of the end of the synthetic electric vector is an ellipse, namely:

$$\left(\frac{E_x}{A}\right)^2 + \left(\frac{E_y}{B}\right)^2 - \frac{2E_xE_y}{AB}\cos(\delta_y - \delta_x) = \sin^2(\delta_y - \delta_x) = \sin^2\delta \tag{8.2}$$

It indicates that the motion trajectory of the end of the electric vector is an inscribed ellipse of a rectangle depicted by the A, B values in the x, y direction respectively, as shown in Fig. 8.7.

When $\delta = 0, \pi$, it is linearly polarized light; when $\delta = \pi/2$, it is circularly polarized light. The rotation direction of the end of the vector depends on the range of δ values. When $0 < \delta \leq \pi$, it was left-handed; when $-\pi < \delta \leq 0$, it was right-handed. It can be seen that the shape and orientation of the ellipse can be determined by the three parameters of amplitude A, B and phase difference δ, so that the polarization state can be determined.

2. Jones vector method

In 1941, Jones first proposed to use a simple vector—a matrix of two rows and one column to represent polarized light, this vector is called Jones vector. It is a binary vector that describes the polarization state represented by the complex amplitude

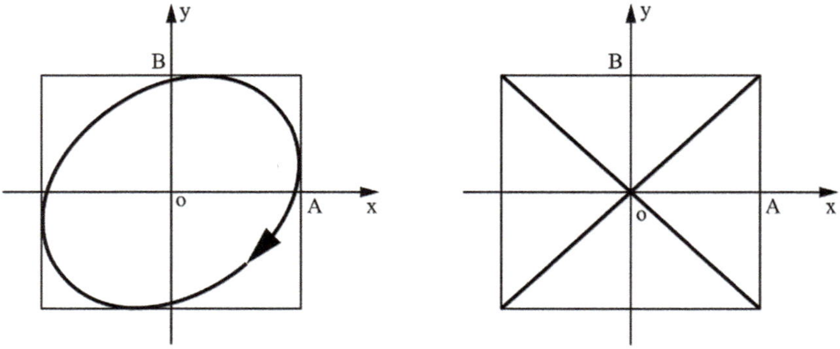

(a) Left-hand elliptically polarized light (b) Linearly polarized light

Fig. 8.7 The elliptically polarized light and linearly polarized light

component at a certain point along the light propagation direction [6, 7]. It is defined as:

$$\begin{bmatrix} E_x \\ \overline{E_y} \end{bmatrix} = e^{i(\omega t - kz)} \begin{bmatrix} Ae^{i\delta_x} \\ \overline{Be^{i\delta_y}} \end{bmatrix} = e^{i(\omega t - kz)} e^{i\delta_x} \begin{bmatrix} A \\ \overline{Be^{i\delta}} \end{bmatrix} \tag{8.3}$$

where $\delta = \delta_y - \delta_x$, Due to the periodic motion of the phase 2π, the main value range of the specified phase difference δ is: $-\pi \leq \delta \leq \pi$. When $-\pi < \delta < 0$, it indicates that the component y lags behind the component x. Conversely, when $0 < \delta < \pi$, it indicates that the component x lags behind the component y. When $\delta = 0$, it means that the component y and the component x are in the same phase; when $\delta = \pm\pi$, it means that the component y and the component x are out of phase.

If the coefficient $e^{i(\omega t - kz)} e^{i\delta_x}$ is Omitting, a beam of polarized light can be described as a Jones vector:

$$J = \begin{bmatrix} A \\ Be^{i\delta} \end{bmatrix} \tag{8.4}$$

The light intensity can be obtained from the following equation:

$$I = A^2 + B^2$$

The normalized Jones vector is $J_n = \begin{bmatrix} \cos\gamma \\ \sin\gamma \; e^{i\delta} \end{bmatrix}$, and the intensity is the unit of 1.

For linearly polarized light, $\delta = \pm n\pi$; for circularly polarized light, $\delta = \pm(2n + 1)\pi/2$. When $A = B$ and $n = 0, 1, 2, \ldots$, the corresponding normalized vectors for linearly polarized light and circularly polarized light are:

$$\begin{bmatrix} \cos\gamma \\ \pm\sin\gamma\, e^{i\delta} \end{bmatrix} \quad \text{and} \quad \frac{1}{\sqrt{2}}\begin{bmatrix} \mu i \\ 1 \end{bmatrix}$$

where $-i$ represents the right-handed circularly polarized light, and $+i$ represents left-handed circularly polarized light.

3. Stokes vector notation

We know that the general form of polarized light is elliptically polarized light, and to fully describe it requires three independent parameters, such as amplitude E_x, E_y and phase difference δ (or the long and short axes a, b of the ellipse and the orientation angle ψ of the ellipse). However, the three most commonly used macroscopically measurable independent parameters are the Stokes parameters, which were proposed by the British physicist Stokes in 1852. Stokes proposed to describe the polarization state of polarized light by a set of parameters with the same physical dimension. Stokes parameters can not only represent fully polarized light, but also partially polarized light.

The Stokes vector is written in the following matrix form with four parameters, this set of parameters is called the Stokes parameters.

$$S = \begin{bmatrix} S_0 & S_1 & S_2 & S_3 \end{bmatrix}^T$$

The four Stokes parameters marked with S_0, S_1, S_2 and S_3 are defined as [8]:

$$S_0 = <\widetilde{E}_x^2(t)> + <\widetilde{E}_y^2(t)> \tag{8.5}$$

$$S_1 = <\widetilde{E}_x^2(t)> - <\widetilde{E}_y^2(t)> \tag{8.6}$$

$$S_2 = <2\widetilde{E}_x(t)\widetilde{E}_y(t)\cos\delta> \tag{8.7}$$

$$S_3 = <2\widetilde{E}_x(t)\widetilde{E}_y(t)\sin\delta> \tag{8.8}$$

Among them, $\widetilde{E}_x$, $\widetilde{E}_y$ is the component of the electric vector along the x and y directions in the selected coordinate system, δ is the phase difference of the two vibration components, and the symbol "< >" represents the average over time. The state of any polarized light, including the degree of polarization, can be represented by this four-dimensional vector. In the above formula, the parameter S_0 represents the intensity of polarized light, S_1 represents the linearly polarized light component in the horizontal direction (x-axis direction), S_2 represents the linearly polarized light component in the 45° direction, and S_3 represents the right-handed circularly polarized light component.

From the Stokes vector, the polarization state information of any light can be obtained as follows:

$$\psi = \frac{1}{2}\arctan\frac{S_1}{S_2} \tag{8.9}$$

$$P = \sqrt{S_1^2 + S_2^2 + S_3^2}/S_0 \tag{8.10}$$

Among them, ψ is the azimuth angle of the ellipse, that is also the orientation of the ellipse; P is an important parameter describing the degree of polarization of partially polarized light, and its value varies from 0 in the case of unpolarized light to 1 in the case of fully polarized light. For partially polarized light, the value takes the middle value.

4. Scattering properties of light waves

The scattering of sunlight by the atmosphere is a very important optical phenomenon, which is the result of the interaction between light and matter. When the refractive index of the medium has a non-single structure, light passes through it, and the phenomenon of scattering occurs. The physical process of light scattering can be briefly described as follows: scatterers such as aerosol particles and atmospheric molecules in the atmosphere, under the irradiation of light, due to the action of oscillating electromagnetic waves by the irradiating light, the scatterers are polarized and oscillating electromagnetic polypoles are induced. The electromagnetic oscillation generated by the electromagnetic multipole of the scatterer radiates electromagnetic waves in all directions, forming a light scattering process. At the same time, scatterers such as aerosols not only scatter part of the energy of the irradiated light, but also often absorb part of the irradiated light energy and convert it into heat energy, which is the absorption effect of scatterers. The scattering process often changes the polarization state of light, causing the sky scattered light to have a certain polarization distribution. For scatterers, scattered radiation is divided into primary and multiple scattering (Fig. 8.8), Rayleigh scattering (Fig. 8.9) and Mie scattering [9] the difference.

Regarding the scattering of light, the theory and formula have been relatively complete, so this book only summarizes the theory, and does not repeat the detailed derivation process. Readers who interested the details can refer to the relevant literature and the book "Polarization Remote Sensing Physics" [4].

8.1.3 Mathematical Model of Polarized Skylight Field and Its Measurement Principle

1. Mathematical expression of sky polarized field

In astronomy, the sky of our top is called the celestial sphere. The celestial sphere is an imaginary sphere with an infinite radius at the center of the earth. All celestial bodies are projected onto the spherical surface of the celestial sphere regardless of their distance from the earth. To determine the position of a celestial body is

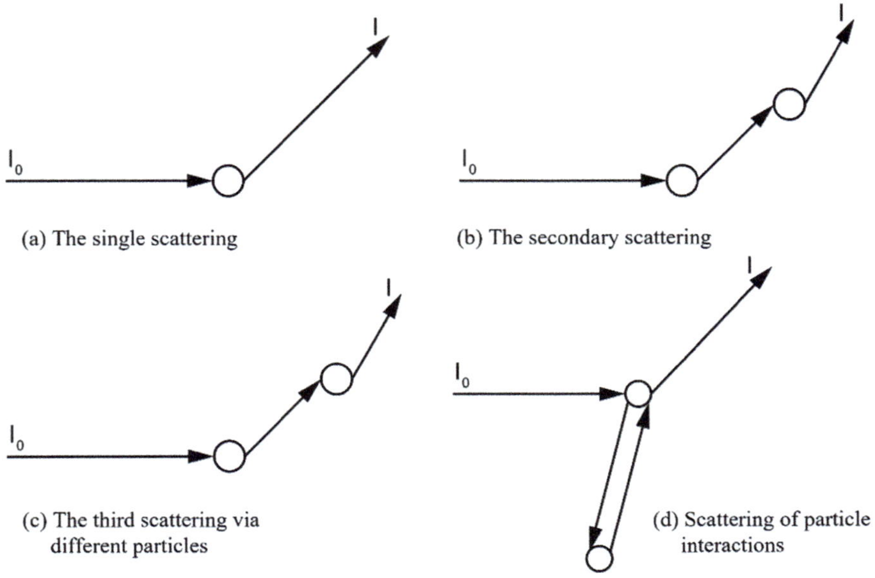

Fig. 8.8 The schematic diagram of multiple scattering

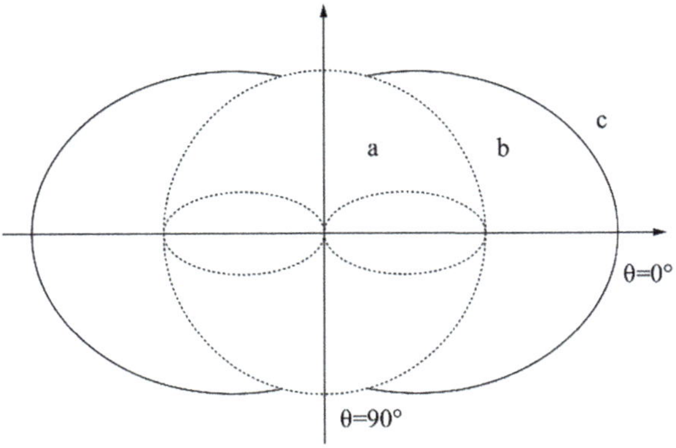

(a) The electrical vector scatters light intensity component parallel scattering plane
(b) The component of the scattering intensity of the electrical vector perpendicular to the scattering plane
(c) The intensity of total scattered light

Fig. 8.9 The angular distribution of Rayleigh scattered light intensity when the incident light is natural light

to determine the position of a celestial body on the celestial sphere. In addition to being represented by a rectangular coordinate system, it can also be represented by the following coordinate systems: ecliptic coordinate system, hour-angle coordinate system, equatorial coordinate system, and horizon coordinate system [10, 11].

The horizon coordinate system is used to establish the sky polarization model diagram. In the horizon coordinate system, the great circle where the horizon where the observer is located infinitely expands and intersects with the celestial sphere becomes the horizon circle, and the countless great circles that pass through the zenith and nadir and are perpendicular to the horizon circle are horizon longitude circle, and the small circle parallel to the horizon circle becomes the horizon latitude circle. The position of the celestial body is represented by the azimuth and elevation angles in the horizon coordinate system. The so-called azimuth is the angle between the meridian circle and the meridian circle passing through the celestial body, starting from the meridian circle (positive south), negative from south to east to north, and positive from south to west to north, such as the sun. In due east, the azimuth is $-90°$, in due north it is $\pm 180°$. The so-called altitude angle is the angle between the line connecting the center of the celestial body and the location of the observer and the horizon, with the horizon as the starting point, positive from the horizon to the zenith, and negative from the horizon to the nadir. When the center of the celestial body directly hits the local meridian, its azimuth angle is 0°, and the hour angle is also 0°. This moment is the moment when the celestial body is on the mid-heaven.

As shown in Fig. 8.10, in the horizon coordinate system, S represents the sun, OP represents the direction of observation in the sky, and Z represents the zenith. Polarization azimuth is defined as the angle between the *E*-vector vibration direction of point P in the viewing direction and the meridian (arc ZP) passing through point P in the sky. And the *E*-vector vibration direction of point P is perpendicular to PS, $\varphi = 90° - \angle ZPS$.

According to the definition of the degree of polarization, the distribution of the degree of polarization in the sky is:

$$P(\Theta) = \frac{1 - \cos^2 \Theta}{1 + \cos^2 \Theta} = \frac{\sin^2 \Theta}{1 + \cos^2 \Theta} \tag{8.11}$$

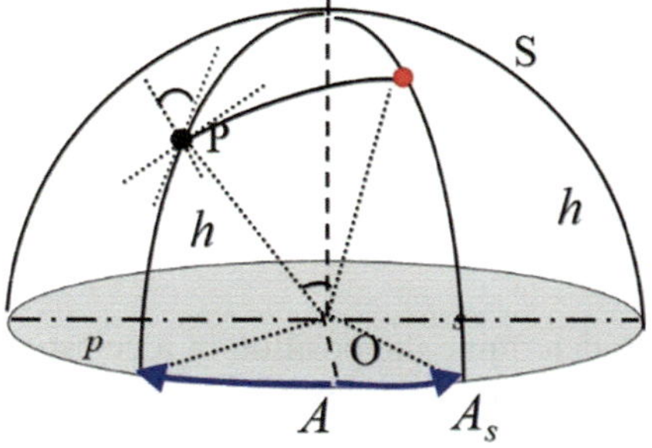

Fig. 8.10 The E vector polarization information at point P in the sky

where Θ is the scattering angle. The polarization azimuth φ can be calculated by the following formula:

$$\cos\varphi = \frac{\sin(A_s - A_p)}{\sin\theta}\cos h_s \quad (8.12)$$

Here, θ is the angle between the incident light direction OS and the observation direction OP, h_s is the elevation angle of the sun, A_s is the azimuth angle of the sun, h_p is the elevation angle of the observation point P in the spherical sky, and A_p is the azimuth angle of the observation point P. Here, due south represents an azimuth of 0°, and due west represents an azimuth of 90°.

The polarization degree and polarization azimuth angle of the beam in the observation direction in the sky can be calculated from Eqs. (8.11) and (8.12), as shown in Fig. 8.11. The direction of the black dotted line in the figure represents the polarization direction, and the width of the dotted line represents the degree of polarization. The shaded part represents the part of the sky that can be seen by the human eye. It can be seen that the polarization pattern map in the sky is closely related to the sun altitude angle. The polarization pattern map distribution, the intensity and direction of the polarized light corresponding to different sun altitude angles are different. When the sun altitude angle is low, the intensity of polarization is relatively large, which means that the intensity of polarization in the sky is greater in the evening or in the morning than at noon. It can also be seen from the figure that the polarization pattern diagram has two sharp lines of symmetry. One is a great circle with a 90° angular distance from the sun or anti-sun point, where the degree of polarization is the largest. Another is the sun and anti-sun point. On this line of symmetry, the polarization angle is 90°.

2. Basic Principles of Polarization Measurement

For any beam of quasi-monochromatic plane wave from the detected target, its polarization state can be characterized by four Stokes parameters S_0, S_1, S_2, S_3. In polarization remote sensing testing technology, the Stokes vector $[S_0, S_1, S_2, S_3]^T$

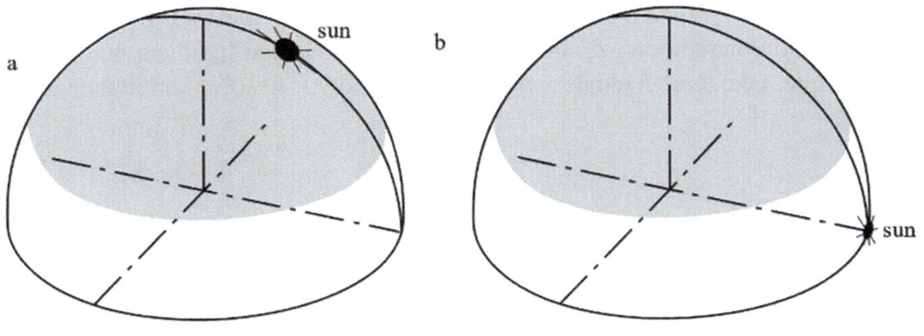

Fig. 8.11 Patterns of polarization in the sky at different times

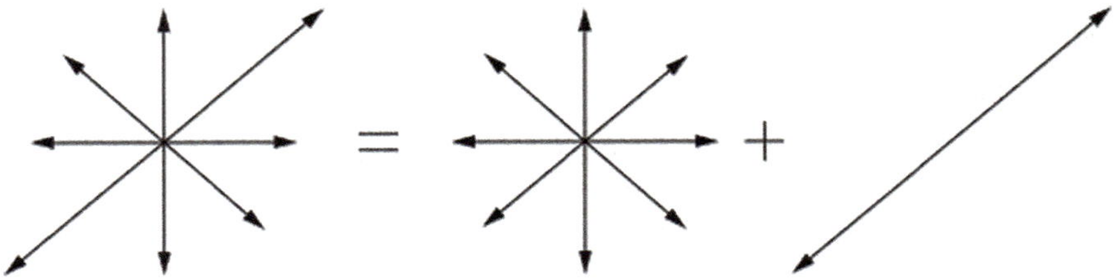

Fig. 8.12 The decomposition diagram of real light

is usually written in another form, i.e., $S = \left[I,\ Q,\ U,\ V\right]^T$. Where I is the intensity of unpolarized light, Q, U represents linearly polarized light in two directions, and V represents circularly polarized light. In polarized light detection, the V component in the Stocks component is considered to be $V = 0$. In this case, the light we observe is considered to be a combination of linearly polarized light and natural light, as shown in Fig. 8.12.

Assuming that the intensity of natural light is I_g, the intensity of linearly polarized light in its polarization direction is $I_{\max}$. In the actual detection, the observation light is made to pass through a polarizer in a certain direction, because the light intensity of natural light in all directions is the same, so the light intensity of natural light after passing through the polarizer becomes 1/2 of the original. At the same time, according to Marius' law, if the angle between the polarization direction of the linearly polarized light and the transmission direction of the polarizer is α, the light intensity of the linearly polarized light after passing through the polarizer becomes $I_{\max}\cos^2\alpha$. Then, in the actual observation, let the observation light pass through the polarizers in the directions of 0°, +45°, +90° and +135°, to obtain the light intensity $I(\alpha_i)$ of the linearly polarized light after passing through the polarizers in the four directions, and then simultaneously obtain the four Stokes vector parameters of I, Q, U and V. At the same time, the value of the polarization degree P and polarization azimuth angle ψ of the polarized light can be simultaneously obtained.

In actual detection, because the value of the V component in the Stocks component is smaller, it is considered that $V = 0$. Therefore, as long as the light intensity $I(\alpha_i)$ of linearly polarized light at three different angles is measured, the three parameters of the Stokes vector I, Q and U can be obtained simultaneously, and then the polarization degree P and polarization azimuth angle ψ of the polarized light can be obtained. For example, take three fixed directions 0°, 45°, and 90° to solve simultaneously.

8.2 Polarization Line of Force 1—Static Measurement of the Parametric Distribution of Polarized Light in the Sky

Polarized field lines of force are directly related to polarized field navigation, and thus firstly introduced in this section. The polarization field line of force is divided into the basic part, including the static part and the periodic part. This section introduces the static part, the dynamic part will be introduced in the next section.

During the actual measurement, install a fisheye lens in front of the digital camera (to ensure that its field of view is 180°), and install a polarizer in front of the lens to pass four different polarizations (0°, 45°, 90°, 135). The measurement of the azimuth angle can measure the light intensity $I(\alpha)$ of polarized light at 4 different angles. The first three components of the Stokes parameter can be calculated, i.e., I, Q and U, and the corresponding polarization degree P and polarization azimuth ψ can be obtained.

8.2.1 Comparison of Sky Images Before and After Adding Polarizer

As shown in Fig. 8.13a, b, the original pictures without polarizer and with polarizer are presented. These two pictures were taken on the 5th floor rooftop of Peking University Remote Sensing Institute at 1:00 pm on May 19, 2008. Figure 8.13c, d are the distribution of sky brightness before and after adding polarizers. Comparing the two images in the figure, it can be clearly seen that the brightness distribution of the sky after adding the polarizer is quite different from the brightness distribution without adding the polarizer.

Comparing the original images, it is found that the two images reflect the same area; the only difference is that the area taken after adding the polarizer is slightly smaller. However, after brightness grading, it is found that there is a big difference between the two images. The brightness values of the points other than the background in the original image are all greater than 0.8, while the points with the brightness value greater than 0.8 after adding the polarizer are much less, which means the light intensity decreased after adding the polarizer.

8.2.2 Measurement of the Parameter Distribution of Stocks

Figure 8.13 is the original image taken from the same area of the sky after the polarizer added, and the image after luminance segmentation. During the shooting, the

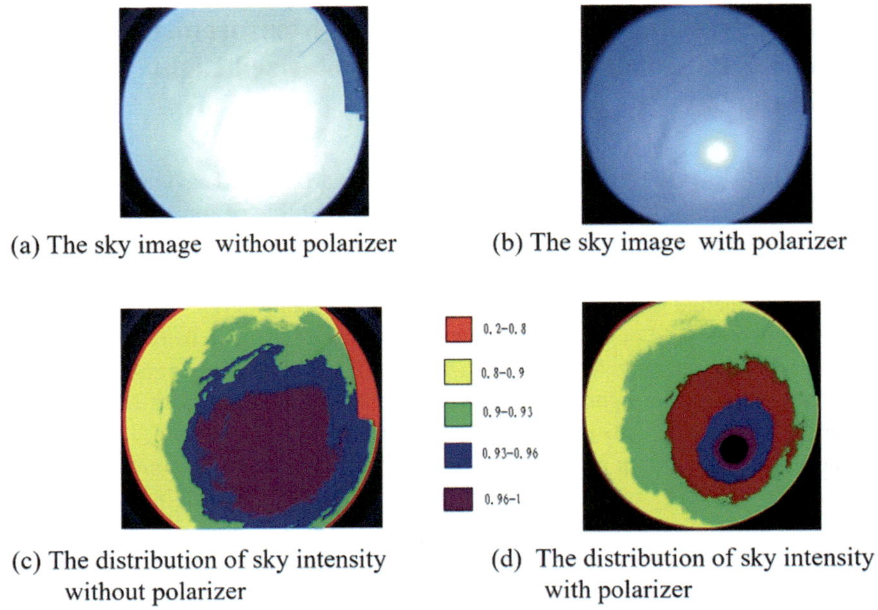

Fig. 8.13 The comparison of original images and distribution of sky intensities

polarizer was rotated 45° clockwise after each shot, and some noticeable differences can be seen from the luminance segmented images (Fig. 8.14).

In the previous chapter, the brightness value of any point of the image taken after adding the polarizer can be written as $0.5I_g + I_{\max}\cos^2\alpha$ (α representing the angle between the polarization direction of the linearly polarized light and the direction of

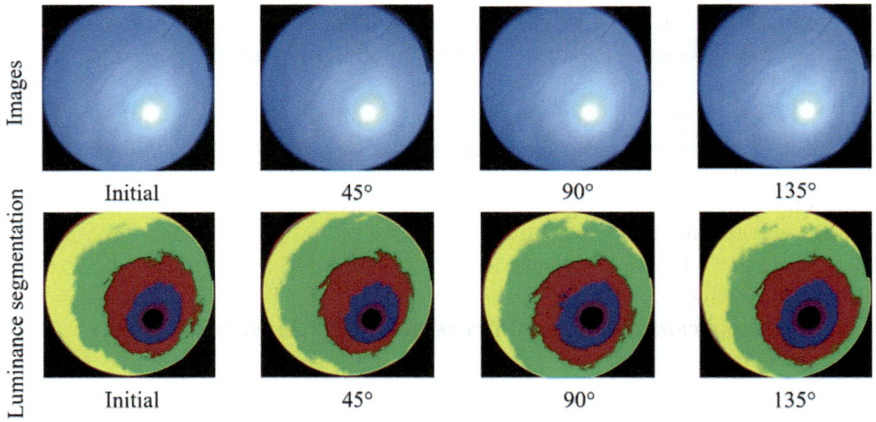

Fig. 8.14 The original images and luminance segmentation of sky intensities

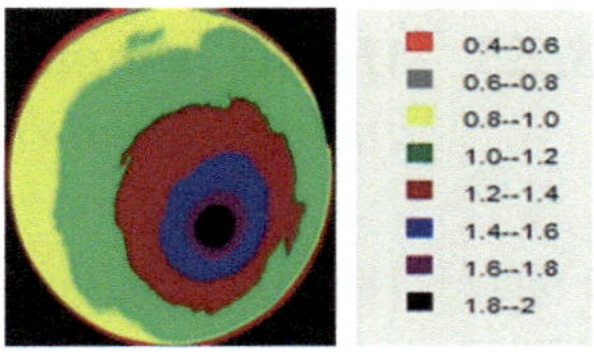

Fig. 8.15 I components

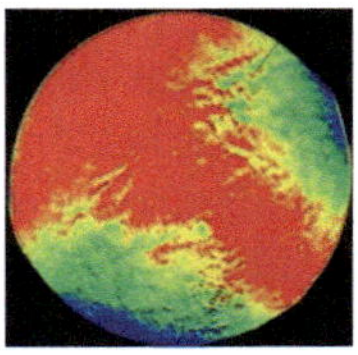

Fig. 8.16 |Q| components

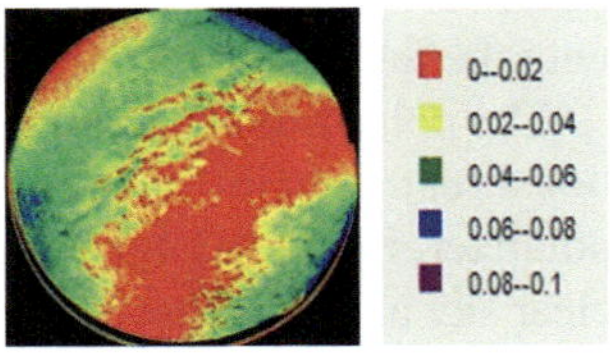

Fig. 8.17 |U| components

the polarizer), and thus the difference between any two images at a certain point is $I_{\max}(\cos^2\alpha - \cos^2\beta)$, which indicates the existence of $I_{\max}$. Some large differences on the image also show that the value of is not small. Therefore, in the measurement range of the whole sky, the value of linear polarization degree is not negligible. Figures 8.15, 8.16, and 8.17 are the distribution diagrams of the Stocks parameters I, Q and U, calculated from the above four images, respectively. By comparison, it is found that the distribution of the components is generally similar to the distribution of the luminance segmented image of the original image.

According to the calculation formula of Q and U components:

$$Q = I(0^\circ) - I(90^\circ),\ U = I(45^\circ) - I(135^\circ)$$

It is found that the values of Q and U are actually changed with the angle of the polarizer at the beginning, so the distribution of these two components is irregular. At the points near the sun, due to overexposure, the brightness values of the pictures taken at all angles are very similar, so the values of Q and U components in these places are relatively small.

8.2.3 Measurement and Research of Polarization Degree and Polarization Azimuth Parameter Distribution

Figure 8.18 is a distribution map of the polarization degree in the whole sky. The red area in the figure is the overexposed part of the sun and the side. It can be seen from the figure that the distribution of polarization degrees in the whole sky is roughly annular. Theoretically, the polarization of the whole sky is symmetrically distributed. The actual measurement results also generally show some symmetry relations of the polarization degree distribution. Of course, we can also see the symmetrical distribution from the polarization distribution maps taken at other time.

Figure 8.18 is the distribution of polarization degree after processing the data taken at 4:00 pm on May 7, 2008 and 1:00 pm on May 21, 2008 on the 5th floor of the Remote Sensing Institute. The distributions of the polarization degrees in the two figures are also roughly symmetrical, and this symmetrical distribution is consistent with the theory.

The polarization azimuth angle of a point refers to the angle between the polarization direction of the linearly polarized light at the point and the initial direction of the polarizer. Figure 8.19 shows the distribution of polarization azimuths in the sky for clear sky and cloudy weather. It can be seen that the distribution of the polarization angle under clear sky conditions is relatively regular, while the polarization azimuth angle under cloudy state is affected by the multiple scattering of cloudy weather, and its distribution changes relative to the clear sky. In general, the properties of its symmetrical distribution have not changed.

In order to obtain the distribution law of polarized light in the whole sky over a period of time, a continuous observation experiment was carried out on the roof of the 5th floor of the Remote Sensing Building from 11:00 to 17:00 on May 30, 2008. In the experiment, the whole sky was photographed every 1 h.

Figure 8.20 is the real-time photo and the distribution comparison of polarization degree and polarization azimuth angle.

In this experiment, the sky was in good condition with almost no clouds. Among them, a small amount of processing was done on the polarization distribution map at 17:00, and the influence of the cloud layer in the left half of the image was eliminated. From the experimental results, we can obtain some rules about the distribution of polarized light in the whole sky: First of all, it can be seen from the distribution of

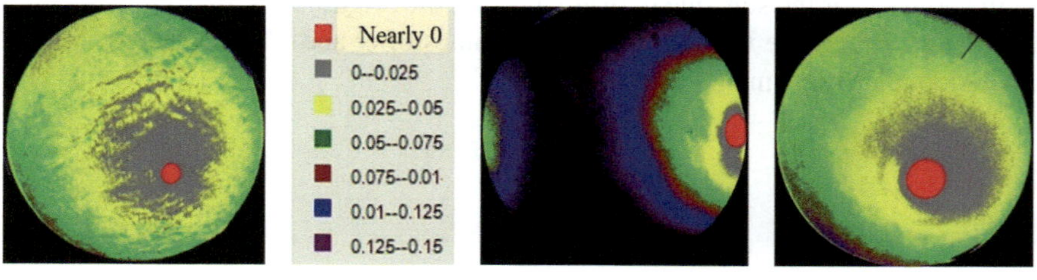

Fig. 8.18 The distribution of polarized degree

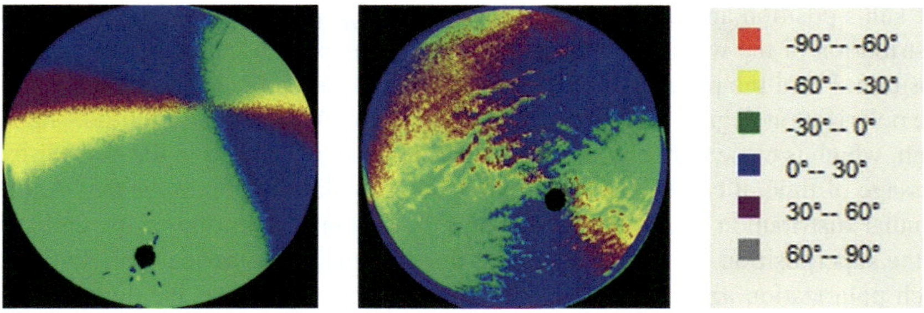

Fig. 8.19 The distribution of polarized azimuth angles

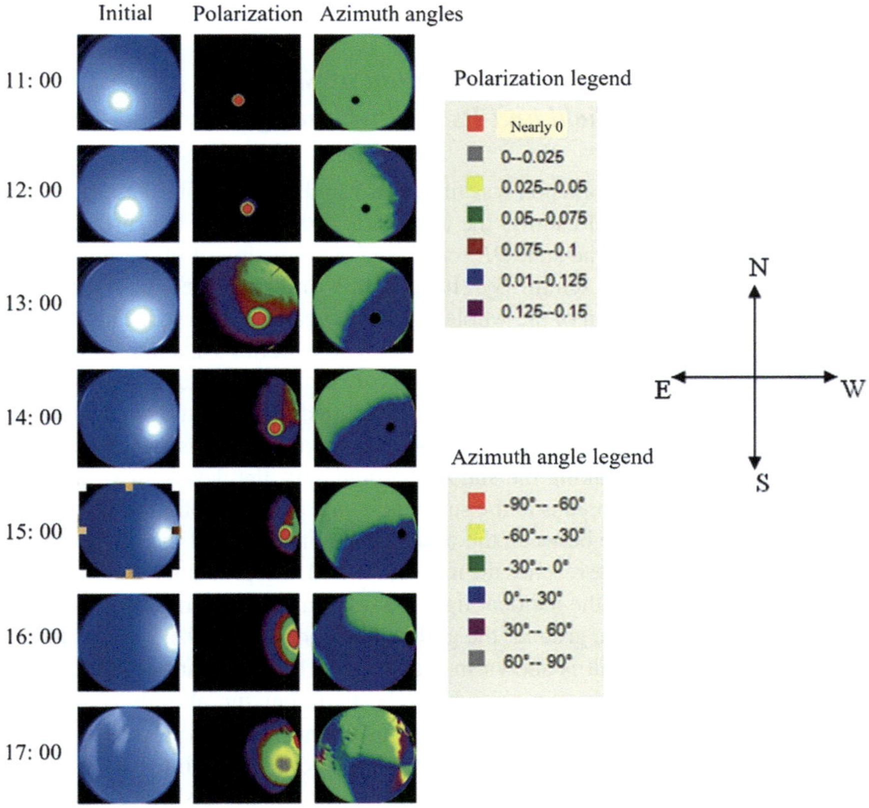

Fig. 8.20 Continuous experimental renderings

the sun's position and polarization degree at each time that the polarization degree distribution of the whole sky is not fixed, but changes with the change of the sun's position. From the polarization degree distribution map, it can be clearly seen that the polarization degree is distributed in a ring shape. When the sun altitude angle is high, which is close to noon, a ring distribution can be observed in the sky. With the passage of time, it can be found that there is a ring distribution in the sky. Another annular distribution is obtained, and the polarization distribution of the entire sky is the superposition of these two annular distributions. Comprehensively looking at each polarization azimuth distribution map, it can be seen that the distribution of polarization azimuth also has a certain distribution law, which is mainly related to the position of the sun.

8.2.4 Single-Point-Based Observation of Sky Light Polarization Field and Its Multi-Point Synthesis

The sky polarization phenomenon was first discovered in the ground-based observation of the sky. Figure 8.21a shows the observation geometry of the surface-to-sky observation. *S* represents the sun, *OP* represents the direction of observation in the sky, and *Z* represents the zenith. By changing the different directions of the *OP*, the polarization distribution of the whole sky locally can be observed. The team has conducted ground observations in Finland, Hungary, Beijing, Dalian, Zhuhai and other places in China. These local observations are highly consistent with the model simulations. As shown in Fig. 8.21b, Shaw, et al. conducted ground observations in the United States. Where the degree of polarization of the atmosphere varies with the position of the sun. Taking the sun as the field axis, the degree of polarization increases gradually as it spreads to the surroundings. Taking the observation point as the origin, when the angle between the observation direction and the sun is close to 90°, the polarization degree reaches the maximum, and then gradually decreases until the mirror image point of the incident light is 0. Based on this condition for vertical upward observation on the ground, the global atmospheric polarization distribution is simulated, and the result is shown in Fig. 8.21c. The polarization pattern in the sky is closely related to the sun altitude angle. Different sun altitude angles have different degrees of polarization, and the intensity and direction of polarized light are different. When the sun altitude angle is low, the polarization intensity in the sky is relatively large, that is to say, the intensity of polarization in the sky is greater in the evening or in the morning than at noon. In different seasons, such as vernal equinox, summer solstice, autumnal equinox and winter solstice, there are regular changes in its distribution. We attribute this slow year-to-year periodic variation as a static feature of the polarized lines of force, with no direct impact in terms of real-time navigation, relative to the variation of the polarization field during the daily solar rise and fall.

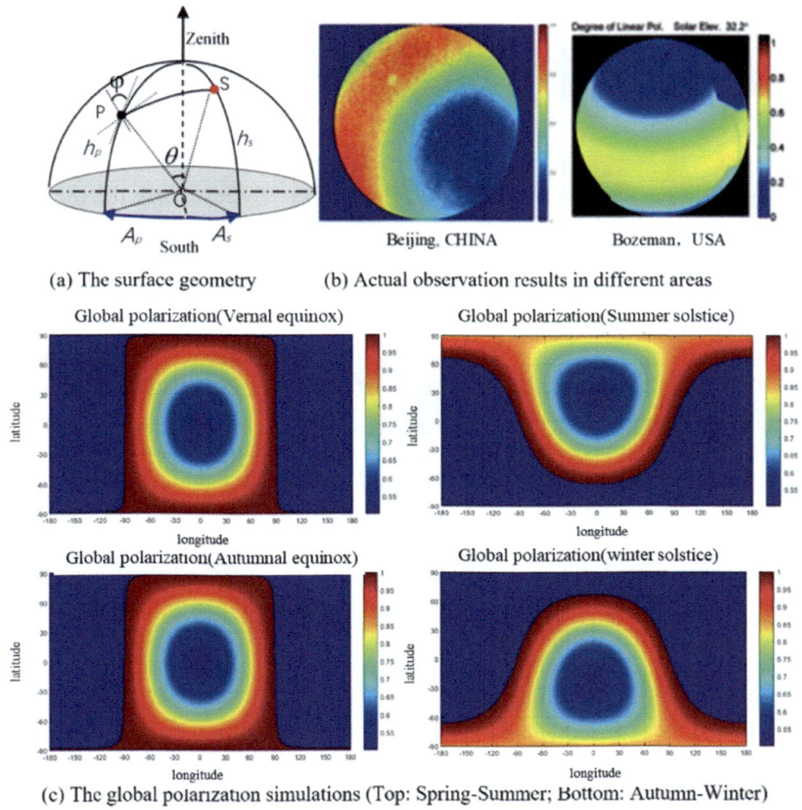

Fig. 8.21 Global polarization field theoretical simulation and single point sky polarization degree verification

8.3 Polarization Line of Force 2—Dynamic Characteristics of the Distribution of Polarized Light in the Sky

Half a century ago, Frisch discovered that bees can use polarized light of visible in the sky for orientation [12]. Although some other animals, including some vertebrates, have also been found to be sensitive to polarized light, but they can only use this information to determine a reference direction relative to the slow-moving sun. In terms of location, this has certain limitations for insects that migrate long distances. Studies have shown that insects, such as bees and pigeons can use the dynamics of polarized light distribution to determine the location of the celestial pole in a short period of time. And use it to provide specific navigation information, such as latitude, local true solar time, and true (geographical) compass direction. In addition, long-distance migratory birds can also use the position of the maximum polarization

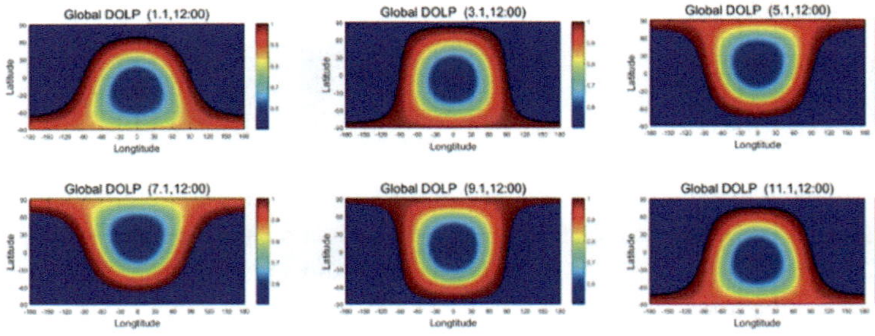

(d) Annual variation of global atmospheric polarization (12: 00 GMT)

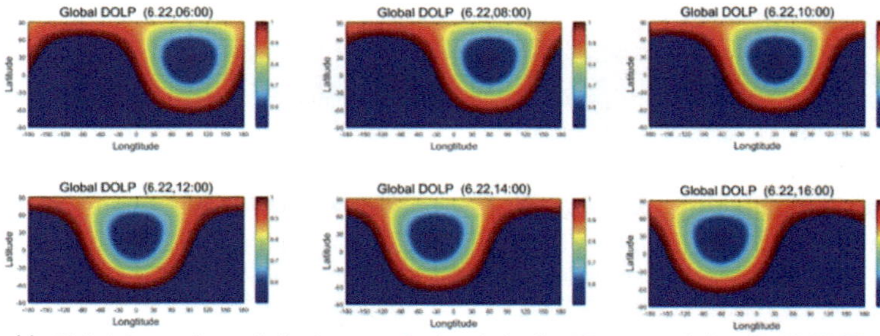

(e) Global degree-day variation in atmospheric polarization (Summer solstice, 12: 00 GMT)

Fig. 8.21 (continued)

line at sunrise and sunset to correct the direction information of the geomagnetic navigation system to further improve the navigation accuracy.

8.3.1 Important Properties of Celestial Poles

In most cases, sunlight scattering is mainly Rayleigh scattering. In this case, the direction of the *E*-vector of polarized skylight is completely predictable. On the surface of the celestial sphere, the *E*-vector direction of all points in the sky relative to the line connecting the observer and the sun can be conveniently defined. At any point in the sky, the *E*-vector direction is tangent to the line connecting the observer and the sun. All points on any circle are equidistant from the sun and have nearly the same degree of polarization. Although the maximum degree of polarization is related to the wavelength of light in the sky and atmospheric conditions, it usually occurs at an angular distance of 90° from the sun. Since the polarization patterns are closely related to the position of the sun, they appear to rotate around the celestial pole at a rate of 15° per hour, as shown in Fig. 8.22.

The celestial pole carries an important property, and the entire *E*-vector distribution rotates around it. Some animals are sensitive to the polarization pattern of the entire sky rotating around the celestial pole at a rate of 15° per hour, and they

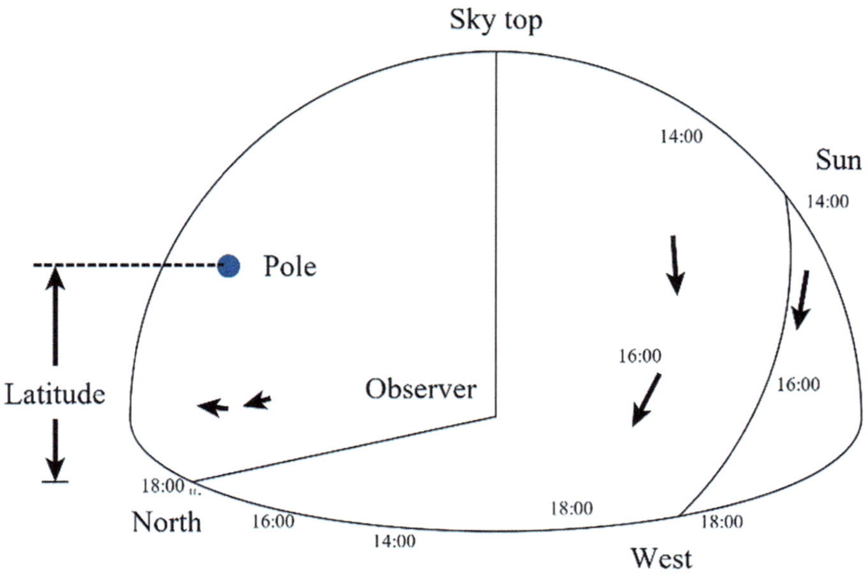

Fig. 8.22 The dynamic distribution of polarized light in northern hemisphere sky

determine the location of the pole by observing the line of maximum polarization rotated around the pole. At this time, the azimuth of the celestial pole is the direction of the polarized compass, and the height of the celestial pole is the latitude of the observer. As Wheatstone pointed out in 1984, the *E*-vector direction at the celestial poles is the local true solar time.

If this important navigation information exists in the sky polarization pattern map, an important biological question is whether humans can develop a relevant sensor to detect the polarization pattern map rotating at a rate of 15°/h. If it can be detected, how long does the detection time take for different accuracy requirements? The amazing ability of pigeons or other birds to return to their nests is partly thought to be the detection of the extreme position. In theory, birds can determine their own direction and distance from the nest by observing the polarization pattern of the sky during the process of returning to the nest. If the bird sees the pole as a beacon lower than it, it flies towards the beacon and vice versa. The longitude can be determined by the difference between the solar time determined by the direction of the *E*-vector at the pole and the time difference in memory. This theory requires that returning birds have the ability to detect rapid rotations in the direction of the *E*-vector. The pigeons in the experiment showed the exact direction of homecoming, with a difference of only 1°–2° between latitude and longitude. Moreover, experienced pigeons can determine the direction of homecoming within one minute, and the *E*-vector rotates only 0.25° around the pole in one minute. That is to say, the bees were able to detect a 0.25° *E*-vector orientation change if navigating with polarized light alone.

Obviously, the detection of the dynamic properties of the polarization pattern map for navigation requires a certain precision. From Frisch and his colleagues' work, honeybees use polarized light from the sky to determine their orientation with great precision. For example, if a large area of the sky (including the sun) is visible, the bees can determine their orientation with an accuracy of 0.2°. When collecting information from a large swath of the sky or given sufficient time, the resolution of the *E*-vector direction by bees is greater. Since bees have the ability to detect the direction of the *E*-vector, if we can design a sensor that is extremely sensitive to the direction of polarized light, and can recognize subtle changes in the direction of the *E*-vector like bees, it is a new breakthrough for polarization navigation, and also an important direction for our follow-up research.

8.3.2 Application Analysis of Maximum Polarization Line of Force

Another potential feature of polarized skylight for navigation is the line of maximum polarization. At the vernal or autumnal equinox, the maximum polarization line 90° away from the sun just passes through the pole, and the distance between the maximum polarization line and the pole at other times is within ±23.5°, the angular distance from the maximum polarization line to the pole is the declination of the sun. Therefore, the animal only needs to identify with a relatively small error of the point around the maximum polarization line rotate of celestial pole. Since the line of maximum polarization coincides with the direction of the *E*-vector above, the direction of the line of polarization can also be used to determine the solar hour angle. Of course, combining the *E*-vector direction and the degree of polarization to determine the location of the poles works is better.

Numerous studies have shown that sky information at sunrise and sunset can also provide important directional information for migrating birds [1]. This type of literature argues that maps of polarized light patterns in the sky can be used to infer geographic directional information. At the vernal and autumnal equinoxes, the maximum polarization line 90° away from the sun just passes through the pole, and the intersection of the maximum polarization line with the ground plane at sunrise and sunset happens to be the north–south meridian. Therefore, using the intersection of the maximum polarization line and the ground plane at sunrise and sunset will not bring systematic errors to correct other compass systems. Except for the vernal and autumnal equinoxes, the intersection of the maximum polarization line and the ground plane at sunrise and sunset is symmetrical with the north–south meridian. By averaging the intersection of the maximum polarization line and the ground plane at the same location, the north–south line can be inferred, as shown in Fig. 8.23. In figure (A, B), the three-dimensional representation of the position of the maximum polarization line relative to the sun at sunrise and sunset is presented. In figure (C, D), the corresponding two-dimensional representation is illustrated. In figure (E),

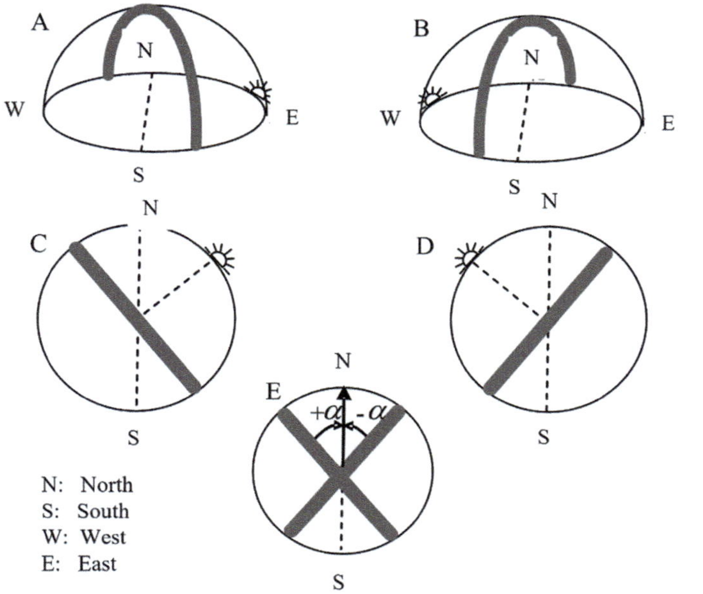

Fig. 8.23 The diagram of using the maximum polarization line of sunrise and sunset to determine the reference direction

represent the line position of the maximum polarization in the average sunrise and sunset time to determine the north–south reference direction. The angular distance between the maximum polarization line and the reference line is determined by the latitude and time of the surveyor. In this way, once the orientation of the line of maximum polarization at sunrise and sunset is sensed, the magnitude can be deduced, and the position of the line of maximum polarization at sunrise or sunset can be used alone to correct other compass systems. However, if other compasses are calibrated using only the position of the line of maximum polarization at sunrise or sunset, periodic updates are required to overcome seasonal and latitude changes.

Although there is currently no direct evidence that all animals use the dynamic properties of polarized skylight, its potential applications are enormous. Since it can not only provide direction information, but also can be used by other navigation systems to correct heading parameters. Therefore, it is necessary to study it more deeply.

8.3.3 The Time-Varying Periodicity of the Force Line Vector of the Perceived Polarization Intensity by the Bionic Polarization Navigation Instrument

Biologists have found that a variety of insects have the bizarre ability to sense the azimuth of polarized light in the sky and use it for navigation to help them complete behaviors, such as foraging, homing, and migration. Sand ants and other insects can use the polarization pattern of the sky for navigation and positioning because there is a pattern of polarized light in the sky, and it has a visual nervous system that is extremely sensitive to the polarization direction of sunlight. Based on the above-mentioned goniometer, a polarized light bionic navigation sensor is formed, and the polarized light field navigation effect at different times of the day is realized. Figure 8.24a, b are the polarized light distributions and the theoretical simulation in the sky from daytime and nighttime observed at sea on May 17, 2016, respectively. It can be seen that the sky polarization field is dominated by the sun during the daytime, and gradually transitioned to moonlight (reflected sunlight) at night, all of them have achieved effective navigation, which proved the periodicity of the sky polarization field centered on the sun. While the nighttime observations show that, the incident light indirectly takes the moon as a refracting body, refracting the complete characteristics of the polarized light field and the 24-h cycle stability and measurability of the day and night at any point on the earth.

8.4 Polarization Field Axis—Atmospheric Polarization Neutral Point Region

Under the condition of single scattering, the polarization of sky light is generally positive, but the multiple scattering of atmospheric particles causes negative polarization. Atmospheric neutrality is created when the degree of sky polarization is zero at the intersection of positive and negative polarizations. The line connecting the center point and the observation center constitutes the polarization field axis. In the sky polarization field, the polarization line of force is centered on the field axis and spreads vertically to the far end. The polarization degree of the neutral point of atmospheric polarization and its adjacent area is 0, which is a blind area for polarization navigation, so the polarization field axis and atmospheric polarization neutral dots are important elements of polarized fields.

8.4.1 Characteristics of the Polarization Field Axis

The polarization field axis has two meanings. First, in the actual atmosphere, due to the existence of multiple scattering, the polarization field axis is not a single straight

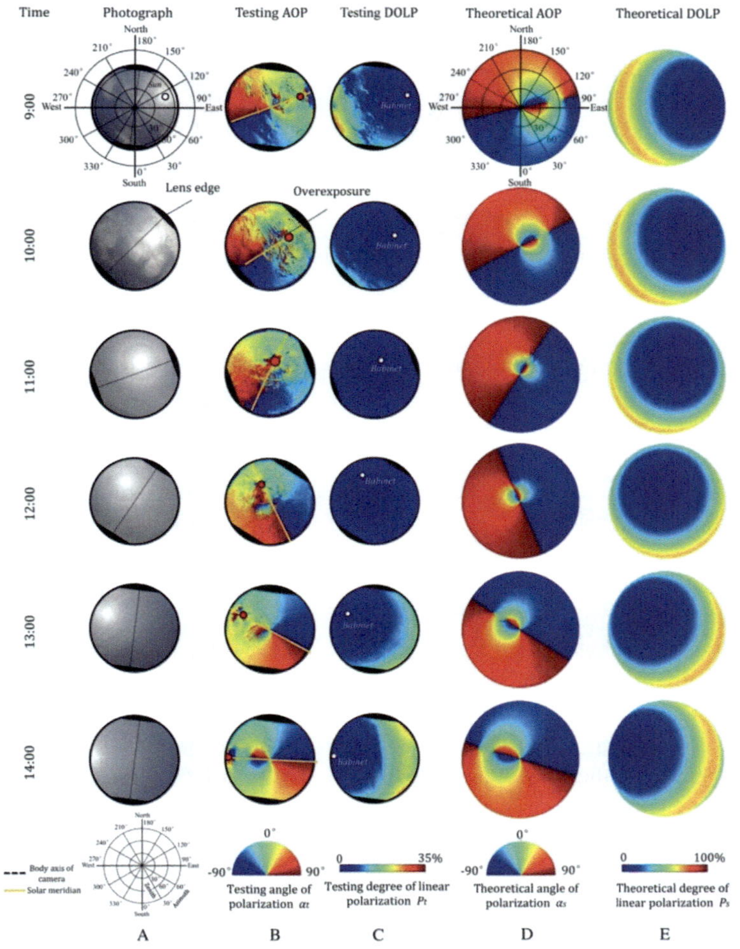

(a) Polarized light distribution at sea observation and simulation results between 9:00 and 14:00.

Fig. 8.24 Polarized light distribution at sea observation and simulation results

line; it's an area that the sky polarization degree of zero (or close to zero). Secondly, the polarization field axis is not a fixed area in the sky, it's a conical area starting from the observation point and facing the neutral point area. In this conical area, the polarization degree at any height is zero.

For ground observations, the amount of information for polarization image observed at the neutral point (the right side of Fig. 8.25, the neutral point of Babinet is the direction of the polarization field axis) is much greater than that of the polarization image observed at the non-neutral point (the left side of Fig. 8.25). Especially, this character is obvious for long-distance objects.

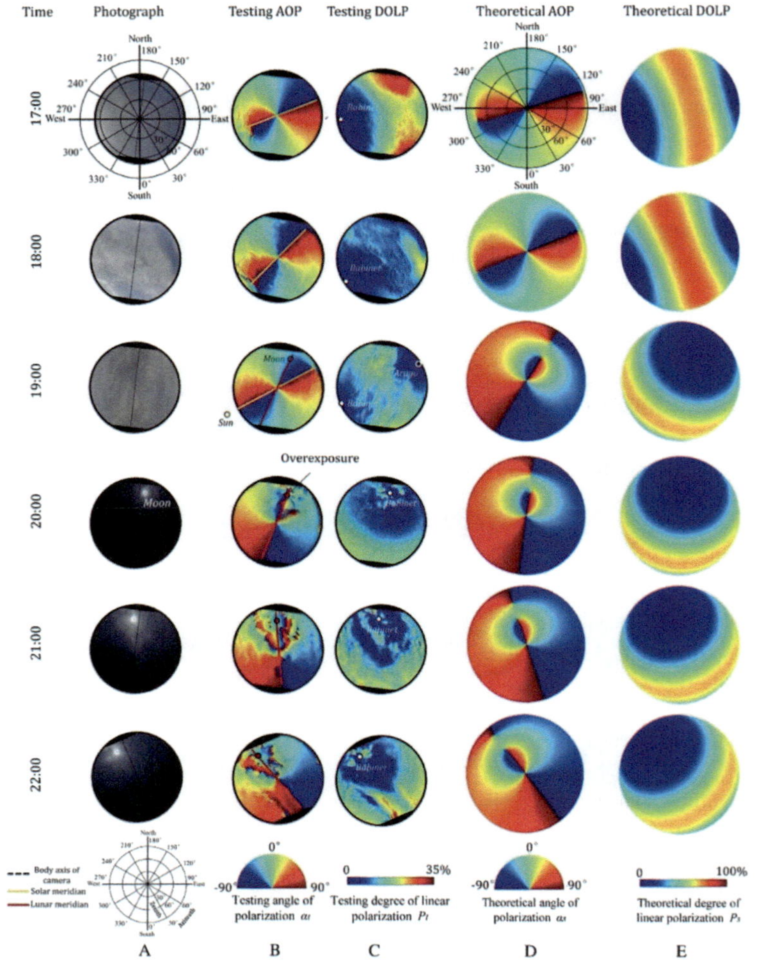

(b) Polarized light distribution at sea observation and simulation results between 17:00 and 22:00.

Fig. 8.24 (continued)

On the polarization image observed at the neutral point (right), whether the nearby Foxiang Pavilion of Longevity Mountain in the Summer Palace (about 3.1 km away from the shooting point) or the distant Xishan (about 6–8 km away from the shooting point in a straight line) are clearly received. Especially, the bare soil (or road) information on the mountain is in good agreement with the non-polarized image. While on the polarized image observed at the non-neutral point (left), the nearby ground object information can be clearly received (as described in the Buddha Incense Pavilion), and the information on the distant mountains is much weaker, which only the outline of the mountain can be vaguely displayed. It means that the polarization information

Non-neutral observation Neutral observation

(a) No polarization images

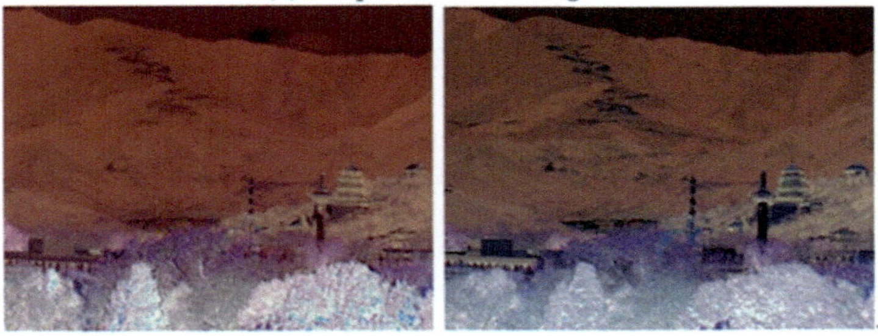

(b) Q component polarization images

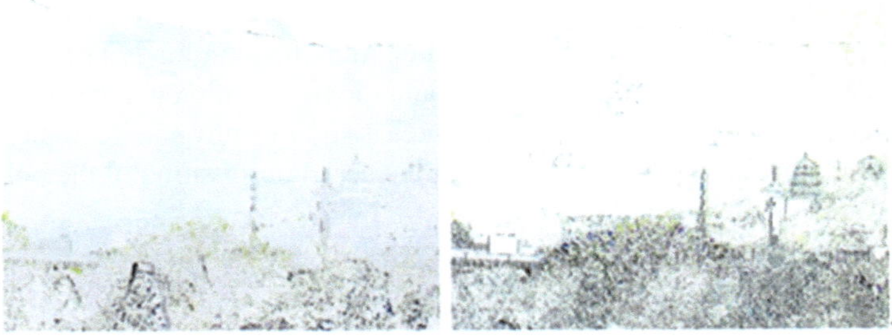

(c) U component polarization images

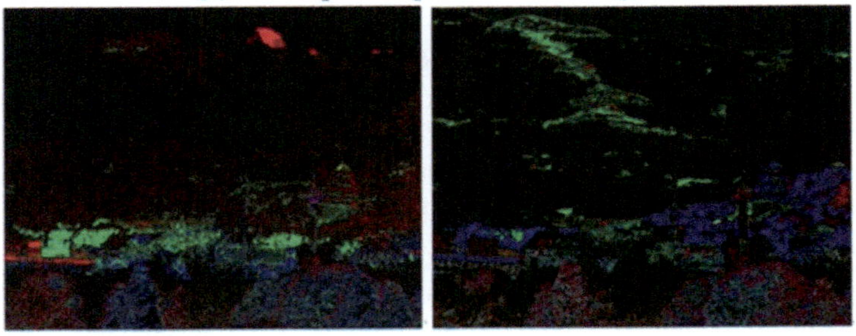

(d) Polarization images

Fig. 8.25 The local ground polarization remote sensing images from ground observations of distant mountains cannot be effectively obtained when observing from non-neutral points.

In aerial observation (Fig. 8.26), the polarization degree of images in the polarized field axis area is mostly less than 0.4, while most of the polarization degrees in the non-polarized field axis area are greater than 0.5, and the details of the objects in the polarized field axis area are significantly more than that of the unpolarized field axis area. Since atmospheric polarization is a strong polarization, this shows that the effect of atmospheric polarization in the polarization field axis region is small. It can be seen from the three-band polarization degree image grading chart of the R, G, and B aerial experimental images that the polarization degree of the red wave band and the green wave band is smaller than that of the blue wave band, and the details of the image information are more than that of the blue wave band. For example, the building in the slashed part of the image, the blue wave band image is partially blank, and some information is missing. The red wave band information is also slightly more than the green wave. In the image grading diagram of the R, G, B three-band polarization degree in the non-polarized field axis area, we can also draw the same conclusion as in the polarized field axis area. It can be found that the wavelength band with longer wavelength is also more conducive to observe the polarization field axis region.

It can be seen from the sky to the surface, the polarization field axis area can be also observed, and the remote sensing observation of ground objects in the polarization field axis area, the proportion of ground object information is larger than that in the non-polarization field axis area, and the polarization field axis area. Compared with the image in the non-polarized field axis area, the image in the polarized field axis area has more ground object information, and the polarization degree concentration distribution area in the polarized field axis area is significantly smaller than that in the non-polarized field axis area. It shows that the aerial observation of the polarization field axis area is effective.

8.4.2 Distribution of Atmospheric Polarization Neutral Points

The point in the atmosphere where the degree of linear polarization is zero is called the neutral point of atmospheric polarization. The polarization neutral point is a blind spot for navigation. Whether the polarization degree can be used for navigation, the polarization degree is within the detectable range should be firstly concerned. If the polarization degree is zero, it is impossible to use the polarization light navigation. There are three atmospheric polarization neutrals in a clear cloudless sky, named Arago, Babinet, and Brewster neutrals [1, 13–16]. The Arago point is located 20°–30° above the anti-solar point, and its position relative to the sun makes it relatively easy to observe. The Babinet neutral point is above the sun on the solar meridian, and the Brewster point is 25°–30° below the sun. The part between the sun and the horizon is the brightest area of the sky, and the haze near the horizon often produces low polarization values, so Brewster's point is the hardest to see. In theory, there will also be a fourth neutral point of 4TH, below the antisolar point, which was recently

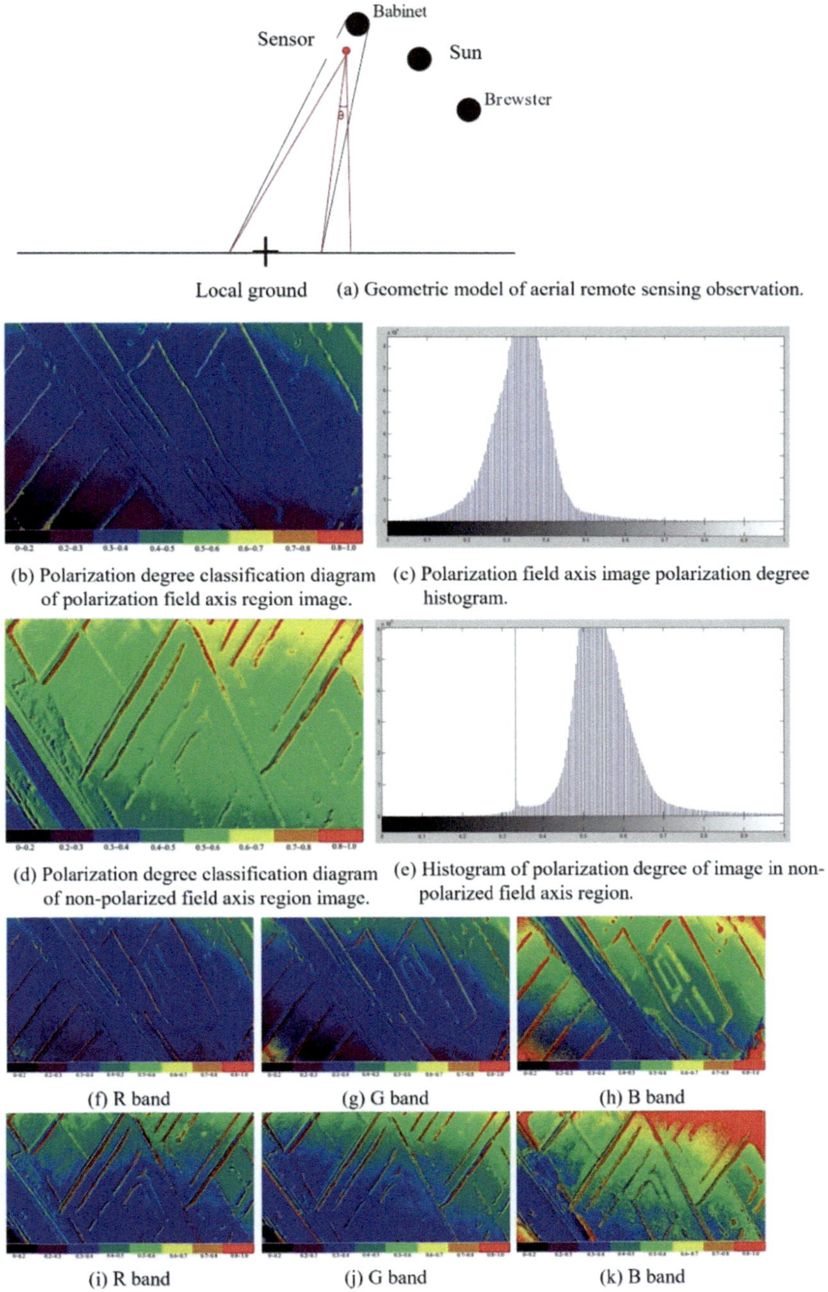

Fig. 8.26 The local ground polarization remote sensing images from aerial observations

discovered by scientists through sounding balloons. Some neutral points also appear in some special cases, such as water specular reflection, turbid atmosphere after volcanic eruption, and total solar eclipse. Figure 8.27 shows where each polarization neutral point of sky light occurs.

Under normal atmospheric conditions, only two points-Babinet and Arago or Babinet and Brewster can be seen at the same time. When the sun is higher, the Babinet and Brewster points are closer together, and when the sun reaches the zenith they almost merge into a neutral point. As shown in Fig. 8.28, the center of the gray area is the polarization neutral point, and the Babinet point and Brewster point can be clearly seen. The Babinet point can be seen in Fig. 8.29 (taken at 08:30); the neutral point seen in Fig. 8.30 (taken at 17:00) is the Brewster point. At the same time, according to the distribution of the polarization degree, another neutral point, the Babinet point, also affects the entire polarization degree distribution.

As shown in Fig. 8.31, the distribution of the polarization degree in the whole sky for continuous observation is presented. It can be seen from the distribution of the sun's position and the polarization degree at any time that the distribution of the polarization degree and the position of the neutral point in the whole sky are not

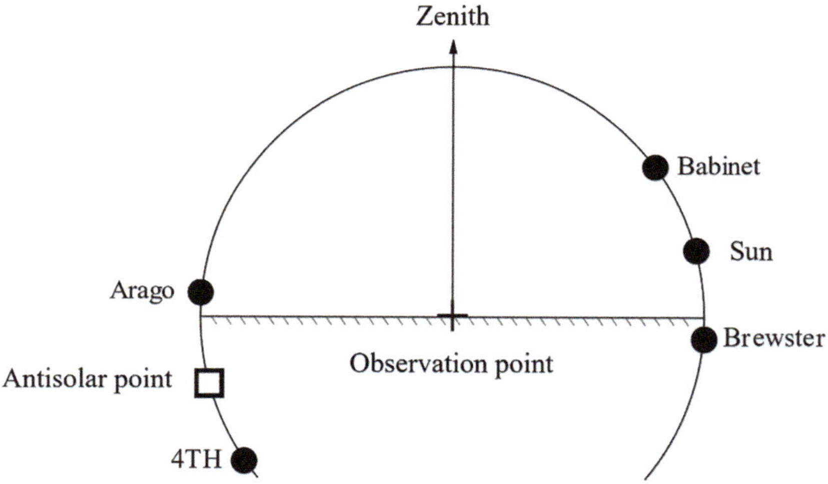

Fig. 8.27 The diagram of polarizing neutrals in the sky

Fig. 8.28 Babinet point and Brewster point

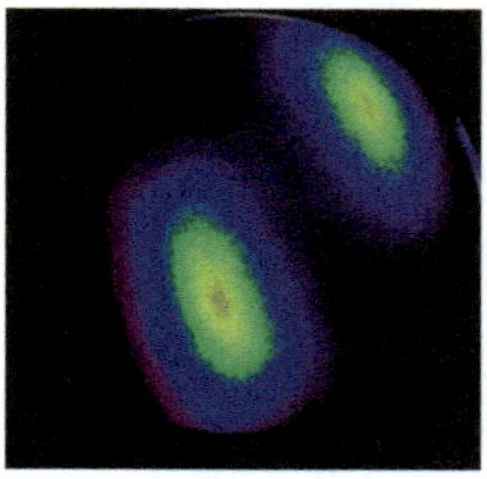

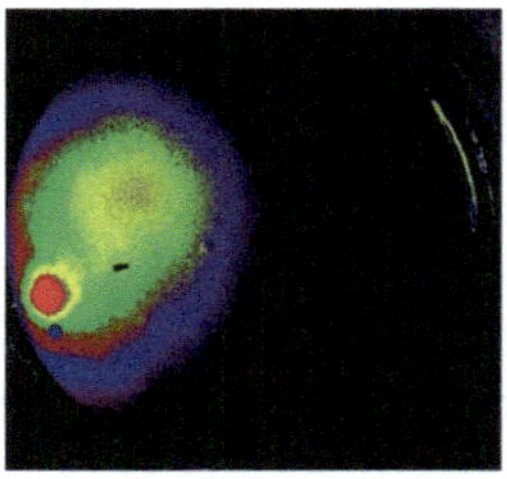

Fig. 8.29 Babinet point

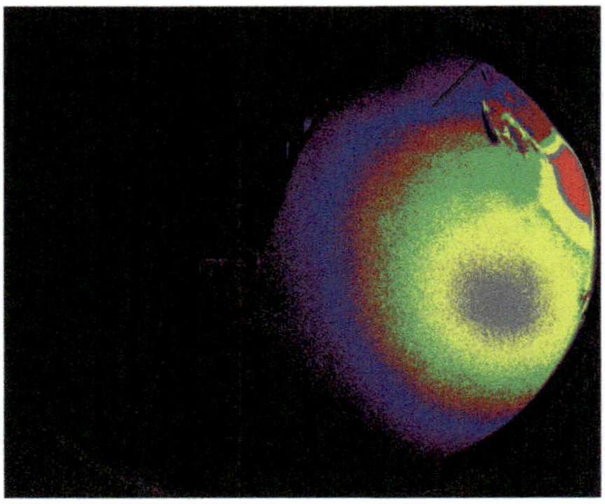

Fig. 8.30 Brewster point

fixed, but changed with the change of the position of the sun, which confirmed the altitude angle to the sun closely related.

From the polarization degree distribution map, it can be clearly seen that the polarization degree is distributed in a ring around the polarization neutral point. When the sun altitude angle is high and close to noon, a ring distribution can be observed in the sky. With the height of sun decreases, another annular distribution

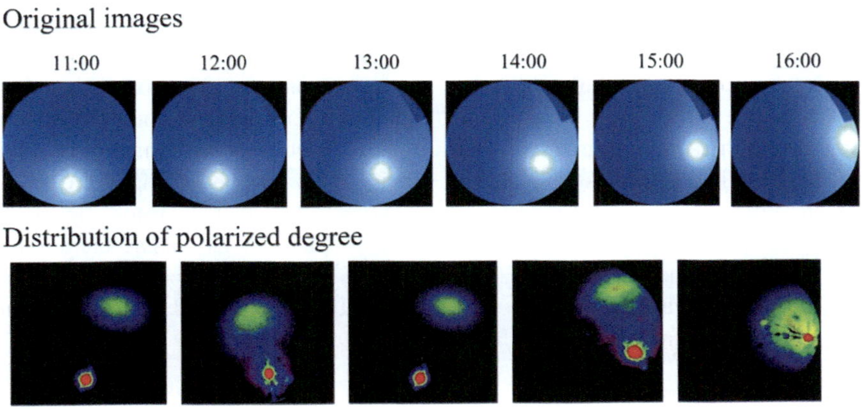

Fig. 8.31 Continuous observations of the polarization distribution across the sky

appears, and the polarization distribution of the entire sky is the superposition of these two annular distributions. Figure 8.32 also shows that the polarization degree of the sky sometimes shows an irregular distribution due to changes in the state of the atmosphere at different times. The red part in the picture is the sun, and the displayed polarization degree is zero, but the area where the sun is located in the image is often overexposed and distorted. The gray part is close to zero, that is, the position of the neutral point. Due to the existence of multiple scattering, the point where the polarization degree of zero is generally difficult to appear, which is closely to zero.

From Figs. 8.32, 8.33, 8.34 and 8.35, it can be seen that the atmospheric polarization window exists, and the angular distance between the polarization neutral point and the antisolar point (or the sun) varies in the range of about 15°–25°. Under the condition of single scattering, the visible polarization neutral points in the sky are the Babinet point and the Arago point; the polarization neutral point appears on the connecting line between the sun and the antisolar point, and the polarization degree is regularly distributed around the polarization neutral point. In this way, when the earth is observed at the position of the neutral polarization point, since the linear polarization degree of this point is zero, the measurement at this point can remove the polarization effect of the atmosphere, so that the polarization information of the surface target can be directly obtained [17]. However, for polarization navigation, the atmospheric polarization neutral point is a blind spot, and the polarized skylight is detected in the direction of the polarization neutral point. Since the degree of polarization is close to zero, the distribution of the polarization angle cannot be detected, and the navigation required orientation information cannot be obtained. Of course, the research on the atmospheric polarization neutral point still needs to further solve the problems of the polarization neutral point theory under the condition of multiple

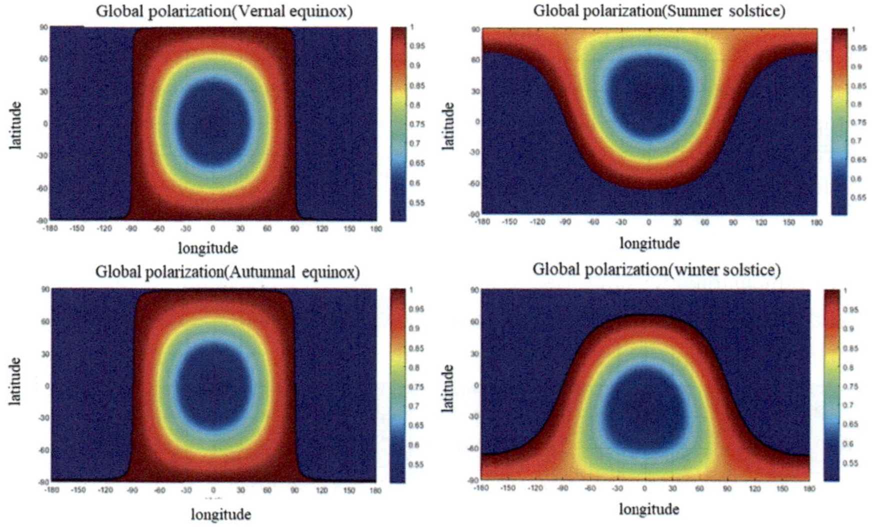

Fig. 8.32 Simulation of global polarization

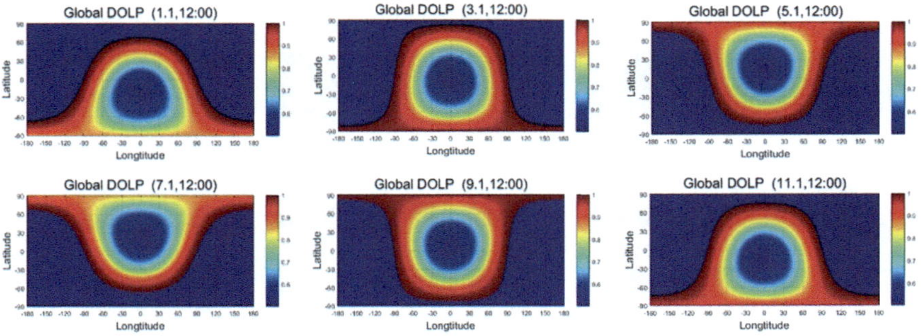

Fig. 8.33 Annual variation of global atmospheric polarization

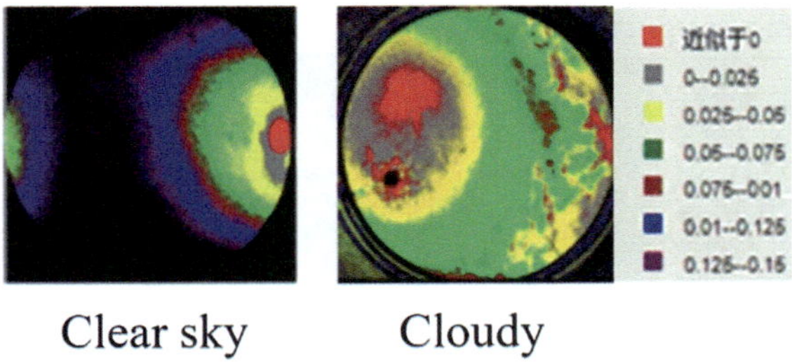

Fig. 8.34 The distribution of polarization degree under different weather conditions

scattering, the simulation of the all-sky atmospheric polarization neutral point, and how to use the polarization neutral point to combine with the polarization sensor. Therefore, the hypothesis about the distribution of atmospheric polarization degree still needs to be further analyzed theoretically and verified experimentally.

8.5 Synthetic Distribution of Polarization Vector Field and Comparison with the Known Natural Fields

The cooperation of polarization field force line and the polarization field axis forms the polarization field of the sun entering the sky, which is the most important manifestation of the stellar energy entering the earth.

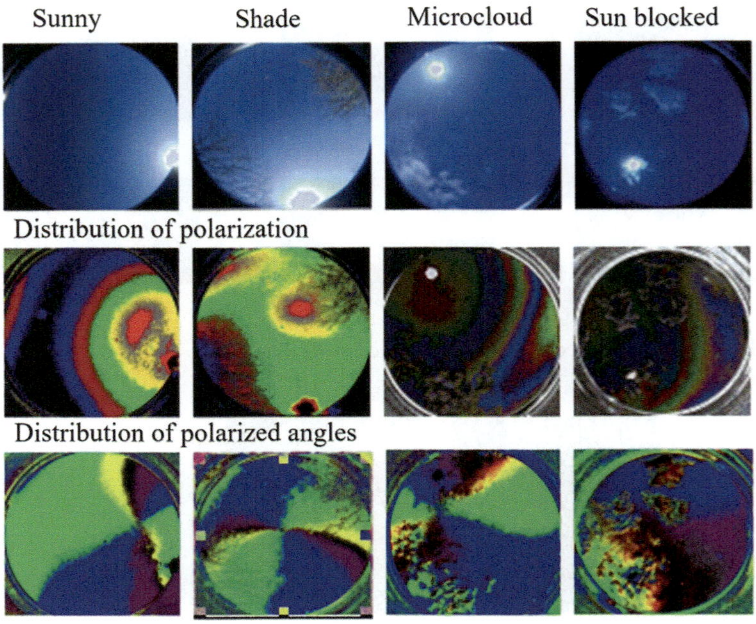

Fig. 8.35 The distribution of polarized light in the sky under different conditions

8.5.1 Distribution of Polarization Degree Under Different Conditions

At a certain time and position of the day, the sky has a relatively stable polarization mode. However, because many factors affecting the atmospheric polarization distribution, such as the number of atmospheric particles, particle size, weather conditions, and measurement bands, etc. The atmospheric polarization mode map has a greater impact, so when using the sky polarization mode map for more in-depth research or applications, such as using the polarization mode map to obtain the direction information required for polarization navigation, more influencing factors need to be considered to make the results closer to the actual conditions.

The polarization degree of the atmosphere varies with the position of the sun. Taking the sun as the field axis, the polarization degree increases gradually as it spreads to the surroundings. Taking the observation point as the origin, when the angle between the observation direction and the sun is close to 90°, the polarization degree reaches the maximum, and then gradually decreases until the mirror image point of the incident light is 0. Based on this principle, we simulated the global atmospheric polarization distribution under the condition of vertical upward observation on the ground, and the results are shown in Fig. 8.1c. The polarization pattern in the sky is closely related to the sun altitude angle. Different sun altitude angles have different polarization degrees; the intensity and direction of polarized light are different. When the sun altitude angle is low, the polarization intensity in the sky is different. It is relatively large, that is to say, the polarization intensity in the sky is greater in the evening or morning than at noon. In different seasons, such as vernal equinox, summer solstice, autumnal equinox and winter solstice, there are regular changes in its distribution.

According to the theoretical distribution of the polarization degree of light in the global sky at 12:00 GMT on the 1st of each month, its annual change (Fig. 8.32), the point with the lowest degree of polarization is theoretically the point of direct sunlight, which moves between the tropic of cancer and a year is a cycle. The point with the lowest polarization degree is theoretically the point of direct sunlight. It changes with the alternation of day and night, and the movement of the sun's position in the hemisphere during the day (Fig. 8.33).

1. The distribution of the polarization degree of the sky under different weather conditions

Since the distribution of polarized light in the sky is inseparable from the scattering of sunlight by atmospheric particles, the distribution of sunny and cloudy days needs to be considered (Fig. 8.34). The red area in the figure is the position of the neutral point of atmospheric polarization, which is the point in the sky where the degree of atmospheric polarization is zero. It can be seen from the image of the cloudless state that the atmospheric polarization degree shows a regular annular distribution around the polarization neutral point. The polarization degree gradually increases from the neutral point to the outside, and then gradually decreases to another neutral point (due to the angle of view, this neutral point is not represented in the figure). The image in the cloudy state on the right has certain similarities with the image in the cloudless state. The polarization degree of the sky is also distributed in a ring shape along with the neutral point, but this distribution is not very regular. The low gray area, which is another polarization neutral point, is also not shown here. The color of the entire image is mostly yellow or green, which means that the degree of polarization is much smaller than the cloudless state. Due to the depolarization effect caused by the multiple scattering of the atmosphere in cloudy conditions, the polarization degree of the image is greatly reduced.

How suitable is polarized skylight for navigation? Figure 8.35 shows the distribution of polarized skylight obtained by the full polarization measurement method under four conditions: clear, shaded, slightly cloudy, and cloudy (the sun is blocked by clouds). The average value of the polarization degree distribution in the figure is presented. It can be seen that although the weather conditions have a great influence on the distribution of the polarization degree in the sky, the average value of the polarization degree under various weather conditions is greater than 10%, which meets the requirements of polarization navigation. At the same time, the sky polarization angle also has a relatively regular distribution under various conditions, and direction information can be obtained from the distribution of the polarization angle under different weather conditions.

2. Sky polarization distribution at different sun altitudes

As shown in Fig. 8.36, from a single polarization degree distribution image, the atmospheric polarization degree shows a regular annular distribution around the polarization neutral point. The polarization degree gradually increases from the neutral point outward, and then gradually decreases after reaching the maximum black area. However, the shape and size of the degree of polarization distribution are different for different solar altitude angles. When the solar altitude angle is high and close to noon, we can see two polarization neutral points, and the entire polarization degree distribution revolves around the two polarizations. When the sun altitude angle is relatively low, we can only see one polarization neutral point, and the entire polarization degree distribution seems to expand around only one polarization neutral point. As shown in Fig. 8.36, the brightness statistics for each polarization degree image are performed, and the average statistics of the polarization degree of the entire image are calculated in Fig. 8.37. As can be seen from Fig. 8.37, from 10:30 am to 12:00 am, the average value of the polarization degree image gradually becomes smaller, and the average polarization degree reaches the minimum when the sun altitude angle is maximum (12:00). And then, as time goes by, the average value of the polarization image gradually increases. From the results of Figs. 8.36 and 8.37, it can be concluded that the distribution of the sky polarization degree is closely related to the sun altitude angle, and the shape and size of the sky polarization change with the change of the sun altitude angle. When the sun altitude angle is lower, the average polarization degree is larger. Conversely, the polarization average is smaller.

3. Sky polarization distribution in different observation bands

Under the condition of a clear sky, the polarization distributions of the purple, blue and red bands are not exactly the same. The polarization degrees of the sky are distributed regularly around the polarization neutral point (the red area in Fig. 8.38). The position of the polarization neutral point of the three bands is the regions where the maximum polarization degree (black area in the figure) appears are also different, which shows that the visible light band also has a certain influence on the distribution of polarization degree.

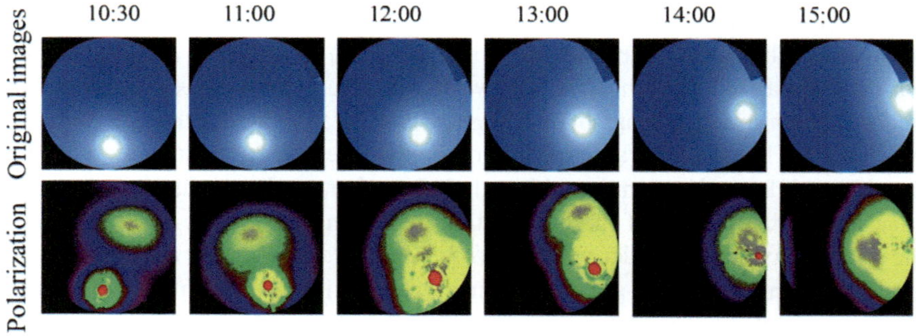

Fig. 8.36 The variation of polarization degree with the sun's altitude angle in sky

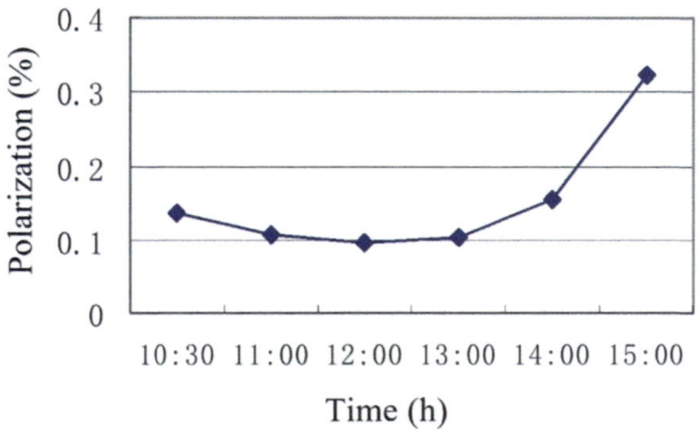

Fig. 8.37 The relationship between polarization and sun's angle

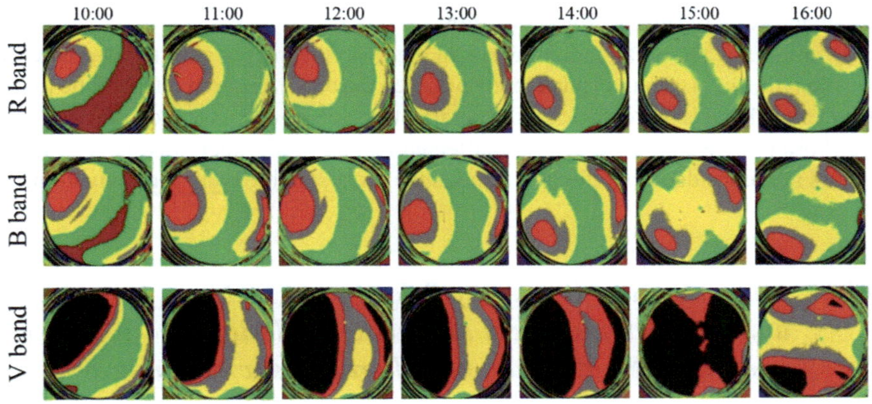

Fig. 8.38 The variation of polarization degree of different bands with the sun's altitude angle in clear sky

As shown in Fig. 8.39, the average statistics of the polarization degrees of the images for each polarization degree are performed. It can be seen from the figure that the polarization degree distribution in the sky is closely related to the observation band. The longer the visible band, the greater the average polarization degree. On the contrary, the average of polarization degree is smaller. The figure also shows that, from 10:00 to 12:00 AM, the polarization degree of the same observed band is decreased, and the average polarization degree reaches a minimum when the sun altitude angle is maximum (12:00). With the passage of time, the average polarization degree image gradually becomes larger, and becomes smaller again at 14:00 and 15:00, which has a certain deviation from the theoretical analysis. This should be caused by the atmospheric turbulence caused by the strong wind at that time, and the atmospheric turbulence has a greater impact on the short waveband and less

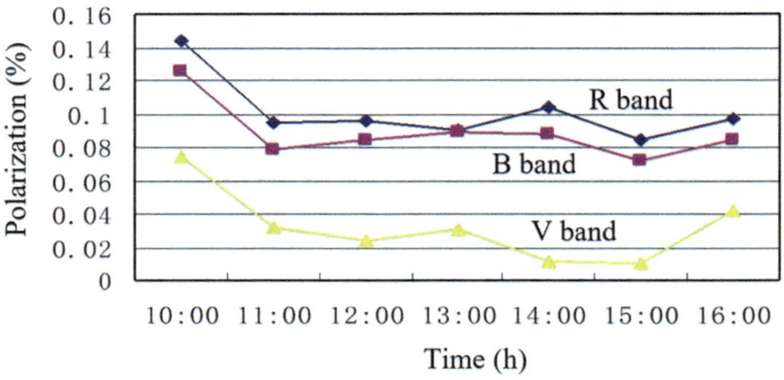

Fig. 8.39 Plot of the mean value of the polarization degree against the altitude angle of the sun

impact on the long waveband. This shows that in clear weather conditions, the long waveband is more suitable for navigation.

In summary, the polarization degree of clear sky decreases with the decrease of wavelength, but in turbulent and cloudy weather, the depolarization effect of the long-wavelength band is more obvious, the polarization degree decreases accordingly, and the shorter waveband is basically unaffected or has little effect. Taking the above factors into consideration, the blue band is the most suitable band for navigation.

8.5.2 Comparison of Earth-Sky Polarization Field and Gravitational Field Magnetic Field Map, and Three-Field Parameter Characteristic Mapping

It is now known that there are two global vector fields on the earth, namely, the gravitational field formed by universal gravitation, and the geomagnetic field formed by the rotation of the earth. Accordingly, they are compared with the Earth-sky polarized vector field formed by the incoming radiant energy of the sun.

The comparison between the polarized fields of light in the earth's sky, the gravitational field, and the earth's magnetic field are performed. The gravity field map is a gravity field map drawn by GOCE satellite data released by ESA (http://www.esa.int/SPECIALS/GOCE/SEMY0FOZVAG_0.html), and the geomagnetic field map is the 11th generation International Geomagnetic Reference Field diagram, which is provided by the International Association of Geomagnetism and Aeronomy Working Group V-MOD.

At any point on Earth, the polarized field of light in the sky has magnitude and direction, and can be measured in many ways. The physical quantities that describe the magnitude and direction of the polarized field are the polarization degree and the angle of polarization. The polarized field is a function of time and space. The basic polarization field is affected by the sun's height angle in a day, and the sky light polarization field at the same location changes with a short period of one day, a period of one year, and a long period of one year.

It can be seen from the sky polarization field, like the earth's gravitational field and geomagnetic field, is a global feature, and there are many similarities. As an objective representation of solar energy incident on the earth, it has wide application prospects. The combination of the three fields is expected to become a perfect representation and regular basis of the interaction between universal gravitational force, the earth's rotation and solar energy on the earth.

8.6 Summary

The regular distribution of polarized light in the sky is the premise and guarantee of polarization navigation. The measurement and analysis of the polarized light distribution in the sky is the focus of this chapter. In this chapter, the cause of the polarized skylight and its theoretical model is firstly analyzed. And then, by experimental measurement, the actual distribution of the polarized skylight is verified to be consistent with the theory. The second half of this chapter describes the dynamic characteristics of polarized light in the sky and its application value, and makes preliminary assumptions about the distribution of atmospheric polarization. The distribution of the polarization degree in the whole sky is the superposition of several annular distributions around the neutral point of atmospheric polarization, and the correctness of the hypothesis is preliminarily verified by measurement experiments.

In this chapter, the all-sky polarization measurement method is used to conduct an experimental study on the distribution of polarized skylight in all-band and red, blue, and purple bands under different weather conditions. The measurement results show that the atmospheric polarization degree shows a regular annular distribution around the polarization neutral point; the image of cloudy state is similar to the image of cloudless state, but its distribution is not very regular, this is because the effect of multiple scattering of atmospheric particles, the degree of polarization in the cloudy state is smaller than that in a clear sky; the polarization degree distribution of sky light is closely related to the sun's altitude angle. With the change of sun altitude angle, the shape and the size of polarization degree are change. The greater the sun altitude angle is, the smaller the average value of the sky light polarization degree. And vice versa, under the same conditions, the sky light polarization degree distribution of different bands is different, and the polarization degree of the clear sky becomes longer with the observation wavelength. However, under cloudy weather conditions, the depolarization effect of the polarization degree in the longer wavelength band is more obvious. The average polarization degree measured under various conditions is greater than 5%, so it is considered that the polarized skylight can be used as an information source for polarization navigation. Considering various factors, it is considered that the blue band of visible light is the most suitable band for navigation.

The polarization angle of sky light is extremely regularly distributed in clear weather and depends entirely on the position of the sun. If the sun is blocked by clouds or vegetation, the polarization angle distribution of sky light will be affected by multiple scattering, but from the polarization angle distribution measured under various conditions, the distribution is relatively regular. The measurement results show that the distribution of the polarization angle of the sky light is relatively stable under the conditions of clear weather, shade and rainy weather, which can be used to obtain the direction information required for navigation.

References

1. Muheim, R., Phillips, J. B., & Akesson, S. (2006). Polarized light cues underlie compass calibration in migratory songbirds. *Science, 313*(5788), 837–839.
2. Yan, L., Guan, G. X., Chen, J. B., et al. (2009). Bionic orientation mechanism in the skylight polarizatin pattern. *Journal of Peking University (Natural Sciences), 45*(4), 616–620.
3. Ye, Y. T., Rao, J. Z., et al. (2005). *Optics course.* Tsinghua University Press.
4. Lei, Y., et al. (2014). *Polarization remote sensing physics.* Science Press.
5. Li, J. Z. (1986). *Optical manual.* Shaanxi Science Press.
6. Wang, Z.-J. (1987). *Optical technical manual* (pp. 291–292). China Machine Press.
7. Wei, G.-H. (1995). *Matrix optics* (pp. 144–186). Higher Education Press.
8. Acham, R. M. A., & Bashara, N. M. (1956). *Ellipsometry measurements and polarized light (in Chinese).* Beijing: Science Press.
9. Mie, C. (1908). Beitrage zur opitic truber medien. *Ann physick, 25*, 377–445.
10. Yang, J. X., & Zhang, S. X. (1988). *Fundamentals of navigation.* Harbin: Harbin Shipbuilding Institute Press.
11. Fang, C. Z., & Ning, X. L. (2006). *Principles and applications of astronomical navigation* (pp. 1–352). Beihang University Press (in Chinese).
12. Frisch, K. V. (1949). Die Polarisation des Himmelslichtes als orientierender Faktor bei den Tanzen der Bienen. *Experientia, 5*(4), 142–148.
13. Chandrasekhar, S., & Elbert, D. (1951). Polarization of the sunlit sky. *Nature, 167*(4237), 51–55.
14. Wu, T., & Zhao, Y. (2005). The bidirectional polarized reflectance model of soil. *IEEE Transactions on Geoscience & Remote Sensing, 43*(12), 2854–2859.
15. Zhao, Y. S., Wu, T. X., Hu, X. L., et al. (2005). A preliminary study on the quantitative relationship between multi-angle polarization reflection and binary reflection. *Journal of Infrared and Millimeter Waves, 24*(6), 4.
16. Lee, R. L. (1998). Digital imaging of clear-sky polarization. *Applied Optics, 37*(9), 1465.
17. Guan, G. X., Yan, L., Chen, J. B., et al. (2010). Characteristics of atmospheric polarization neutrals under single scattering conditions. *Journal of Harbin Engineering University, 31*(4), 4.

Chapter 9
Bionic Navigation and Positioning Method of Polarization Vector Field

Over 3.5 billion years of evolution, about 2 million kinds of organisms have gradually formed. These organisms have different forms and abilities, but what they have in common is that they have tissues, organs and structures, which are adapted to the growth environment. This feature provides unlimited imagination for the technical innovation of human beings. For example, in the Sahara desert, there is an ant called Cataglyphis owns amazing navigation and positioning skills. The study has indicated that the ants with a brain of only 0.1 mg can arrive at hundreds of meters away from the nest to hunt food. Once a suitable prey is found, they can back to the old nest accurately on a straight path (as shown in Fig. 9.1).

Their journeys are thousands of times than their body lengths, there are also no landmarks to refer to in the desert, and the dry environment makes it impossible for them to navigate using any scents, while they have no developed brains to make judgments.

Compared to ants, bees and crickets are no less capable of navigation. According to calculations, a bee needs to collect the powder of 2 million flowers to make 1 kg of honey and fly 4.5×10^5 km back and forth, but they never get lost. The magical navigation ability of these types of insects has attracted the attention of researchers for a long time. As early as the early twentieth century, Pieron, Cornetz, and Santschi et al. have begun to explore how these insects perceive and calculate during the navigation process. Preliminary research has showed that they navigated by using polarized light from the sun. An interesting experiment found that, sand ants in the desert came out of their nests to randomly search for food, once they obtained, they could directly return to the nest. This phenomenon that following a new path approximately a straight line quickly returns to the nest instead of repeating the circuitous route traversed during the food searching, indicating that animals can be changed directions randomly during the foraging process, but they are aware of the direction of each step of their own with memory function, which is called vector calculation.

L. Yan et al., *Mathematical and Physical Fundamentals of Navigation and Positioning with Earth's Natural Fields*, Navigation: Science and Technology 13,
https://doi.org/10.1007/978-981-97-5524-0_9

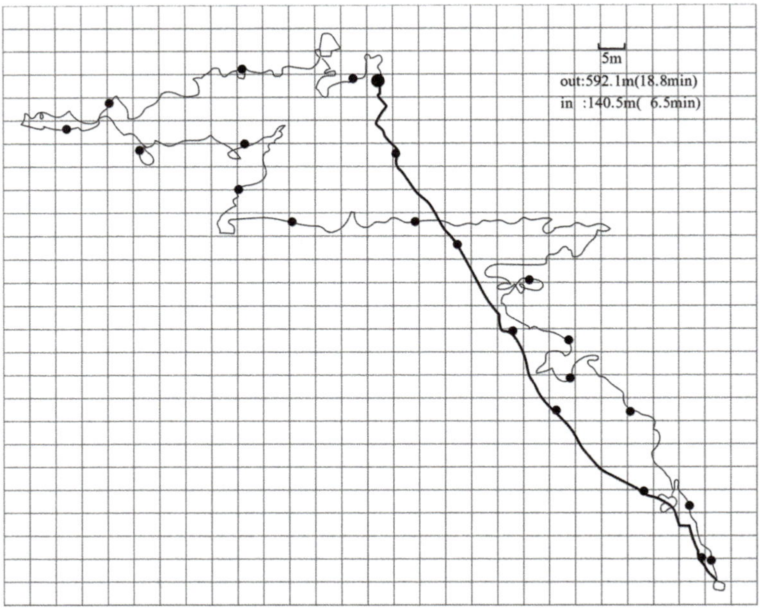

Fig. 9.1 Sand ants back to their home using polarized light navigation

Another example is that the sand ants are divided into three groups, the legs of group A are sawed off, the legs of group B are lengthened by adding a seesaw, and the group C remains the same.

(1) When they walking back to the nest from a same starting point, for group A, they stopped to look for them before the nest arrived. For group B, they stopped to look after walking through the nest. For group C, they stopped just after walking back to the nest.
(2) After training for 3–5 times, these ants can be directly arrived at the nest without searching. The above situation firstly proves that the biology is a complete navigation by superimposing the vector step size, at the same time, this vector calculation of the organisms has the self-learning ability, thus formed the ability of self-learning and self-correcting for the biological polarized light navigation.

With the rapid development of precision experimental instruments and the continuous improvement of processing technology, researchers have made major breakthroughs in the research on the neural tissue structure and behavioral physiology of these polarization-sensitive insects. At the same time, computer vision imitating insect polarization vision has also achieved great progress.

9.1 Mechanism Analysis of the Polarization Navigation for Sand Ants

Neurobiologist studies have shown that insects such as bees, crickets, spiders and sand ants have very similar polarized visual systems [1–5]. This system can be divided into two layers, the retinal layer and the nerve center layer. The retinal layer perceives the intensity and direction distribution pattern of the polarized light of the sun in the sky, and the nerve center layer comprehensively processes its response results. Research on spider's shows that spiders have three pairs of compound eyes, one of them is extremely sensitive to polarized light and responds to different degrees of polarized light in different directions. Figure 9.2 shows the response to stimulation of spider compound eyes corresponding to 0°, 45° and 90° polarized light.

Meanwhile, studies on sand ants have shown that an area of the sand ant's retina, located at the highest dorsal border of the eye, is extremely sensitive to polarized light, called the POL area. Sand ants use the special distribution of optic nerve rods in the POL area to complete the navigation task. The optic nerve rods in the POL area are spatially arranged in a hierarchical structure, as shown in Fig. 9.3.

How does the retinal layer sense the direction of polarized light? Studies on sand ants have shown that the original retina is composed of hundreds of optic nerve rods facing in different directions, and each rod is composed of 8 unidirectional photoreceptors intersecting and perpendicular to each other, as shown in Fig. 9.4.

This structure makes each sensor rod show a high degree of polarization sensitivity, with the maximum response when the polarized light is parallel to it, and the minimum

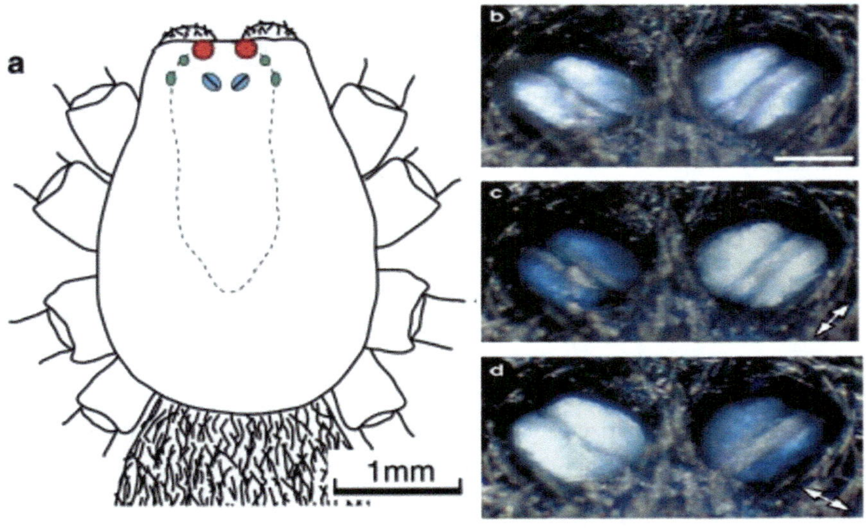

Fig. 9.2 The compound eyes of spiders and their response to polarized light

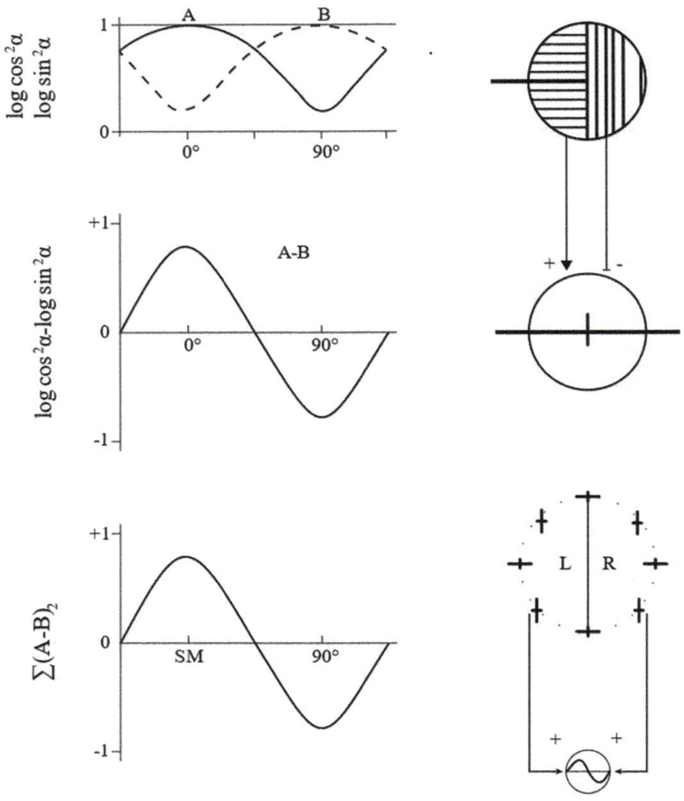

Fig. 9.3 Hierarchy of optic nerve sensory rods in the insect POL region

Fig. 9.4 The rod bundle arrangement of Cataglyphis retinal sensory

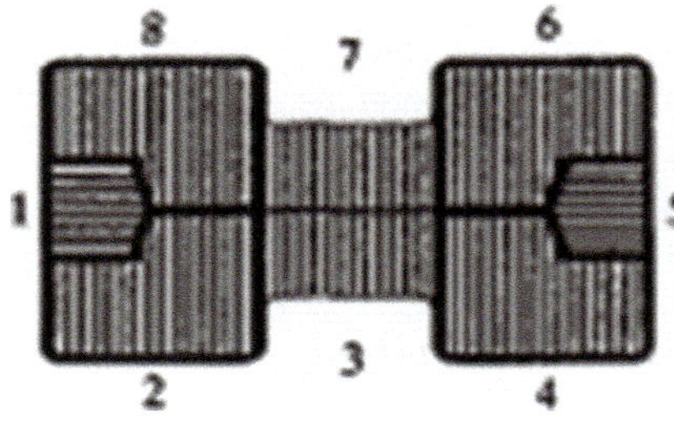

response when the polarized light is perpendicular to it. Moreover, the stimulus–response of each sensor rod to polarized light in a specific direction is a sine curve, forming a "+" cross_photoreceptor structure. This configuration makes each sensor rod a strict directional analyzer, as shown in Fig. 9.5.

And all the polarization-sensitive nerve rods simulate the pattern of polarized light in the sky to form a polarization-sensitive array on the retina, as shown in Fig. 9.6.

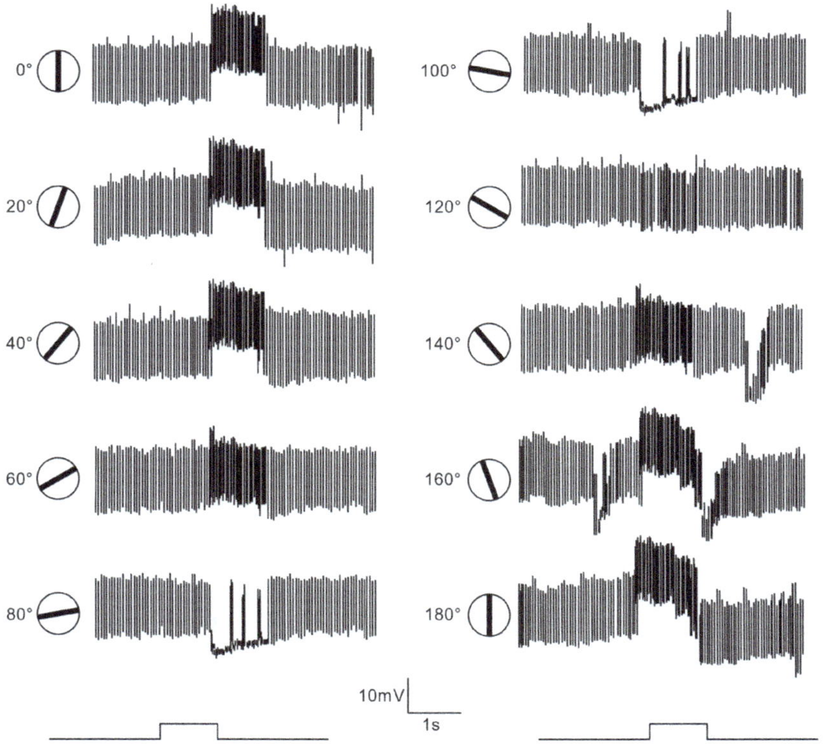

Fig. 9.5 The response of the polarization orthogonal element

In this way, when the sand ant turns its body around its head, the neural response changes systematically. Once the body axis of the sand ant coincides with the solar meridian, the response of the retinal neurosensory rod reaches an extreme value, as shown in Fig. 9.7.

Fig. 9.6 The diagram of polarized-sensitivity

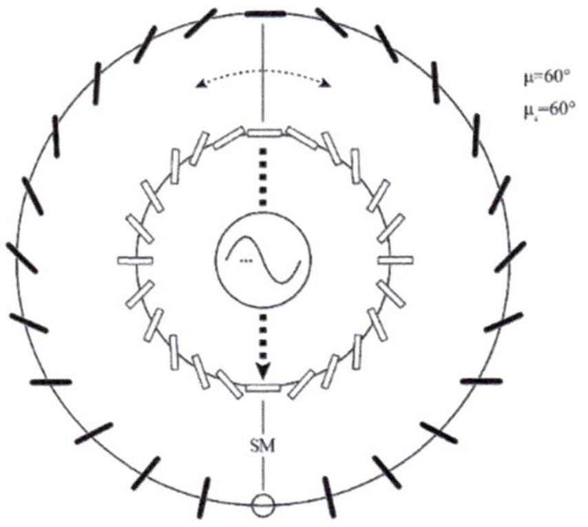

Fig. 9.7 The response of body axis paralleled to the solar meridian

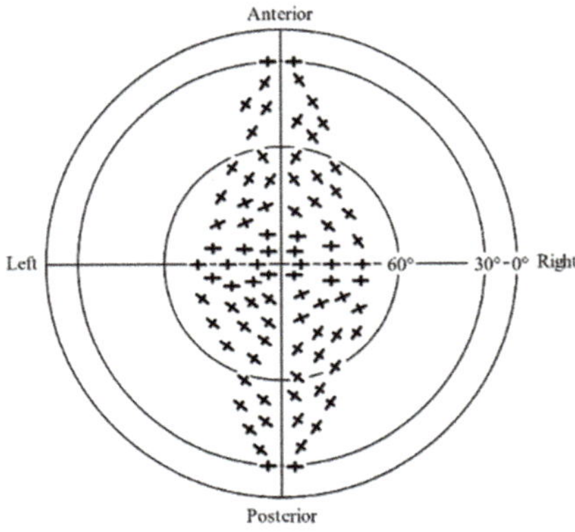

The responses of retinal polarization-sensitive rods converge as input to the optic lobe portion of the central nervous layer. The interneurons in the optic lobe receive input from the optic rods on the retina whose directions of greatest sensitivity are perpendicular to each other. Their responses to neural input to the retina are also sinusoidal, reaching maximum response values when the angle between the body axis of the sand ant and the solar meridian is 0°, 60°, and 120°, respectively, as shown in Fig. 9.8. The response result of the interneuron is used as input to the compass neuron in the sand ant's brain. After calculation and decoding, the angle between the sand ant's body axis and the solar meridian can be obtained, so as to realize its navigation function.

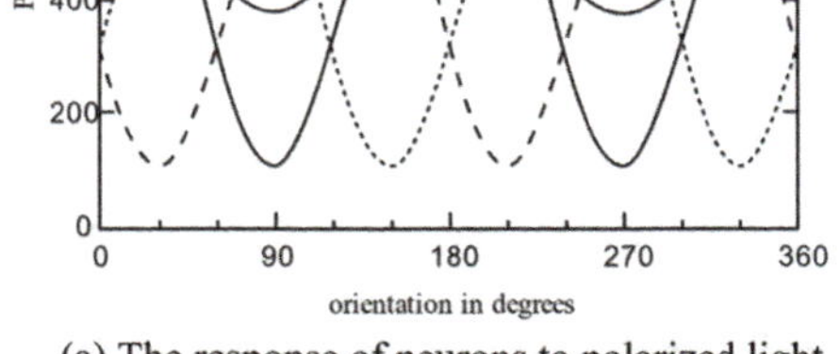

(a) The response of neurons to polarized light.

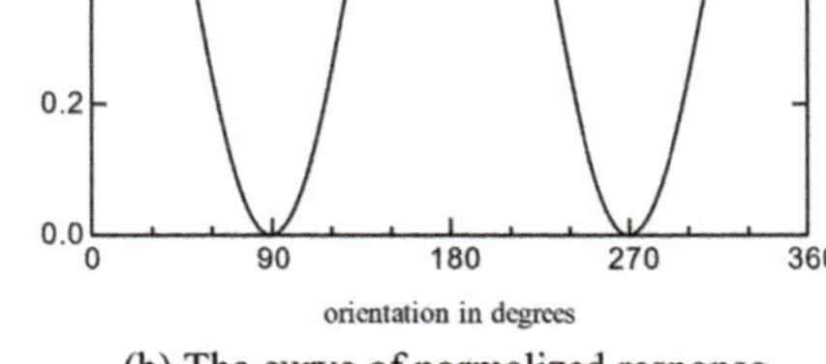

(b) The curve of normalized response.

Fig. 9.8 The response curve of neurons to polarized light

9.2 Bionic Polarization Navigation Model and Heading Angle Determination

Polarization navigation models can be divided into scanning and concurrent types [5]. On the basis of these two models, if a more accurate polarization navigation model was studied, a more accurate navigation could be realized.

9.2.1 *Scanning Model*

For the scanning model, the solar meridian is first discovered by using the directional sensitivity of the polarization orthogonal unit to the sky polarization pattern, and then this direction is used as the zero-degree reference direction of the body system. In the basic scanning model, only one orientation of polarization-sensitive quadrature elements is required to determine the orientation of the navigator. When the navigator rotates around its own body-long axis to scan the sky, the polarization quadrature unit will produce the greatest response when the navigator's body axis is parallel to the symmetry line of the sky polarization pattern map (solar or anti-solar meridian).

$$p(\varphi_{\max}) \geq p(\varphi), \forall\varphi \in [0^\circ, 360^\circ], \varphi_{\max} \in \{\varphi_{sol}, \varphi_{asol}\} \tag{9.1}$$

where φ is the output angle of the current direction of the polarization orthogonal unit relative to the solar meridian. This function has two local maxima, which correspond to the solar and anti-solar meridians, respectively. The solar meridian and anti-solar meridian can be determined by the directional light intensity sensor installed on the head [8]. This model has a simple structure, which can be easy to implement photo-electrically. However, a defect is also existed, because the response curve near the local maximum value is very flat, the detection of the maximum value position is imprecise. The solution by Lambrinos et al. is to use the signals extracted by two other polarization quadrature units to finely modulate the region of local maxima. Here, the response directions of the maximum E-vectors of the three fundamentally polarized orthogonal elements are 0°, 60°, and 120°, respectively. The response curve of the scanning process can be expressed as:

$$f(\varphi) = p_1(\varphi) - \|p_2(\varphi) - p_3(\varphi)\| \tag{9.2}$$

The basic model (response curves of POLOP1, POLOP2, POLOP3) and the improved model of Lambrinos et al. (response curves of dotted line) are shown in Fig. 9.9.

Obviously, the response curve of the improved model is sharper at the local maxima at 0° and 180°, and f(φ) is a quasi-linear function, which can be easily realized and calculated for the model system. Even so, there are still many angle values close to the local maxima, which is difficult to accurately search for the

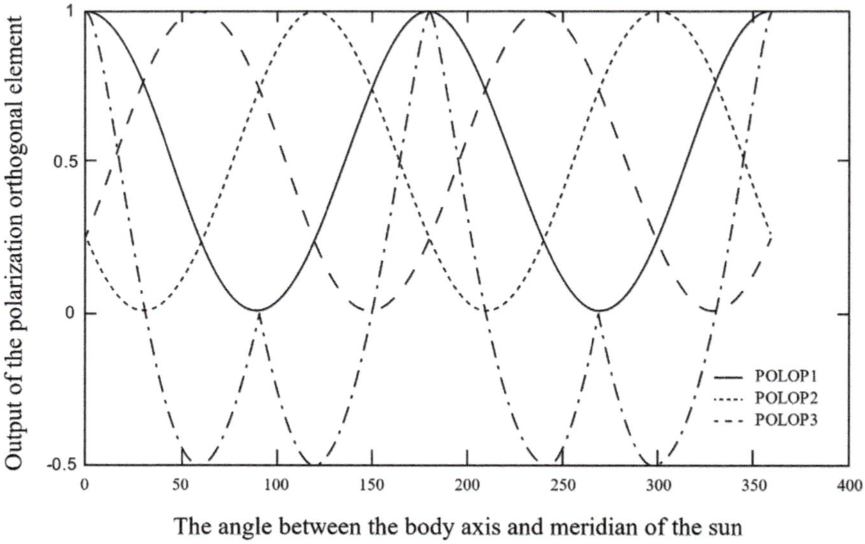

Fig. 9.9 The scanning model improved by Lambrinos

maxima. The research team of Professor Chu Jinkui in China found that the intersection points of the two response curves of POLOP2 and POLOP3 are correspond to angles of 90°, 180°, 270° and 360°, respectively. Is this a coincidence? Can we use this feature of the response curves of POLOP2 and POLOP3 to search for the location of the maximum value more accurately? Hence, their team made such a derivation, let $p_2(\varphi) = p_3(\varphi)$, and the following description can be obtained:

$$\lg\left[\frac{1 + d\cos(2\varphi - 2\pi/3)}{1 - d\cos(2\varphi - 2\pi/3)}\right] = \lg\left[\frac{1 + d\cos(2\varphi - 4\pi/3)}{1 - d\cos(2\varphi - 4\pi/3)}\right] \tag{9.3}$$

After calculation, a relation can be expressed as:

$$\cos\left(2\varphi - \frac{4\pi}{3}\right) = \cos\left(2\varphi - \frac{2\pi}{3}\right) \tag{9.4}$$

When $\varphi \in [0, 2\pi]$, the possible values $\varphi = 90°$, $\varphi = 180°$, $\varphi = 270°$, $\varphi = 360°$ can be obtained. Thus, the scanning model can be further improved, during the process of three polarized orthogonal units rotating 360° around the body axis, when the output responses of $p_2(\varphi)$ and $p_3(\varphi)$ are equal, the angles between the body axis and the solar meridian are $\varphi = 90°$, $\varphi = 180°$, $\varphi = 270°$, $\varphi = 360°$, respectively. By comparing the output responses of the polarization quadrature cells, the reference direction can be accurately found, as shown in Fig. 9.10.

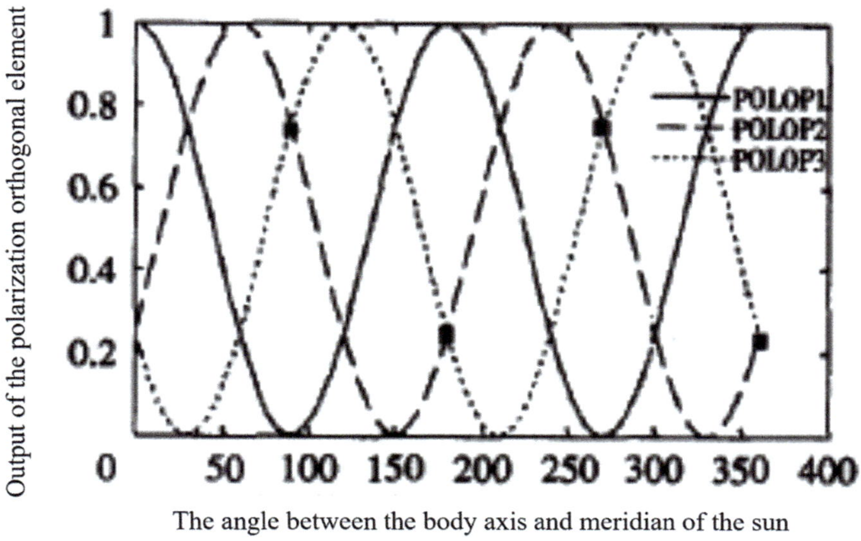

Fig. 9.10 The schematic diagram of research responses of $p_2(\varphi)$ and $p_3(\varphi)$

9.2.2 *Concurrent Model*

For the concurrent model, the angle between the long axis of the object's body and the solar meridian needs to be compared with the output results of the current three polarization sensors and the look-up table generated at the beginning of the motion, and the two optimal ones are found. After the matching value is judged by the adjacent light intensity sensor, the corresponding angle can be determined. To match the lookup table, it will undoubtedly increase the amount of calculation and reduce the accuracy. Lambrinos et al. improved the original concurrency model. The algorithm they proposed does not require a look-up table, and the reference direction can be determined by further analysis and calculation on the basis of the original algorithm. The output of a polarization unit can be described as:

$$s(\varphi) = KI[1 + d\cos(2\varphi - 2\varphi_{\max})] \tag{9.5}$$

In the formula, I is the total light intensity; d is the degree of polarization; φ is the current direction of the long axis of the guided object relative to the solar meridian; $\varphi_{\max}$ is the direction that makes $s(\varphi)$ the maximum value; K is a constant. Thus, the output equation of the polarization quadrature unit is:

$$p_1(\varphi) = \lg\left[\frac{1 + d\cos(2\varphi)}{1 - d\cos(2\varphi)}\right] \tag{9.6}$$

$$p_2(\varphi) = \lg\left[\frac{1 + d\cos(2\varphi - 2\pi/3)}{1 - d\cos(2\varphi - 2\pi/3)}\right] \tag{9.7}$$

$$p_3(\varphi) = \lg\left[\frac{1 + d\cos(2\varphi - 4\pi/3)}{1 - d\cos(2\varphi - 4\pi/3)}\right] \tag{9.8}$$

In the formula, $p_1(\varphi)$, $p_2(\varphi)$, and $p_3(\varphi)$ are the output values of the polarization orthogonal unit when the polarization axis is modulated to 0°, 60°, and 120°, respectively. Because the output of the polarization quadrature unit is log-scaled, its value is independent of light intensity, but closely related to the degree of polarization. Studies in the previous chapters have shown that the degree of polarization varies with the position of the sun as well as weather conditions, and this change directly affects the output of the polarization quadrature unit. The solution proposed by Lambrinos et al. is to remove the effect of the degree of polarization by normalizing the output of the polarization quadrature unit. This method first applies the sigmoid function for inverse logarithmic transformation, and obtaining:

$$\frac{1}{10^{p(\varphi)} + 1} = \overline{p}(\varphi) \tag{9.9}$$

So that, the three output equations are transformed to be:

$$1 - 2\overline{p}_1(\varphi) = d\cos(2\varphi) \tag{9.10}$$

$$1 - 2\overline{p}_2(\varphi) = d\cos(2\varphi - 2\pi/3) \tag{9.11}$$

$$1 - 2\overline{p}_3(\varphi) = d\cos(2\varphi - 4\pi/3) \tag{9.12}$$

Among them, d and φ are unknowns, and there are two unknowns in the three equations, the unknowns can be obtained by any two of them. If the first two equations are used, we have:

$$d = \frac{1 - 2\overline{p}_1(\varphi)}{\cos(2\varphi)} \tag{9.13}$$

Let the Eq. (9.13) inserting into (9.11), we can obtain

$$\tan(2\varphi) = \frac{3 - 2\overline{p}_1(\varphi) - 4\overline{p}_2(\varphi)}{\sqrt{3}(1 - 2\overline{p}_1(\varphi))}$$

And the following derivation can be obtained

$$\varphi = \frac{1}{2}\arctan\left\{\frac{\overline{p}_1(\varphi) + 2\overline{p}_2(\varphi) - 3/2}{\sqrt{3}\left[\overline{p}_1(\varphi) - 1/2\right]}\right\} \tag{9.14}$$

Because $p_1(\varphi), p_2(\varphi), p_3(\varphi)$ are known, which can be read out from the sensors. While $\overline{p}_1(\varphi), \overline{p}_2(\varphi), \overline{p}_3(\varphi)$ can be derived from $p_1(\varphi), p_2(\varphi), p_3(\varphi)$, so that the value of the direction angle φ can be determined according to the sensor readout. Obviously, the calculated value φ is independent of the degree of polarization, and only two of the three equations are used to calculate the value, so what does the third equation do? We perform the following functional transformation on the outputs $p_1(\varphi), p_2(\varphi), p_3(\varphi)$ of the three polarization orthogonal units at 0°, 60°, and 120°:

$$1 - \frac{2}{10^{p(\varphi)} + 1} = \tilde{p}(\varphi) \tag{9.15}$$

So that, the following relations can be obtained.

$$\tilde{p}_1(\varphi) = d\cos(2\varphi) \tag{9.16}$$

$$\tilde{p}_2(\varphi) = d\cos(2\varphi - 2\pi/3) \tag{9.17}$$

$$\tilde{p}_3(\varphi) = d\cos(2\varphi - 4\pi/3) \tag{9.18}$$

When the $d \in (0,\ 1]$,

$$\tilde{p}_1(\varphi)/\tilde{p}_2(\varphi) = \cos(2\varphi)/\cos\left(2\varphi - \frac{2\pi}{3}\right) \tag{9.19}$$

$$\tilde{p}_1(\varphi)/\tilde{p}_3(\varphi) = \cos(2\varphi)/\cos\left(2\varphi - \frac{4\pi}{3}\right) \tag{9.20}$$

$$\tilde{p}_2(\varphi)/\tilde{p}_3(\varphi) = \cos\left(2\varphi - \frac{2\pi}{3}\right)/\cos\left(2\varphi - \frac{4\pi}{3}\right) \tag{9.21}$$

In the above derivation process, $\tilde{p}_1(\varphi), \tilde{p}_2(\varphi), \tilde{p}_3(\varphi)$ can be directly obtained from the output values $p_1(\varphi), p_2(\varphi), p_3(\varphi)$ of the three polarization orthogonal units [6]. In this way, the value of φ can be obtained by any one of the three Eqs. (9.19), (9.20) and (9.21), which is the same as that the polarization angle is independent of the degree of polarization. At the same time, the three Eqs. (9.19), (9.20) and (9.21) are added simultaneously, after transformation, it is easy to obtain the following relationship:

$$\tilde{p}_1(\varphi) + \tilde{p}_2(\varphi) + \tilde{p}_3(\varphi) = 0 \tag{9.22}$$

This conclusion can be used for accuracy verification.

It should be noted that the output of the sensor is a function of the period π, which means that there are two φ values (φ and $\varphi + \pi$) are satisfied with Eqs. (9.6)–(9.8). These are the clip angles between the direction of carrier motion and the solar meridian, and the angle between the direction of motion of the carrier and the anti-solar meridian, respectively. To determine the angle between the moving direction of the carrier and the solar meridian, the ambient light sensor for auxiliary judgment is necessary. According to the output of the ambient light sensor, the relationship between moving direction of the carrier and the solar meridian can be roughly estimated.

9.2.3 Determination of Heading Angle of Bionic Polarization Navigation

The angles calculated by the above two models are the angles between the motion direction of the navigator and the solar meridian, but the solar meridian is constantly changing with the movement of the sun. There are certain limitations for long-time navigation, so it is proposed to use the current azimuth of the sun to correct the angle calculated above, and thus the direction of the navigation object relative to the geographic north and south can be obtained, which will make polarization navigation more applicable.

Here involves the problem of finding the azimuth angle of the sun. In astronomical navigation, the right ascension and declination of the celestial body are often obtained by querying the “Navigational Astronomical Calendar” or by calculation. The location of the up-projection point GP-longitude and latitude are corresponded (as shown in Fig. 9.11). According to solve the navigation triangle (as shown in Fig. 9.12), the altitude and azimuth of the celestial body can be obtained.

As shown in Fig. 9.12, in the navigation triangle, the sun elevation angle $h_s(90° - z)$ can be calculated as:

$$\sin h_s = \sin \phi \sin \delta + \cos \phi \cos \delta \cos t \tag{9.23}$$

And the sun azimuth can be calculated as:

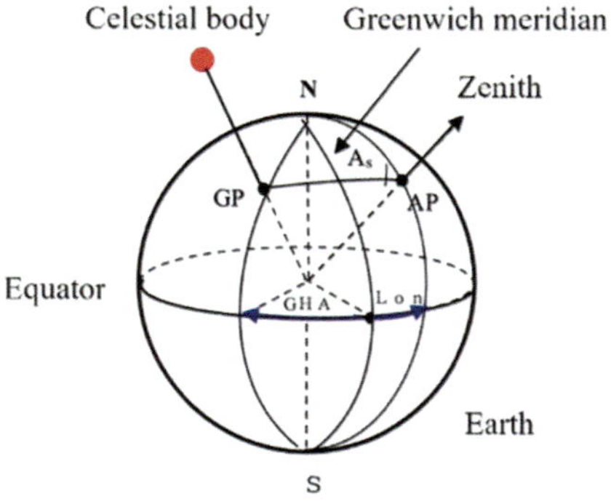

Fig. 9.11 The schematic diagram of projection and navigation triangle

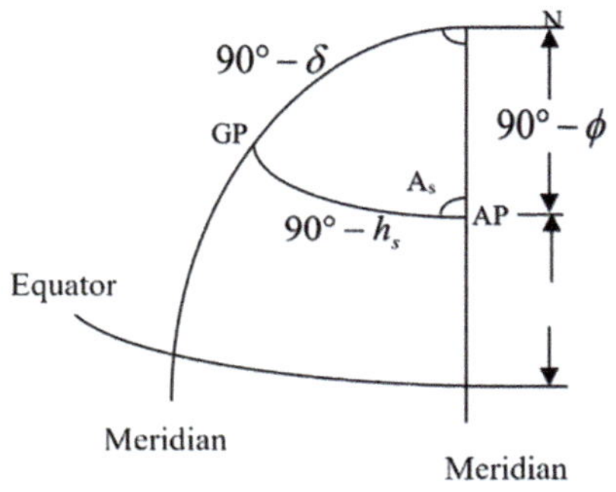

Fig. 9.12 The schematic diagram of navigation triangle

$$\sin A_s = \frac{\cos\delta}{\cos h_s}\sin t \tag{9.24}$$

It is negative when the sun is east and positive when it is west. If the value calculated by the above formula is greater than 1 or its absolute value is small, the following formula can be utilized to calculate:

$$\cos A_s = \frac{\sin h_s \sin\phi - \sin\delta}{\cos h_s \cos\phi} \tag{9.25}$$

where, h_s is the solar altitude angle; A_s is the solar azimuth; t is the local solar hour angle; ϕ is the geographic latitude; δ is the solar declination.

Due to the diurnal apparent motion of the sun, the sun rotates from east to west relative to the earth, and its trajectory is parallel to the celestial equator, as shown in Fig. 9.13. Taking the horizon coordinate system as the navigation coordinate system, if the initial position of a point is known, the position (altitude and azimuth) of the sun at a certain moment relative to the point can be calculated and determined according to formulas (9.23) and (9.24), combined with the light distribution of sky polarization, the degree and angle of polarization at any point in the sky are determined according to the formula, which means that at a certain time and position on a certain day, there is a relatively stable distribution pattern of polarized light in the sky. There is a fundamental line of symmetry (solar meridian) for polarized light distribution patterns. At the same time, studies have shown that insects such as sand ants and bees can sense the angle between the long axis of the body and the solar meridian through their own polarized visual systems [7].

In this way, if the azimuth angle A_s of the sun and the angle β between the long axis of the insects' body (such as sand ants and bees) and the meridian of the sun are known, the angle $\alpha = A_s + \beta$ between the current direction of motion and the north–south direction can be determined according to these two angles. That is, the heading angle α is determined at any time. As long as the speed is sensed by other sensors, the position at the next moment can be calculated. When the position of the next moment is known, the relative position of the sun is determined, and the position of the sun determines the distribution pattern of polarized light in the sky, so

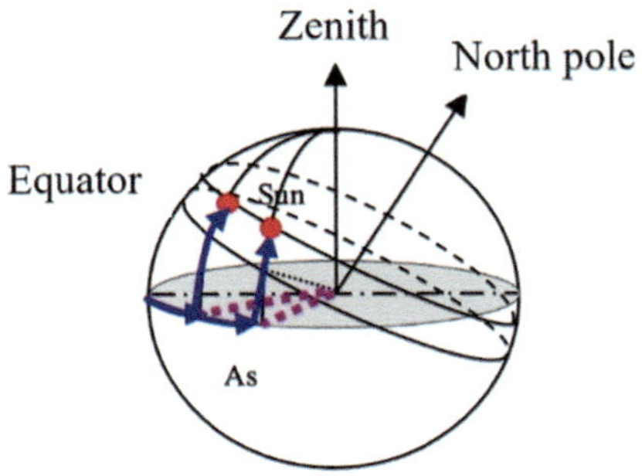

Fig. 9.13 The diagram of the position of the sun

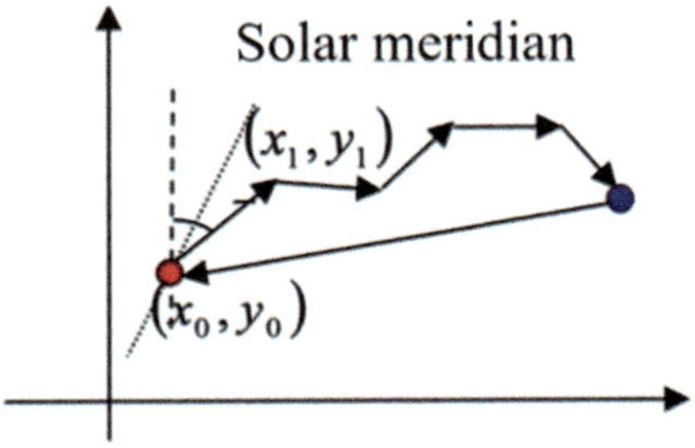

Fig. 9.14 The diagram of navigation and positioning

the insect can sense the distance between the current long axis and the solar meridian through its own polarized vision system. In this way, iterative formula (9.26) can complete the task of navigation and positioning, as shown in Fig. 9.14.

$$\begin{cases} x_{i+1} = \int v_i \cdot \sin\alpha \, dt + x_i \\ y_{i+1} = \int v_i \cdot \cos\alpha \, dt + y_i \end{cases} \quad i = 0, 1, 2 \cdots \tag{9.26}$$

where, (x_i, y_i) and v_i are represent the position and speed at the moment i, and α is the heading angle.

When calculating the position of the sun [8], a key parameter is involved-the local hour angle, which is the angle between the celestial time circle and the local meridian. Similar to the Green hour angle, this angle can be measured from the local meridian to the west (0°–360°), and the angle measured in this way is called the local hour angle *LHA*. The local hour angle *LHA* is equal to the algebraic sum of the celestial body's Green hour angle *GHA* and the observer's longitude *Lon*. In order to ensure that the angle thus calculated is within the defined range, the following rules should be used in the calculation:

$$LHA = \begin{cases} GHA + Lon \\ GHA + Lon - 360^\circ \end{cases}, \text{ when } \begin{cases} GHA + Lon < 180^\circ \\ GHA + Lon > 180^\circ \end{cases}$$

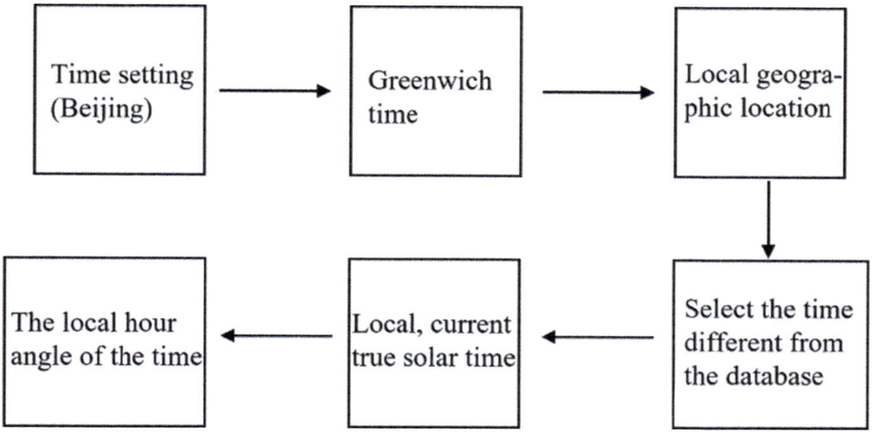

Fig. 9.15 The local hour angle calculation process

The local hour angle is the time difference between the true solar time at a certain moment in the local area and the mid-heaven moment of the local sun (the moment when the center of the sun passes through the local meridian). Converting this difference into an angle is the local hour angle. The local meridian of different places will be different due to different geographic longitudes, as shown in Fig. 9.15, the flow chart of local hour angle calculation.

Example analysis: Calculate the solar hour angle at 11:25 in the morning of November 21, 2006 (Beijing time) in Beijing at 39°92′ north latitude and 116°46′ east longitude. First, the declination and time difference of the sun at that time must be calculated. Since the declination of the sun is given in the astronomical almanac according to the universal time, whether it is to consult the astronomical almanac or substitute into the formula for calculation, it is necessary to convert the calculation time into the universal time. After calculation, the world time corresponding to 11:25 Beijing time is 3:25, that is, 3.42 h. Querying the almanac database, the sun's declination is −19°53′ at this time, and the time difference is 14 m 10 s. Since the time given by general clocks is Beijing time, and the east longitude of Beijing area is 116°46′, when Beijing time is 11:25, the local time in Beijing area is:

$$m = 3.42 + (116 + 46/60)/15 = 11.204\ (h)$$

The local true solar time at this time is:

$$m\Theta = 11.204 + (14 + 10/60)/60 = 11.44\ (h)$$

The local time angle in Beijing area at this time is:

$$(11.44 - 12) \times 15 = -8.4\ (°)$$

9.3 The Test Platform of Bionic Polarization Navigation Angle Measurement

Studies by biologists have shown that the reason why sand ants have magical navigation and positioning skills is that the optic nerve structure of the compound eyes of sand ants is extremely sensitive to the distribution of polarized light in the sky [9, 10]. With the increasing demand for navigation and positioning, the research and development of bionic polarized navigation angle sensors based on the principle of polarized light navigation and positioning of sand ants' compound eyes has highlighted its significance. To develop a bionic polarized navigation goniometer sensor, the corresponding research on the polarized light test of sand ants must be carried out first. The overall design of the polarization navigation angle measurement test platform is shown in Fig. 9.16. The whole system consists of photoelectric receiving module, signal processing module, data acquisition module and angle calculation module.

The photoelectric receiving module imitates the polarization-sensitive compound eye structure of sand ants, perceives the direction and intensity of polarized light in the sky, and converts the perceived light signal into an electrical signal; the signal processing module is responsible for converting the current signal output by the photoelectric receiving module into a voltage. The data acquisition module mainly completes the conversion of analog signals to digital signals; the angle calculation module uses the relevant navigation model to calculate the heading of the carrier to complete the final goal of the system.

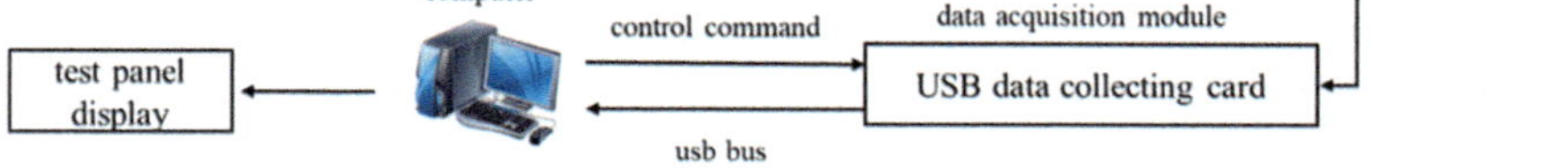

Fig. 9.16 Structure block diagram of polarization navigator

9.3.1 Simulation and Implementation of Polarization Orthogonal Unit

1. Structure design of polarized light detection

In nature, most of the polarized light is linearly polarized light, which can be quantitatively described by the measurement of linear polarizers. When the polarizer is placed in front of the eyes and rotated, the changes of light intensity in different directions can be seen. This change in light intensity is produced by polarized light. In this way, polarized light in different directions can be measured. A special method of cutting the polarizer in a triangle can be utilized to obtain the polarized light intensity in different directions simultaneously without rotating the polarizer (for the specific principle, please refer to the author's monograph on polarization remote sensing "Polarization Remote Sensing Physics").

The structural design of the system polarizer is shown in Fig. 9.17. In addition to meeting the functional requirements, the design of the polarizer should also consider the overall structure. Its shape requirements are the same as those of ordinary polarizers on the market, requiring a diameter of 77 mm and threaded openings at the top and bottom, as shown in Fig. 9.18.

2. Photoelectric receiving module

To carry out the test and research on the polarized light of sand ants, the first step is to use the photoelectric conversion element to convert the optical signal into an electrical signal. According to the research of biologists, different insects are sensitive to different wavelength bands of polarized light in the sky. Therefore, a photodiode that is sensitive to this band must be selected.

The polarized light detection platform is composed of six triangular polarizers, and the system needs to place a photodiode under each polarizer. Considering the requirement for the smallest possible system volume, place the photodiode in the center of each polarizer, as shown in Fig. 9.19.

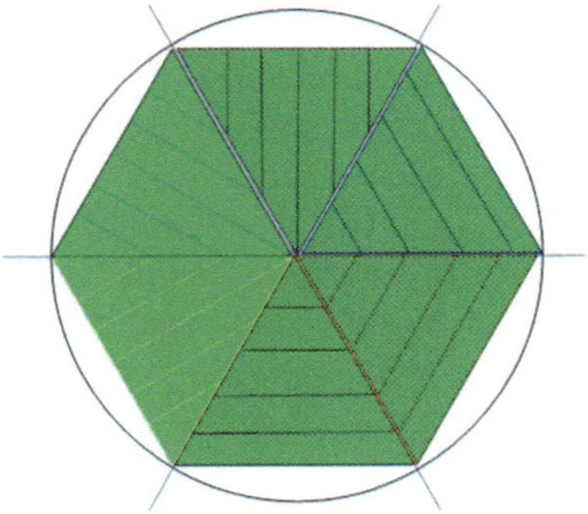

Fig. 9.17 The polaroid arrangement

Fig. 9.18 Image of polarizer

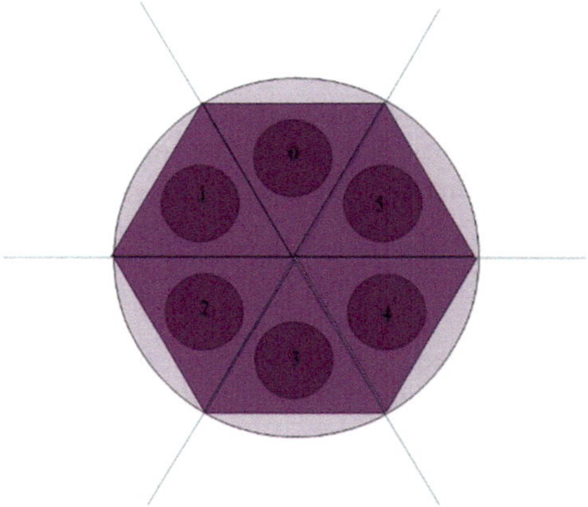

Fig. 9.19 The schematic diagram of the position of the photosensitive element

The sleeve is also an important part of the optical path system design. On one hand, the optical path system should have a certain field of view. On the other hand, the photodiode should not be affected by stray light and light leakage. At the same time, the sleeve height as an influencing factor needs to be demonstrated experimentally. That is to say, the height of the sleeve should be considered in the design, and the height of the sleeve should be finally determined by changing the height of the sleeve during the test and comparing the test results.

The key to the sleeve design lies in the design of the mechanical structure. The design must meet the requirements for installing filters and polarizers, and can be correctly installed on the circuit board. At the same time, try to ensure that the light radiation received by the six polarization sensors at the same time The amount is basically the same, the field of view of the sleeve is 60°, the diffuse reflection is as small as possible, the blue filter and the polarizer in six directions are installed.

3. Signal processing module

The signal output by the photoelectric receiving module is a current signal. The value of the current signal is extremely small (about 1 μA–1 mA). It spans an order of magnitude of 10^3, which cannot be directly converted to analog-to-digital. At the same time, the general data acquisition module is to collect the voltage signal. Therefore, for the needs of subsequent data acquisition work, the system adopts the precision logarithmic operational amplifier LOG102 produced by Texas Instruments (TI) in the United States. Each logarithmic operational amplifier LOG102 converts the two photosensitive current signals into voltage signals after operation, and outputs them. This not only realizes current–voltage conversion for subsequent A/D acquisition, but also performs logarithmic ratio operation on the signal to achieve the purpose of enhancing the polarization contrast of incident polarized light.

The proportional operation of LOG102 can eliminate the influence of external light intensity changes on the calculation angle, and the logarithmic operation can also greatly reduce the variation range of useful signals, providing convenience for subsequent data acquisition. In addition, the logarithmic operational amplifier LOG102 adopts advanced integrated circuit technology. When the input current or the ratio of the two input currents changes within the range of 100 dB, the total output error can be guaranteed to be 0.37% of the full-scale output voltage. Below, the deviation from the ideal logarithmic relationship is not more than 0.1%. Since the chip integrates amplifiers, logarithmic transistors and low-drift thin-film resistors, the gain can be changed without external components, which brings great convenience to users.

The output voltage of the logarithmic operational amplifier LOG102 is:

$$V_{out} = K \log \frac{I_1}{I_2} \tag{9.27}$$

In the formula, K is the magnification, namely $K = 1 + \frac{R_2}{R_1}$; I_1 and I_2 are the current signals generated by the two photodiodes respectively. By adjusting the size of the variable resistors R_1 and R_2, the magnification of the output voltage can be changed.

4. Data acquisition module

The voltage signal processed by the photoelectric conversion and logarithmic proportional operation circuit is a continuous analog signal in time, which cannot be directly recognized by the digital system for related calculation and analysis. Therefore, a corresponding data acquisition module is required to complete the analog signal to digital signal conversion.

The azimuth information output by the test system varies in the range of −180°–+180°, and the resolution is required to be no less than 0.05°, so the number of digits of the digital-to-analog conversion chip is required to be no less than 13. At the same time, the digital-to-analog conversion chip itself has factors such as linearity error and temperature drift that affect the accuracy. Therefore, the actual performance of the digital-to-analog conversion chip is difficult to reach the standard value, which is usually lower 1–2 bit than the standard value. In the actual circuit, it will also be affected by other chips, which further reduces the actual accuracy.

5. Angle calculation module

The angle calculation module takes the PC as the processing platform, analyzes and calculates the data sent by the data acquisition module according to the concurrent navigation model, and calculates the azimuth angle required by the navigation through the program.

According to the above design, a test platform for polarized compass is built. The experimental platform adopts the sleeve structure to realize the test function. On one hand, the sleeve structure meets the requirements of the experimental system. On the other hand, it also tries to ensure that the amount of optical radiation received by the six channels at the same time is basically the same. There is a central axis in the center of the sleeve, and the interior is evenly divided into six channels by the partition plate, the angle between the polarizers placed in each channel is 60°. The two opposite channels are respectively placed with polarizers whose light transmission axes are perpendicular to each other, and a photodiode is placed on the bottom plate under each polarizer. Just like a direction analyzer for the experimental system. We define the direction of one of the polarizers as the 0° polarization channel, and the other two are the 60° and 120° polarization channels in turn. Meanwhile, the polarization direction of the 0° polarization channel is defined as the main direction of the sleeve. In this way, the polarized light enters the six photosensitive amplifiers to obtain six amplified voltage signal outputs, and the two photosensitive signals in opposite directions are output after logarithmic operation. Six photosensitive signals and three logarithmic signals are converted into digital signals through the 16-channel A/D acquisition module, and then collected into the computer through USB, and the azimuth angle information can be solved by referring to the algorithm in the theoretical analysis section.

9.3.2 *Bionic Polarization Navigation Angle Measurement*

The optoelectronic model of the polarization orthogonal unit is established by simulating the visual neurons of the insect, and the symmetrical distribution of the sky polarization pattern is perceived through the model to obtain the navigation reference direction. The photoelectric conversion adopts a photodiode with a blue filter, and three photodiodes (with linear amplifier function) are placed on the metal box; the photodiode is surrounded by a metal cylinder whose inner wall is coated with a diffuse reflection layer, with a field of view of 60° and the angle between the optical axis and the zenith is 25°; two of the photodiodes are equipped with polarizers (horizontal polarization and vertical polarization), placed on the top of the cylinder, and

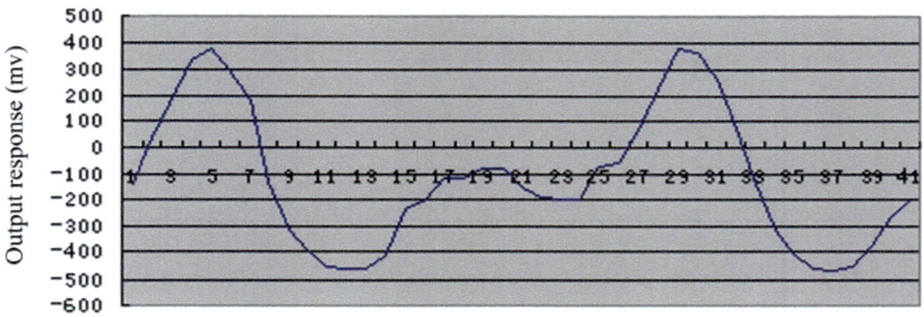

Fig. 9.20 The response curve of single channel polarization orthogonal element

the transmission directions are perpendicular to each other; the other photodiode acts as a light intensity detector. The output of the horizontal polarization channel and the vertical polarization channel after being amplified by the logarithmic ratio is the curve of the polarization orthogonal unit in response to the polarized light.

A DC motor is used to drive the rotation axis, and each scan rotates 360° to obtain the response curve of the polarization orthogonal unit relative to the distribution of polarized light in the sky. The curve has the greatest response in the solar meridian and anti-solar meridian directions, and the solar meridian direction, which is the reference direction for navigation, can be determined by the output of the light intensity detector [1].

In the actual measurement, the DC motor (ZGA20RU1100i) is used to control the bracket to rotate at a constant speed of 30°/s, and the sleeve is driven to complete a scan (12 s) every 360°, and each rotation starts from the south direction. Figure 9.20 is the data collected on April 1, 2010 at 10:18 am.

It can be seen from the measurement results in the figure that the output response of a single channel exhibits an obvious periodic variation law, and in each cycle, the output response has two maxima, representing the direction of the solar meridian and the anti-solar meridian respectively, which is consistent with the theoretical It is completely consistent, indicating that the bionic polarization navigation angle measurement model designed by us is correct, and can complete the response of different intensities to the stimulation of polarized light in different directions, and can determine the reference direction according to the response intensity.

The polarizer is added in the experiment, and the waveform changes of the three-way signals generated by the continuous rotation of the sleeve around the axis are shown in Figs. 9.21, 9.22 and 9.23.

It can be seen from the figure that when the incident light is polarized light, the three output signals generated when the goniometer platform rotates around the axis all show a certain trend of sinusoidal response. However, due to the limitation of actual conditions, it can only rotate at most two times at a time instead of continuous rotation, and because the sleeve rotates unevenly, the output response of each channel is not as smooth as ideal. But in general, the waveforms of the output signals of the three channels are basically in line with the response output of the theoretical model,

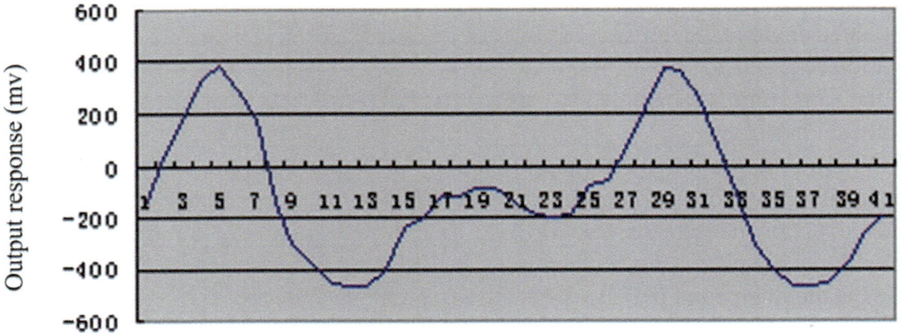

Fig. 9.21 The response curve of polarization orthogonal element at 0°

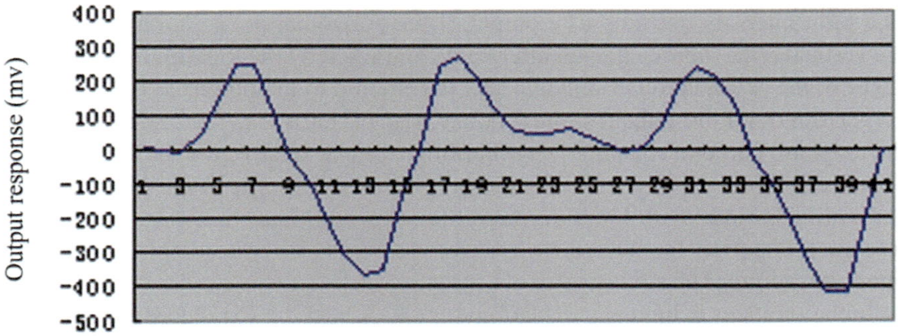

Fig. 9.22 The response curve of polarization orthogonal element at 60°

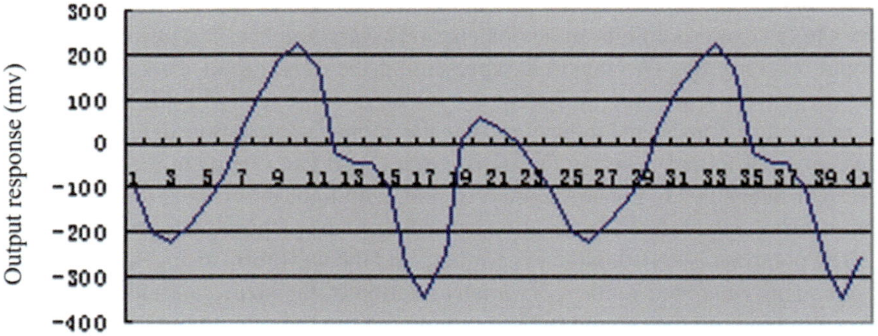

Fig. 9.23 The response curve of polarization orthogonal element at 120°

which can complete the response to the stimulation of polarized light in different directions to produce different intensities.

9.3.3 Three-Channel Polarization Orthogonal Unit Test

In the experiment, a polarizer is added to make the incident light into linearly polarized light, and the waveform changes of the three-way signals generated by the continuous rotation of the sleeve around the axis are shown in Figs. 9.24, 9.25 and 9.26, respectively. The output responses of the three channels of the goniometer platform were collected at 10:00 am, 10:30 am, and 11:00 am on April 1, 2010, respectively.

It can be seen from the figure that when the incident light is polarized light, the output responses of the three channels generated by the rotation of the sleeve around the axis basically show a sinusoidal variation law, and there is a relative relationship between the output responses of the three channels. fixed phase difference. The results show that the output responses of the three channels basically conform to the

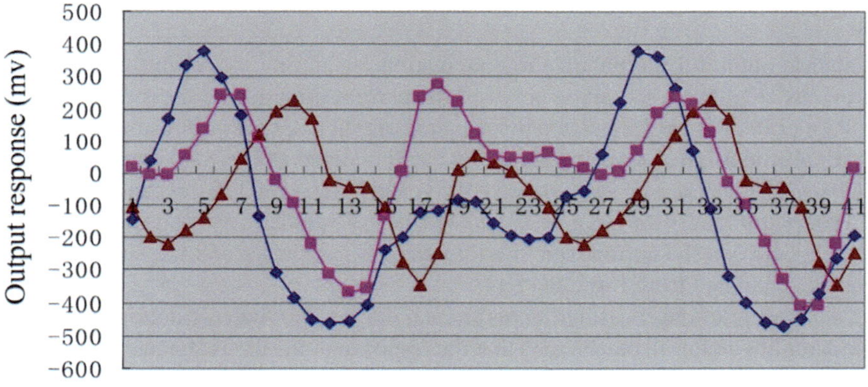

Fig. 9.24 The response curve at 10:00, Apr. 1. 2010

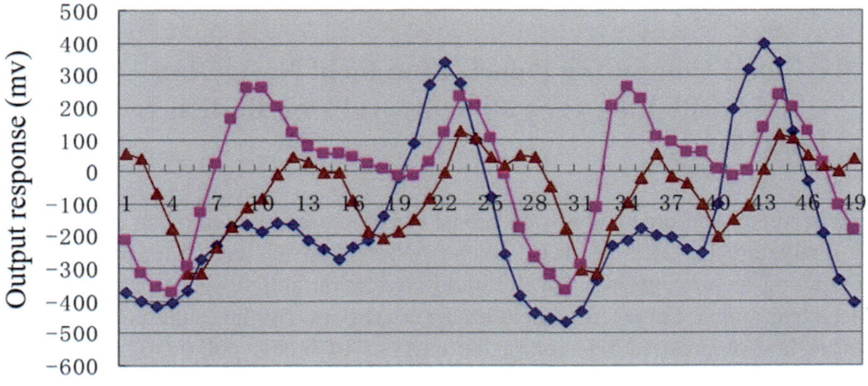

Fig. 9.25 The response curve at 10:30, Apr. 1. 2010

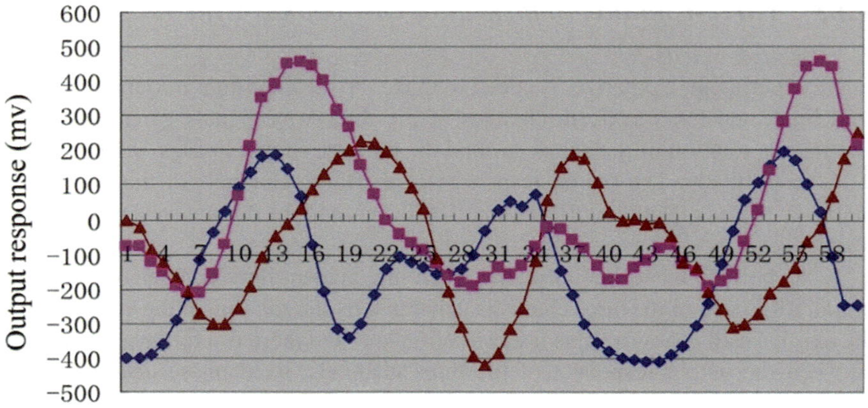

Fig. 9.26 The response curve at 11:00, Apr. 1. 2010

response output of the polarization navigation concurrent model introduced in the previous chapter, so the angle between the current direction of the system and the solar meridian can be calculated by using the output responses of the three channels.

Example analysis: According to the data collected at 10:18:40 a.m. on April 1, 2010, $p_1(\varphi) = 334.4\,\text{mv}$, $p_2(\varphi) = 51.9\,\text{mv}$, and $p_3(\varphi) = -179\,\text{mv}$ can be obtained. $\overline{p}_1(\varphi)$, $\overline{p}_2(\varphi)$, and $\overline{p}_3(\varphi)$ can be calculated from formula (5.9), and the angle $\varphi = 18.72°$ between the current running direction of the carrier and the solar meridian can be calculated from formula (5.14).

According to the measured experimental curves and the calculation analysis of the sampling points, it can be seen that the response of the three-channel polarization orthogonal unit model to polarized light basically conforms to the response law of the sand ant's visual nervous system to the direction of the polarized light in the sky, which verifies the polarization validity of the light goniometric model.

9.4 Cloud Computing Polarization Field Navigation and Positioning 1: Spatiotemporal Distribution Law and Model

The accuracy and practicability of bionic polarization navigation can be improved by analyzing the error of single bionic polarization navigation and using cloud computing technology to assist the multi-navigation unit system to perform high-speed operations. This section includes the construction of the spatiotemporal distribution law of polarization modes, the analysis of bionic polarization navigation errors, the spatiotemporal resolution model of polarization mode graphs based on cloud computing, the analysis of multi-cell measurement errors, and the introduction of related parallel algorithms.

9.4.1 Analysis of Bionic Polarization Navigation Error and Multi-navigation Unit Cloud Computing Support System

As shown in Fig. 9.27, this system mainly includes: sky polarized light distribution model and its characteristic analysis, implementation of bionic polarization navigation and single navigation error analysis, proposal of multi-navigation unit cloud computing support system and parallel algorithm analysis; information source time and space domain Establishment of error model; random error of time and space domain measurement system, system error analysis and model establishment, and parallel algorithm in the overall operation process of bionic polarization navigation under the condition of cloud computing support system.

1. Analysis of Bionic Polarization Navigation Error and Proposition of Multi-Navigation Unit Cloud Computing Support System

The all-sky polarization measurement method is used to analyze the relationship between the sky polarization pattern and the time or position of the carrier, and the error of the polarization navigation angle measurement platform designed in the previous stage is analyzed. The cloud computing and parallel algorithms to improve the accuracy of polarization navigation are proposed. Specifically, it includes the sky polarized light distribution model and its time and space characteristics analysis, the realization of bionic polarization navigation and the analysis of single navigation

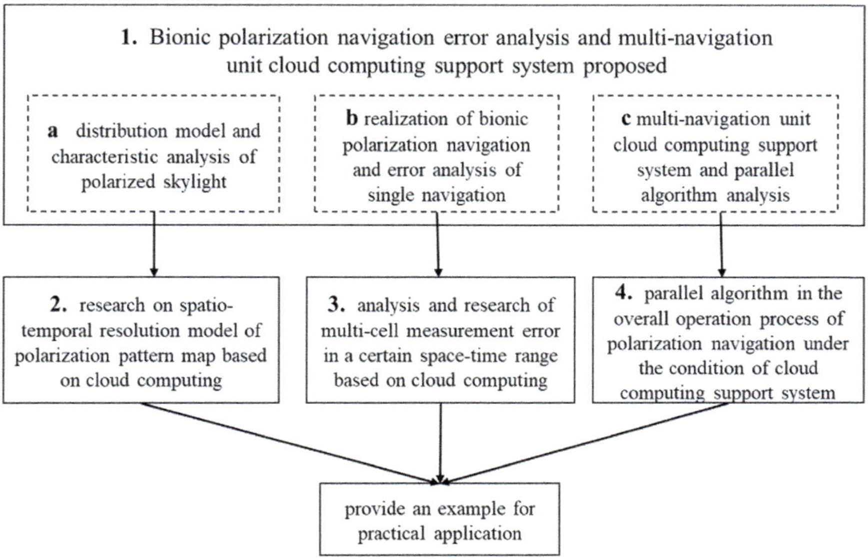

Fig. 9.27 Cloud computing support technology block diagram

error, the proposal of multi-navigation unit cloud computing support system and the analysis of parallel algorithm.

2. Research on the spatiotemporal resolution model of polarization mode map based on cloud computing

Divide the range (or time period) in which different measurement positions (or moments) within a given accuracy can obtain the same polarization pattern result. The goal is to establish a common interval and general polarization pattern map based on the multi-navigator cloud computing method. period model. Specifically, it includes the establishment of the information source spatial error model and the establishment of the information source time domain error model.

3. Research on the multi-cell measurement error model based on cloud computing

Establish an optimal estimation algorithm for multi-cell navigation measurement sequences under the same polarization pattern map for time domain and regional changes-time-domain and regional Kalman filter recursive algorithm, analyze and determine the measurement deviation of multi-cell polarization navigation in the space–time range and its bias suppression model. Specifically, it includes the random error of airspace measurement system, systematic error analysis and model establishment, random error of time domain measurement system, systematic error analysis and model establishment.

4. Parallel algorithm in the overall operation process of bionic polarization navigation under the condition of cloud computing support system

It provides specific computing means for the study of the temporal and spatial resolution model of the polarization pattern map and the study of the multi-unit navigation error model. Specifically, it includes the parallel computing structure design of the information source spatial domain error analysis model, the parallel computing structure design of the information source time domain error analysis model, the parallel algorithm for determining the carrier measurement error analysis in the space domain, the parallel algorithm for determining the carrier measurement error analysis in the time domain, Parallel computation of multi-unit Kalman filtering and measurement estimation.

In order to make the bionic polarization navigation system operate automatically and with high precision in the actual field, a multi-unit, long-term repeated operation "measurement-calculation-search-analysis-description-improvement" research scheme was adopted to analyze the error model of the information source time domain, space domain, and measurement error of the carrier in the specific time domain and space domain is analyzed and studied, the information processing algorithm and its effect of suppressing the error are experimentally verified.

Taking the 10-m accuracy of polarization navigation as an example, the contribution rate of the information source error does not exceed 10% (less than 1 m) for practical use. The error of the information source is mainly caused by the change of the sky polarization pattern with time and space. This topic firstly studies in which area the change of the sky polarization pattern causes the error of the information source is within one meter. Within a given range, over what time interval does the change in the sky polarization pattern cause the source of information to be within 1 m of error? The research idea is to first put 100 samples in the same direction in a small area, and calculate the statistical error of 100 samples at the same time. If it is less than 1 m, expand the area and calculate the error until it reaches the maximum allowable information source error, so that The constant space area A is determined. Then, 10 samples in the same direction are placed in the above determined area, and the information is sampled at different time intervals and the error is calculated until the time interval within the allowable error range is found, so that the sampling interval T in a specific space area is determined. This part of the research content can provide reference for the practical use of polarization navigation.

The measurement error is the main error of the carrier. The error of the information source can be regarded as a constant in a certain area and a certain time interval. So how does the measurement error of the carrier change? How to extract it? The research idea is to first place 100 samples in the same direction in a specific area A, sample the information multiple times within a certain time interval T, and eliminate the single random error through Kalman filtering, so that the measured value with systematic error can be obtained. Then put 100 samples in the same direction in the above determined area, and collect the measurement values of 100 samples at the same time, because the systematic errors of each sample are irrelevant, the quasi-true value that eliminates the measurement error can be obtained by Kalman filtering. After then, simulate the mathematical model of bionic polarization navigation, and compare the simulation results with the quasi-true value to get the system error of each monomer.

The above experiments involve 100 single measurement systems, 10,000 sampling moments, 100 repeatable measurements and calculations, and the recursive operation of the improved Kalman filter. Both the information source error and the carrier measurement error involve time and space issues, the geographical range and sampling interval continue to change, and the amount of calculation in the middle is huge, each cell needs to use Kalman filtering to eliminate random errors, and multiple cells need to perform time-domain and spatial-domain search evaluation calculations and result parameter statistics at the same time. On one hand, these problems are interrelated, computationally complex, and applied in parallel or serially in multiple single and multiple repetitive experiments. On the other hand, various calculations have a large number of identical structures and models, so the reconfigurable and concurrent features of the computing unit array can be fully utilized to divide tasks and improve computing efficiency. In addition, the above-mentioned error analysis and calculation array structure based on multi-cell filtering has strong redundancy and fault tolerance. It does not care about the calculation error of a certain unit or even the loss of information, and the system can be reconstructed at any time, ensuring the real-time and effective calculation. Under such a multi-unit and multi-process computing array system close to the cloud computing structure, it is very necessary to use cloud computing technology and parallel algorithms to jointly solve the problem through the changing multi-navigation unit network, as shown in Fig. 9.28.

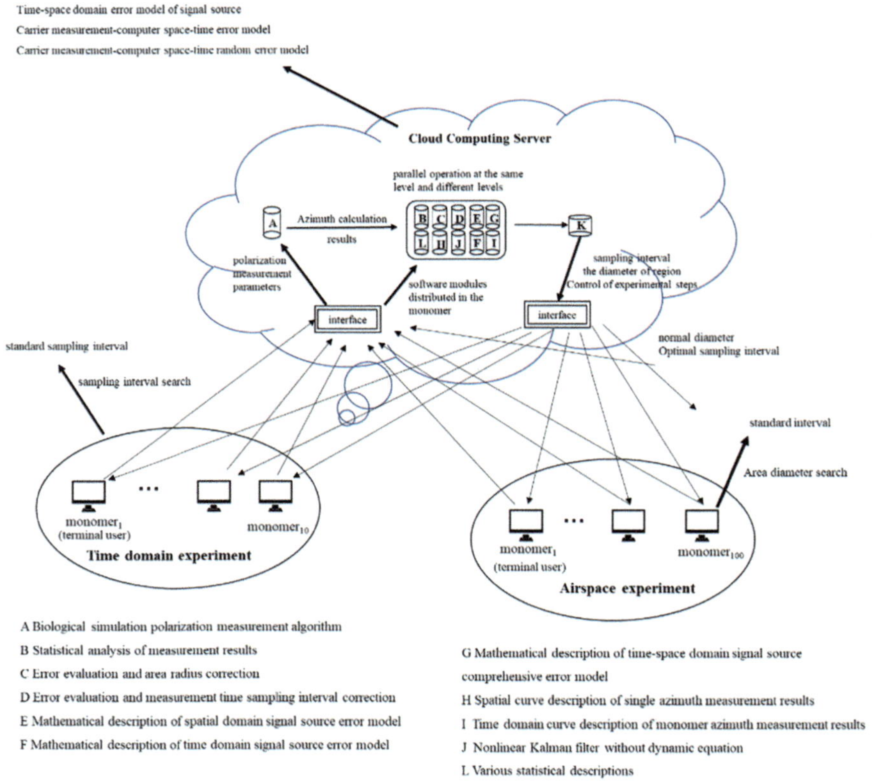

Fig. 9.28 The cloud computing support system for information source error and measurement error model analysis

9.4.2 *Research on Spatiotemporal Resolution Model of Polarization Pattern Map Based on Cloud Computing*

In order to improve the accuracy of the cloud computing-based bionic polarization navigation measurement in the field, it is first necessary to establish a universally applicable spatial domain and time domain error model of the information source. The following takes the group movement composed of 100 monomers, the single movement composed of 10 monomers, and the allowable error of the celestial information source system at a distance of 1 M (that is, the azimuth distance error of the earth's surface) as an example to describe the errors in the space domain and time domain model building process, respectively. The measurement error of the carrier is mainly caused by the random error caused by noise, thermal effect, and the systematic error caused by the mechanical structure. This part of the content is to analyze and establish a model for multi-unit time domain, spatial random error, and systematic error within the space–time range confirmed by the above experiments.

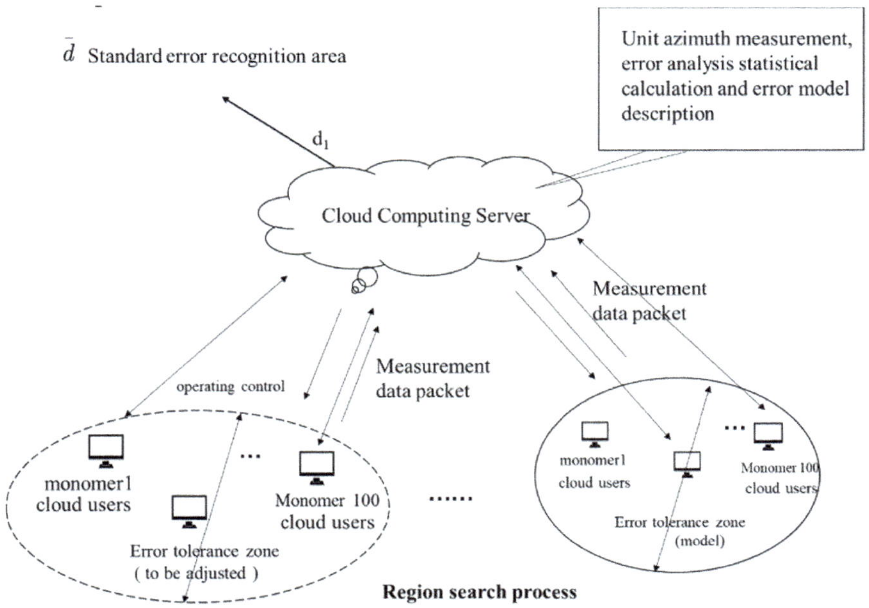

Fig. 9.29 The schematic diagram of the cloud computing system

1. Establishment of information source airspace error model

As shown in Fig. 9.29, the azimuth equivalent length test is carried out for 100 monomers in the area to be measured, and the statistical average error is obtained. The area diameter is evaluated and adjusted until the critical value of error tolerance is reached, thereby determining the error tolerance area diameter of the information source space domain.

Step 1: Randomly distribute monomers, randomly distribute (fill the circular surface) 100 moving monomers on a circular surface with a diameter of dM (d starts from 100 and increases in steps of 10) and make them in the same direction and stand still during the experimental process.

Step 2: Measurement and calculation of the azimuth angle of a single unit. Let each unit performs a measurement and calculation of the azimuth angle of a single unit at the same time (the same sky pattern). Assuming that the measurement results are the azimuth angles α_{kj}, $k = 1, 2, \ldots, j = 1, 2, \ldots, 100$ (k is the number of experiments, and j is the monomer number).

Step 3: Calculation of the statistical error. Calculating all the difference between the azimuth angles of all monomers and the average value of all monomer azimuth angles, and convert it to an equivalent geographic distance, denote it as ΔL_{kj} (unit is M), and calculate the root mean square error: $\Delta L_k^* = \sqrt{(\sum_{j=1}^{100} \Delta L_{kj})/100}$.

Step 4: Evaluate and adjust *d*, evaluate the diameter of the area *d*: if $\Delta L_k < 10$, increase d by one step, and let k = k + 1.

Step 5: Repeat the adjustment of d, repeat steps 1–4 until it reaches $\Delta L_{kj}^* \geq 10$, and record the previous area diameter d1,l = 1, 2, … as the number of experiments.

Step 6: Repeat the experiment, let l = l + 1, repeat steps 1–5 (the first step of each experiment is to re-distribute the above 100 monomers randomly) until d1,l = 1, 2, …, 100.

Step 7: Statistical average, calculate the statistical average of d1, $\overline{d} = \frac{\sum_{l=1}^{100} d_l}{100}$, and d1 is the desired value.

2. Establishment of information source time domain error model

As shown in Fig. 9.30, the average value of the azimuth equivalent length statistical error is evaluated for the 10 monomers in the above standard error tolerance area until the error reaches the tolerance critical value, so as to determine the best measurement sampling that meets the signal source error tolerance range interval.

Step 1: Monomer distribution, randomly distribute 10 moving monomers in the area d1 determined by the above experiment (filling the entire area).

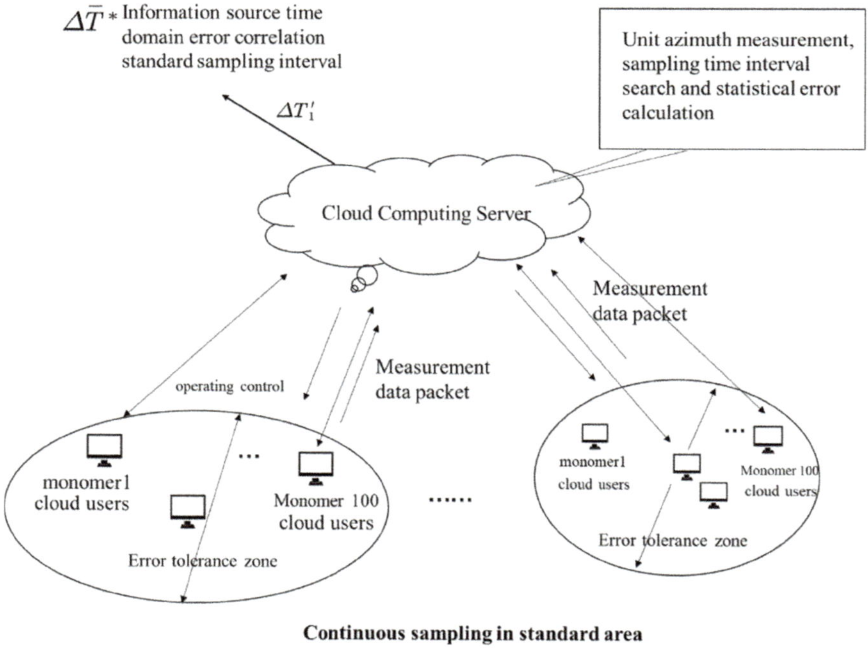

Fig. 9.30 The schematic diagram of the cloud computing system

Step 2: Single unit azimuth angle measurement and calculation, at the time T_k, k = 1, 2, …, 10,000, when the interval time is ΔT_1, make all monomers perform single unit azimuth angle measurement and calculation at the same time, and obtain 10,000 groups, each group contains 10 azimuth angles $\{\alpha_{kj}\}$, k = 1, 2, …, 10,000, j = 1, 2, …, 10.

Step 3: Statistical error calculation, calculate the statistical error of all the monomers obtained above, convert it into an equivalent geographic distance, record it as ΔL_{kj} (unit is M), and calculate the statistical error $\Delta \overline{L}_j^* = \frac{\sum_{k=1}^{10000} |\Delta L_{kj}|}{10000}$, where j = 1, 2, …, 10.

Step 4: Evaluation of ΔT, if $\Delta L^* < 10$, ΔT should be increased by 1 h, and repeat the steps 2–3 until $\Delta L^* > 10$ appears at the first time. Recording the sampling time interval $\Delta T'$ of the previous moment, and calculate the average value $\overline{T}_1' = \frac{\sum_{j=1}^{10} \Delta T_j}{10}$, where l is the number of experiments.

Step 5: Repeat the experiment. Repeat the steps 2–4 to 10,000 times, and obtain $\Delta T'l$, l = 1, 2, …, 10,000.

Step 6: Calculate the statistical average sampling interval. Calculating the statistical results of the above 100 experiments (where i is the number of experiments), $\Delta \overline{T}^* = \frac{\sum_{l=i}^{100} \Delta \overline{T}_i'}{100}$, ΔT^* is the sampling interval required by the system.

3. The system random error of airspace measurement, analysis of systematic error and model establishment

As shown in Fig. 9.31, using the sampling interval and area diameter specified by the information source error model, measure the azimuth angle of multiple monomers at the same time, calculate the statistical average value of their equivalent length errors, and use the improved Kalman filter and statistical calculation to analysis the and final model description for systematic and random errors.

Step 1: Randomly distribute monomers, randomly distribute (fill the circular surface) 100 moving monomers on a circular surface with a diameter of dM (obtained in 2.2.2) and keep them in the same direction and stationary during the experiment.

Step 2: Measure and calculate the azimuth angle of the single unit. Let each unit in the circular plane performs an azimuth angle measurement and calculation at a certain time at the same time, and converts the measurement result into the geographic length L_{jl}. Where the unit serial number j = 1, 2, …., 100, and the number of experiments l = 1, 2, …, 100. Every experiment gets a geographic length curve (observation curve) L_{jl}, j = 1, 2, …, 100.

Step 3: Measure and calculate the high-precision value of the azimuth angle of the monomer, and obtain the high-precision azimuth angle equivalent length quasi-true value curve L_{jl}, j = 1, 2 …, 100 of the above-mentioned monomer by the simulation method.

Step 4: Separation of the system error in the space domain of the equivalent length of the single azimuth angle, design a non-dynamic equation nonlinear Kalman filter algorithm suitable for the measurement conditions and environment

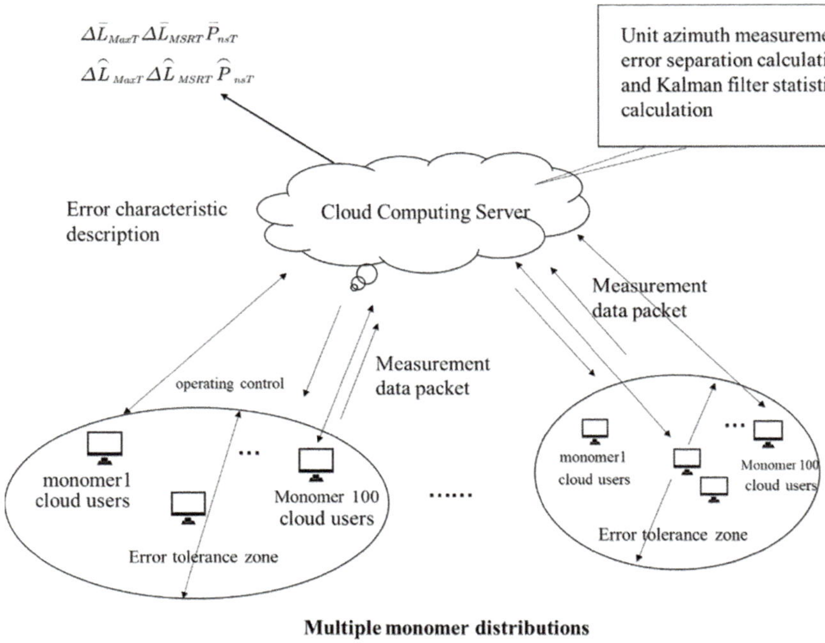

Fig. 9.31 The schematic diagram of the cloud computing system

faced by this experiment, and apply it to L_{j1}, j = 1, 2, ..., 100 to obtain a measurement (estimation curve) with sufficiently small random errors and dominant systematic errors, denoted as $\widetilde{L}_{j1}$ (j = 1, 2, ..., 100). On this basis, the quasi-system error curve $\Delta\widetilde{L}_{j1} = \widetilde{L}_{j1} - L'_{j1}$ is calculated, and the maximum value, average value and root mean square value of this error are calculated as: $\Delta\widetilde{L}_{Max} = MAX(\Delta\widetilde{L}_{j1})$, $\Delta\widetilde{L}_{Mean} = \frac{\sum_{j=1}^{100}\Delta\widetilde{L}_{j1}}{100}$ and $\Delta\widetilde{L}_{MSR} = \sqrt{\frac{\sum_{j=1}^{100}(\Delta\widetilde{L}_{j1})^2}{100}}$.

Step 5: Repeat the calculation to find the statistical average of the system error description parameters, let l = l + 1, repeat steps 1–4 (the first step of each experiment, the above 100 monomers are re-randomly distributed), until l = 100, so as to obtain the average value of the above-mentioned parameters. With these parameters as the evaluation criteria, the source analysis of the system error of the instrument and the environment can be carried out, so as to provide a basis for suppressing the systematic error of the measurement and calculation.

Step 6: Separation of the random error in the space domain of the equivalent length of the single azimuth angle, calculate the quasi-random error curve $\Delta\hat{L}_{j1} = L_{j1} - \hat{L}_{j1}$, j = 1, 2, ..., 100, and calculate the maximum value of this error. The mean value and root mean square are: $\Delta LMax = MAX(\Delta L_{jl})$, $\Delta\hat{L}_{Mean} = \frac{\sum_{j=1}^{100} \Delta\hat{L}_{j1}}{100}$.

Step 7: Repeat the calculation to find the statistical average of the random error description parameters, let $l = l + 1$, repeat steps 1–4 (the first step of each experiment, the above 10 monomers are re-randomly distributed), until $l = 100$, so as to obtain the average value of the above-mentioned parameters. Taking these parameters as the evaluation criteria, the random error source analysis of the instrument and the environment can be carried out, thus providing an evaluation basis for the calculation of the suppression measurement.

4. Time-domain measurement system random error, systematic error analysis and model establishment

As shown in Fig. 9.32, using the sampling interval and area radius specified by the information source error model, the long-term azimuth angle of multiple cells in the area is carried out. According to the measurement and use improved Kalman filter and statistical calculation to analyze and model systematic error and random error.

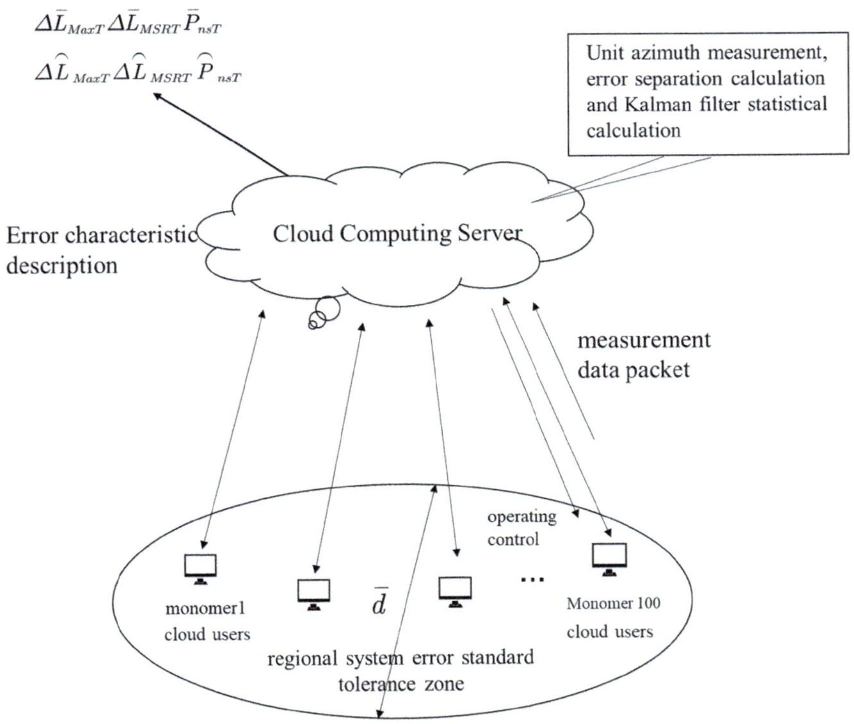

Fig. 9.32 The schematic diagram of the cloud computing system

Step 1: Randomly distribute monomers, and randomly distribute 100 moving monomers within the standard error tolerance area determined by the above experiments (filling the entire area).

Step 2: Measurement and calculation of the azimuth angle of the monomer, during the time interval ΔT^* obtained above under the moment of T_k, k = 1, 2, ..., 10,000, perform the measurement and calculation of the azimuth angle of the monomer simultaneously for all the monomers. Obtaining the azimuth equivalent length observation curve L_{kjl} of 10,000 groups, each group of 100 monomers (single number j = 1, 2, ..., 100, and the experiment number l = 1, 2, ..., 10,000).

Step 3: The measurement and calculation of the high-precision value of the azimuth angle of the monomer, the equivalent length L_{kjl}, j = 1, 2, ..., 100 of the 100 monomers obtained at the moment of k, perform spatial Kalman filtering to obtain the position (i.e., Equivalent length estimate curve L_{kjl} as a function of monomer number). Since any time series L_{kjl}, k = 1, 2, ..., 100 is uncorrelated with any space sequence L_{kjl}, j = 1,2,...,100, therefore, L_{kjl} can be considered as the quasi-true of L_{kjl} at time k value, which is used as an approximate true value in the following experiments.

Step 4: Separation of the time-domain system error of the single azimuth equivalent length, design a non-dynamic equation nonlinear Kalman filter algorithm suitable for the measurement conditions and environment faced by this experiment, and apply it to L_{kjl}, k = 1,2,...,10,000, so as to obtain the measurement result with sufficiently small random error and dominant systematic error, the dominant systematic error of the estimated value curve $\tilde{L}$ is the single azimuth equivalent length quasi-system error curve: $\Delta\widetilde{L}_{kj1}$, j = 1,2,...,100. On this basis, the quasi-systematic error curve $\Delta\widetilde{L}_{kj1} = \widetilde{L}_{kj1} - L'_{kj1}$ of the single azimuth equivalent length with the systematic error as the main error is calculated, and the maximum value, mean value and root mean square of this error are calculated as $\Delta\widetilde{L}_{MaxT} = MAX(\Delta\widetilde{L}_{kj1})$, $\Delta\widetilde{L}_{MeanT} = \frac{\sum_{j=1}^{100}\Delta\widetilde{L}_{j1}}{100}$ and $\Delta\widetilde{L}_{MSRT} = \sqrt{\frac{\sum_{j=1}^{100}(\Delta\widetilde{L}_{j1})^2}{100}}$.

Step 5: Repeat the calculation to find the statistical average of the system error description parameters, let l = l + 1, repeat steps 1–4 (the first step of each experiment, the above 100 monomers are re-randomly distributed), until l = 100, so as to obtain the average value of the above-mentioned parameters. With these parameters as the evaluation criteria, the source analysis of the system error of the instrument and the environment can be carried out, so as to provide a basis for suppressing the systematic error of the measurement and calculation.

Step 6: Separation of random errors in the time domain of single azimuth equivalent length, and calculate the quasi-random error curve ($\Delta\widetilde{L}_{kj1} = L_{kj1} - \widehat{L}_{kj1}, k = 1, 2, \ldots, 10000,\ j = 1, 2, \ldots 100$) of measurement dominated by random errors. And calculate the maximum value, mean value and root mean square value of this error.

Step 7: Repeat the calculation to obtain the statistical average of the time-domain random error description parameters, let l = l + 1, and repeat steps 1–4 (the first step of each experiment is to re-randomly distribute the above 100 monomers), until l = 100, the average value of the above parameters is obtained. Taking these parameters as the evaluation criteria, the random error source analysis of the instrument and the environment can be carried out, thus providing an evaluation basis for the calculation of the suppression measurement.

9.5 Cloud Computing Polarization Field Navigation and Positioning 2: Parallel Computing Method

The overall operation of polarization navigation is a very complex set of algorithms. In the process of establishing the error model and error analysis in the space and time domains, because it involves several complex models and algorithms, and the algorithms need to share resources and execute repeatedly. Therefore, a huge amount of calculation and data processing is generated. To improve the computing speed, we choose to perform parallel computing at the algorithm level and support it with cloud computing systems. The parallel computing at the algorithm level ensures that the operations between the units are independent of each other, and their operations can be executed in parallel. Based on the aforementioned error analysis and modeling experiments involving multiple monomers, the related parallel computing design is described below.

9.5.1 A Parallel Algorithm for Determining the Error Analysis of Carrier Measurements in the Space–Time Domain

The parallel computing structure system of the information source spatiotemporal error analysis model is divided into six layers, as shown in Fig. 9.33. The first, second, and third layers perform the operations of random distribution of cell positions, one azimuth measurement calculation for each cell, equivalent length conversion calculation, and signal source spatial domain error statistics calculation. The operations of each layer are composed of 100 A single unit participates and executes concurrently at the same time. The unit measurement task is completed independently by each unit, and after obtaining the raw data, it is sent to the cloud server in the form of data packets. The fourth, fifth and sixth layers are error evaluation, standard mode diameter statistical value calculation, logic control of 100 repeated experiments and overall calculation operation control of statistical calculation, which are completed by the cloud server itself.

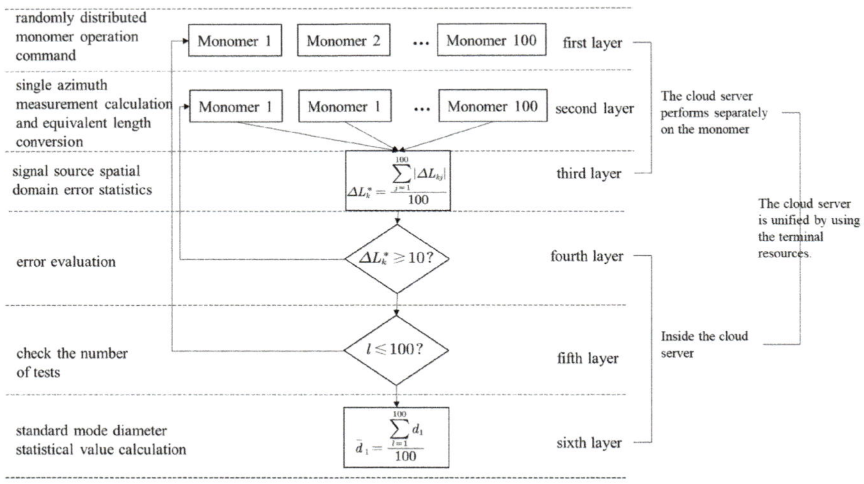

Fig. 9.33 Parallel system structure of spatial error analysis and modeling of information source

The parallel calculation of the error analysis model in the time domain of the information source is performed by 10 cells each at 10,000 sampling times. The entire parallel system is divided into nine layers, which perform random distribution of cell positions and azimuth measurement at one sampling time. Calculation, logic control of repeating 10,000 measurements according to the sampling time, evaluation of signal source time domain error and adjustment of sampling interval, single operation step inspection and control, time domain statistical average of single signal source time domain error. Nine parallel calculations and operations, such as the logic control of 100 repeated experiments and the statistical average of the results of repeated experiments, as shown in Fig. 9.34. All measurement and data transfer operations are controlled by the cloud server. The logical process of the time-dependent progressive cycle, the search cycle and the repetitive experiment cycle of each monomer is controlled and executed by the cloud server depending on the completion of the cycle.

9.5.2 *Determining Parallel Algorithms for Carrier Measurement Error Analysis in the Space–Time Domain*

The parallel algorithm for determining the carrier measurement error analysis in the space domain is divided into seven layers, which realizes the random distribution of the cell position, the measurement calculation of the cell azimuth angle, the logic control of the cell experiment steps, the Kalman filter in the space domain, and the systematic error of obtaining the equivalent length of the azimuth angle. Obtaining the

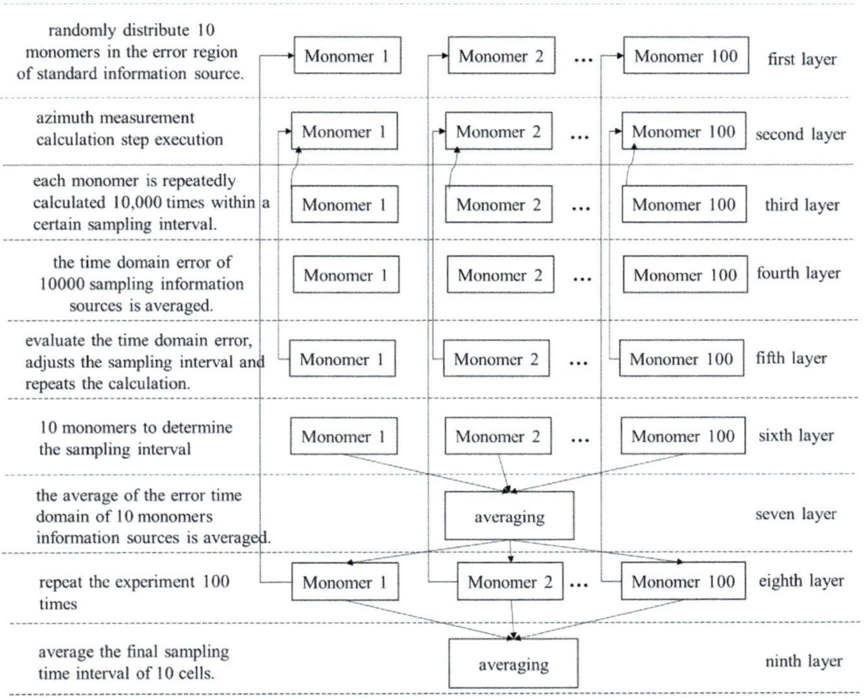

Fig. 9.34 Parallel computing technique for error analysis of information source in time domain

random error of the equivalent length of the azimuth angle, the high-precision value of the azimuth angle by simulation, the statistical parameters of the systematic error, the statistical parameters of the random error, and repeat the experimental logic control etc., as shown in Fig. 9.35. The first and second layers involve the parallel operations of 100 units, and the distributed computing resources are called by the cloud server for simultaneous execution. Airspace Kalman filtering and calculation of high-precision azimuth angle, systematic error and random error of azimuth equivalent length, and statistical parameter calculation of the two errors are carried out simultaneously at the same layer.

The parallel algorithm for determining the carrier measurement error analysis in the time domain is divided into eight layers, which realizes the random distribution of cells. Using the sampling interval obtained in the above experiments to calculate the equivalent length of the azimuth angle, logical control of the number of cell measurements, Kalman filtering in the time domain, and spatial domain. Kalman filtering to obtain the quasi-true value, to obtain the systematic error of the azimuth equivalent length, the random error of the azimuth equivalent length, the statistical parameters of the systematic error, the statistical parameters of the random error, and the logic control between the cells and the logical control of the number of multi-unit experiments, as shown in Fig. 9.36. Among them, the first, second, third, and sixth

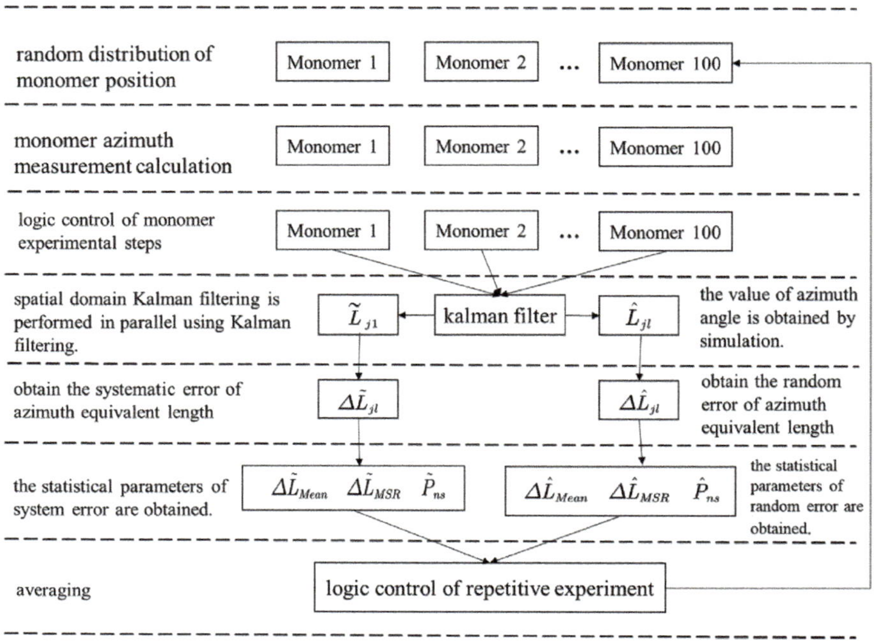

Fig. 9.35 Parallel calculation of spatial domain stochastic and systematic error analysis modeling

layers are 100 units of parallel computing, and the cloud server invokes computing resources for simultaneous execution. In addition, the time-domain Kalman filtering and the spatial-domain Kalman filtering to obtain the true value, to obtain the systematic and random errors of the azimuth equivalent length, and to obtain the statistical parameters of the systematic and random errors are all calculated simultaneously in the same layer.

9.5.3 Parallel Computation of Multi-unit Kalman Filtering and Measurement Estimation

In a specific area, 100 samples in the same direction need to eliminate their own random errors through Kalman filtering. At the same time, these 100 samples in the same direction need to collect their respective measurement values at a certain time and eliminate each monomer through Kalman filtering to remove the system error. The basic linear univariate Kalman filtering process calculates the situation at time k + 1 from the observation value at time k, the error matrix and the prediction of the observation value. On this basis, 100 monomers at the same time need to perform Kalman filtering at the same time. The parallel structure of the entire Kalman filtering is divided into five layers, as shown in Fig. 9.37. A recursive process of linear

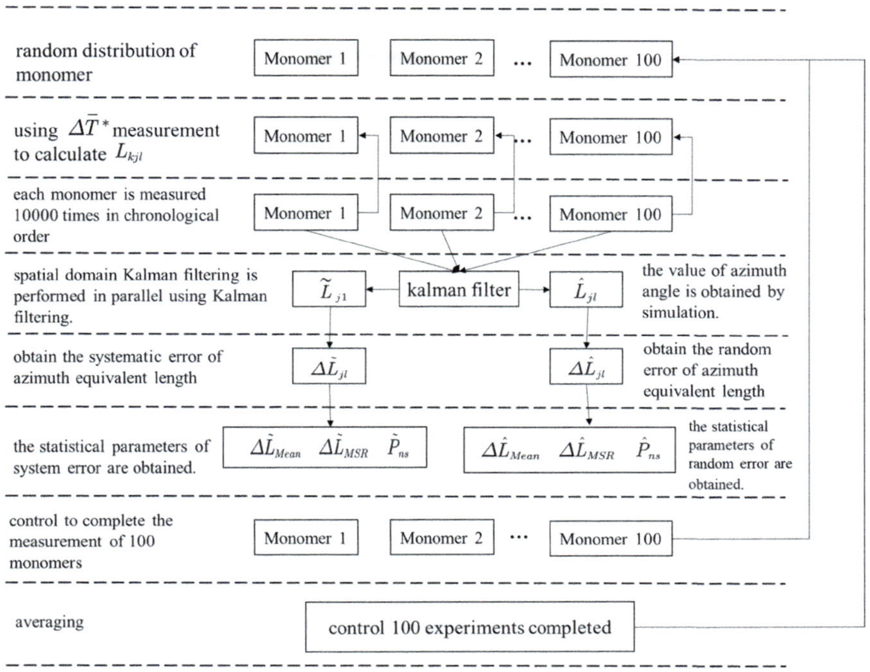

Fig. 9.36 Parallel calculation of stochastic and systematic error analysis modeling in time domain

univariate Kalman filtering is performed from the first layer to the fifth layer, and in each layer, the operation tasks are the same, and the estimated value at the next moment is continuously calculated from the value calculated at the previous moment, so as to realize continuous The recursive loop of, and the calculation of the estimated value at each moment can be performed in parallel.

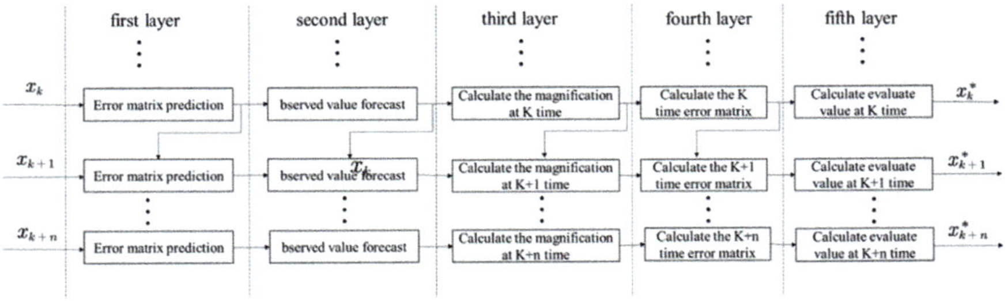

Fig. 9.37 The linear univariable Kalman filter parallel algorithm structure

9.6 Summary

This chapter first takes sand ants as an example to briefly introduce the mechanism of biological polarization navigation. Then, it describes two models currently used in bionic polarization navigation, which is called scanning model and concurrent model. Since the angles calculated by the two models are constantly changing. The author believes that this is not practical for long-term navigation, so use the current azimuth of the sun to correct the angle calculated by the model proposes was proposed, and finally obtain the angle between the movement direction of the object being navigated and the geographic north–south.

Based on the polarization navigation principle of sand ants, this chapter firstly proposes the overall design scheme of the bionic polarization navigation angle measurement test platform, according to the modular design idea, the design and implementation of photoelectric receiving module, signal processing module, data acquisition module and the angle calculation module are briefly introduced. After then, the actual measurement was carried out using the designed goniometric verification platform. The measurement results show that when the incident light is polarized light, the response of the three-channel polarization orthogonal unit model to polarized light basically conforms to the response law of the sand ant's visual nervous system to the direction of polarized light in the sky, which verifies the polarized light angle measurement. The correctness of the model can therefore be used to calculate the angle between the current direction of the system and the solar meridian using the output responses of the three designed channels. Finally, the error generated by the bionic polarization navigation angle measurement verification platform is analyzed, and it is considered that the validity of the navigation model is the most important for the accuracy of the calculation results. Of course, the installation of the polarizer and the sensitivity of the photosensitive device also have a great influence on the angle measurement accuracy.

To express sensitive laws more precisely, there are still many problems to be solved:

(1) The models are critical. The description of the polarization sensor ($s(\varphi) = KI[1 + d\cos(2\varphi - 2\varphi_{\max})]$) determines the goniometric accuracy of the entire system. The light received by the two polarization sensors of a single channel has a certain difference. To reduce the difference in light intensity, the only way is to reduce the volume of the bionic polarization navigation polarization sensor as small as possible.

(2) Mechanical structure of polarizer. The polarization axes of the two polarization sensors in each polarization orthogonal unit are perpendicular to each other to imitate the cross-vertical structure of the sensing rods in the compound eye of insects. The stimulus response to polarized light in a specific direction is required to be a sinusoid, which is a strict direction analyzer. Therefore, the mechanical design of the polarizer should ensure that the vibration transmission directions of the two opposite polarizers are perpendicular to each other, and each polarizer is an ideal linear polarizer. In addition, it is also necessary to ensure that the

angle between the transmission directions of the adjacent channel polarizers is 60°.

(3) For differences in photosensitive devices, ensure that the parameters of each photosensitive device are consistent as much as possible.

(4) If the sleeve rotates unevenly, try not to use switching power supply or transformer for power supply, so as to avoid the 50 Hz power frequency interference caused by AC power.

References

1. Labhart, T. (1999). How polarization-sensitive interneurones of crickets see the polarization pattern of the sky: A field study with an optoelectronic model neurone. *The Journal of Experimental Biology, 202*, 757–770.
2. Kalayjian, Z. K., Andreou, A. G., Wolff, L. B., & Sheppard, N. (1999). A polarization contrast retina that uses patterned iodine-doped PVA film. In *Proceedings of the international joint conference on neural networks* (pp. 2329–2332).
3. Pomozi, I., Horvath, G., & Wehner R. (2001). How the clear-sky angle of polarization pattern continues underneath clouds: Full-sky measurements and implications for animal orientation. *The Journal of Experimental Biology, 204*, 2933–2942.
4. Lambrino, D., Kobayashi, H., Rolf, P., Maris, M., Labhart, T., & Wehner, R. (1997). An autonomous agent navigation with a polarized light compass. *Adaptive, 6*(1), 131–161.
5. Esch, H., Zhang, S., Srinivasan, M., & Tautz, J. (2001). Honeybee dances communicate distances measured by optic flow. *Science, 411*, 581–583.
6. Chu, J. K. & Zhao, K. C. (2005). Research on photoelectricity model of polarimetric sensing for compound eye of insect. *Micronanoelectronic Technology, 12*, 541–545.
7. Waterman, T. H. (1979). Polarization sensitivity. In *Handbook of sensory physiology* (Vol. VII/6). Springer.
8. Yu, H. J. (2006). Study on calculation method of solar declination and time difference for meteorological purpose. *Meteorological Hydrology and Oceanographic Instruments, 9*, 50–53.
9. Guan, G. X., Yan, L., Chen, J. B., & Wu, T. X. (2008). Vehicle tracking algorithm based on GPS and map-matching. In *Measurement theory and systems and aeronautical equipment* (Vol. 7128, pp. 542–547).
10. Collett, M., Collett, T. S., Bisch, S., & Wehner, R. (1998). Local and global vectors in desert ant navigation. *Nature, 394*(6690), 269–272.

Chapter 10
Electrostatic Field Navigation and Positioning Basis and Technical System

The characteristic of inertial navigation technology is that it does not need to provide the position data of the carrier without external reference and signal, and its core is precision and autonomy. Figure 10.1 shows the various inertial sensor—gyroscope accuracy and application areas. Since the 1950s, inertial technology has gradually dominated the navigation field. In the integrated navigation system started in the 1980s, inertial technology is still dominant, but inertial navigation technology has always been troubled by setting the initial value and decreasing accuracy over time, and the price is expensive.

The "Navigation Satellite Timing and Ranging Global Positioning System(GPS)" was completed in the 1990s in the United States and the GLONASS system in Russia can continuously provide high-precision three-dimensional position, three-dimensional speed and time information, with precision independent of time. For example, the three-dimensional positioning accuracy of GPS is in the order of 10^2m in civilian services (Standard Positioning Service referred to as SPS), and less than 16m in military password protection services (Precision Positioning Service referred to as PPS). In terms of price, an SPS differential navigator costs about $6,000, a non-differential SPS device costs less than $3,000, and a PPS device costs about $10,000 to $20,000.

Therefore, inertial technology is challenged by GPS. But the weakness of GPS is that it is essentially an alternative navigation, and its technical control is in the hands of the Americans, so it cannot replace the autonomy of the inertial system. From a technical point of view, GPS is affected by celestial bodies and sometimes interrupts the signal; the GPS signal update rate is 50 times lower than that of the inertial system, and cannot be used for flight control weapons with high real-time performance; GPS is not yet able to perform high-precision attitude measurement of moving carriers. These technical weaknesses are precisely the strengths of inertial systems. Therefore, even in the United States, where GPS is widely used, the aircraft inertial navigation system (Inertial Navigation System, abbreviated INS) with a positioning accuracy of 1 nautical mile/hour priced at $200,000 still has a large market. The above situation

L. Yan et al., *Mathematical and Physical Fundamentals of Navigation and Positioning with Earth's Natural Fields*, Navigation: Science and Technology 13,
https://doi.org/10.1007/978-981-97-5524-0_10

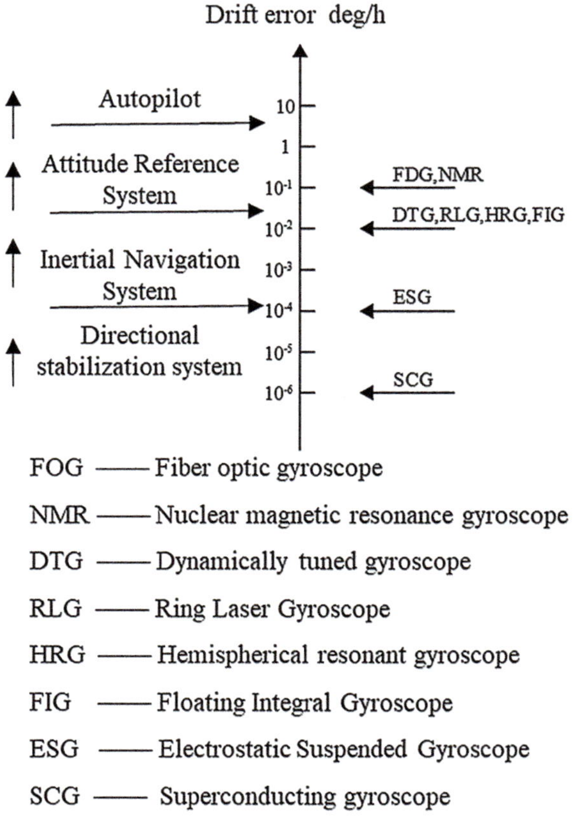

Fig. 10.1 Accuracy and application areas of new gyroscopes

brings opportunities for the development of inertial technology. Therefore, INS can be developed from two aspects:

1. The combination of medium-precision INS (with a drift rate of 0.1°~0.01°/h gyro) and GPS is developed to improve the function, reduce and expand the scope of use. The INS is calibrated by GPS, thereby reducing the requirements for inertial sensors; when the GPS signal is lost, the INS provides memory navigation and helps the GPS receiver to re-acquire satellites.
2. In terms of highly autonomous and highly concealed military uses such as nuclear submarines, long-range ballistic missiles and strategic bombers, INS continues to develop towards high-precision. In these occasions, it is not allowed to transmit radio signals to the outside world, nor to receive external radio signals, so it is necessary to rely on INS autonomous navigation to achieve directional positioning. Obviously, this kind of high-precision inertial sensor requires stable accuracy, long reset period and very low drift rate .For example, the missile with a range of 8000km in the United States is launched by an ESG nuclear submarine, the error is 100m, and the ESG readjustment period is 15 days, then the compensation error drift rate $\triangle \theta_r = \triangle d / \triangle t$,, here r is the missile range, $\triangle d$

is the diameter of the error circle, and $\Delta\ t$ is the readjustment period, unit week. Therefore,

$$\Delta\theta_{\min} = \frac{\Delta d}{rt} = \frac{100m}{8000km \times 2circle} = 6.25 \times 10^{-4\circ}/circle = 3.47 \times 10^{-8\circ}/h \tag{10.1}$$

As can be seen from Fig. 10.1, the inertial sensor that meets the above-mentioned high precision is currently only ESG, and the superconducting gyroscope SCG under development is expected to have higher precision than ESG.

Looking forward to my country's aerospace technology in the 21st century, ESG and SCG will be a unique inertial space orientation element. Because they automatically meet the high vacuum conditions in space, the weightlessness makes the bearing not have the problem of overcoming gravity, and the low temperature makes the expansion coefficient of the ball and the bowl tend to zero, which in principle almost eliminates the main source of interference torque. ESG research in the United States shows that the accuracy is as high as 4.4×10^{-2} arcseconds/year (ie $1.40\times 10^{-10\circ}$/h) under this condition. In this regard, ESG applications are more profound.

10.1 Problems and Challenges Related to Electrostatic Field Suspension BearingTechnology

For Electrostatic Suspension System (ESS), it must contain ESG. Because the ESG body structure is a part of the bearing system, it forms the displacement signal sensor of the ESS system, and its rotor is a second-order pole in the ESS open-loop system; the "control" and "sensing" links are closely related to the ESS, and their most of the functions can be replaced by ESS. Therefore, in a sense, the application research of ESG, first and foremost, is the problem of ESS engineering.

ESS is mainly used in the core components of the following four products:

1. Quartz flexible electrostatic accelerometer [1, 2], the measurement accuracy is $3 \times 10^{-6}g$, the problem is the linearization of the electrostatic torquer, and the electrostatic feedback force changes with the gap.
2. Electrostatic gyro test platform [3, 4]. It is an experimental device based on the advantages of force feedback magnetic field torque application [5] and servo method test [6, 7].
3. Electrostatic gyroscope to find the north instrument. Robert C (1978)[8] initially discussed the North-finding instrument scheme, and A.P.Truban (1979) [9] proposed a rapid measurement accuracy index: the accuracy of 3 arcseconds within 25 minutes. For this project, the overall structure, ESG ontology and ESS circuit engineering prototype were considered.
4. Solid ball solution for electrostatic gyro monitor. The *Wire and Cable Product Sample* has carried out experimental research on the electrostatic gyro rotor mass

unbalanced signal device, and the angle measurement accuracy is less than 27 arcseconds, which is the key problem of the solid sphere scheme. Other issues have not yet been covered. According to the existing data, the annunciators are divided into three categories [10]: large-angle optical annunciators, photoelectric small-angle annunciators, and rotor quality unbalanced annunciators. They are now summarized in Table 10.1 by technical characteristics.

The eight-electrode solid sphere scheme needs a comprehensive study, because the electrode marking is simple, the signal angle measurement is completed by the ESS, and the centroid offset creates conditions for the constant speed of the ESS electric field to eliminate the rotor deformation caused by the constant speed of the eddy current magnetic field. From the perspective of the rotor, there is no line marking and no fear of surface damage. Therefore, it can withstand breakdown and electrode rubbing accidents, which greatly improves reliability and practicability. From the ESS point of view, the single-frequency constant current beaing is only suitable for the application of eight electrodes; and the eight electrodes are easy to realize the fault tolerance control.

Problems faced by the working beaing circuit: measuring the electrical bridge and the high-voltage transformer, the former determines the acquisition of the displacement signal, and the latter determines the strength of the beaing electric field. The link between the two is the ESS control mode, which determines the dynamic stiffness and quality of the ESS. At present, almost all researches on ESS circuits involve the work of high-voltage transformers. Since the passband required by ESS system

Table 10.1 Annunciator scheme

Type of annunciator	Optical wide angle measurement			Small angle photoelectric measurement	Mass Unbalance Modulation annunciator
	Great arc	Cosine	Equatorial triangle		
Precision	7' (70s) <15"(US)		1' (two quadrature annunciators) (US) 1~2" (differential compensation of two pairs of quadrature annunciators) (Russia)	within tens of cents 0.5"(China)	20" (US) 27"(principle experiment) (China)
Additional conditions	A full set of sensing and measuring circuits, the rotor is marked with lines, the rotor cannot have breakdown spots, and the electrodes cannot be rubbed.			Complete set of sensor measuring circuit rotor scribe line. The rotor can not have breakdown spots, can not rub the electrode	No special sensing circuit, only the ESS itself. The rotor is radially eccentric, without marking lines, and it is not sensitive to surface scars

is greater than 800Hz, it is necessary to increase the carrier frequency to 1.74KHz. In fact, when the carrier is wave greater than 12KHz, the modulation output can better reproduce the 800KHz signal. However, due to no breakthrough in the afterburner transformer, the power current reaches 1 ~ 2A, which must be solved.

The supporting object, the main body of the gyro, is a part of the ESS. The balls and bowls together form the gap capacitor bridge of the ESS. The generated measurement signal is passed through the ESS control circuit to form a high voltage, which is then applied to the ball. Therefore, the process structure of the ball and bowl and the corresponding vacuum, constant speed, acceleration, centering, and angle measurement are closely related to ESS [11]. However, no systematic research has been carried out in this regard. The problems it faces include:

1. The vacuum problem. Zhang Yanzhong (1981) [12] pointed out that the vacuum pump ion current is less than 10 μA when the initial levitationis started. However, the reason and basis are not pointed out, and there is no specific research on the vacuum degree of reliable work. In recent work and research, it has been found that low vacuum is an important reason for the current ESG reliability reduction and large drift. The vacuum first determines the reliability of the high-voltage support and initial levitation. When the vacuum is good, the gap between the ESG balls is not easy to produce breakdown discharge. Secondly, the vacuum itself does not produce source movement, but the quality of the vacuum determines the magnitude of the constant velocity torque of the magnetic field, which determines the rotor eddy current induced temperature rise and the size of the gap deformation, thus also determines the ESS stability, dynamic stiffness and deformation drift error. Ren Zhong (1993) [13] provided valuable theoretical basis and technical realization means for ESG vacuum estimation for the first time. The literature also proposes to use the rotor side-swing effect to measure the speed, speed decay and so on. Therefore, it is necessary to use this achievement to estimate the vacuum and propose the structure scheme of the vacuum pump.
2. Constant speed and temperature deformation effect. Zhang Yanzhong (1981) [12] listed the relationship between the optimal rotational speed of the ball rotor and the drift rate

   As shown in Fig. 10.2, curve a: drift caused by gravitational inertial force field and electrostatic field interference sources related to odd harmonics; b: drift caused by stray magnetic field interference sources; c: drift caused by interference related to even harmonics; d: total drift rate. The literature has completed the design of the reference frequency given circuit with variable frequency, and the adaptive frequency selection circuit design of the band-pass frequency changing with the given speed change. The adjustable frequency range is 0~486Hz, and the minimum adjustment jump variable is 2Hz. This is useful for choosing the best speed, but the article does not consider the magnetic.
3. The new drift effect caused by the rotor temperature deformation effect caused by the constant speed of the field torque. The same is true of the findings of G.D. Sammons (1979)[14].

Fig. 10.2 Rotation speed W and drift rate Ω curve

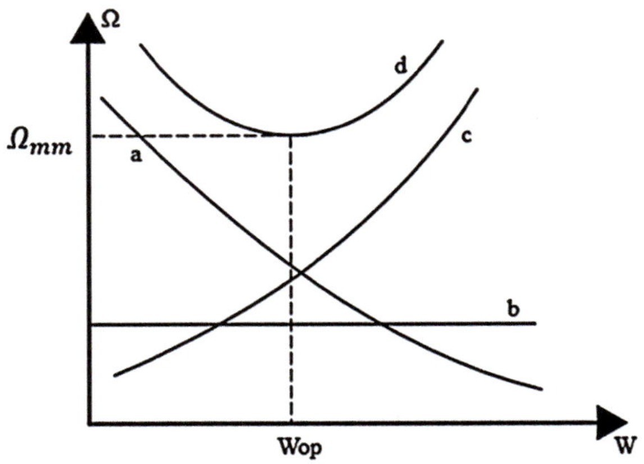

Kamptrek (1986) [15] studied the temperature-induced deformation and constant temperature control, and pointed out that when the average gap changes due to temperature changes, the support stiffness fluctuates, and the support voltage changes the position of the center of mass, resulting in unpredictable interference forces; The temperature change makes the rotor non-ideal spherical, and the electrostatic interference torque changes. The literature points out that the bowl temperature is controlled at $82.2° \pm 1.1°C$ abroad, but the rotor temperature cannot be controlled. In the literature, the temperature rise of the rotor is indirectly measured by the constant bowl temperature and the variation of the ball and bowl gap. However, the heat transfer between the balls can only be done by radiation, and its thermal constant is as high as 7.2 hours under the support condition, which is almost unrealistic for rapid measurement engineering applications. Therefore, eliminating the magnetic field eddy current thermal effect of rotor deformation is the fundamental way out. That is to say, a new solution should be sought for the constant velocity of the magnetic field.

Li Qingliang (1985) [16] discussed the electric field constant velocity scheme. The electric field does not produce eddy current thermal effects, but the premise of constant speed is that the rotor has a slight radial eccentricity. It is traditionally believed that radial eccentricity does not introduce interference torque, but Russian scholars raised doubts about this at the Sino-Russian Inertial Technology Conference in 1991. In addition, the amount of eccentricity required for constant speed is related to the vacuum, so determining the relationship between the allowable amount of eccentricity and the degree of vacuum under the current conditions is a problem that must be solved first for the constant speed of the electric field. Li also developed an ESS that measures the swing period T of the ball rotor, and calculated the equivalent centroid offset by measuring T

$$e = \frac{J}{mg}\left(\frac{2\pi}{T}\right)^2 \tag{10.2}$$

where mg: rotor weight, J: moment of inertia. The literature points out that the current processing ball rotor can reach a swing period of 60-70 seconds, and the corresponding center of mass offset is $0.2\mu m$, while the foreign hollow beryllium ball T=120s, the equivalent $e = 0.05\mu m$, and its unbalanced moment is less than $2 \times 10^{-9} kg/m$, the drift error is $10^{-3\circ}/h$.

Chen Defang et al. (1985) [17] conducted a study on the static balance measurement of the rotor by the electric suspension method, and the device used a bell jar structure. A vacuum of 1.33×10^{-5}Pa is required, and the minimum working vacuum of the gap is 1.33×10^{-3}Pa. It is pointed out in the literature that the rotor balance error measurement is divided into static (measurement of the period T in the suspended state, derived from e) and dynamic (e from the radial runout of the altimeter rotor or the fluctuation of the supporting force); The problem of dynamic balance between the axis of symmetry and the main axis of inertia, so the dynamic and static measurements are equivalent. The literature believes that this kind of disturbance torque comes from two aspects: one is that when the center of mass deviates from the center of the sphere, it causes axial drift (the drift rate of the diagonal momentum $h = 5 \times 10^{-3} kgm^{2/\sec}$ reaches $4°/(h.g.\mu)$,while the coupling of radial migration with ESS anisotropy or external vibration also produces drift; second, aspherical error ΔR produces interference torque with supporting force. As far as domestic processing technology is concerned, $\Delta R \leq 0.3\mu m$, corresponding third harmonic $< 0.05\mu m$, centroid The offset is limited by the measurement and correction level and $> 0.2\mu m$, which is the main reason for ESG and gravity related drift. The literature proposes that the regular octahedron electrode, the support system has equal stiffness in all directions, which is conducive to the uniform distribution of the electric field force.

10.2 Electrostatic Support System

This section summarizes the basic knowledge of the electrostatic support system. First, the electrostatic attraction force and its calculation method are introduced. The maximum support force of the electrostatic field is analyzed and explained according to the capacitance change between the electrode and the rotor. Finally, the principle of active electrostatic support control loop is introduced.

10.2.1 The Concept and Calculation Method of Electrostatic Attraction

If there is an ideal sphere with no mass, the center of which is placed at the origin O of the space Cartesian coordinate system OXYZ, and if there is a pair of forces acting on the sphere along the three coordinate axes with the following relations, then the sphere is balanced by these three pairs The force is supported in a stable position in the center of space:

$$\begin{aligned} F_{x+} &= -F_{x-} \\ F_{y+} &= -F_{y-} \\ F_{z+} &= -F_{z-} \end{aligned} \tag{10.3}$$

If these three pairs of forces are all forces acting on the sphere without contacting the sphere, for example, they are all electrostatic attraction, the sphere will be stabilized at the center of the suspended space by the balance of the three pairs of electrostatic attraction. When the ideal sphere rotates at a high speed around the fixed point O, and the outer surface of the sphere still maintains the ideal spherical shape, the three pairs of electrostatic attraction will suspend the balance support of the ball rotor in a stable position in the center of space, which is the basic concept of electrostatic suction suspended ball rotor.

The suspension function of the electrostatic gyroscope is to use the electrostatic force generated by the strong electric field with a field strength of (3.0~4.0)$\times 10^5$V/cm to support the spherical rotor under the ultra-high vacuum condition of 10^{-8} mbar. The ultra-high vacuum of 10^{-8} mbar is maintained by a closed inner spherical cavity composed of two hemispherical ceramic bowls, as shown in Fig. 10.3.

The surface of the ceramic inner spherical cavity is a regular hexahedron electrode, and the uniform gap d between the inner spherical surface of the electrode and the surface of the metal ball rotor is very small, with a typical value of 70μm. When the high voltage V is applied to the electrode, and the ball rotor is at zero potential, a uniform electrostatic field is formed in the gap between the electrode and the ball rotor, and its field strength is E=V/d. been supported.

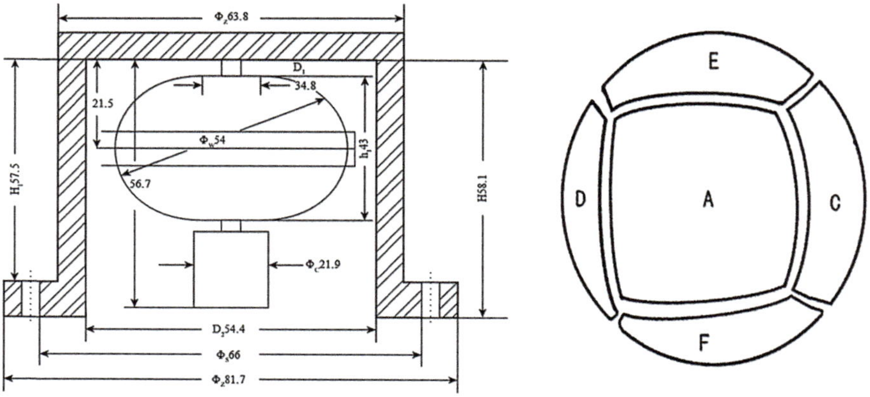

Fig. 10.3 The closed inner spherical cavity composed of two hemispherical ceramic bowls (left) and its six equal-area electrodes (right)

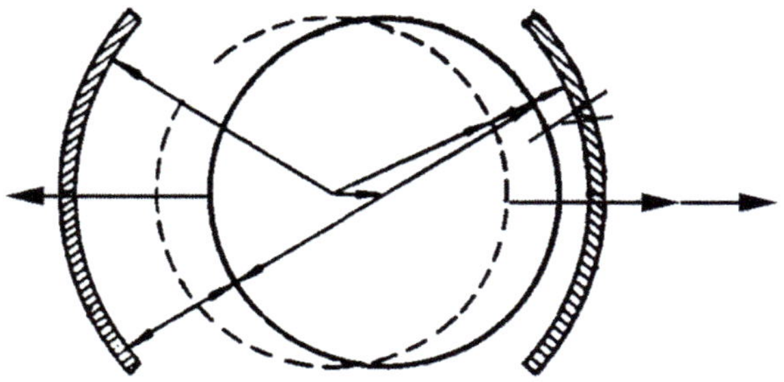

Fig. 10.4 Electrostatic attraction between parallel electrode plates

The electrostatic attraction effect of the electrostatic field formed in the gap between the electrode and the ball rotor is the same as that of the electrostatic field generated between a pair of parallel metal electrode plates, as shown in Fig. 10.4.

Through derivation (for the detailed process, please refer to the author's other monograph *Electrostatic Suspension Control System*), the electrostatic attraction calculation formula applicable to all six electrodes can be written

$$F = \frac{0.83}{2}\varepsilon_0\left(\frac{V}{d}\right)^2 S \tag{10.4}$$

where ε_0 is the vacuum dielectric constant, $\varepsilon_0 = 8.86 \times 10^{-12} F/m$, V is the potential difference between the two polar plates, d is the gap between the two polar plates, and S is the area of the two polar plates.

The formula of electrostatic attraction between spherical electrodes and flat electrodes is similar in form, but the former is multiplied by a coefficient of 0.83, which is easy to understand from the physical concept.

In the actual electrostatic support gyroscope, considering the cutting groove between each electrode and the electrode area occupied by the light-passing hole and the ventilation hole, the effective area of the electrode is reduced by about 13%, so in fact, each electrode in the regular hexahedral electrode is the effective area should be:

$$S = \frac{1}{6} \times 4\pi r^2(1 - 13\%) = 1.82r^2 \tag{10.5}$$

10.2.2 Automatic Adjustment of Support Voltage and Maximum Support Force of Electrostatic Field

The capacitance change ΔC between the electrode and the rotor has an approximate linear relationship with the displacement Δx, so a bridge measuring circuit can be used to sense this capacitance change, and the output voltage signal of the sensitive

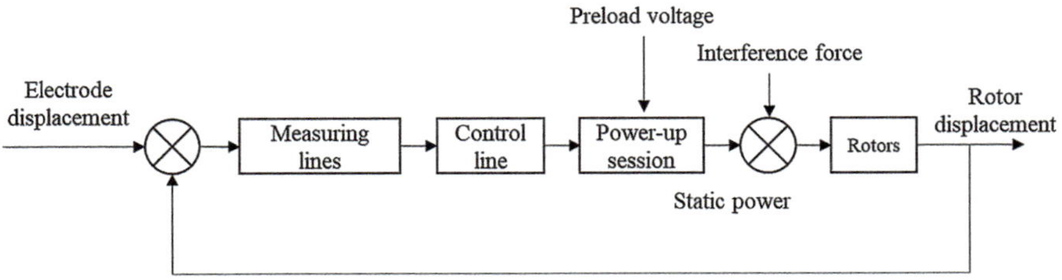

Fig. 10.5 The principle block diagram of automatic adjustment of electrostatic support voltage.

bridge can be amplified through the amplification circuit. The voltage ΔU has a linear relationship with the capacitance change ΔC, that is, it has a linear relationship with the displacement of the rotor Δx, that is

$$\Delta U = k_c \Delta C \tag{10.6}$$

or

$$\Delta U = k_c \frac{C_0}{\delta_0} \Delta x = k \Delta x \tag{10.7}$$

where k_c is the amplification factor of the measurement amplification circuit; $k = k_c \frac{C_0}{\delta_0}$ is the amplification factor of the support circuit

If the voltage is used as the control voltage, and the preload voltage of the two electrodes is added or subtracted respectively, the stable support of the rotor can be realized. Figure 10.5 shows a block diagram of the automatic regulation of the electrostatic support voltage.

Only when the external force acting on the rotor is less than the maximum supporting force generated by the electrode, can the normal supporting effect on the rotor be guaranteed. When the electrostatically supported gyroscope is installed on the vehicle and is in the gravitational field or acceleration field, the gyro rotor is subjected to the action of gravity or inertial force. If the rotor with mass m is in the gravitational field g, the magnitude of the gravity on the rotor is W=mg; if the acceleration of the vehicle is a, the magnitude of the inertial force on the rotor is F=ma. If the acceleration a is represented by the multiple n of the gravitational acceleration g, that is, a=ng, the inertial force on the rotor can be expressed as F=nmg. According to the condition that the inertial force acting on the rotor should be less than the maximum supporting force, it can be obtained,

$$n < \frac{F_m}{mg} = \frac{F_m}{W} \tag{10.8}$$

The coefficient n is called the carrying capacity of the electrostatically supported gyroscope, which is an important parameter of the electrostatically supported gyroscope.

The carrying capacity is related to the following factors:

1. Electric field strength. It can be seen from the above analysis that the bearing capacity is proportional to the square of the electric field strength. In order to improve the load-carrying capacity, it is necessary to try to increase the voltage on the electrode and reduce the gap between the electrode and the rotor to obtain a higher electric field strength. However, when the electric field strength is 4.0×10^5V/cm, the electron emission between the electrode and the rotor will increase significantly, and a large current will appear, causing burn marks on the surface of the rotor. If reading scheme of the shaft position is used, which is marked on the rotor surface, the reading accuracy of the angle will be reduced. In addition, when the electric field strength exceeds 5.7×10^5V/cm, arc discharge even occurs and the support fails. Therefore, the electric field strength should generally not exceed the breakdown field strength of 4.0×10^5V/cm. Obviously, when the control voltage applied to the electrode is the maximum value, that is, $\triangle\ U = U0$, the electric field intensity $E = 2U0/\delta\ 0$ will reach the maximum, so the electric field intensity $U0/\delta 0$ generated by the preload voltage should generally not exceed 2.0×10^5V/cm.
2. The structure of the ball rotor. For the hollow ball rotor, the bearing capacity is inversely proportional to the thickness of the spherical wall. In order to improve the bearing capacity, the thickness of the spherical wall must be appropriately reduced, that is, the rotor is made into a thin-walled structure. For a solid ball rotor, the bearing capacity is inversely proportional to the rotor radius. In order to improve the bearing capacity, the rotor radius must be appropriately reduced, that is, the rotor is made into a small ball structure.
3. Power supply voltage waveform. The voltage waveform applied to the electrodes also has a great influence on the load-carrying capacity. For a sine wave voltage, the rms voltage that produces the maximum support force is only $1\big/\sqrt{2}$ its maximum voltage value (peak voltage). For the square wave voltage, the RMS voltage that produces the maximum support force is about 0.9 to 0.95 times the maximum voltage. Therefore, when calculating the maximum support force, if the maximum value of the sine wave voltage is substituted into the calculation, it should be multiplied by an effective coefficient α=0.5. If the maximum value of the square wave voltage is substituted into the calculation, it should be multiplied by an effective coefficient α=0.8~0.9. It can be seen that under the support voltage of the same amplitude, the maximum support force and bearing capacity of the square wave support is about twice as large as that of the sine wave support.

The above analysis is the case of the regular hexahedron. For regular octahedral electrodes, the analysis method is similar. When the electrostatically supported gyroscope is used on different sailing bodies, the sailing acceleration should be different, and the requirements for the bearing capacity are also different. For example, the acceleration of a rocket may reach 20g, or even greater, while the acceleration of a ship or submarine is smaller, so the electrostatically supported gyroscopes used on different vehicles have two so-called "high g" and "low g". a working state. The "high g" working state is to add a high preload voltage to the electrode to obtain a large carrying capacity. The "low g" operating condition is to apply a low preload voltage to the electrodes to obtain a small power loss.

10.2.3 Static Support Stiffness and Active Electrostatic Support Control Loop

When the electrostatic support works, the resultant force of the electrostatic attraction of the two electrodes is the support force of the electrostatic field on the rotor, so that the:

$$F = -K\Delta x \tag{10.9}$$

In the formula (10.9),

$$K = 1.66\varepsilon_0 U_0 k \frac{S}{\delta_0^2} \tag{10.10}$$

K is called the stiffness coefficient of the electrostatic bearing or simply the bearing stiffness. It represents the supporting force generated by the electrode when the rotor is displaced per unit. The larger the support rigidity is, the smaller the displacement of the rotor when the external force acts, so that the rotor can be easily supported at the center of the electrode.

The principle of the active electrostatic bearing control loop is shown in Figure 10.6. It is an axial bearing control loop, and the other two axial bearing control loops are the same.

The capacitances between electrode A and electrode B and the ball rotor are CA and CB, respectively, and they form the bridge arm of the sensitive bridge with two fixed capacitances C1 and C2. e1 is the low-voltage high-frequency power supply for this sensitive bridge, and its operating frequency is usually 1~2MHz. C3 and C4 are high-voltage isolation capacitors, which isolate the low-frequency high-voltage applied to the two electrodes from the sensitive bridge circuit.

When the ball rotor is in the center of the two electrodes, the capacitance between the two electrodes and the rotor is equal, that is, CA=CB=C0, the sensitive bridge has no signal output, and the electronic circuit of the control loop has no control voltage output. At this time, the low-frequency power supply e2 of the control loop gives a fixed voltage V0, which is added to the center tap of the secondary pole of the synthetic transformer Bc. The voltage is amplified by the powers A and B to drive the high-voltage output transformers BA and BB. The low-frequency high-voltage U0 output by the transformers BA and BB is respectively applied to the electrode A and the electrode B as the preload voltage through the high-frequency isolation network C5, L5 and C6, L6.

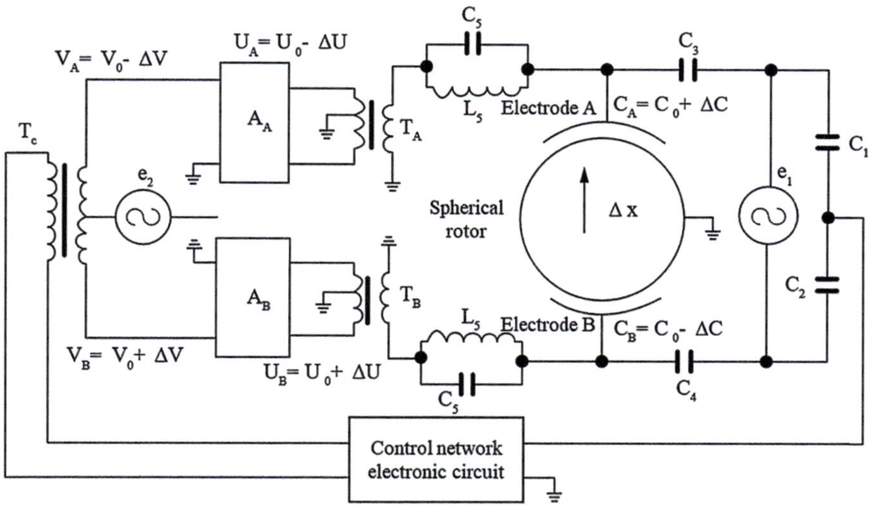

Fig. 10.6 Schematic diagram of active electrostatic support control loop

When the disturbance force makes the rotor produce a small displacement, such as displacement △ x to one side of electrode A, the capacitance between electrode A and rotor becomes $CA = C0+ \triangle C$, and the capacitance between electrode B and rotor becomes $CA = C0- \triangle C$, at this time, the sensitive bridge is out of balance and there is a signal output. The high-frequency signal is subjected to high-frequency AC amplification, demodulation, filtering, DC amplification, network correction, low-frequency modulation, band-pass filtering and low-frequency AC amplification through the electronic circuit of the control loop, and then a low-frequency control voltage ΔV is output and added to the synthetic transformer BC. 's primary. The frequencies of ΔV and V0 are the same, these two voltages are synthesized in the synthesis transformer BC, and the output voltages of the secondary poles of BC are respectively

$$VA = V0 - \Delta V, VB = V0 + \Delta V$$

After the voltages VA and VB are amplified by the power amplifiers A and B, respectively, they drive the high-voltage output transformers BA and BB. The outputs of the secondary poles of transformers BA and BB are respectively

$$UA = U0 - \Delta U, UB = U0 + \Delta U$$

They are respectively added to A and electrode B, so that the voltage on electrode A decreases, thereby reducing the electrostatic attraction to the rotor; the voltage on electrode B increases, thereby increasing the electrostatic attraction to the rotor. In this way, the electrostatic support force overcomes the interference force and support the rotor.

The advantages of the active electrostatic support system are that the working frequency is wide, the support stiffness and the bearing capacity are large. The disadvantage is that the circuit is more complicated, and the use of high-voltage output transformers increases the size of the structure. This form is suitable for moving bodies with large impact acceleration.

10.3 The Mechanism of Electrostatic Suspension Carrier and Theory of Transformer Capacitive Load

The electrostatic levitation servo control uses the carrier medium to transmit the information of the passband of 800Hz, and uses the afterburner transformer as the actuator of the suspension support. Therefore, the carrier is closely related to the ESS passband and rotor suspension deviation. On the other hand, the working band of the afterpower transformer must be suitable for the carrier.Therefore, studying the relationship between the two, determining the theoretical basis for the design of the afterburner transformer, completing the mathematical analysis of its production, and the special theoretical problem of the afterburner transformer capacitive load have special significance for the realization of the afterburner of the suspension support. This chapter only summarizes the important conclusions that connect the context. For the detailed derivation process, interested scholars can refer to the author's monograph *Electrostatic Suspension Control Theory*.

10.3.1 Electrostatic Levitation Carrier Mechanism

(1) Theoretical relationship between carrier and passband

The passband $f_{\Omega m}$ of ESS is 800Hz,, which requires the system to provide a high voltage of the corresponding frequency band and apply it to the ESG to resist various external interference (frequency f_{Ω}) within 0 to 800Hz. Due to the requirement of electrostatic field force for high voltage (1.3KV at 8g), the ESS open-loop gain link requires a large ratio high-voltage transformer to convert the follow-up control low-voltage signal (a dozen volts) into high voltage. However, the introduction of the transformer makes it difficult for the low-frequency signal f_{Ω} within a few tens of Hz to be effectively transmitted directly in the ESS to cancel the low-frequency interference. This contradiction is solved by modulating the low-frequency signal f_{Ω} by the high-frequency carrier f_c. The modulated signal is applied to the ESG electrodes via a high-voltage transformer and converted into rotor displacement, which is a demodulation process.

According to the sampling quantification, if the low-frequency signal f_Ω is reproduced without distortion during demodulation, the carrier frequency f_c should satisfy

$$f_c \geq 10 f_{\Omega m} \tag{10.11}$$

That is, the operating frequency of the high-voltage transformer $fc \geq 10f_{\Omega m} = 10 \times 800Hz = 8kHz$, but the existing support system uses a iron core transformer made by silicon steel sheet, $f_c = 1.74KHz$, that is, $fc = 2f\Omega m$ just satisfies the sampling theorem "reproducible original signal" the lowest limit. Obviously, the original $f_c = 1.74kHz$ cannot meet the requirements of the ESS passband B=800Hz, and the modulated signal can be reproduced to a certain extent when $f_\Omega \leq 232Hz$, For the 333Hz rotational speed frequency, it is a disturbance signal to the ESS, and only when $f_c \geq 6KHz$ can the gap displacement effect be well controlled. Therefore, for the ESS with B=800Hz, $f_c \geq 15KHz$ should be selected, at least $f_c \geq 6KHz f_c \geq 6KHz$

(2) The relationship between carrier and suspension rotor displacement deviation.

In order to quantitatively analyze the influence of f_c on ESS, let the low-frequency signal output by the PID correction link be $\upsilon\Omega$, the preload DC v_0 adder-subtractor amplification factor K_j, the multiplier modulator gain K't[V/V^2], and the modulation gain $Kt = K'tv_c$ the carrier υ_c, the amplification factor of the power amplifier and the transformer is KgNg, after deduction, The influence of carrier fc on rotor displacement is $\frac{\Omega^2}{2(2\omega_c-\Omega)^2}$ times of the controllable deviation of ESS,If the ESS passband $B = f_{\Omega m}$, the maximum influence of f_c on rotor displacement is $\frac{\Omega^2}{2(2\omega_c-\Omega)^2}$ times of the controllable deviation of ESS.When $\omega_e \geq \Omega$

$$Y(t) = \frac{2K'}{m\Omega^2}(1 - \cos \Omega t) \tag{10.12}$$

That is, ω_c does not cause additional bias at this time, which is exactly what ESS needs.

10.3.2 Basis for Design of Electrostatic Suspension Afterburner Transformer

It can be seen from the above that increasing the carrier frequency f_c is an important means to reduce the displacement error of the ESS system. The experiment shows that the sine peak clipping and flat-top waveform is more suitable, which is close to the situation where the upper and lower edges of the square wave are not very steep, as shown in Fig. 10.7. When ϕ_o is large, it not only has a relatively gentle commutation process, but also inherits the advantages of large square wave output. But there are still richer harmonics. Also, some kind of clipping or absorbing circuit

Fig. 10.7 Square wave support waveform and parameters

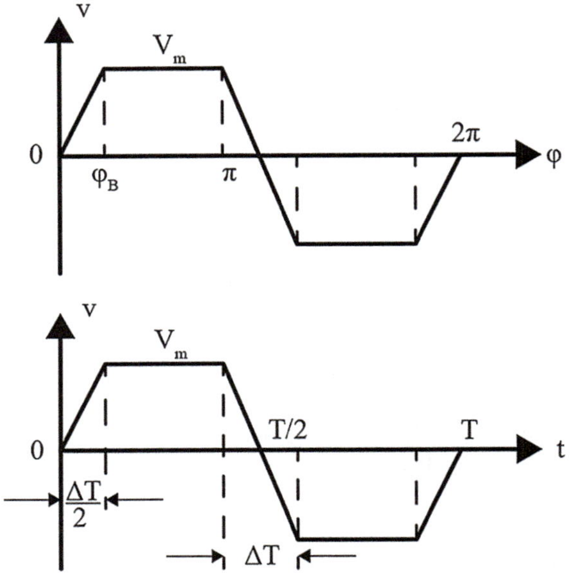

can be used for excessive shocks, but this additional device increases the point of failure. In this sense, therefore, sinusoidal support is advantageous. Because the higher the sine degree, the weaker the harmonic effect, the analysis also shows that the electric shock is almost negligible, the maximum instantaneous current is only 1.15mA, and the maximum instantaneous power is only 8w. The sine wave is thus selected.

When the frequency is changed from 1.74kHz to 2kKHz, ω_c increases by 10 times, such a large power consumption is not allowed. Therefore, the operating frequency point must be selected at the parallel resonance or near resonance of the transformer secondary inductance and the load capacitance. However, this also puts forward requirements for the high-frequency natural frequency of the transformer, that is, it can make the ESG body on the transformer street, the resonant frequency f_{c0} of the equal-adjustment capacitor and the auxiliary pole are in the audio frequency range, and have a certain bandwidth B_2, so as to ensure that the carrier wave f_c is selected in the audio frequency range. At this point, the width of 800Hz on the left and right sides (ie, the passband B of the ESS system) is within B_2.

The technical requirements are: static—overload capacity of 8g, maximum support voltage of 1.3kV; dynamic—adaptive carrier frequency fc in the range of 5kHz ~ 33kHz, bandwidth Bc $\geq$ 4.8kHz. The secondary inductance L resonates with the capacitive load, which reduces the power consumption and can adapt to the change of the capacitive load in a wide range. Power consumption and volume are as small as possible.

In order to reduce power consumption, the special approach that should be taken is: (1) The secondary inductance and the load capacitance form a resonant circuit to reduce reactive power consumption; (2) Reduce the stray capacitance and inductance of the supporting transformer to improve power conversion; (3) Reduce or even remove the high-voltage resistance R to reduce active power consumption; (4) Use a supporting transformer core suitable for high frequency to reduce iron losses such as eddy current.

In order to reduce the volume, the technical approaches that should be adopted are: (1) increase the carrier frequency, (2) change the iron core suitable for high frequency to reduce the number of secondary turns required to generate high support voltage; (3) change the skeleton structure For example, the skeleton structure with the original secondary windings is used to improve the breakdown resistance and make the cushion layer as little as possible.

10.4 Physical Basis for Removing Stray Capacitance of Spherical Bowl of Gyro Rotor on Secondary Side of High Voltage Transformer

The purpose of electrostatic support engineering research is to reduce power consumption and achieve miniaturization on the basis of improving performance indicators. Reducing power consumption is an important technical requirement for ESG to work for a long time, and it is suitable for applications such as satellites and other occasions where the power supply capacity is strictly limited; miniaturization is to install the ESS on the ESG body, cancel the connector, and improve the reliability of the ESG. Electrostatic booster actuator is one of the key components of ESS. The passband of ESS is wider and the support voltage is higher, so it is necessary to use a support transformer that generates high voltage as the actuator element of the afterburner. The feature of the afterburner actuator is that a resonant circuit needs to be used to generate a supporting high voltage, and a higher carrier frequency should be used for this purpose.

In the engineering research of ESS, a hollow rotor with a diameter of φ38 and a gap of 40~70 μm is used, and a solid rotor with a diameter of φ10 and a gap of 7 μm may be used in the future. If twelve electrodes are used, the series gap capacitances formed by them are 21~38pF and 17pF, respectively, and for six electrodes, they are 42~72pF and 34pF, respectively. Therefore, the load capacitance of the afterburner actuator is about 17~72pF. This capacitive load has the following characteristics:

1. The load capacitance value is smaller than the conventional transformer stray capacitance;
2. During the start-up process, the dynamic capacitive load changes greatly, that is, the corresponding capacitive reactance varies from MΩ to 0;
3. The three-way capacitive load must have no mutual coupling effect on each of the executive elements;
4. System passband $\geq$ 800HZ, rotor displacement measurement frequency is 500kHz, the lower limit of carrier frequency is required to be more than 6 times of 800HZ, and the upper limit is 1/10~1/15 of 500kHz or less;

5. The high-voltage support terminal should be isolated from the rotor measurement circuit.

As above, the impedance matching, passband, power consumption, and volume of the forcing element must be well resolved.

10.4.1 Choice of Iron Core

In order to increase the carrier frequency and reduce the number of turns, that is, to reduce the influence of spurious factors, three core materials are selected as shown in Table 10.2.

10.4.2 Framework Design of Ferrite Core and Calculation of Stray Capacitance

The key to the design of the skeleton structure of the afterburning element is to reduce the stray capacitance of the secondary side Li Qingliang, [14]. The general structure of the coil skeleton of the forcing element is that the primary winding is inside and the secondary winding is outside, the axial length is c, the number of segments is x, the thickness of the segmented baffle is d, and each gear is wound with N layers.

For the force components made according to the number of winding segments, see Table 10.3) for the on-load resonance frequency measurement, and the ferrite air gaps are all 0.15mm. From this, two types of ferrites and four types of winding skeleton structures that can be used are shown in Table 10.4), and the calculated and measured values of their stray capacitances are shown in Table 10.5.

In the above structure, the stray capacitance C_2, the load inductance C_L and the resonant frequency f_{co} should satisfy the experimental formula:

$$f_{co} = \frac{1}{2\pi\sqrt{L_m(c_{2'} + c_{L'}}} = \frac{1}{2\pi\sqrt{L_2(c_2 + c_L}} \tag{10.13}$$

where

$$c_{2'} + c_{L'} = \left(\frac{w_2}{w_1}\right)^2 (c_2 + c_L), L_2 = \left(\frac{w_2}{w_1}\right)^2 L_m \tag{10.14}$$

Two points can be obtained by changing C_L

$$(c_{L1}, f_{co1}), (c_{L2}, f_{co2}) \tag{10.15}$$

Table 10.2 Selection of iron core of high frequency afterburning element

Type number	country of origin	Material	Model	$\mu_i(4\pi \times 10^{-7} H/m)$	Bs (T)	Br (T)	Hc (A/m)	volume (mm^3)	adaptation frequency (Hz)
I	Japan	Ferrite	H7C1-EC28	2500	0.51	0.117	12	$28 \times 28 \times 10$	$\leq 0.5M$
II	China	Ferrite	R2KB	2000 ± 500	0.40	0.12	24	$25 \times 25 \times 18$	$\leq 0.3M$
III	China	High silicon steel	0.3mm Thick rolling	15000	1.7	/	/	$30 \times 20 \times 13$	$4K \sim 20K$

Table 10.3 On-load resonance frequencies of different winding segments forcing elements

Number of winding segments	1	2	3	4	5	6	7	10
Resonant frequency(KHz)	6.8	8	/	13	16	18	19	19.5

Table 10.4 Magnetic core and winding skeleton structure of force element

Type number	Ferrite Core	Skeleton		Number of segments	Gap state
		Single	Double		
I-1-0	H7C1-EC28	√		1	Untunable
I-1-1	H7C1-EC28	√		2	Untunable
II-2-2	R2KB short		√	6	Adjustable
II-1-1	R2KB long		√	7	Adjustable

Table 10.5 Design value and measured value of stray capacitance under different structures

Parameters model	W_2 (turn)	X	Design value				Measured value
			$C_1(pF)$	$C_m(pF)$	$C_n(pF)$	$C_2(pF)$	$C_2(pF)$
I-1-0	2400	1	20.95	0.11	0	21.06	24.8
I-1-1	2400	2	5.04	14.03	2.20	21.27	12.7
II-2-2	3000	6	0.16	0.54	1.52	2.22	2.49
II-1-1	3000	7	0.25	1.61	1.03	2.89	3.41

10.4.3 Analysis of Test Results of Secondary Side Stray Capacitance After Structure Design

For II-type $g = 0$ that is no interlayer absolutely paper force element, winding interlayer stack, only the number of wires in the b direction for the number of layers N,then the layer width e wire number is $\frac{W_2}{XN}$, the total number of wires in each span $\frac{W_2}{X} = (\frac{W_2}{XN})N$, The total number of wires in each span is $\frac{W_2}{X} = (\frac{W_2}{XN})N$, their relationship with b and e is $\frac{W_2}{XN/N} = \frac{e}{b}$, then the layer width e, layer number N, equivalent layer gap f are:

$$e = \frac{c+d}{x} - d, N = \sqrt{\frac{w_2}{x}\frac{b}{e}}, f = \frac{b}{N} - \phi \tag{10.16}$$

(1) C_m is small but C_1 is large when X = 1 ; when X = 2, C_1 decreases to 1/4 times, but C_m increases suddenly;Because the innermost layer of X=1 has a minimum potential difference of $0 \sim \frac{U}{N}$, and the innermost layer of X = 2 has a higher potential difference of $0 \sim \frac{U}{2N}, \frac{U}{2} \sim \frac{U}{2} + \frac{U}{2N}$,relative to the location, the corresponding storage charge effect is large, so that C_m increases suddenly. Figure 10.8 shows the potential distribution of the innermost coil when X = 1

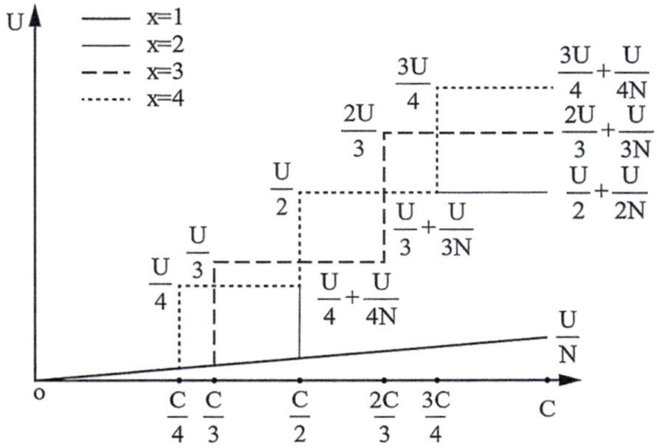

Fig. 10.8 Potential distribution of innermost coil when X is different

~ 4. Obviously, X increases and C_m decreases, When X is large enough, C_m has a minimum $\frac{c-xd+d}{f+g_i}\frac{\varepsilon_0\varepsilon_r}{3}\alpha$.

(2) There is little difference in C_2 between x = 1 and x = 2.. The former is to increase the edge symmetry, which is beneficial to the ESG load.
(3) In order to reduce c_1 but not increase c_m, the distance g_j between the primary and secondary windings can be increased. Therefore, the use of different skeletons for the primary and secondary windings reduces the process and improves the withstand voltage, c_2 is greatly reduced.
(4) After X≥6, $c_2 \leq \frac{c_L}{10}$, the effect on the load frequency characteristic is very small, so X=6 is selected as the 20kHz afterburning element.
(5) Combining with Table 10.3, it is known that the X≥4 skeleton structure should not be used for the force element working at 6~10KHz, because even if the air gap is 0 under load, the value of $c_2 + c_L$ is not enough to make $f_{c0} < 10KHz$. If the parallel capacitor C_p makes f_{c0} decreases, reactive power on C_p needs to be provided and active power consumption is increased.
(6) For the I-type afterburning element, the process requires the "dense-dense-dense-dense" process between the layers, which actually increases the layer spacing and reduces the interlayer area, thereby reducing the stray capacitance.
(7) For the design of the forcing element, firstly, the resonant frequency f_{c0} under load is estimated by calculating C_2, and then the corresponding skeleton structure is used to make.

10.5 Electrostatic Force Actuator Sample and Performance

In Sect. 10.4, the physical basis for removing the stray capacitance of the gyro rotor on the secondary side of the high-voltage transformer is analyzed. On the basis of the theoretical analysis, this section designs the key hardware of the system, the electrostatic force component, as an example. Analyze sample properties.

10.5.1 *Performance Test and Analysis of Ferrite Forcing Element*

The horizontal bar in Figure 10.10, 10.11 and 10.12 is the number of test individuals, of which No. 6 to No. 11 are the number of turns on the side of 1200 × 2, and the rest are 1000 × 2; No. 1 to 6 have large air gaps, and the rest are small (Fig. 10.9).

1. Temperature, frequency and repetition time characteristic test and analysis

Figure 10.10 shows the influence of temperature variation parameter characteristics. The voltage increases with the increase of the temperature T, and the resistance increases slightly with the increase of T, that is, it has inherent thermal noise and iron loss, and the stray capacitance C has no obvious relationship with T.

Figure 10.11 examines the repeatability of inductance measurements over time. The repeatability is good when the number of turns is small, and the repeatability is good when the gap is larger.

Figure 10.12 studies the effect of different measurement frequencies f on each parameter. Among them, the inductance decreases slightly with the increase of f, indicating that the stray capacitance effect increases when the f increases, which offsets part of the inductance; the resistance increases with the increase of f, which approximately satisfies

$$R \geq 1 \tag{10.17}$$

According to this, the change of active power consumption caused by the change of carrier fc can be estimated by referring to the above formula; C2 decreases slightly

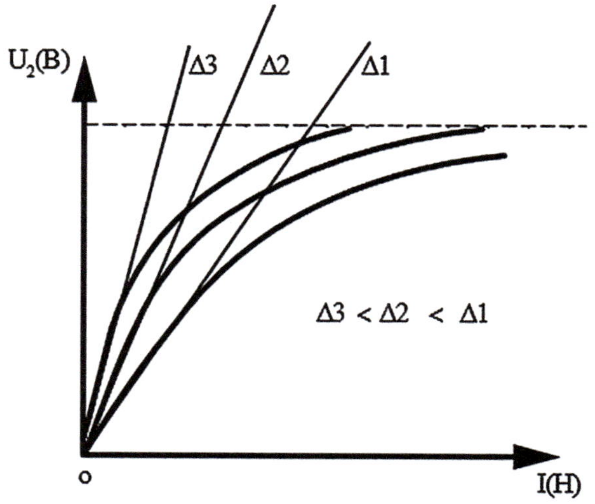

Fig. 10.9 The relationship between U2-I or B-H as a function of gap Δ

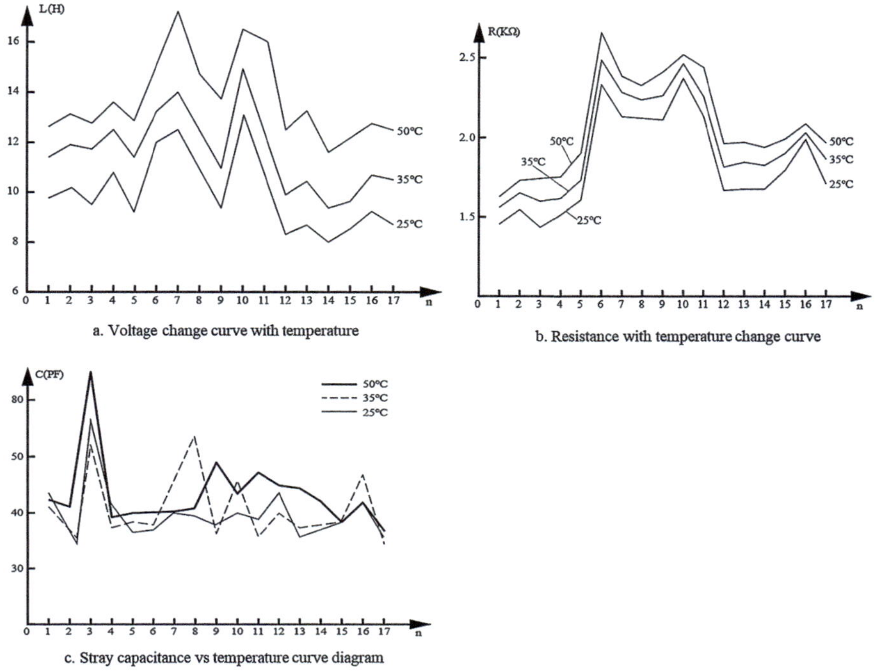

Fig. 10.10 Experimental characteristics of high-voltage transformers at different temperatures

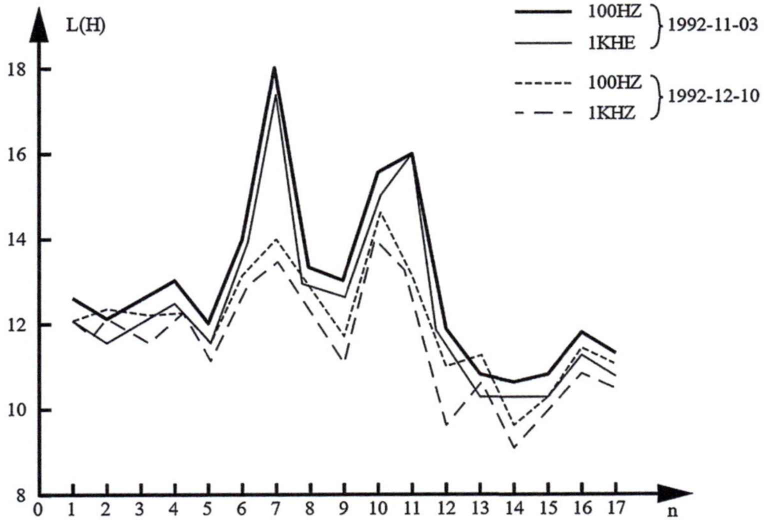

Fig. 10.11 Experimental characteristics of transformer inductance measured at different times

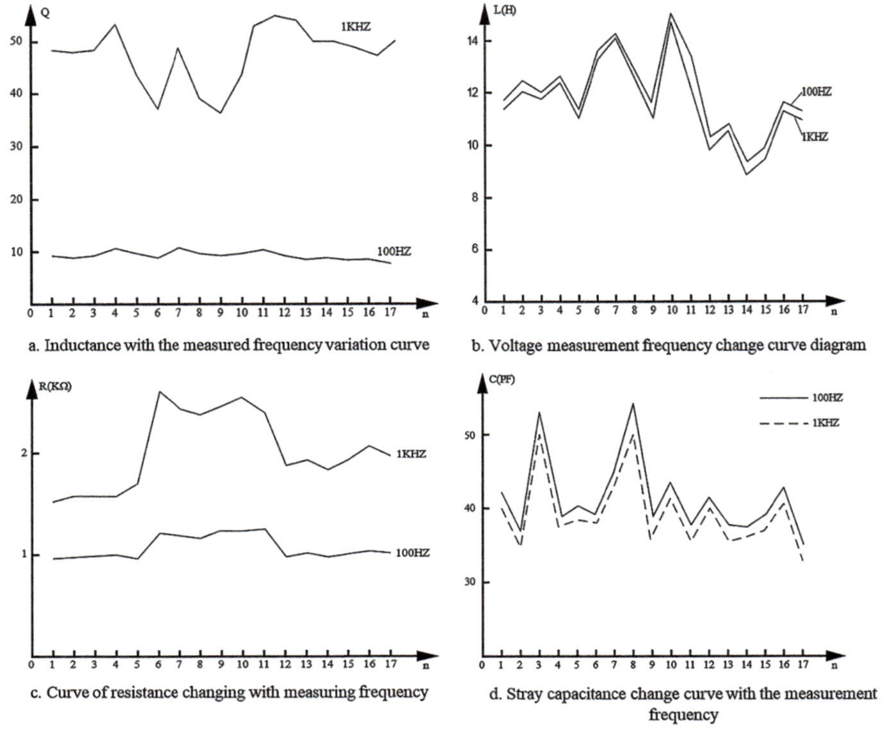

a. Inductance with the measured frequency variation curve
b. Voltage measurement frequency change curve diagram
c. Curve of resistance changing with measuring frequency
d. Stray capacitance change curve with the measurement frequency

Fig. 10.12 Experimental characteristics at different frequencies

with the increase of f, so the increase of fc is beneficial to C, and the quality factor Q approximately satisfies

$$Q = \frac{kf}{\lg f}.$$

2. Comparison with the original different afterburning element

For low-silicon steel forcing components (Table 10.6), the temperature characteristics are relatively stable, but the inductance value drops sharply with the increase of f, which is the eddy current effect that hinders high-frequency applications. For the old ferrite, the stability characteristics are better, because the plane of the magnetic core is relatively rough to form a gap effect, but the product consistency is poor; Yang Wuqiang, [18] once pointed out that the resonance frequency of the same set of afterburning components differs by 1~ 2kHz, which is not allowed in the application. The new ferrite afterburner has good consistency and almost no difference in resonant frequency. In addition, the number of secondary turns of the old ferrite (2800×2) is twice that of the new ferrite, but the inductance is very small, which seriously hinders its practicality.

It can be seen that the new ferrite afterburner has better high frequency characteristics than the low silicon steel afterburner, and higher magnetic permeability than the old ferrite, so it is smaller in size and less winding, resulting in stray capacitance smaller.

3. Frequency characteristics of afterburner element

The frequency characteristic curve of the I-1 forcing element is shown in Fig. 10.13, which was measured by the frequency spectrum analysis method at the Experimental Vibration Center of Tsinghua University.

4. Experimental study on characteristics of resonant frequency f_{c0}

Figure 10.14 shows the characteristics of the input current I1 and the secondary voltage V_2 to f_c of the type II-2-2 forcing element under the condition of constant input, in which 0.63kV corresponds to the preload value of 4g.

Figure 10.15 is the curve of the resonant frequency f_{c0} when the C_L of the type II forcing element changes, which corresponds to the fco of about 20KHz at C_{Lo} = 32pF. Obviously f_{c0} is sensitive to the change of C_L, indicating that the stray capacitance C_2 is very small and does not play a role, thus providing a possibility for the ESS resonance operating characteristics, especially the selection of the best resonance forcing negative feedback characteristic point.

Figure 10.16 is the relationship curve between f_{c0} secondary inductance L and gap Δ when the II-2-2 type of forcing element with C_L=32pF. Since $l\alpha\Phi_{m,}$ for the determined U, if Δ increases, L decreases, Φm decreases, and fco increases. This characteristic ensures that f_{c0} is adjustable in a considerable range, and for the determined Δ and f_{c0}, the time-frequency characteristic is only translated without shape change, corresponding to the passband$\geq$ 5KHz.

Table 10.6 Comparison of Three High Voltage Transformers for ESS

Temperature		25 °C			35 °C			65 °C		
Model	Parameter Measurement frequency	L (H)	r (Ω)	C (pF)	L (H)	r (Ω)	C (pH)	L (H)	r (Ω)	C (pF)
Type I-1	100HZ	9.32	970	35.5	10.73	997.3	34.7	12.59	1032	36.1
	1KC	9.12	1670	35.2	10.59	1780	33.2	12.55	1800	35.7
old ferrite	100HZ	1.88	1357	38.3	1.89	1380	39.5	1.91	1475	38.9
	1KHZ	1.86	1610	38.9	1.87	1630	39.1	1.89	1710	38.8
Low silicon steel sheet	100HZ	45.2	4237	29.0	48.8	4612	27.1	47.98	4705	25.6
	1KHZ	41.4	1810	28.7	43.4	1912	27.0	43.0	1920	25.1

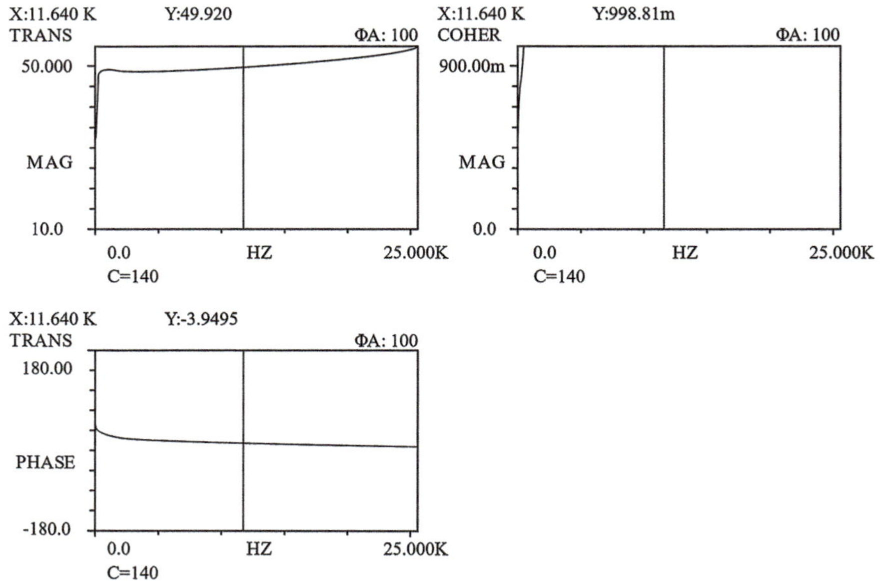

Fig. 10.13 Experiment of frequency characteristic of I-1 type supporting transfor

Compared with the past afterburner, the carrier frequency f_c of ESS is no longer limited by the characteristics of the afterburner, but is determined by the good working state of ESS. The f_{c0} of the afterburner is adjusted $\triangle$ to be consistent with f_c, so that the afterburner is subject to the system (ESS).

Table 10.7 shows the input and output characteristics of the forcing component under a certain air gap and load. Apparently fco is about 17KHz. Passband $\geq$ 6KHz.

5. Experiment on the relationship between the passband B_c, the secondary high voltage resistance R and the quality factor Q under the carrier frequency $f_{c0} = 20KHz$.

The test result of the bandwidth B_c of the afterburner element when the secondary side is connected to the high-voltage resistor R under the condition of 20KHz. According to the following formula

$$Q = \frac{f_c}{B_c} \tag{10.18}$$

The value of the quality factor Q available.

Obviously, the secondary resistance R greatly affects the Q value and the bandwidth. In order to resonate and reduce power consumption, it is better to have a larger Q, but it cannot meet the requirements of the system passband B: Bc$\geq$6B; if Bc is too large, the frequency selection is low, the resonance characteristics are not used, and the power consumption is greatly increased. Therefore, the rational selection

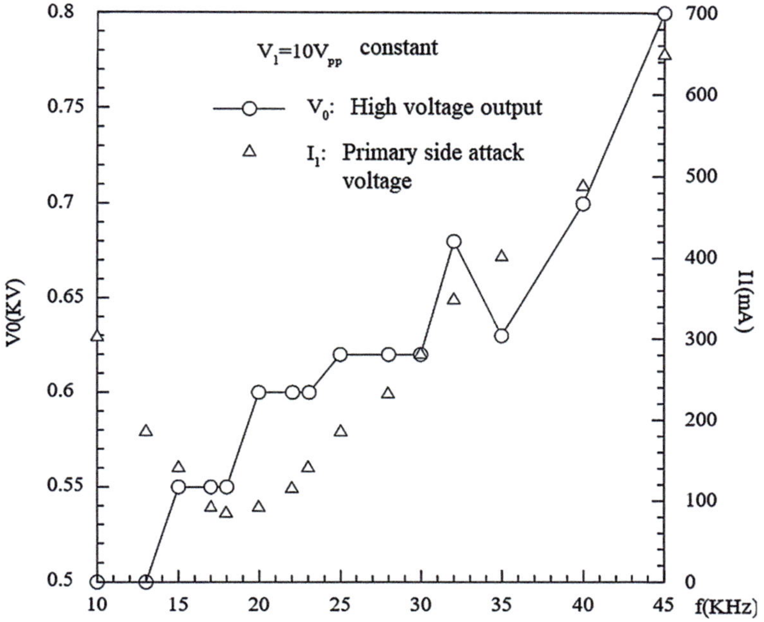

Fig. 10.14 Characteristic curve of II-2-2 type transformer

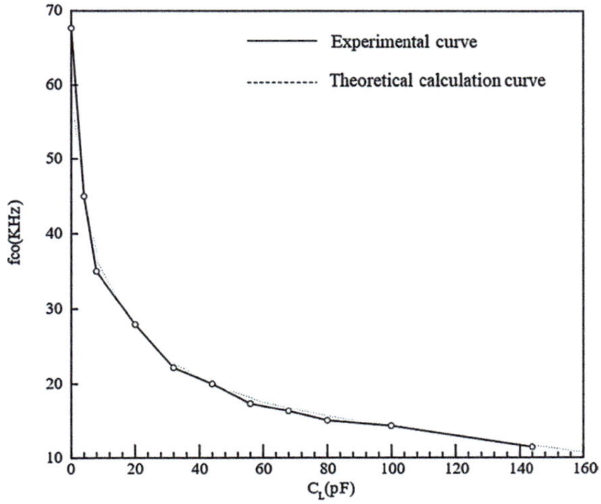

Fig. 10.15 II-2-2 type transformer load C1 and resonant frequency Fm solid line curve

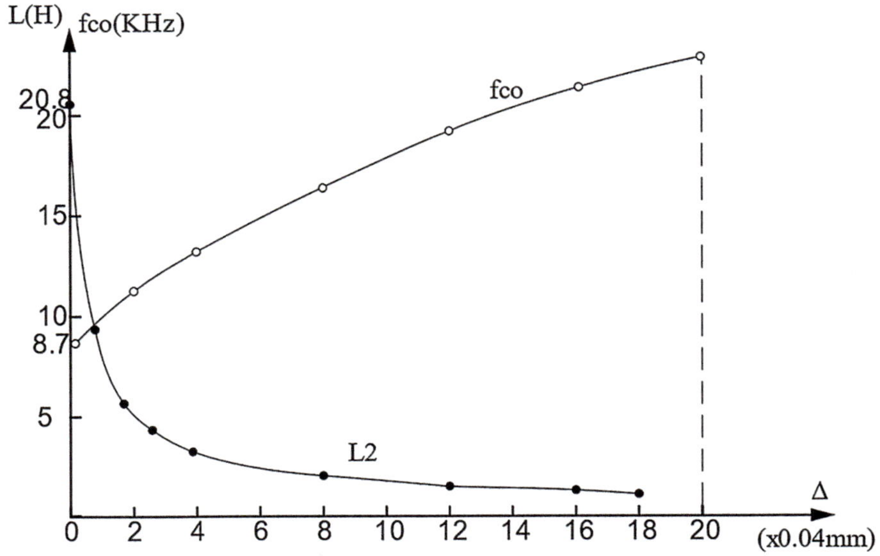

Fig. 10.16 The relationship between the secondary inductance and the load resonance frequency Fdm and the gap △

Table 10.7 Input and output characteristics at = 33pF, Δ=0.24mm

f (KHz)	15	17	20	23	25	28	30	35
V_2 (KV)	1.2	1.2	1.2	1.24	1.2	1.24	1.24	1.30
I_1 (mA)	140	80	180	280	350	440	500	600

of R is an important part of the resonance characteristics of the forcing component Table 10.8.

Table 10.8 High voltage resistance R and passband at 20KHz experiment

R(KΩ)	0	2	4	6	8
B_c (KHz)	1.4	2.0	5.0	8.0	13.0
Q	14	10	4	2.5	1.5

10.5.2 Afterburning Elements and Samples of High Silicon Steel Cores

The iron core developed by the author's team has a high magnetic permeability (Table 10.2), and can work in audio frequency. The main power consumption is related to the coil winding coefficient (generally up to 0.85).

Figure 10.18 shows the developed forcing element sample. From left to right, they are I-1-0, I-1-1, III-1, II-1-1, II-2-2, and the corresponding codes are shown in Tables 10.2 and 10.5. They all meet the technical requirements of 10.4.1, and have undergone high-voltage load experiments. Type I has undergone 6-10KHz single and three-axis support experiments; Type II has been used in new 20KHz ESS circuits, and its withstand voltage test is > 3.5KV without shock Put on.

The actual product shows that it is necessary to use the afterburner actuator (high-voltage support transformer) to realize the resonance characteristics of the secondary inductance and the capacitive load. To this end, the carrier frequency must be increased, and the operating frequency of 8 to 33KHz can be applied. This technical approach also helps to reduce the volume of the supporting transformer.

The developed 10~35KHz working frequency support transformer adopts domestic R2KB ferrite core, the corresponding high-voltage skeleton is separated from the low-voltage skeleton, the withstand voltage is high, the stray capacitance is about 3pF, and the winding is simple; it adopts the standard manufacturer's standard process, and the price is cheap ; The volume of the finished product is $35 \times 30 \times 20mm^3$, the gap can be adjusted after potting, so as to ensure that the resonance frequency reaches the required value, and the current minimum power consumption can be ≤ 0.5W. The authors recommend the use of such a transformer.

The developed 6~10KHz working frequency support transformer has a stray capacitance of about 12pF. It adopts Japan's TDK-H7C1-EC28 ferrite core and self-made skeleton, which is suitable for the ESS circuit with large capacitance (72pF) load in the future, but the coil The winding process is strictly required; the use of a low transformation ratio can reduce the load effect, which is conducive to reducing reactive power consumption. In addition, the magnetic core used in the supporting transformer has high quality, good product consistency and high magnetic permeability. The author recommends purchasing similar cores of smaller size and making corresponding high and low voltage coil bobbins to improve the electromagnetic characteristics of the supporting transformer and reduce the size and power consumption.

The working frequency band of the developed high-silicon steel core supporting transformer is 8-12KHz. The support transformer has high magnetic permeability, but the process is complicated, and the power consumption rises rapidly after deviating from the resonance point, so it is not practical. The author recommends further research on it.

From the perspective of ESS applications, if you customize smaller-sized magnetic cores and skeletons, the volume can be reduced by half. Power consumption can be

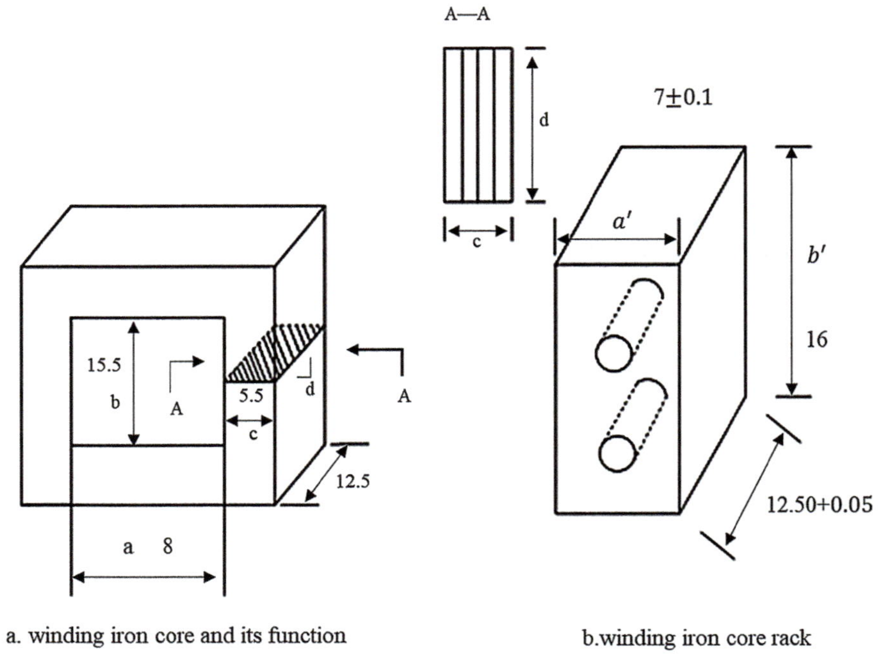

Fig. 10.17 Schematic diagram of processing dimensions of high silicon steel core

further reduced by reducing the number of turns and using the load capacitance tuning feature to improve the figure of merit (Fig. 10.17).

10.6 Control Circuit of Electrostatic Support System

In the ESS, the two ends are the rotor displacement measurement circuit and the afterburner actuator, and the two ends are connected by the control circuit, including: start-off correction, preload, deviation synthesis and modulation, and power amplification. The following will be researched and determined:

(1) The carrier frequency of the electrostatic force device;
(2) Power amplifier, aiming at the fastness of the driving circuit, high back pressure and low power consumption;
(3) Resonance start-up correction, aiming at solving the problem of breakdown discharge;
(4) AC small preload modulator, aiming at the modularization of the circuit and the low drift of ESG.

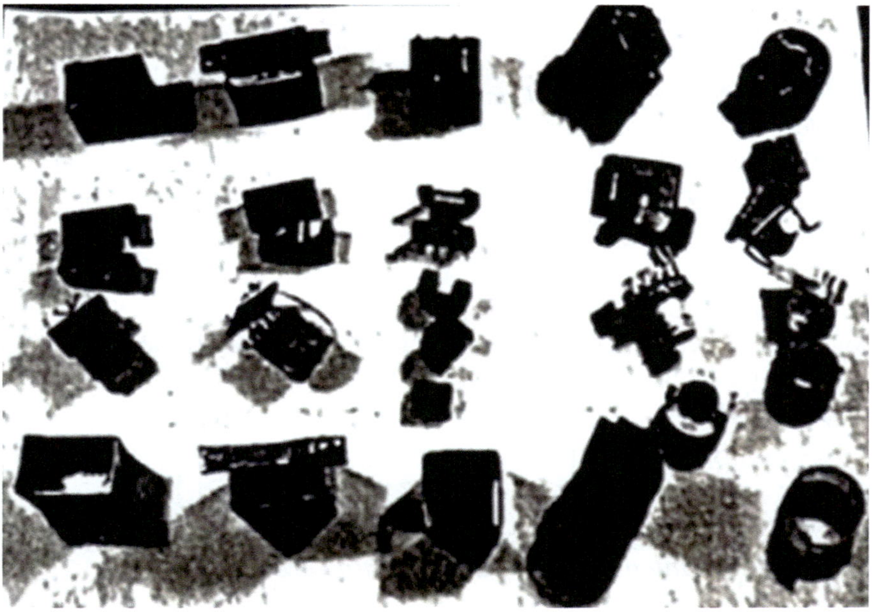

Fig. 10.18 Sample of afterburning element developed

10.6.1 Afterburner Carrier Frequency

The selection of the carrier frequency fc should be considered from five aspects: (1) the passband and measurement frequency of the system; (2) the working frequency band of the afterburner actuator; (3) the start-up performance; (4) the carrier interference; (5) the power consumption.

(1) In order to ensure that the passband of the system is ≥800Hz, fc≥6.4KHz should be selected. In order to make the measurement frequency fr not affect the carrier frequency fc, take

$$f_c \le \frac{f_r}{15} = 33KHz \tag{10.19}$$

(2) The allowable working frequency band of the actuator is 10k~35KHz.
(3) Starting characteristics. Known the inductance L of the afterburner and the resonant frequency of the auxiliary capacitor C:

$$f_{c0} = \frac{1}{2\pi\sqrt{LC}} \tag{10.20}$$

For a gyro with a gap of 456, C = 35pF, and the gap of the upper electrode before the starting is 26, corresponding to C = s, there is

$$f_c(\delta_0) = \frac{f_c(2\delta_0)}{1.19} \quad (10.21)$$

Considering $f_{c\max} \leq 33KHz$ in (1), let it correspond to $f_c(2\delta_0)$, then from Eq. (10.21), $f_c \leq 27KHzf_c \leq 27KHz$

(4) The effect of the carrier frequency f_c on rotor displacement measurement is $\frac{1}{2(2K-1)^2}$ times the ESS controllable deviation, where

$$K = \frac{F_C}{B} \quad (10.22)$$

(5) In order to reduce the power consumption, through experimental research, it is found that the inductance of fc is larger at 15-22KHz, and the power consumption is the smallest.

To sum up, the working frequency fc=20KHz is determined.

10.6.2 Experiment of Resonant Pick-up Circuit

See Fig. 10.19 for the starting principle of the single-circuit support circuit, in which the inductance L2 of the afterburner actuator is a constant. In the ESG used in the experiment, the upper and lower electrode gaps and the corresponding capacitances before the support are: $d_1 = 140_{\mu m}$, $d_2 = 0$, $C_{o1} = 21_{pF}$, $C_2 = \infty$. the process of initial levitation, the capacitance of the upper electrode increases and the resonant frequency decreases; the opposite is true for the lower electrode. When the start-up is completed, the upper and lower capacitances and resonant frequencies are the same.

According to the formula (10.21), $f_{lm} = 24KHZ$, that is, when the support is started, the resonance frequency of the upper electrode decreases from 24kHz to 20kHz, and the corresponding lower electrode increases from 0Hz to 20kHz.

In order to ensure that no electric field breakdown occurs when the support is started, the quality factor of the resonant circuit should not only ensure that the

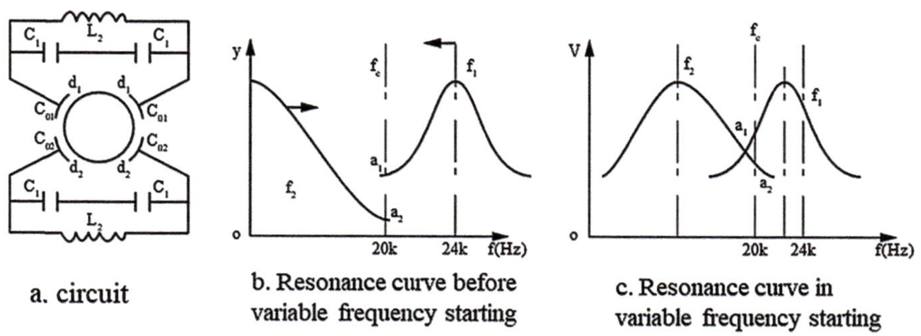

Fig. 10.19 Principle of Single Support Resonance Erection

upper electrode has sufficient output, but also ensure that the lower electrode does not break down. For example, in the experiment, the carrier frequency of the afterburner actuator is 20KHz, the secondary inductance L2 = 2H, and the total resistance R2 $\approx 50\ K\Omega$, which is calculated by

$$Q = \frac{\omega_c L_2}{R_2} \tag{10.23}$$

$$B_c = \frac{f_c}{Q} \tag{10.24}$$

Get Q=5~6, the bandwidth $B_c \approx 4KHz$ after loading. From the point of view of and fm=24KHz, it can meet the requirements of resonance start-up support.

1. Static simulation experiment of resonance support

The experimental circuit is shown in Fig. 10.24, but it is necessary to adjust CL1 and CL2, and use variable capacitors to simulate the upper and lower gap capacitances of ESG. As shown in Fig. 10.20, there are two curves measured by the frequency characteristic experiment of the resonant frequency shift. The center frequency is set at 20KHz and 24KHz by adjusting the air gap of the transformer, and then the frequency characteristics are tested respectively. It can be seen from the figure that the passband Bc $\geq$ 5KHz of the supporting transformer with capacitive load, and the Bc and frequency characteristic curves have nothing to do with the center frequency; it is only related to the distance of the center frequency, so the voltage at fc when the support is started can be regarded as f_1, f_2 as the center of the waveform translation at the value of the intersection at f_c,f_c, thus explaining the dynamic characteristics of the start-up process. It can be seen from the $f_{co} = 24KHZ$ curve that when the deviation from the center frequency is greater than or equal to 15KHz, the V_2 is very small, then the f_c curve is translated to take 0KHz as the center, which corresponds to the resonant frequency curve before the ESG lower electrode is supported, and the a2 point is extremely low. When starting the support, $\Delta V2 = V_{a1} - V_{a2}$ is enough to lift the ball without causing breakdown. If f_c is small, it is impossible to have the characteristics shown in the figure, that is, the a_2 point is very high, which is easy to cause breakdown.

Figure 10.21 is the curve of changing the resonant frequency caused by changing the simulated load. Converting the capacitance value to the gap variable describes the relationship between the upper and lower electrodes $f_{0\alpha}$ and $f_{0\beta}$ during the start-up process. It can be seen from the figure that with $\delta = 70\mu m$ as the center, the resonant frequency change caused by the gap change is always the one with the gap reduction (corresponding to the f_{0b} curve), which has a larger change rate, which means that the voltage of the gap reduction loop decreases. The small rate is greater than the increase rate of the gap increase, so as to ensure that the gap is small and will not break down and discharge; the above characteristics exist in the start-up and system operation, so it is not only suitable for the start-up, but also for the working state; The resonance characteristics ensure the static frequency selection characteristics

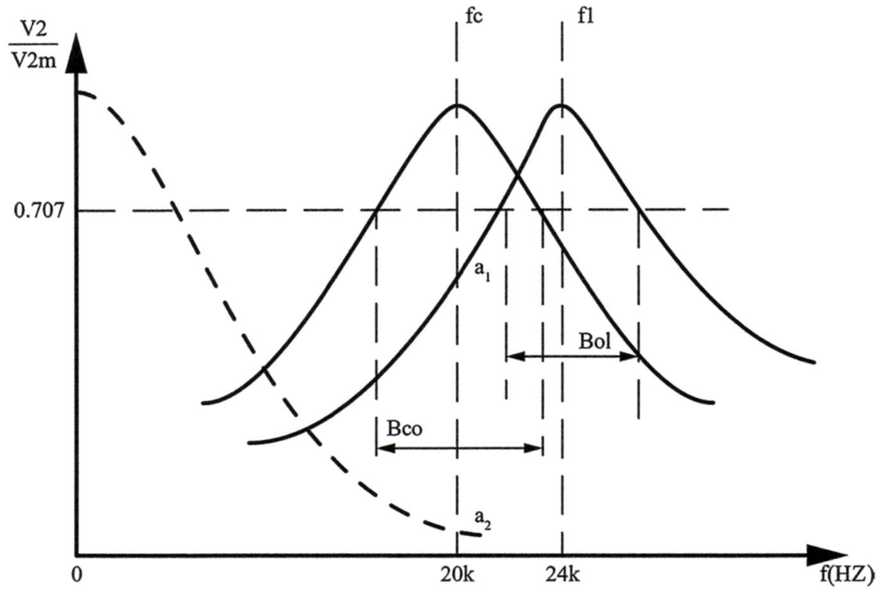

Fig. 10.20 Experiment of frequency characteristics with stable resonance frequency

of the high-voltage circuit, that is, the resonance support characteristics, and the control afterburner plays a role in rectifying the interference. Therefore, the system combined with the two methods can not only work effectively at the optimal static operating point, but also ensure sufficient rigidity; in order to reduce the secondary resistance R to reduce the active power consumption, Q should be increased, and the frequency selection requirement should be increased, as In this study, the frequency conversion start-up circuit is studied, see derails [4, 5], which is omitted in this paper.

2. Influence of position feedback loop in ESS

It should be pointed out that while the resonance plays a role in the support characteristic, the ESS position feedback loop still plays a role in the support. The ESS circuit is under the condition of large deviation before the start of the branch, in which the rotor measurement circuit K_L and the multiplier modulator K_t are in a non-linear working state, and the transformer is no longer linear due to its tuning characteristics, as shown in Figure 10.22. Therefore, the analysis of nonlinear systems can be divided into three steps:

(1) The PID correction is only the ID correction, and the proportional link P is removed, which is equivalent to the disconnection of the position feedback loop and does not constitute a closed loop, so the above-mentioned tuning static characteristics are consistent with the dynamic resonance.
(2) By removing the nonlinear tuning link of the transformer, the high-voltage resistance can be increased to more than $10k\Omega$, and the bandwidth $B_t \geq 15KHz$

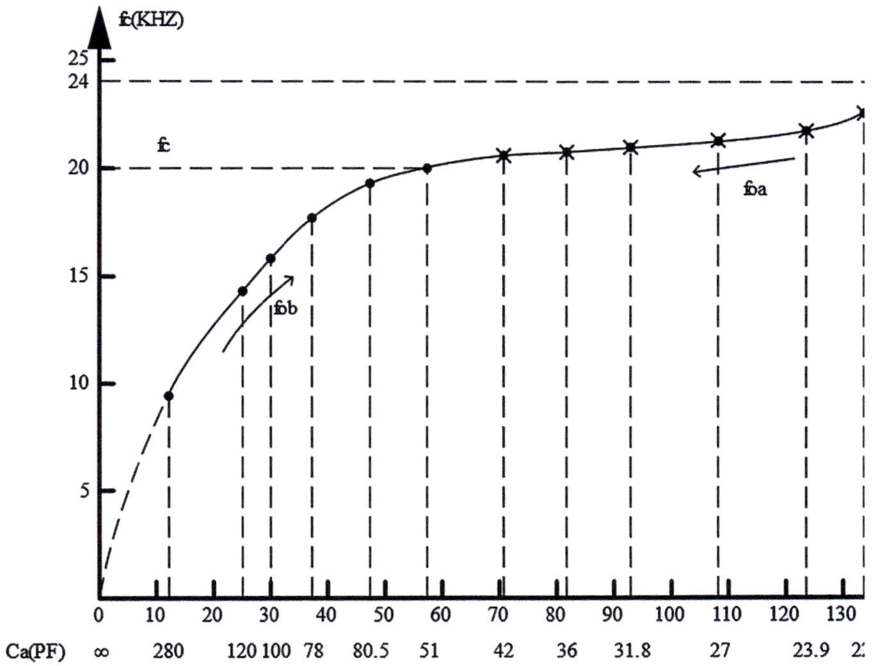

Fig. 10.21 Experiment of resonant frequency changing with gap capacitance

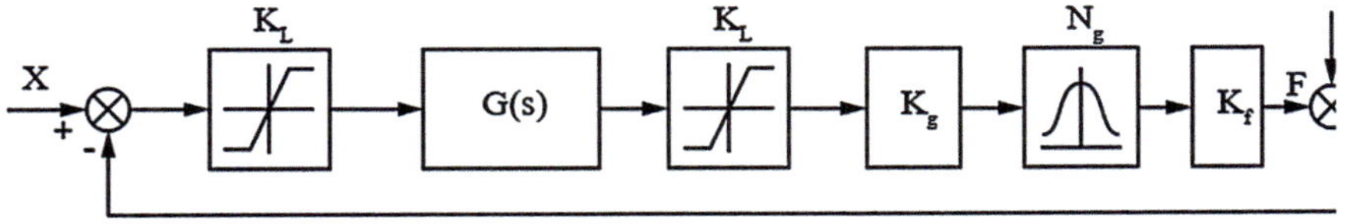

Fig. 10.22 Structural diagram under the condition of large deviation of ESS before the start of support

so that the transformer is still regarded as a linear link N_g, so as to analyze the ESS characteristics of the large deviation state. At this time, try not to saturate the kt link, that is, the ESS only has the measurement link K_L saturated, so the analysis is relatively simple. As far as the actual circuit is concerned, only one link of K_L is saturated, so there is a study on the improvement of K_t in the next section

(3) Combining the two analyses, the combined support characteristics of the resonance support and the control force can be obtained. This research is yet to be furthered. But in any case, the resonant start-up under the high-frequency carrier greatly improves the start-up situation。

10.6.3 Experiment of AC Preload Circuit

Using the characteristics of MOSFET voltage controlled resistance (VCR), the AC preload circuit of VCR modulation can be realized. Figure 10.23 is the principle circuit, where V_y is the 20KHz carrier and VGS is the modulated signal, then

$$V_0 = -\frac{R_f}{R_l} V_y - K R_f V_y V_{GS} \tag{10.25}$$

The modulator integrates DC preload V_0, adder-subtractor, blocking positive feedback, and 2208 multiplier, which is beneficial to secondary integration and engineering. The precision matching adder and subtractor are omitted, and the DC preload is replaced by an AC carrier. The whole circuit is shown in Figure 10.24 [4], and the related comparison is shown in Table 10.9. In addition, a small preload can be adjusted to the optimum point, thereby reducing system drift [19].

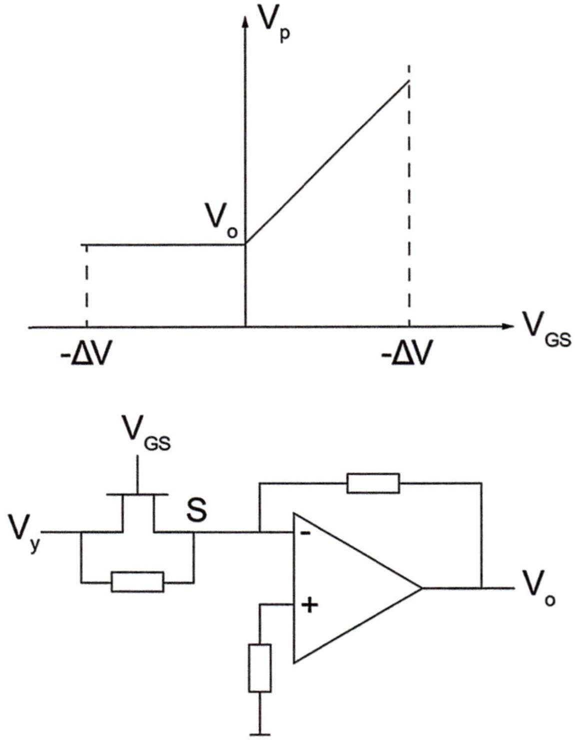

Fig. 10.23 Principle circuit of AC preload modulator

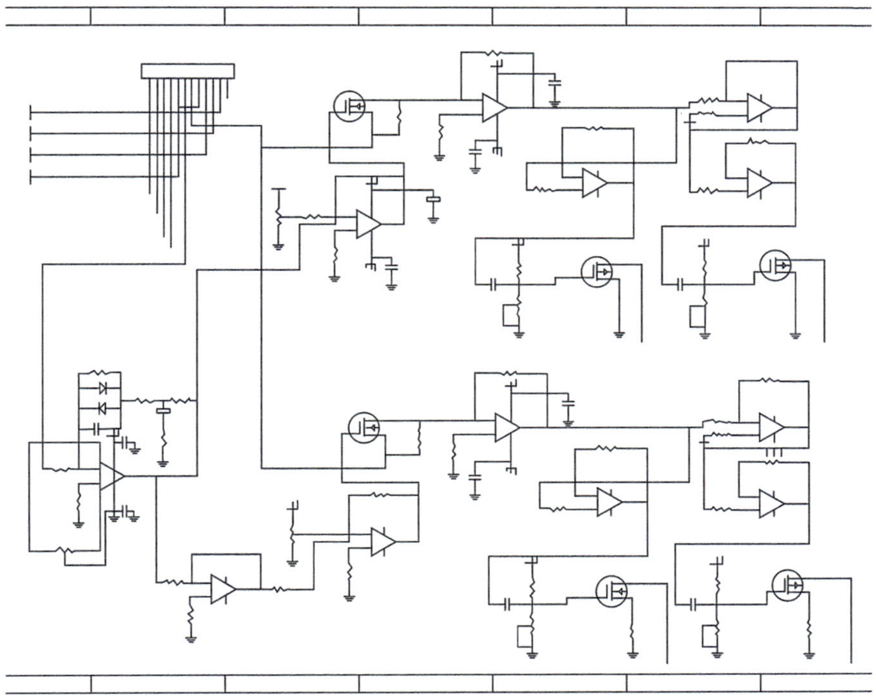

Fig. 10.24 AC preload modulator experimental circuit

Table 10.9 Comparison of MOSFET modulator and DC preload multiplier

Component model	3DJZG series MOSFET circuit	2208 circuit
DC preload voltage	unnecessary	Need
Preload range	>2.0V	<0.5V
Carrier voltage	<0.2Vpp	大
The impact of positive feedback	none	Yes, need to add blocking positive feedback circuit

10.6.4 Key Points of Electrostatic Support System Control Technology

1 Considering ensuring the passband of the ESS, reducing the carrier interference and ensuring the accuracy of the rotor displacement measurement, and at the same time adopting the principle of resonance support to reduce power consumption and eliminate support breakdown, the carrier frequency of the ESS is determined to be 20KHz.

2. In this chapter, VDMOS power transistors are used as power amplifiers. The experimental results show that: in order to ensure that the single-tube class B power amplifier has a good sine wave output, the operating frequency of the transformer with capacitive load $f_c \geq \frac{R_2}{4L_2}$ should be guaranteed, where R_2 and L_2 are the total resistance and inductance of the auxiliary transformer supporting the transformer.

   The above characteristics and conditions apply to all transformers with capacitive loads. Using this achievement, this chapter realizes the circuit of single-tube VDMOS driving a pair of supporting electrodes, with good sinusoidal waveform, lower power, and half of the power amplifier and driving circuit components compared with the traditional circuit, which is conducive to integration.
3. The analysis results show that: in order to realize the variable frequency resonance start-up support, the passband of the support transformer and the quality factor of the secondary resonance circuit should be reasonably designed. The experimental results show that when the carrier frequency is 20KHz, the resonance start-up can completely eliminate the electric field breakdown phenomenon.
4. Using the characteristics of MOSFET voltage-controlled resistance, this chapter proposes and implements a single op amp AC preload modulator circuit, which can replace the original DC preload source, adder (subtractor), blocking positive feedback and multiplying modulator of the ESS. Four links, it is conducive to the integration of the control circuit. The preload of this circuit can be adjusted, which is beneficial to reduce system drift.

10.7 Summary

This chapter first describes the electrostatic gyroscope and its supporting principle. The gap between the spherical rotor and the bowl electrode in the spherical electrode bowl of high vacuum is very small. When the electrostatic force on the spherical rotor is balanced, the spherical rotor is supported. Through the quantitative analysis and single test of the error of the ESS rotor displacement measurement circuit, the separation method and result of each component of the circuit in the total drift error are expounded. On this basis, the methods and approaches to improve the precision index are summarized, a fully shielded and modular structure circuit is developed, and the gyro body is installed.

The high-precision rotor displacement capacitance resolution calibration device developed by the author can provide capacitance changes from $1.25 \times 10^{-3} pF/\mu m$ to $0.2 pF/\mu m$, which cover the capacitance changes corresponding to ESG gap changes of 0.0015 μm to 0.15 μm quantity. The device provides conditions for testing the accuracy and stability (zero point drift) of the rotor displacement measuring circuit. The experimental results show that the developed rotor displacement measurement circuit has good performance, the capacitance resolution and instability are less than or equal to 0.012pF, and the equivalent ESG displacement is 0.01 μm

The high-voltage support transformer, a dedicated electrostatic force actuator, has been successfully developed, which realizes the parallel resonance characteristics of the secondary inductance and the ESG gap capacitance. The volume of the transformer is 35×30×25mm, which can be installed in the gyro body, and can meet the working requirements of different capacitive loads of φ38 hollow rotor and φ10 solid rotor. Resonance characteristics can eliminate the breakdown of ESS support and reduce power consumption. The realization of the resonance characteristic depends on the reduction of the secondary stray capacitance of the supporting transformer. The author puts forward the stray capacitance model and engineering calculation formula of the high-voltage coil bobbin, the calculation formula of the gap and resonant frequency, and the gap and electromagnetic parameters. Their calculated values are consistent with the measured values of the product. This achievement has application value for the design and capacitive load matching of small audio transformers.

In order to drive the capacitive load, a VDMOS power amplifier is developed, the quiescent current is ≤20mA, and the volume with heat sink is 20×20×10mm. The experimental results show that the push-pull power amplifier circuit can be replaced by the single-tube Class B working condition under the condition of ensuring good sine wave. The necessary conditions are: the transformer has capacitive load, and the transformer working frequency $f_c \geq \frac{R_2}{4L_c}$, where R_2 and L_2 are the total resistance and inductance of the transformer pay side. This result applies to all transformers with capacitive loads. The paper realizes a circuit of a single-tube VDMOS driving a pair of supporting electrodes for the twelve-electrode ESG. The sine waveform is good, the power consumption is reduced, and the components of the power amplifier and driving circuit are reduced by half, which is conducive to miniaturization and modularization.

Based on the above, it lays a foundation for the navigation and positioning of the electrostatic vector field and the verification of Einstein's relativity effect excited by the solar light in the sky.

References

1. Liu, W. L., (1992). Research on development and characteristics of new type ferrite high voltage transformer. *Ph.D. Thesis, Tsinghua University.*
2. Yan, L., & Teng, Y. H., (1993). The research and manufacture of high volfage transformer with ferrite core. *Proc. of the International Conference on Modern Design, Manufacture and Measurement, International Academic Pub., Beijing, China.*P358-363
3. Wang, S. W. (1990). *Application of VDMOS Field Effect Transistor in Handzhou.* Science Press.
4. Chen, J. (1993). Real-face research on electrostatic gyroscope with high frequency, small preload and variable frequency start-up. Ph.D. Thesis, *Tsinghua University.*
5. Li, X. Y. (1993). Experimental research on signal power supply for electrostatic gyro support. *Ph.D. Thesis, Tsinghua University.*
6. Compiled by the Ministry of Machinery Industry. (1985). Wire and cable product samples (1). *Machinery Industry Press.*

7. Atkinson, J. L. (1975), Electrostatic bearing sensing and control circuit, U.S. patent, No.3891285.
8. Robert, C.S. (1978). Adaption bias for electrically gyroscope, U.S. patent, No. 4078436.
9. Truban, A. P. (1979) Application of instrument rotation in the N73 standard inertial navigation system, *Proc. of the IEEE 1979, Nat. Aeros. & Elect. Conference Voll Nav. Technology*, P1179.
10. Yang, Y. J. (1992) Experimental study of eight-electrode electrostatic support circuit. *Ph.D. Thesis, graduation thesis of Tsinghua University.*
11. The Central Scientific Research Institute. (1992). ESG with a solid rotor. *Proc. of the 3rd China-Russian Symposium on Inertial Technology (Russion Part), Nanjing, China, Oet.* P18–27.
12. Zhang, Y. Z., & Teng, Y. H. (1981). Several problems of High-precision electrostatic support system in inertial navigation. *Proceedings of the Academic Report of the 1980s Society of Aeronautics and Astronautics of the Chinese Institute of Electronics.*
13. Ren, Z. (1993). Fabrication and experimental research of high-precision rotor displacement capacitance resolution measurement device. *Ph.D. Thesis,Tsinghua University.*
14. Sammons, G. D. (1979). ESG producibility:IEEE-NAECON. P1193~1198.
15. Kamptracker, V. W. (1986). Ferrite cores. *Science Press.*
16. Li, Q. L. (1985). Small transformer design. *People's Posts and Telecommunications Press.*
17. Chen, D. F., & Jin, S. S. (1985). Manufacturing and application of electrical epoxy insulation materials. *Machinery Industry Press.*
18. Yang, W. Q. (1988). Research on electrostatic gyro support system with variable structure and variable preload. Ph.D. Thesis,Tsinghua University.
19. Teng, Y. H. (1992). A Space-Stabilized platform with ESG. *Proc.of China-Russian Symposium on Inertial Technology (China part), Nan jing, China, Oet.*

Chapter 11
Electrostatic Field Ultra-Precision Navigation and Localization Methods and the Basis of Relativistic Effect Verification

The core of the electrostatic gyro is to levitate the rotor in the spherical cavity by the electrostatic force in the electrostatic field, which causes the change of the gap between the rotor and the spherical cavity due to the change of the external field environment, gyro attitude, etc., thus causing the change of the gyro output signal for navigation and positioning. This chapter first compares and elaborates on the electrostatic gyro support system, and then introduces the typical application of electrostatic field navigation—Gravity Probe B experiment to verify Einstein's theory of relativity, systematically.

11.1 Theory of Capacitive Loading of Transformers Supported by Electrostatic Suspension

The electrostatic field is obtained through a high-voltage transformer electrostatic high voltage and electrostatic gyroscope with three vertical direction ball bowl connected, through the ball bowl bearing the vacuum electrostatic metal rotor to form the gap capacitance, through the high-speed rotation of the metal ball rotor to generate the electrostatic field navigation of the fixed-axis effectiveness, and accurately determine the micron-level difference between the different directions of the gap to reduce the incoming motion. This section discusses the theory related to the transformer capacitive load in the support system, derives the elliptic load line equation, and demonstrates it by the characteristics related to the output waveform of a single-tube Class B amplifier.

L. Yan et al., *Mathematical and Physical Fundamentals of Navigation and Positioning with Earth's Natural Fields*, Navigation: Science and Technology 13,
https://doi.org/10.1007/978-981-97-5524-0_11

11.1.1 Derivation and Theoretical Proof of Elliptic Load Line Equation for Transformer Capacitive Load

The transformer as the load of T_1, the amplifier circuit can be equated to two paths of AC and DC, where r_1 is the equivalent internal resistance of the original side of the transformer, C' is the equivalent capacitance is the ratio of the paid side of the transformer W_2 to the original side W_1, and

$$C' = N_g C \tag{11.1}$$

$$N_g = \frac{W_2}{W_1} \tag{11.2}$$

where C is the equivalent capacitance of the ESG measurement circuit, $\mathrm{C} = \frac{C_{01}+C_1}{2} = \frac{42p+22p}{2} = 32\mathrm{pF}$. And $X'c = \frac{1}{\omega C'} = \frac{1}{\omega N_g^2 C} = \frac{X_c}{N_g^2}$, when $f_c = 10KH$, the smaller N_g is, the larger $X'c$ is, the smaller the output i_{cm} is. And when $X'c \gg r$

$$I_{cm} = \frac{E}{X'_c} = \frac{EN_g^2}{X_c} = \frac{U_2 N_g}{X_c} \tag{11.3}$$

$$E = \frac{U_2}{N_g} \tag{11.4}$$

$$Q_p = I_{ce}E = \frac{U_2^2}{Xc} \tag{11.5}$$

$$P = I_{cm}(T_e + T_i)^2 \tag{11.6}$$

Under the premise of ensuring the same negative terminal voltage U_2, the smaller N_g, the smaller I_{cm}, the larger E, the reactive power remains the same, but the active power P is much reduced, and the parametric changes at different N_g are shown in Table 11.1.

Figure 11.1 also shows that $r = r_i + r_e$ forms the DC load line at the output of the amplifier with a slope

Table 11.1 Variation of ESS circuit parameters at different times

	Ng	V2 (kv)	E (v)	X'c (Ω)	Icm (mA)	P (w)	Q (w)	Cs (pF)	Vs (mH)
A	100	V_{20}	24	50	500	$I_{cm}^2 r$	Q_0	C_{so}	L_{so}
B	50	V_{20}	48	200	250	$I_{cm}^2 r$	Q_0	$N_g^2 C_{so}$	L_{so}
$\frac{B}{A}$	0.5	1	2	4	0.5	0.5	1	0.25	1

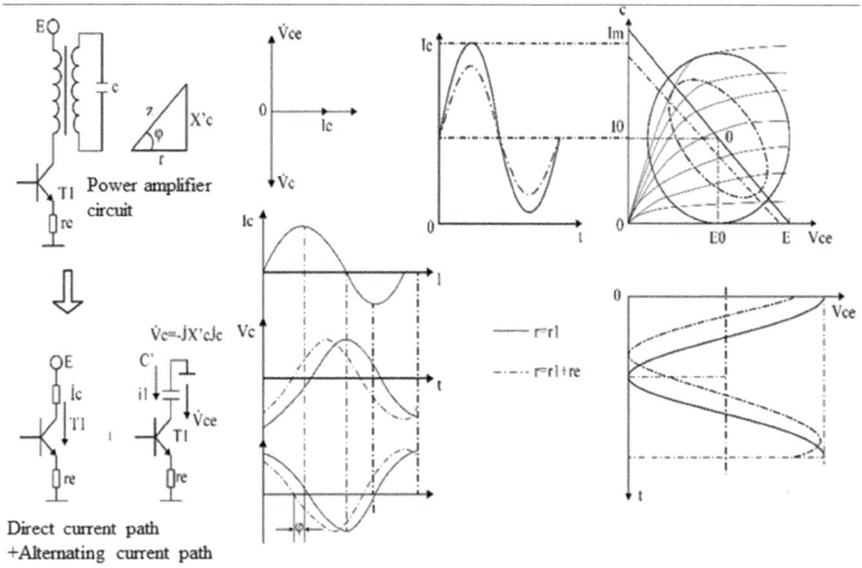

Fig. 11.1 Elliptical load line formed by capacitor C on power tube T_1. **a** Ac-dc path decomposition. **b** Phasor waveform diagram. **c** Output ellipse load line

$$k = \frac{-1}{r} = \frac{-1}{r_i + r_e} \tag{11.7}$$

When $r_c = 0$, k increases and the slope becomes steeper. The equivalent capacitive resistance X_c' forms an elliptical AC load line, where V_{cc} is about 90° ahead, so that the elliptical load line forms a trajectory in an anti-clockwise direction.

$$i_c = I_0 sinw_c t = \frac{E_0}{z} sinw_c t \tag{11.8}$$

$$\varphi = tg^{-1}\frac{x'_c}{r} = cos^{-1}\frac{r}{z} \tag{11.9}$$

$$\begin{aligned} v_{ce} &= E_0 cos(w_c t + \varphi) \\ &= E_0 cosw_c t \times \frac{r}{z} - \frac{E_0 x'_c}{z} sinw_c t \\ &= \frac{rE_o}{z} cosw_c t - x'_c i_c \end{aligned} \tag{11.10}$$

Then

$$v_{ce}^2 = \frac{r^2}{z^2} E_0^2 cos^2 w_c t + x'^2_c i_c^2 - 2\frac{r}{z} E_0 x'_c i_c cosw_c t$$

$$= \frac{r^2}{z^2} E_0^2 cosw_c t + x'^2_c i_c^2 - 2(v_{ce} + x'_c i_c) x'_c i_c$$

After sorting out

$$v_{ce}^2 + z^2 i_c^2 + 2x'_c i_c v_{ce} = \frac{r^2}{z^2} E_o^2 \tag{11.11}$$

By the general quadratic equation

$$Ax^2 + Bxy + cy^2 + Dx + Ey + F = 0 \tag{11.12}$$

Denote

$$\Delta = \begin{vmatrix} 2A & B & D \\ B & 2C & E \\ D & E & 2F \end{vmatrix}, \quad \delta = B^2 - 4AC$$

Then
$A = 1$

$B = 2x'_c$

$C = z^2$

$D = E = 0$

$F = -E_0^2 \frac{r^2}{z^2}$

$$\Delta = \begin{vmatrix} 2 & 2x'_c & 0 \\ 2x'_c & 2z^2 & 0 \\ 0 & 0 & -E_0^2 \frac{r^2}{z^2} \end{vmatrix} = -4E_0^2 \frac{r^2}{z^2} (z^2 - x'^2_c)$$

$$= -4E_0^2 \frac{r^4}{z^2} \neq 0, \delta = 4x'^2_c - 16z^2 = -4(3x'^2_c + 4r^2) < 0 \tag{11.13}$$

That is

$$(A + C)\Delta(1 + z^2)(-4E_0^2 \frac{r^4}{z^2}) < 0 \tag{11.14}$$

It proves that Eq. (11.11) is an ellipse. Obviously, the larger x is, the smaller the region enveloped by the ellipse. For class B amplifier, the positive and negative half-week two amplifier tubes are on, then there is Fig. 11.2, where $E_0 = E$.

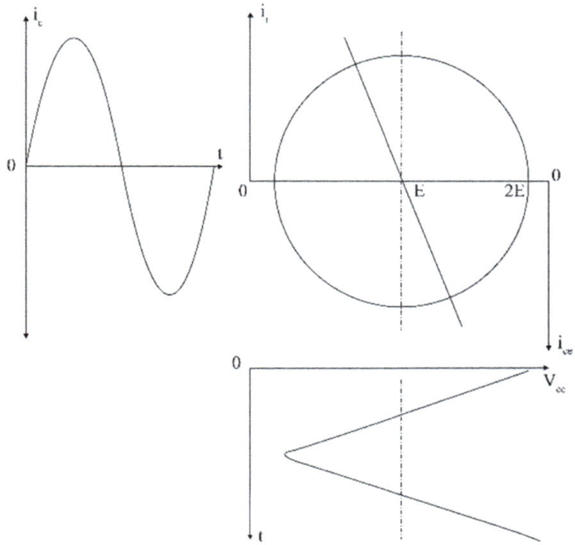

Fig. 11.2 Elliptical load line formed by class B push–pull amplifier

11.1.2 Sine Wave Output Mechanism of Single-Tube Class B Amplifier with Capacitive Load and Analysis of Anti-impact Stiffness

Single-tube Class B amplifiers have been successfully applied to ESG suspension supports. The rationale is that transformer capacitive loads have circular load lines. As in Fig. 11.3, in the positive half of the voltage V_i, the single-tube amplifier T conducts, and the power current i_{dl} passes through the primary side L_1 of the transformer T to provide energy to its pay-side capacitor C. At this time, the magnetic energy of the transformer pay-side inductor L_2 is converted into the electrical energy of the capacitor C, i.e. C is charged. In the negative half of V_i output, the single-tube amplifier T cuts off, the primary current idl is zero, and the power supply E_c no longer provides power to the transformer load C; Thus the electrical energy of C is released and supplied to L_2, thus generating a negative half-cycle voltage at L_1. Therefore, from the single-tube Class B amplifier side, that is, the original side of the transformer, V_{dl} is the sine output, i_{dl} is the half-wave output. This realizes the sinusoidal output of single-tube amplifier voltage V_{dl}, load voltage V_2, and load current i_2. As the amplifier tube T only conducts in the half-cycle, and outputs positive half-cycle sine wave, so the amplifier tube is in the class B working state.

(a) Circuit Principle (b) V_1 Positive half-cycle equivalent circuit(c) V_1 Negative half-cycle equivalent circuit.

The key to the practical use of capacitive load single-tube class B amplifier sine output theory for ESS lies in the ability to withstand various disturbances in the passband $B = 800\,\text{Hz}$ range. Specifically, the circuit should ensure that in each negative half cycle of the amplifier tube does not conduct (20 KHz frequency), relying

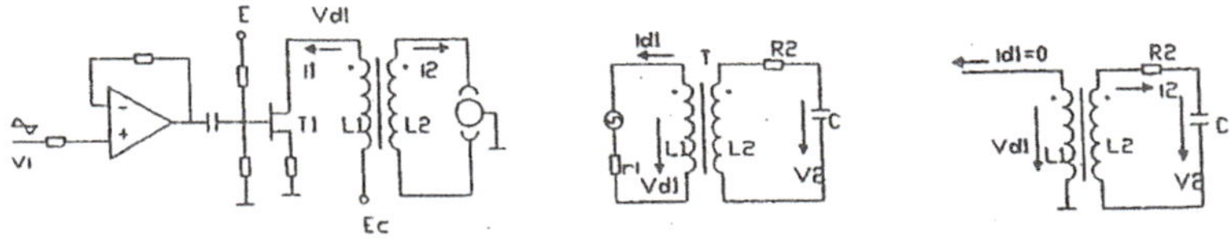

Fig. 11.3 Single-tube Class *B* amplifier sine output power analysis. **a** Circuit Principle. **b** V_1 Positive half-cycle equivalent circuit. **c** V_1 Negative half-cycle equivalent circuit

on its own capacitive load storage energy, to resist the shock or disturbance within 8g in 0–800 Hz . The following is a demonstration of the mechanism of the practical application of this circuit theory in terms of waveform coefficients and interference.

(1) Considering the passband B_c and the waveform coefficient K_b of the sine wave.

Waveform of a 20 KHz signal driving a single tube Class *B* amplifier: The upper half of the cycle provides energy to the VDMOS forward channel, and the lower half of the cycle maps the load capacitance energy to the primary side when converting magnetic energy to the transformer pay-side inductor. where the passband *B* pay-side woe reduction constant

$$\tau = \frac{4L}{2R + R_0} \tag{11.15}$$

R_0 is the negative side AC internal resistance, $R_0 = 8\text{k}\Omega$, R is the high voltage resistance, L is the negative side inductance, $L = 2H$ and the waveform coefficient K_b indicates the degree of waveform attenuation in a carrier period T.

$$K_b = \frac{U_m - U_m(T_c)}{U_m} \tag{11.16}$$

Take the peak of the negative half-cycle waveform

$$U_{-m} = \frac{U_m + U_m(T)}{2} \tag{11.17}$$

Here U_m is the positive half-cycle output amplitude caused by the original side excitation, U_m is the output amplitude after the decay of the period T, is the negative half-cycle amplitude of the decay of the pay-side energy storage, then K, satisfies

$$K_b \leq 1 - e^{-\frac{T_c}{\tau}} \tag{11.18}$$

And

$$T_c = \frac{1}{f_c} \tag{11.19}$$

Then Table 11.2 is available.

Since *Kb* is the decay factor after one week T_c, the decay rate for $T_c/2$ cycles is

$$K_{-b} = \frac{U_m - U_{-m}}{U_m} = \frac{k_b}{2} \tag{11.20}$$

Take $K_b \le 10\%$, then $B \le 5KHz$, $K_{-b} \le 5\%$ that is, for the passband $Bc \le$ 5 KHz, $K_b \le 10\%$ corresponding to the negative half-cycle amplitude to meet the engineering requirements.

(2) Considering interference.

① Figure 11.4 shows the 20 KHz carrier and $2B = 17$ KIz waveform analysis curves, which both have the same support voltage U_m.

For 1.74 KHz sine Um can withstand 8*g* of acceleration, for 1.5 KHz square wave U_m can withstand 20*g* of acceleration, for 20 KHz, its negative half-cycle has external energy input, this half-cycle width corresponds exactly to the 15° width of 1.7 KHz. Therefore, in the 15° range of 1.74 KHz, to find the maximum rate of change of the waveform interval and determine the corresponding amplitude change, and for 20 KHz negative half-period amplitude U_{-m} satisfied

$$U_{-m} \ge \Delta U_m \tag{11.21}$$

Table 11.2 Relationship between *Bc* and τ, *Kb* at $fc = 200$ KHz

Bc (KHz)	1.4	2.0	5.0	8.0	13.0
R(kΩ)	0	2	4	6	8
t (ms)	1	0.67	0.5	0.4	0.33
Kb	0.05	0.07	0.10	0.12	0.14

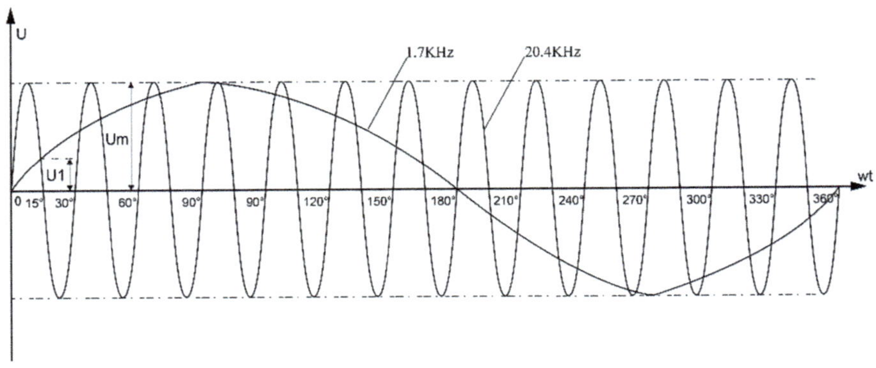

Fig. 11.4 20.4 kHz carrier 2B = 1.7 kHz waveform analysis curve

Then the 20 KHz single-tube class B sine amplifier can withstand the same acceleration of external interference.

② Determine the maximum variation rate interval and variation amplitude ΔU_m for 1.7 KHz and 15° range.

Obviously, the maximum variation of the sine wave is at the crossing of zero, then the value of 0–15° range is taken as the maximum. That is, from $sin 15° \approx 0.25$, we know that $\Delta U_m = 0.25 U_m$.

③ ΔU_m compared to 20 KHz negative half-period amplitude U.

From the previous calculation has been obtained, 20 KHz sine half-period $U_{-m} \geq 0.95 U_m$, then $U_{-m} \geq 0.25 U_m = \Delta U_m$, that is, under $8g$ interference,20 KHz sine negative half-period energy storage is sufficient to resist the voltage changes caused by the interference. In particular, the greater the interference the greater the low component, no harm to the 1.74KHz ESS reliable support, and even less for the 20 KHz carrier with a frequency about 12 times higher than 1.74 KHz. The high frequency component of the above-mentioned low interference decreases rapidly, and the system is insensitive above 800 Hz.

(3) Summary

①) If the interference reduces the gap (e.g. short circuit) at 20 KHz half cycle energy runs off and no voltage, which is exactly what is needed. If the interference increases the gap at 20 KHz half-cycle, then $U_{-m} \geq 0.95 U_m$, as long as the interference is within $B = 800$ KHz can be tolerated. Exactly how much the maximum limit can withstand, need to be experimentally verified.

② Interference rate ≤5 KHz, then the 20 KHz carrier has a cycle of positive circumference from the power supply to provide energy to withstand the interference. Now $B \leq 1$ KHz, corresponding to 20 KHz carrier wave has ≥20 cycles of positive half-period from the power supply to provide energy, fully able to withstand interference.

In the extreme, $\left.\begin{array}{l}\text{1.5 KHz Square wave}\\ \text{1.7 KHz sine}\end{array}\right\}$ can withstand 8 KHz interference, this can test the effect of disturbance on 20 KHz carrier at 10 KHz.

Experiment:5 KHz electrical interference experiment (equivalent to $4g$ acceleration), starting support and working normally, thus proving the correctness of the theoretical analysis.

11.2 Suspension Support Follower Control Technology and Integrated Electromechanical Technology

The electrostatic support system is one of the two main elements of high precision electrostatic gyroscopic inertial guidance device (suspension support, fixed-axis goniometry), which requires the gyroscopic rotor to be absolutely and reliably suspended in the vacuum sphere cavity to withstand acceleration shocks above $8g$, vibration in the range of 800 Hz, and sub-micron rotor centering accuracy (e.g. rotor mis-centering and resolution are $< 0.05\mu$m). These requirements give the actual electrostatic support system a static stiffness or open-loop gain of up to about 106 N/m and a preloaded high voltage V_0 with a single peak of about $1kV$. This subsection discusses how to solve the its problems while satisfying the conditions of high gain and high preload.

11.2.1 *Derivation of Control Mode and Proposed Control Scheme*

(1) Continuous frequency conversion and variable structure starting support principle proposed.

The single branch circuit is shown in Fig. 11.5, where the transformer L is a constant. The upper and lower electrode gaps and the corresponding capacitance before starting the branch are: $d_1 = 140\mu\text{m}$, $d_2 = 0$, $C_1 = 21pf$, $C_2 = \infty$, the total equivalent capacitance of upper and lower electrodes C_a, C_b and their respective resonant frequencies with the corresponding resonant frequencies are

$$C_a = \frac{(C + C_1)}{2} = \frac{(C + \varepsilon_0 \frac{S}{d_1})}{2}$$
$$C_b = \frac{(C + C_2)}{2} = \frac{(C + \varepsilon_0 \frac{S}{d_2})}{2} \tag{11.22}$$

$$f_1 = \frac{1}{2\pi\sqrt{(LC_a)}}$$
$$f_2 = \frac{1}{2\pi\sqrt{(LC_b)}} \tag{11.23}$$

where ε_0: vacuum dielectric coefficient S: effective area with poles,d_1,d_2: upper and lower ball bowl gap.

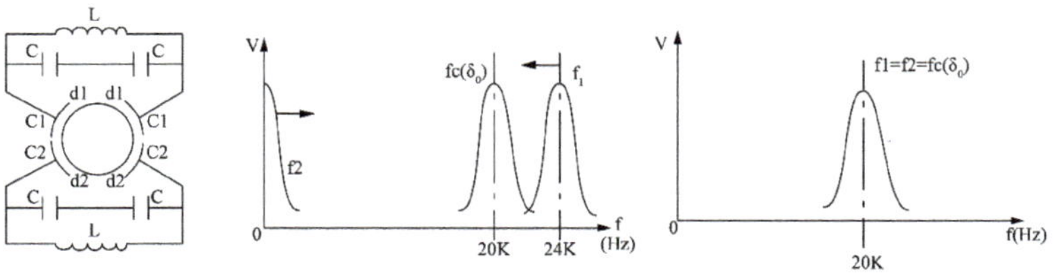

Fig. 11.5 Single-support inverter regulator starting principle. **a** Circuits. **b** Resonance curve before variable frequency start-up. **c** Resonance curve after variable frequency start-up

Smooth switching to overrun hysteresis operation regulator after rotor start.

(2) Circuit implementation of preload regulator for starting support.

The starting preload regulator is used only in the vertical direction, and the implementation circuit to satisfy Fig. 11.6 is shown in Fig. 11.7. The charging starts at the moment of power on, and automatically enters the working preload state after the charging is completed, and the preload regulator for starting support no longer works, and the experimental results are consistent with the analysis of the above study.

(3) Circuit implementation of variable frequency start regulator.

The circuit of the variable frequency starting regulator is shown in Fig. 11.8a, and the corresponding operating curve is shown in Fig. 11.9b.

(4) Principle of the proposed linear preload modulator.

The schematic block diagram is shown in Fig. 11.10a, and its characteristic curve is shown in Fig. 11.10b.

(5) Preload modulator without $|\Delta V| < |\Delta V_0|$ segment linear regulation area and with base voltage $V_p < 0$.

Fig. 11.6 Preload integral regulator for *Z*-axis start

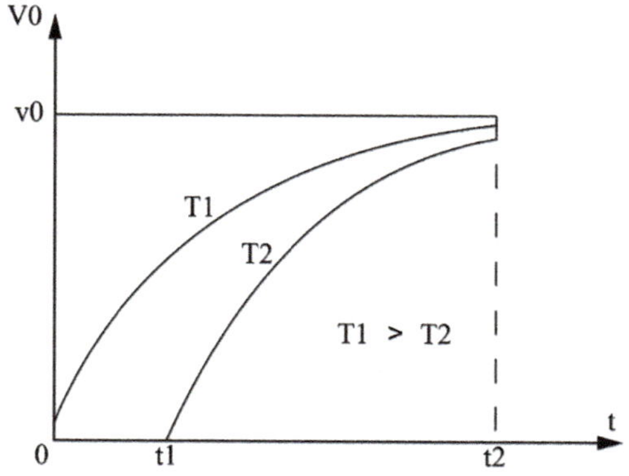

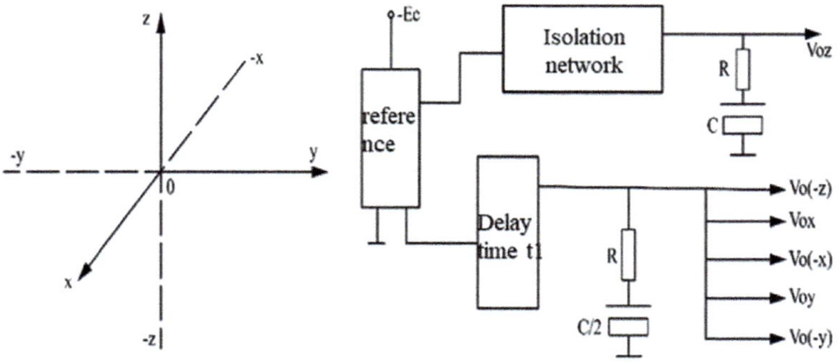

Fig. 11.7 Preload regulator for starting support. **a** Support electrode coordinate system. **b** Circuits

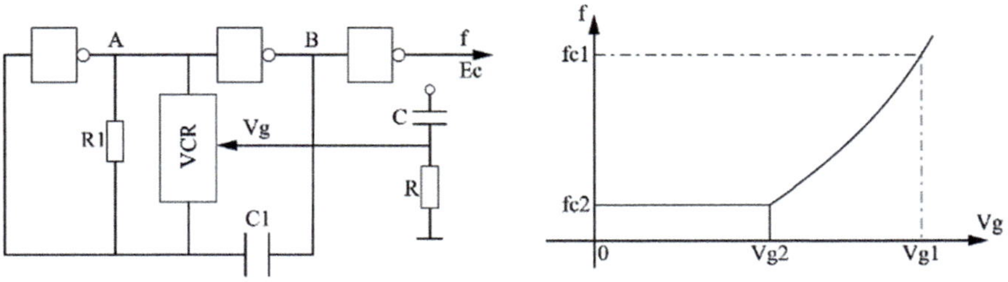

(a) VCF experimental circuit (b) VCF working curve

Fig. 11.8 Variable frequency start regulator

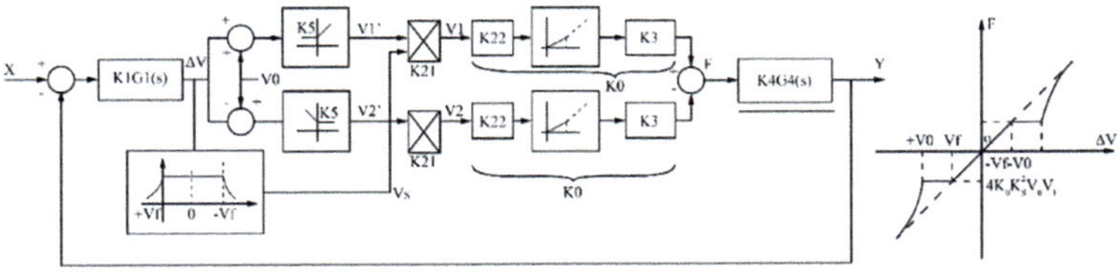

Fig. 11.9 Proposed linear preload modulator. **a** Block diagram. **b** Characteristic curve

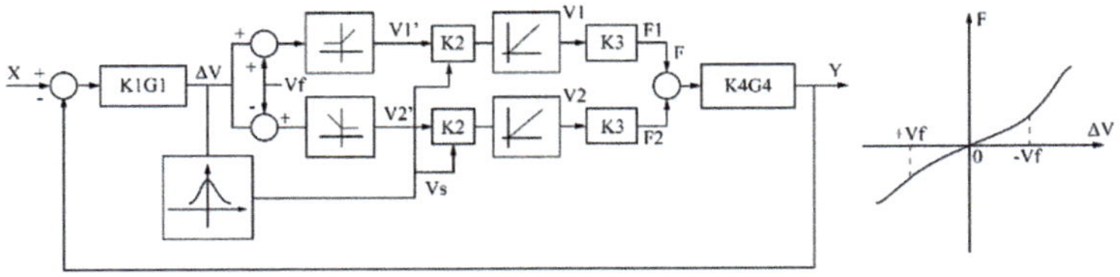

(a) Block Diagram (b) Characteristic Curve

Fig. 11.10 Preload modulator without $|\Delta V| < |\Delta V_0|$ segment linear regulation region and with base voltage $V_p < 0$. **a** Block diagram. **b** Characteristic curve

The structure block diagram is shown in Fig. 11.11a, and its characteristic curve is shown in Fig. 11.10b.

(6) Circuit structure of preload modulator.

The corresponding circuit diagram of the modulator with linear regulation area $|\Delta V| < |\Delta V_0|$ and in the controller is shown in Fig. 11.9, and the circuit diagram is shown in Fig. 11.11a; the corresponding circuit diagram of the modulator without $|\Delta V| < |\Delta V_0|$ segment linear regulation area and containing a base voltage $V_p < 0$ is shown in Fig. 11.11b.

(a) Proposed linear variable structure circuit.
(b) No $\Delta V < V_b$ SDM mode in the linear regulation region.

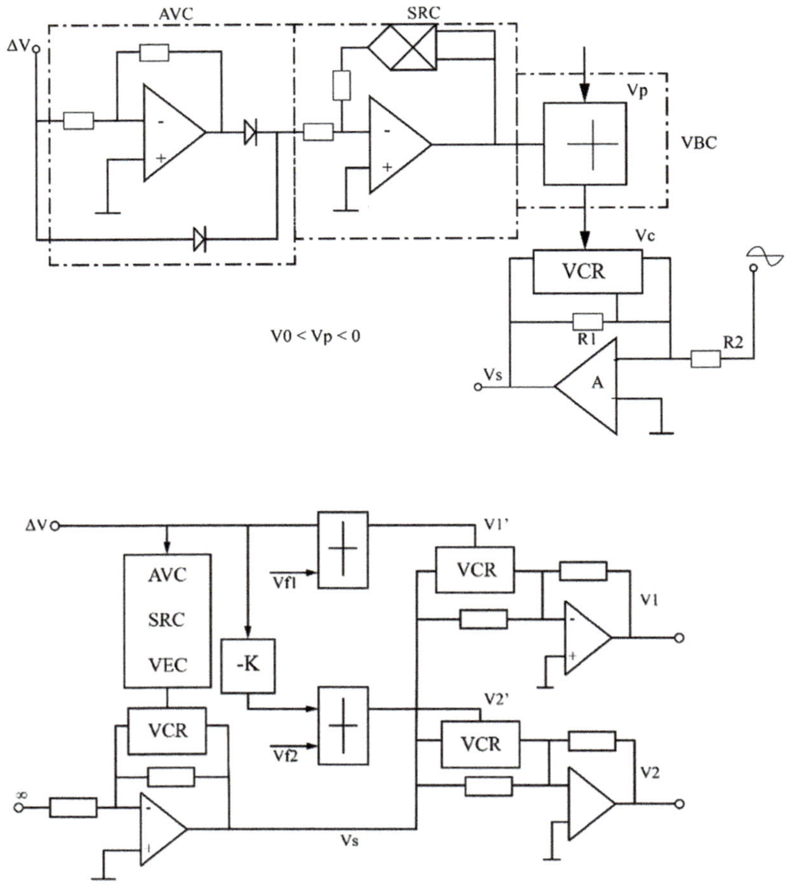

Fig. 11.11 Proposed linear preload modulator circuit implementation. **a** Proposed linear variable structure circuit. **b** No $\Delta V < V_b$ SDM mode in the linear regulation region

11.2.2 Comprehensive Technical Issues of Suspension-Supported Inertial Sensors

(1) General problems faced by various gyro rotors.

The function of gyroscopic navigation is to establish the carrier position and direction, whose angular momentum $\vec{H}$ and rotor vector diameter $\vec{r}$, mass m, linear velocity C satisfy

$$\vec{H} = \vec{r} \times m \vec{v} \tag{11.24}$$

This is the fixed axis of the gyroscope. However, the gyroscopic motion produces deviation, and its drift speed and and disturbance moment M satisfy

$$\vec{w_d} = \frac{\vec{H}}{H^2} \times \vec{M}_d = \frac{\vec{H}}{H} \times \vec{M}_d / \left[rmvsin(\vec{r}, \vec{v}) \right] \tag{11.25}$$

Theoretically, the larger the H i.e. the larger the rotor radius r, mass m and speed n, the smaller the drift m_d for the same disturbance M_d. However, the increase of r and m is constrained by the structure and the increase of n is constrained by the material deformation. This is a universal problem faced by various gyro rotors.

(2) General issues faced by ESG.

The fundamental advantage of ESG is the elimination of mechanical shaft friction and goniometry with a suspension support. The full implication of the suspension support is that the metal sphere is suspended in a metal sphere cavity with a high vacuum p between narrow d and rotates at a high speed n. The narrow d, metal bowl and ball ensure the formation of the electrode capacitance and high electrostatic levitation field E, the very low vacuum pressure p ensure the gap without molecular friction and constant rotor speed, the high speed n ensure the large angular momentum $\vec{H}$ can resist external disturbance M_d, the above quantities are mutually constrained.

Angle measurement is the purpose of ESG, which uses two photoelectric sensors or rotor imbalance signalers to obtain the signal of carrier deviation and obtain the carrier's relative $\vec{H}$ position.

(3) The specificity of the problems faced by ESS.

The ESS uses the electrostatic field E to generate a support force $\vec{F}$ to levitate the rotor m. By the formula

$$\vec{D} = \varepsilon \vec{E} = \varepsilon \frac{U}{d} \frac{\vec{E}}{E} \tag{11.26}$$

$$q = \oint \vec{D} \times d\vec{s} \tag{11.27}$$

$$c = \frac{q}{u} \tag{11.28}$$

$$F = \frac{W_e}{d} = \frac{1}{2d} cu^2 \tag{11.29}$$

where s and d are the effective area and distance of the two parallel pole plates respectively. $\vec{D}$: Electric displacement vector; q, u are the charge, capacitance and voltage of the pole plate respectively; F, Wc: the electrostatic force and the work it does on m moving d distance; $\varepsilon_0 = 8.85 \times 10^{-12}$F/m is the vacuum dielectric coefficient. $\varepsilon = \varepsilon_0 \varepsilon_r$:$\varepsilon_r$ is the relative permittivity, which is 1 in vacuum. Notice that the $\vec{D}$ and $\vec{S}$ vectors are in the same direction, then the above four equations are solved jointly to obtain the electrostatic force in vacuum as

$$F = \frac{\varepsilon_0}{2} S(\frac{u}{d})^2 = 4.42 \times 10^{-12} S(\frac{u}{d})^2 = 4.42 \times 10^{-12} SK_p(\frac{U_m}{d})^2 \tag{11.30}$$

where U_m is the peak value, K_p is the waveform coefficient, and

$$U_m = \frac{U}{\sqrt{K_p}} \tag{11.31}$$

For the sine wave $U_m = \sqrt{2}U = \sqrt{\frac{1}{0.5}}U$, that is, K_p= 0.5. and let the electrode apparent area is S0 then the electrode effective area coefficient Ks, which deducts the electrode division and other factors caused by the area loss and electrode projection effect, that is,$K_s = \frac{S}{S_0}$, then

$$F = \frac{\varepsilon_0}{2} K_s S_0 K_p \left(\frac{U_m}{d}\right)^2 = 4.42 \times 10^{-12} K_s K_p S_0 \left(\frac{U_m}{d}\right)^2 \tag{11.32}$$

Obviously, the electrostatic force is related to the Um provided by the ESS circuit on the one hand, and to the apparent area So, area efficiency K, and bowl gap d generated by the electrode structure on the other hand. In order to increase F, it is necessary to increase the power factor UM and KP on the one hand, and to make the structure and environment factor S0 and KS as large and d as small as possible on the other hand. However, the increase of S0 and KS is constrained by the increase of the structural factor - rotor mass m: the external disturbance force Fa and the impact acceleration a satisfy

$$F_a = ma \tag{11.33}$$

For gravitational directional disturbance forces, also influenced by gravitational acceleration

$$F_{ag} = m(g + a) \tag{11.34}$$

The larger the m, the larger the Fa under the same a, thus offsetting the increase in the electrostatic support force F. On the other hand, the increase of Um, Kp and the decrease of d are influenced by the support environment, i.e. the vacuum pressure P: P, d and breakdown voltage U satisfy

$$U_B = f(Pd) \tag{11.35}$$

Here the function $f(\cdot)$ is used to show that the relationship between them cannot yet be expressed by a simple non-transcendent function, but the law of change is realized by the well-known Paschen curve, which is represented in Fig. 11.13, where it is clear that Pd is inversely related to UR. But in fact $P \propto \frac{1}{U_n}$, only P has a more significant effect on UR than d (d $\neq$ 0). On the other hand, UR is also related to the material between the two pole plates, and the corresponding relationship curves are shown in Fig. 11.12.

Thus, as far as the suspension support is concerned, the reduction of P will greatly improve the support environment, so the reduction of P is an absolutely sought-after indicator, which does not have a negative effect on any parameter. At the same time, d reduction is a sub-absolutely pursued indicator, and as long as Pd decreases together to ensure that UR > > Um, d reduction does not have a negative impact on other parameters. And So, Ks, both affect the electrostatic force F and are constrained by Fa through m.

(4) Support object and environmental factors division and its interrelationship.

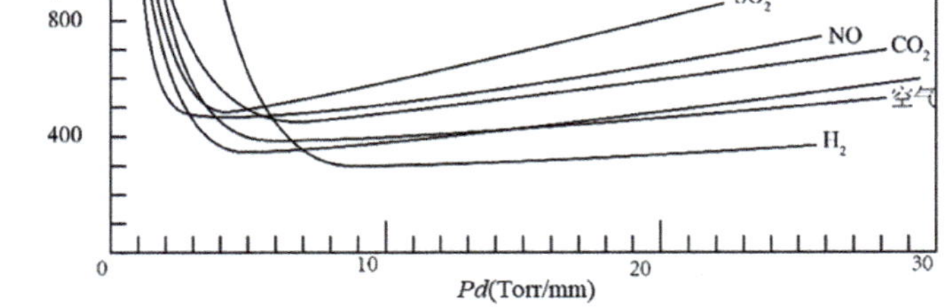

Fig. 11.12 Up = f(pd) experimental curve

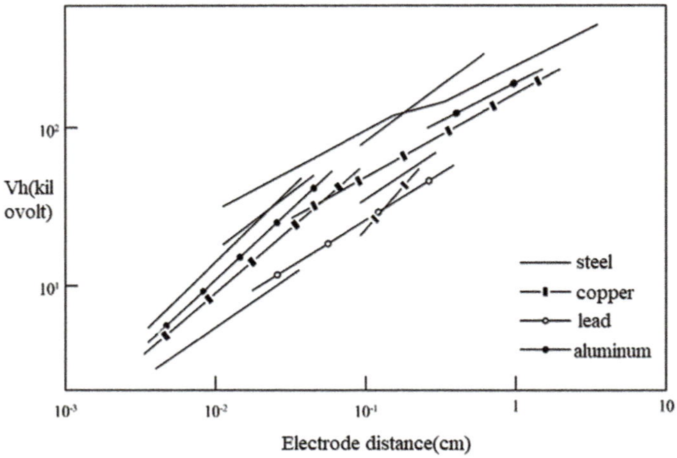

Fig. 11.13 Experimental curve of Up and bipolar plate material

According to the above analysis, it is known that the parameters related to the support object should include: ball mass m, radius r, ball material, ball structure (hollow, solid), ball bowl clearance d, bowl electrode effective area KS0 The parameters related to the support environment should include: rotational speed n, temperature T, vacuum pressure P, signal structure (photoelectric or rotor unbalance signal).

Table 11.3 shows the interrelationship of the parameters. "+" means the covariance increases, "–" means the covariance decreases, indicates that the covariate does not cause a change in the covariate of the corresponding column, and the symbol at the beginning of each horizontal row indicates that the corresponding covariate is set to change according to the desired trend, thus causing a change in the covariates of the other columns of a row to tend to heat up, and if the positive or negative trend is the same as the positive or negative sign of the element at the top of the corresponding column, it means that the factor favors the covariate; If inconsistency indicates that the causal variator changes are unfavorable to the covariate changes. The more covariates change caused, the greater the role of their causal variables in each covariate, and the greater the significance of improvement. The principles of improving the support object and environment are: improving the engineering level of the ESS system; reducing the role of error sources in the ESS.

Table 11.3 Interrelationship of parameters

Hope trends		+	+	+	+	+	–	–	–	–	+	+	+	+	–
Covariant		(1)	(2)	(3)	(4)	(5)	(6)	(7)	(8)	(9)	(10)	(11)	(12)	(13)	(14)
Invariant symbol		H	E	F	$\frac{1}{p}$	N	d	T	m	r	k_sS_o	U_m	K_p	U_B	ω_d
Setting trends	Factor														
+	H		/	/	/	/	/	/	/	/	/	/	/	/	–
–	E	/		+	/	/	–	/	/	/	/	+	/	/	/
+	F	/	+		/	/	/	/	/	/	/	+	/	/	/
+	$\frac{1}{p}$	+	+	+		+	–	–	/	/	/	+	+	+	–
+	n	+	/	/	/		/	+	/	+	/	/	/	–	–
–	d	–	+	+	–	–		+	/	/	/	+	/	–	/
–	T	–	/	/	–	+	–		/	–	/	–	/	+	–
–	m	–	/	/	/	+	/	/		–	/	/	/	/	+
–	r	–	/	/	/	+	/	/	–		-	–	/	/	+
+	K_sS_o	/	/	+	/	/	/	/	/	/		/	/	/	/
+	U_m	/	+	+	/	/	/	/	/	/	/		/	/	/
+	K_p	/	+	+	–	/	/	/	/	/	/	/		/	/
+	U_B	+	+	+	/	+	–	/	/	/	/	+	+		/
–	ω_d	+	+	+	/	+	/	/	–	–	/	+	+	/	

11.2.3 Comprehensive Technical Issues of Suspension-Supported Inertial Sensors

(1) Error source classification

Figure 11.14 shows the main module for realizing free-state ESG goniometry, which is essentially to overcome various kinds of friction to make it rotate at constant speed in inertial space. The friction of ESG can be divided into three categories.

I type is mechanical friction, it is a source of static friction, caused by gravity, is the contact friction of the ball, bowl. When the ESS works reliably, this friction is automatically eliminated and the degree of "freedom" of rotation goes up a notch compared to other mechanical gyros. Using the mechanical friction gyroscope operating condition as the friction index base point1, there is Fig. 11.16 showing the level of friction elimination.

Class II is molecular random friction, which is a source of sliding friction and consists of air molecules. If a high vacuum is achieved, this source of friction will be eliminated and the degree of "freedom" of rotation will rise another notch, corresponding to an increase in the friction index to the order of 10^{-2}.

Class III is a variety of indirect friction, which is a source of combined friction. The level of its elimination determines the degree of improvement of the friction index. This "friction" is applied to the ESG rotation shaft in the form of torque. Fundamental to ESG hardware is the blocking of internal structural conditions that create disturbing moments on the rotation axis, i.e., improving ball and bowl symmetry, sphericity and geometric form center coincidence. As for the external disturbance torque, it is not a problem considered by the ESS hardware itself.

(2) Vacuum Environmental Issues

For vacuum, it is required to systematically examine and solve the current problems of poor vacuum and speed decay, and at the same time eliminate or reduce the possible interference so that it is lower than the influence of the interference torque

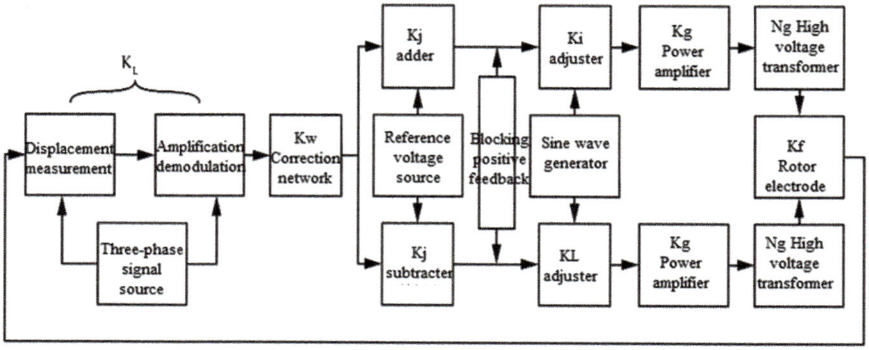

Fig. 11.14 ESS composition

of category III. For example, the change in rotor geometry caused by the vacuum constant speed, so that the rotor plane normal direction axis due to deformation and deviation from the rotor geometry center, the electrostatic force to that point will not pass through the rotor geometry center and thus generate electrostatic interference torque. This introduces the vacuum study.

(3) Constant velocity environmental issues

There are two measures of the advantages and disadvantages of the constant speed device: one is the ability to constant speed, and the other is not to introduce disturbing torque and not to create conditions for the disturbing torque to be generated.

(1) The existing constant speed uses the magnetic field method. It induces eddy currents in the rotor by means of a magnetic field, similar to the induction motor principle, to drag the rotor through the vacuum at a fixed speed. The worse the vacuum level, the greater the rotor-induced eddy currents needed to overcome the resistive torque, and the more severe the thermal effects. This causes rotor temperature rise and thermal expansion and deformation. Due to the skin effect of eddy currents and the specific location of the induced magnetic field coil, the distribution of eddy currents is also not uniform, but into different closed-loop shapes. The resulting temperature rise variation is different and the rotor deformation is inconsistent. Because the heat can only be transferred by radiation, so the accumulated thermal energy is far more than the thermal energy radiated out of the ESG, this cumulative thermal expansion effect on the one hand, the rotor diameter increases until the ball falls, on the other hand, the sphericity becomes worse, the normal vector of the rotor plane deviates from the center of the circle, the electrostatic force under the action of the interference torque, so that the ESG can not work.

The fundamental change to the above situation lies in constant temperature, or no magnetic field constant velocity scheme.

(2) Existing temperature control device features

① Several experiments have shown that the temperature rises during constant speed, deceleration and acceleration. It shows that the ESG is extremely weak in radiating energy, so it should not be given some kind of heat source like a vortex, so the magnetic field constant velocity is fundamentally inappropriate.

② Now the temperature control circuit process is complex, and can only measure the bell temperature rise. However, not the actual gap temperature rise, because the two is to transfer energy between the radiation. Hence, the reflected temperature is difficult to represent the actual gap temperature rise, which is similar to the boiling water bottle effect: outside the cool, inside may have been hot expansion to the extent that it can not work! It can only warm up and not cool down in certain ranges, so it cannot be used for the cooling process in constant speed.

(3) Constant speed is often not used in practical applications, but for high-precision ESG, accuracy is constrained by speed decay.

An approach to temperature measurement using the gap has been proposed in the literature, which has the advantage of directly capturing the required information about the rotor gap temperature and its deformation. However, no specific practical methods for temperature control, especially cooling, are given in the text. In fact, this problem is not easy to solve.

Even if temperature control can be solved, the method is extremely inappropriate: do not want the temperature change, but in order to constant speed and produce temperature change, and in order to constant temperature and temperature control. The device is complicated, and in order to eliminate the class II sliding friction, the more troublesome constant speed method of mechanical thermal deformation is used, and finally the thermal deformation is corrected by the cooling method.

(4) The real elimination of the distortion of constant speed lies in the removal of the method itself that generates this interference, i.e., the removal of the eddy current source and thus the study of a constant speed solution without eddy current geometry distortion.

4. ESS itself controls environmental issues

   ESS requires that no new disturbing moments are generated under reliable support conditions, or that the ESS itself may carry lower disturbing moments than the class I effects. Since the electrostatic force of the ESS points in the direction normal to the rotor and the bowl surface, when the Class III friction source is eliminated and bowl roundness and symmetry are ensured, the electrostatic force will pass through the center of the ball. Therefore the key to the ESS own control is: high resolution gap sensitivity under reliable support conditions, zero stability and accurate and effective addition, so that the ball center is really centered at the geometric center, i.e. the bowl center.

5. Bowl structure issues

   The bowl is the bearer of the ESS support and sensitive signal sensor.

   Ball and bowl geometry center overlap degree, ball and bowl roundness, mechan-ical size symmetry degree, determine the high level of friction index ball bowl me-chanical parameters processing accuracy, is the premise of ESS high precision. Without them, no amount of support control or vacuum can bring the friction index to a higher level. In addition the bowl structure also directly determines the ESS control scheme, and for this reason a study of the bowl structure, especially the electrode division scheme, was conducted to improve the level of bowl symmetry and electrode symmetry to substantially solve the problem of friction sources.

6. Ball structure problem

   The ball is the support object of ESS and one of the electrodes of ESS capacitive sensor. Its material (beryllium, aluminum), mass m, radius r, stiffness and implementation structure (hollow, solid) all determine the level of elimination of the class friction source and thus directly the level of ESS control.

The bowl is the bearer of the ESS support and sensitive signal sensor.

Ball and bowl geometry center overlap degree, ball and bowl roundness, mechanical size symmetry degree, determine the high level of friction index ball bowl mechanical parameters processing accuracy, is the premise of ESS high precision. Without them, no amount of support control or vacuum can bring the friction index to a higher level. In addition the bowl structure also directly determines the ESS control scheme, and for this reason a study of the bowl structure, especially the electrode division scheme, was conducted to improve the level of bowl symmetry and electrode symmetry to substantially solve the problem of friction sources.

11.3 Example of a System with Electrostatic Gyro Support

This subsection gives an example of the overall system of electrostatic gyroscopic support, including the composition structure of the system hardware, the system gain regulation and the system test results.

11.3.1 System Composition and Gain Allocation

Figure 11.14 shows the ESS composition, Fig. 11.15 shows the ESS box plot, and Table 11.4 shows the gain assignments. The static stiffness K_0 and the open-loop gain K are

$$K_0 = K_L K_W K_j K_t K_g N_g K_f \geq 9.81 \times 10^5 \text{ N/m} = 0.981 \text{N}/\mu\text{m} \quad (11.36)$$

$$K = \frac{K_0}{m} \geq 9.81 \times 10^7 l/S^2 \quad (11.37)$$

Where, X (S): support electrode center displacement; Y (S): rotor heart displacement; E (S): rotor center relative to the center of the support plate difference F (S):

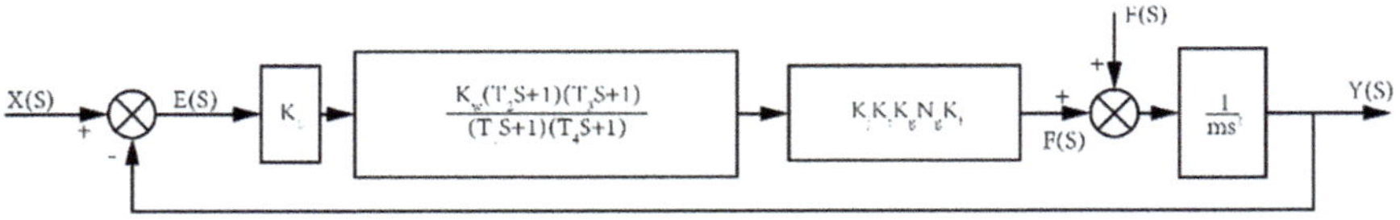

Fig. 11.15 ESS block diagram

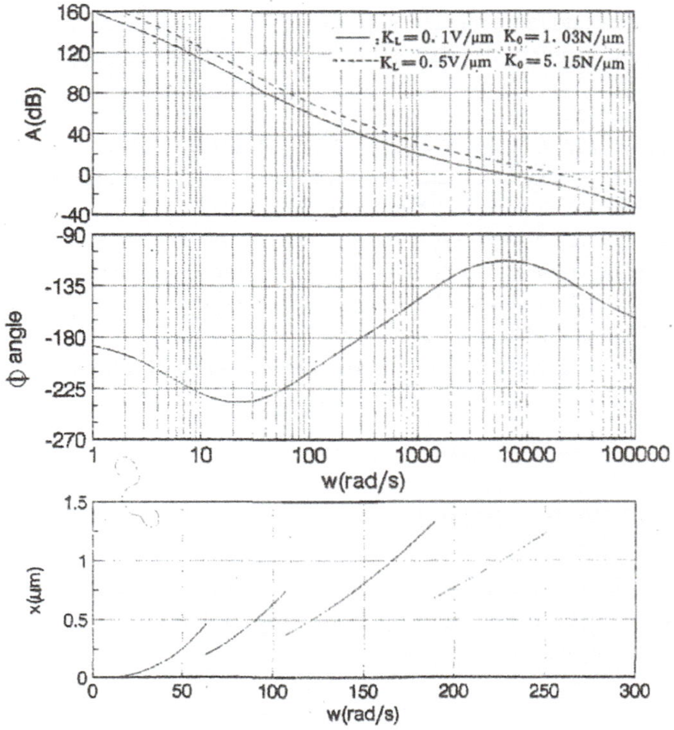

Fig. 11.16 ESS open-loop amplitude and phase characteristics and dynamic error analysis

Fig. 11.17 ESS experiment site

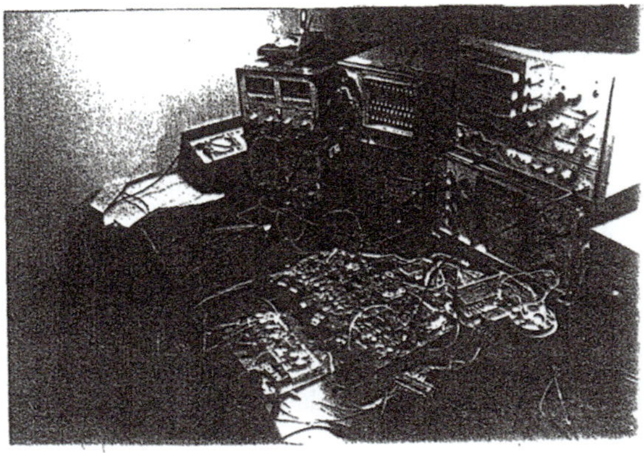

Table 11.4 Gain allocation for each link of ESS

ESS model	F_c (KHz)	Support transformer	Special section	K_L (V/μm)	K_w	K_j	K_L	K_g K_{g1} K_{g2}		NSG	K_f (N/V)	K_0 (N/μm)
A-1	20	II-2–2	KL Smaller	0.1	5	2	0.5	5	2.6	2 × 90	8.79×10^{-4}	1.03
A-2	20	II-2–2	KL greater	0.5	5	2	0.5	5	2.6	2 × 90	8.79×10^{-4}	5.15
B	18.5	II-1 model	No Kg1	0.13	5	2	0.5	×	15	2 × 60	8.79×10^{-4}	1.03
C	8	I-1–1	AC smaller KL larger	0.5	5	×	1	4.6		2 × 50	8.79×10^{-4}	1.01

electrostatic support force; F' (S): external interference force; K_L: rotor displacement electric sensitivity; $K_w\frac{(T_2S+1)(T_3S+1)}{(T_1S+1)(T_4S+1)}$: correction network transfer function; K_d: adder, subtractor amplification; K_l: modulator pass; K_g: power amplifier amplification; N_g: high voltage support variable boost ratio; K_f: support electrode control voltage to electrostatic support force transfer coefficient; $\frac{1}{mS^2}$ rotor dynamic characteristics; m: rotor mass.

Take the hysteresis-overrun correction network parameters $1/T_1 = 6.6l/S, \frac{1}{T_2} = 801/S, \frac{1}{T_3} = 12601/S, \frac{1}{T_4} = 310001/S$, Then the open-loop pre-rate characteristic of the system is expressed as

$$W(jw) = \frac{K(jwT_2+1)(jwT_3+1)}{(jw)^2(jwT_1+1)(jwT_4+1)} = \frac{9.81 \times 10^7(\frac{jw}{80}+1)(\frac{jw}{1260}+1)}{(jw)^2(\frac{jw}{6.6}+1)(\frac{jw}{31000}+1)} \tag{11.38}$$

The transfer function can meet the dynamic and static requirements, let e_m be the rotor center to branch electrode amplitude system dynamic error, then

$$e_m = \frac{A}{|W(jw)|} = \frac{A}{10^{\frac{|w(jw)|(ab)}{20}}} \tag{11.39}$$

Several of the maximum error frequency points wk and error values em are shown in Table 11.5. Figure 11.16 shows the open-loop amplitude-frequency characteristics of the ESS and the allowable values of the system dynamic error in the range of 0–251 rad/s, which shows that the maximum allowable deviation of the ESS is $\leq +2$ μm. The dashed line in the figure is the open-loop gain characteristic of the system design scheme A-2 in Table 11.4, which is 5 times larger than the scheme A-1 open-loop gain K, i.e., 13.9 db, and therefore shifted up by 13.9 db, the amplitude margin Δk increases from 40 to 50 db, and the phase margin Δφ decreases from $67^o(w \approx 6.5 \times 10^3$rad/s) to 58° (w = 1.9 × 104 rad/s), and the system is stable.

A: ESS open-loop amplitude and frequency characteristics.

Φ: ESS open-loop phase frequency characteristics.

X: Maximum dynamic error in the vibration exclusion zone.

Table 11.5 Dynamic error indicators and typical values

Ω (rad/s)	A (μm)	ω_k (rad/s)	$\|W(j\,\omega_k)\|$(db)	em(j ω_k) (μm)
6.28–62.8	1540	62.8	70	0.49
62.8–107	660	107	55	1.17
107–189	330	189	47	1.47
189–251	170	251	43	1.20

11.3.2 Field Tests and Results

The site of the device is shown in Figs. 11.17 and 11.18, where the rotor displacement measurement circuit module and the actuating element of the force-actuating device are mounted in the ESG body box in Fig. 11.21, the ESG is 8105#, and the ESS circuit is the circuit of type A in Table 11.4 and Fig. 11.19.

(1) Parameter value test

The closed-loop parameters are tested in Table 11.6.

The demodulator filter output has a 1 MHz noise width of about 8 mV; the high-voltage common ground has a 20 kHz noise width of about 10 mV. Blocking positive feedback is achieved with diode bias, bias voltage +0.78 V, works well, if removed, the system may be destabilized when subjected to large disturbances.

(2) Shock and vibration

Table 11.7 shows the corresponding peak-to-peak values of the VDMOS amplifier pay-side voltage V for this system at different accelerations. Set VDMOS amplifier power supply Vc = + 10 V, Vdmax = 2Vc = 20Vpp > 8 when the voltage 14.0Vpp, the experiment shows that the maximum output can be up to 20 Vpp, corresponding to 16 g acceleration.

Continuous impact test in three directions of ESS for many times, the oscilloscope shows that Vd reaches 20Vpp and instantaneously saturates at 20Vpp, indicating that

Table 11.6 ESS closed-loop parameter test

	A		B		C		Conditions
Measurement circuit output V1 (mV)	0.8		1.5		2.6		No interference
	0.8		1.5–15		2.6		B 30 Hz 0.32 V interference is added to the PID input
VDMOSE efficacy output (Vdpp)	15	12	14	14	13.5	13.0	No interference
Measurement circuit demodulation output noise (mV)	≤200		×		×		No interference

Table 11.7 VDMOS amplifier output voltage under different acceleration

External acceleration (m/S^2)	1g	2g	3g	4g	5g	6g	7g	8g	16g
High voltage support transformer pay-side voltageV2 (Vp)	446	630	772	892	997	1091	1182	1261	1784
$2N_g = 180, V_d = \frac{V_{2p} \times 2}{180} (V_{pp})$	5.0	7.1	8.6	9.9	11.1	12.1	13.2	14.0	19.8

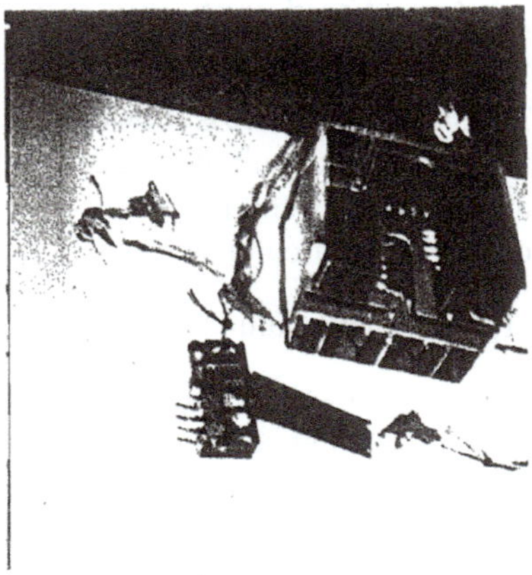

Fig. 11.18 Three-way rotor displacement module with six high-voltage transformers being assembled into the ESG body box

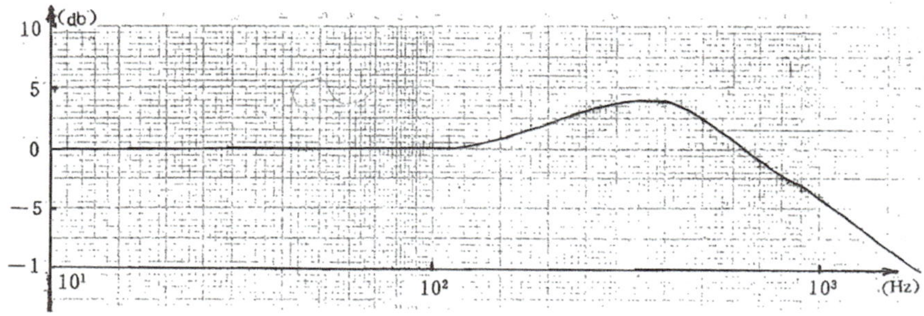

Fig. 11.19 ESS amplitude and frequency characteristics test

the output voltage has the ability to withstand 16 g of load, the impact force is much greater than 8 g, and the support is good.

(3) Frequency characteristics test

Figure 11.20 shows the measured amplitude-frequency characteristics, as seen from the figure: the system passband B = 900 Hz, the system resonant frequency f = 360 Hz, the peak vibration Mr = 5db. Figure 11.24 shows a photo of the unit step response. The overshoot is $\leq$ 30%, the number of oscillations is two to three, and the transition process time is $\leq$ 10 ms. Adjusting the carrier frequency fc finds that there is no significant change in the system in the 18.5 K–23.5 kHz band; i.e., the system has a stable operating bandwidth Br $\geq$ 5 kHz (Fig. 11.21).

(4) Electrical interference experiment

Experiments with electrical interference can be equivalent to ESS withstanding various acceleration shocks. Add interference signal from demodulator wave filter. In order to further equate with the mechanical vibration experiment, the system

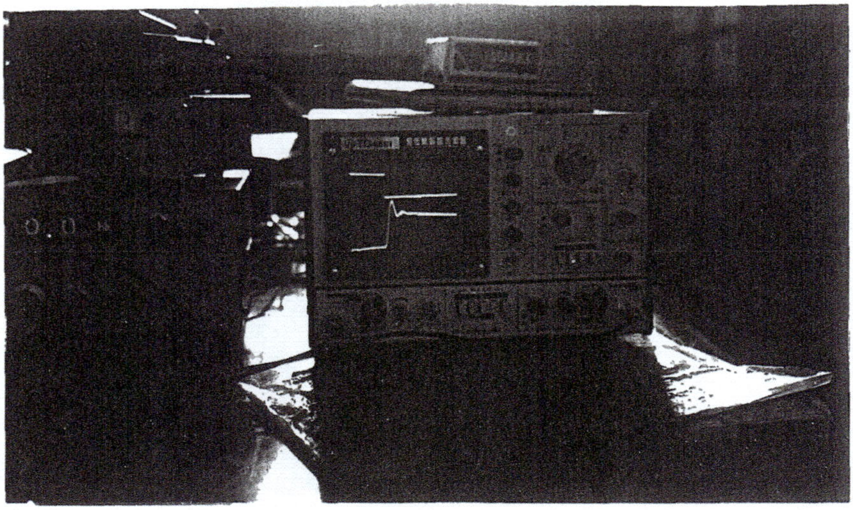

Fig. 11.20 Unit step and ESS response waveforms

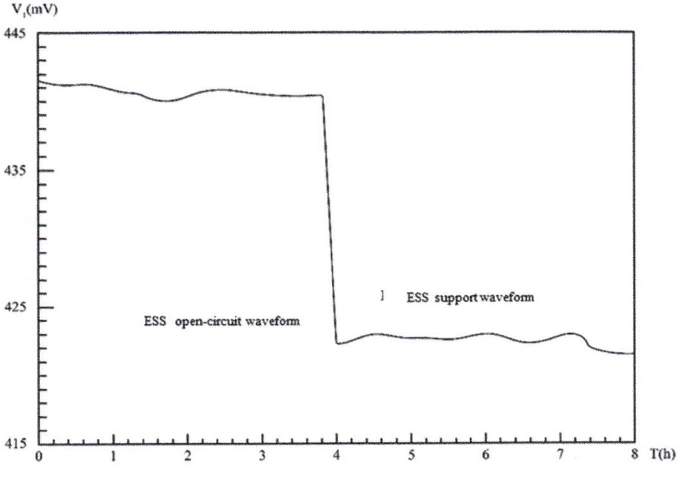

Fig. 11.21 Adder output voltage stability experiment

experiment of Yang [1] is taken for comparison, and the equivalent four-point vibration results are shown in Table 11.8. Where V_R is the disturbance signal, V_L is the measured value, the equivalent disturbance acceleration a vs. V_L is obtained from Yang [1] P95, the rotor amplitude A vs. V_L is calculated from the measurement circuit sensitivity of rotor displacement $K_L = 0.1$ V/μm. Compared with Yang [1], KL is twice smaller due to the same total open-loop gain. There is no change in the

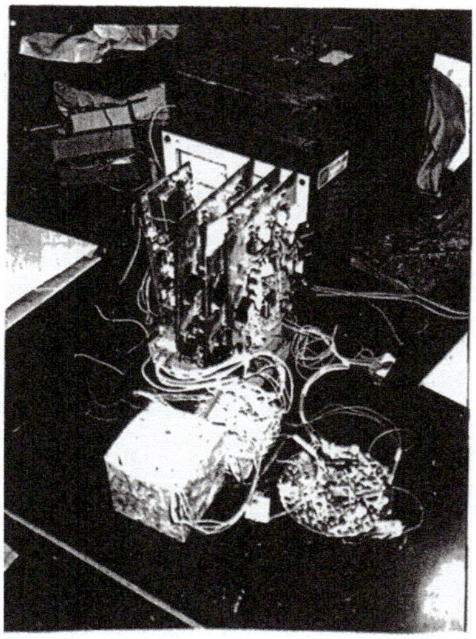

Fig. 11.22 8 kHz test site

Fig. 11.23 Representation of space in Newton's theory (left) and Einstein's general relativity representation of space (right)

vacuum ion flow throughout the experiment, indicating that the system meets the vibration stability requirements with a maximum dynamic error $\leq \pm 3.8$ μm.

Similarly there are high g equivalent vibration experiments. That is, the required VR at high g is obtained from the linear relationship between VR and a in Table 11.8. The results are shown in Table 11.9. The experiments show that the system can withstand 40 Hz × 6g, 30 Hz × 6.6g, 10 Hz×, 7.7 g equivalent mechanical vibration.

Table 11.10 shows the equivalent shock vibration experimental results for the given disturbance signal as a square wave.

Table 11.8 Four-point mechanical vibration results of electrical interference equivalent

Frequency (Hz)	InterferenceV_R (V_{pp})	ResponseV_L (V_{pp})	Open-loop equivalent acceleration(g)	Equivalent rotor amplitudeA(μm)	Durationt (min)
10	0.2	0.2	0.56	1.0	25
17	0.31	0.33	0.70	1.7	15
30	0.55	0.55	1.11	2.7	15
40	0.55	0.55	0.97	2.7	15

Table 11.9 High g vibration results of electrical interference equivalent

Frequency(Hz)	40	30	10
InterferenceVR(Vpp)	3.3	3.3	2.8
Equivalent acceleration(g)	6	6.6	7.7

The vibration experiments of three ESS A, B and C with different frequency interference signal VR input are shown in Table 11.11. The table "/square" means input square wave, the rest is sine, "X" means no input interference. The ESS works normally in this type of experimental case.

(5) Support experiment

(1) Non-interference start experiment

Repeatedly start the branch experiment, are smoothly supported, ion flow in the starting branch can not see the change in the reading, only in the original place chattering a little; power supply current transient shock for 600 mA, is the normal operating current of 500 mA 1.2 times, the control circuit current no change.

(2) Experiment of starting under the condition of electric interference

120 experiments with different frequency electrical disturbances were conducted, where the square wave was equivalent to periodic shock disturbances. Corresponding experimental data are shown in Table 11.12. All branches were successfully started, and there was a 0.05uA change in the starting ion current. The power supply transient shock is 600 mA, and there is no change in the control circuit current.

(6) Stability experiments

The experiments were conducted on the adder output V + . The experiments were divided into two stages: 1) adjusting the PID output to zero without starting the branch, setting a certain DC bias, measuring the adder output, and testing for 4 h; 2) adjusting the PID output to normal operation, starting the branch ESG, and measuring the adder output, and testing for 4 h.

The experimental curve is shown in Fig. 11.21, and it can be seen from the figure: the adder range is 440.4–441.6 mV with stability $\leq 2.7‰$ when non-starting branch; the stability is $\leq 4.74‰$when starting branch is working.

Table 11.10 Electrical disturbance equivalent shock acceleration results

Frequency(Hz)	Square wave interference$V_R(V_p)$	Equivalent impact acceleration (g)	Equivalent pulse width (mS)	Equivalent number of impacts	Phase-sensitive rectification responseV_L(Vp)	Equivalent rotor displacement(μm)
1	0.38	2	500	180	0.38	3.8
1	0.8	4	500	180	0.8	8
2	1.6	7	250	180	1.6	16
2	1.9	8	250	180	1.9	19

Table 11.11 Three-dimensional different frequency interference experiments

	Experimental order number	1	2	3	4	5	6	7	8	9	10
A	Frequency(Hz)/ Waveform	×	800	800	800	800	800	50	50	30	10
	VL Amplitude(Vpp)		0.75	0.75	0.75	0.75	0.75	1.4	1.4	1.0	1.0
B	Frequency(Hz)/ Waveform	330	800	800	200	70	70	60	×	×	×
	VL Amplitude(Vpp)	0.95	0.75	0.75	1.0	1.2	1.2	1.2			
C	Frequency(Hz)/ Waveform	330	×	×	200	×	70	×	×	×	×
	VL Amplitude(Vpp)	0.95			1.0		1.2				

Table 11.12 Test results with interference start

Interference frequency(Hz)	11		110		200		400		800	1.1K
Interference waveform	Square	Sine	Square	Sine	Square	Sine	Square	Sine	Sine	Sine
Interference voltageVR(mVp)	50	50	50	50	50	50	50	50	50	50
ResponseVL(Vp)	40	40	50	50	54	54	70	70	30	28
Power supply transient shock(mA)	600	600	600	600	600	600	600	600	600	600

(7) Power consumption

Table 11.13 shows the system power consumption. 8 g preload power supply power consumption P1 $\leq$ 5w, and the operating voltage is only +10 V; the total power consumption P $\leq$ 11w. 4 g preload and reduced control voltage can make P $\leq$ 6.7w.

The ESS circuit is the B circuit in Table 11.4, in which the voltage-variable ratio Ng = 60, model II-1. no high-voltage resistor R on the pay side, ESC 81–15#, support rate 19 kHz, power supply voltage + 15 to + 24 V. experimental data are shown in Table 11.14 The circuit also successfully achieves three-way support, and the power current is obviously smaller than that of the A circuit, which indicates that the turn ratio is reduced and the pay side Do not resist R, can improve the pay-side quality factor Q and power circuit loss, but the stability is not as good as the former.

To test the high sensitivity KL = 0.5 V/μm, high frequency and measurement circuit integrated and AC small preload circuit practicality, the set of ESS test branch test; in order to test the operation of the type I force actuators, using 8 kHz carrier frequency test site see the photo in Figs. 11.22 and 11.23, the set of circuit corresponds to the set C ESS in Table 11.5. The ESG used is 81–15 #. The high supply voltage is used to further explore the transformer low ratio high voltage scheme. The system small preload modulator D board is used to provide a technical basis for the integrated modularity study. The experimental results are shown in Table 11.15.

The system experiments in this example show that: the experimentally developed 20 kHz carrier ES system passband is 900 Hz, leap response overshoot $\leq$30%, response time <10 ms static time support voltage stability is $\leq$2.7‰, support when

Table 11.13 System power consumption under different operating conditions

Work requirements	Power Supplies			Control circuit power supply			Total power consumption P(W)
	Voltage (V)	Current (A)	Power (W) P1	Voltage (V)	Current (A)	Power (W) P2	
Maximum load 16g	10	0.5	5	± 12	0.2	4.8	9.8
Maximum load 8g	7	0.5	3.5	± 12	0.2	4.8	8.3
Maximum load 8g, control Low voltage for control circuits	7	0.5	3.5	± 8	0.2	3.2	6.7

Table 11.14 Table of relevant parameters of the three-way support experiment

ESGTri-axis placement orientation			A(Vpp)		B(Vpp)		C(Vpp)		Ic (mA)	Remarks
X(A)	Y(B)	Z(C)	Vd+	Vd–	Vd+	Vd–	Vd+	Vd–		
Front	Right	Up	8.2	8.6	8.7	5.5	13.6	5.0	230	
Front	Up	Left	8.4	8.6	14.4	3.7	8.2	7.7	250	
Up	Front	Right	13.0	4.9	7.6	5.8	7.0	6.8	240	
Up	Front	Right	15.0	6.4	8.1	6.0	7.1	66.4	200	Road A has LC1 series network

Table 11.15 ESS test branch experiment at 8 kHz

	Conditions	Power supplies Ec (V)	Turn ratio Ng	VDMOS output (Vpp)		Power currentIc (mA)	Results
				T1	T2		
Three ways	A B C	36	60	30 20 17	10 16 24	380	Ball micro-vibration sliding, starting support
Single circuit	A plug D board	36	60	30	12	200	Vertical pull up ESG
Three ways	B-way plug D board	36	60	30 18 17	10 15 24	370	Ball micro-vibration sliding, starting support

stability $\leq 4.74‰$; experience different frequency vibration, shock interference, good support. The starting support experiment shows that: the circuit uses displacement feedback control and high resonance support characteristics, starting support without breakdown. Under 120 times of various electrical disturbances, the starting branch does not break down the vacuum, the ion flow of titanium pump is constant, and the transient inrush current of the power supply is only 20% larger than that of normal operation. When the power supply is lowered from +24 to + 10V, the power circuit power consumption under 8 g preload is <6W, the control circuit power consumption is $\leq$4.8w, and the full support circuit power consumption is <11W; when the 4 g preload and power supply is +7 V, the power consumption is <3.5W, and the total system power consumption is <8W. The volume of the developed high-voltage support transformer is 35 × 30 × 25 mm. The volume of the developed high-voltage support transformer is 35 × 30 × 25 mm. The single-way VDMOS amplifier circuit is stable and the heat sink is only 10 × 20 × 20 mm in size together with the amplifier, which can be modularized and mounted into the gyro body.

11.4 Navigation Techniques for Electrostatic Gyroscopes with Relativistic Verification Basis

The electrostatic gyroscope, with electrostatic levitation support as the core technology, can be used for inertial navigation and positioning, and is also the most sophisticated core hardware for navigation and positioning. This subsection is based on the electrostatic gyro for inertial navigation, for the classical experiment in academia, the Gravity Probe B (GPB) project of Stanford University, combined with the relevant Stanford information and the data processing results of the author's group under the guidance of Associate Professor Xiaoxia He of Tsinghua University, the electrostatic The progress of the gyroscope in the verification of relativity is described.

11.4.1 Two Predictions of Einstein's Theory of Relativity

Newton believed that space and time were absolute or fixed entities, and that gravity could be expressed as a gravitational force that acts instantaneously in some way between objects at a certain distance. In Newton's universe, the axis of rotation of a perfect gyroscope orbiting the earth would remain fixed forever with respect to absolute space. Einstein believed that space and time were relative entities, intertwined into a "structure" that he called space–time. He also realized that no force—not even gravity—could act faster than the speed of light. In Einstein's universe, the presence of celestial bodies causes spacetime to bend; gravity is the product of objects moving in curved spacetime (Fig. 11.24). Thus, in Einstein's universe, the axis of rotation of a perfectly rotating gyroscope around the Earth would drift (change its direction) with time relative to the distant universe as it follows the curved and twisted curvature of spacetime.

According to Einstein's theory, over the course of a year, the short-range line effect of the Earth's local space–time should cause the rotation axis of each gyroscope to deviate from its initial calibration by a very small angle of 6.606 arc seconds

Fig. 11.24 Observation principle of gyroscope and telescope in GP-B

(0.0018°) in the plane of the spacecraft's orbit. Similarly, the distortion of the Earth's local space–time should cause the rotation axis to move at a much smaller angle of 0.039 arc seconds (0.000011°) in the Earth's equatorial plane—which is about the width of a hair when viewed from a quarter of a mile away.

11.4.2 Gravity Probe B Experimental Design Principles

Gravity Probe B (GP-B) is a NASA physics research mission designed to experimentally investigate Albert Einstein's 1916 theory of general relativity—the theory of gravity. Using four superconducting quantum interferometers (SQUIDs, or gyroscopes) and a telescope mounted on a satellite 642 km (400 miles) from Earth, GB-P is measuring, in a new way and with unprecedented precision, two effects predicted by general relativity: the geodetic effect—the degree to which the Earth distorts the local space–time in which it is located. The frame-dragging effect—the amount of drag that a rotating Earth exerts on its local space–time (first direct measurement). Huge celestial bodies (such as the Earth) drag their own local space–time as they rotate. This effect of a moving "gravitational charge" is often compared to the magnetic field produced by a moving charge, often referred to as gravitomagnetism. It may explain the enormous energy produced by the most massive and explosive objects in the universe (such as black holes).

As in Fig. 11.24, the gyroscope and telescope are placed on a polar orbiting satellite, starting by aligning the telescope and gyroscope rotation axis to a distant reference point, a guide star, IM Pegasii (HR 8703). The telescope has been tracking this very slow-moving star, but the gyroscope is not affected by the star's so-called "intrinsic motion". The displacement of the guide star positioning is subtracted from the gyro drift data to find the difference between the angular displacement of the gyro and the initial position of the guide star, which is the relative drift of the gyro. The telescope was allowed to align with the guide star for one year, during which time the change in the spin axis alignment of each gyro in the orbital plane (short-range line drift) and the orthogonal drift in the Earth's rotation plane (coordinate system drag drift) were measured. The GP-B experiment observes both effects by precisely measuring the drift (displacement) angles of the four gyroscope spin axes over a one-year period and comparing these experimental results with the predictions of Einstein's theory.

Designing a physical experiment involves a fundamental choice: maximize the effective parameters, or make the "noise" as small as possible so that the effective signal is not masked. For the GPB experiment, Einstein's relativistic effect is very significant near black holes, but very small near the Earth. Thus, the experimental design needed to satisfy seven near-zero design constraints - three near-zero constraints on the gyro rotor, which had to be a near-perfect sphere with a uniform superconducting coating and virtually no defects in shape or density, and four near-zero constraints on the original starboard, cryogenic laboratory, which housed the gyro and shielded it from any sources of noise or interference that could distort

Table 11.16 The 7 near-zero constraints of GP-B

Experimental variables	Binding conditions	Actual value
Gyro rotor		
Mechanical Spherical	50 nm	<10 nm
Homogeneity of matter	$3/10^6$	$3/10^7$
Electrical spherical	$5/10^7$	$<5/10^7$
Detector environment		
Temperature	1.95 Kelvin (−271.2 °C or −465.2 °F)	1.8 Kelvin (–271.4 °C or –456.4 °F)
No gravity residual acceleration	Less 10^{-10}g	小于 5×10^{-12} g
Background magnetic field	10^{-6} Gauss	小于 10^{-7} Gauss
Detector pressure	10^{-11} Torr	小于 10^{-11} Torr

the results - solar radiation, atmospheric friction, magnetic fields, electromagnetic signals from Earth, and so on. Table 1.1 summarizes these seven near-zero constraints, listing the required tolerances and the actual tolerances achieved during the mission (Table 11.16).

11.4.3 Gravity Probe B's Gyro and Its Electrostatic Support System

While the concept of Gravity Probe B is relatively simple, conducting this experiment requires some of the most precise and complex technology ever developed. In fact, scientists and engineers from Stanford University, Lockheed Martin and NASA had to invent about 12 entirely new technologies to meet the near-zero limitations of the GP-B. Because most of these technologies simply did not exist when the experiment was first proposed in late 1959-early 1960. To measure the tiny angles predicted by Einstein's theory, the GP-B team needed to build a near-perfect gyroscope—one whose axis of rotation would not deviate from its starting point by more than one hundred billionth of a degree per hour as it rotated. In contrast, the spin axis drift of the most complex Earth gyroscopes found in high-tech aircraft and nuclear submarines is seven orders of magnitude greater (more than 10 million times greater) than what GP-B can allow.

In order to accomplish this particular challenge, the GP-B gyroscope rotor (Fig. 11.1–11.4) had to be perfectly balanced and homogeneous inside. They must be free of any bearings or mechanical supports, and they must work in a near-perfect vacuum. After years of work and the invention of new techniques and processes for polishing, measuring sphericity, and thin film coating, the GP-B research team finally developed a uniform 1.5-inch ball of pure fused silica polished to a perfect smoothness within a few atomic layers [2]. In fact, the GP-B gyro rotor is now listed in the Guinness World Records database as the roundest object ever made; Only neutron stars can topple them into a sphere. The spherical rotor is the core of every GP-B gyroscope. The raw quartz is mined in Brazil and then melted (baked) and refined in a proprietary process at Heraeus Amercil in Germany. The internal composition of each gyro rotor is uniform to two parts per million. On its surface, each gyroscope rotor is less than three millionths of an inch from a perfect sphere. This means that every point on the rotor surface is exactly the same distance from the center of the rotor to within 3 × 10–7 inches. If the GP-B gyroscope's rotor is enlarged to the size of the earth, its highest mountain or deepest trench is less than eight feet!

Finally, the GP-B gyroscope can be detached from any mechanical bearings or supports by suspending the spherical rotor inside a precision machined fused silica shell body cavity. Six electrodes, evenly distributed inside the housing, keep the rotor suspended inside the shell cavity. During the IOC phase of the mission, a stream of pure helium is used to increase the rotational speed of the four gyroscopes to approximately 4000 revolutions per minute. After this, all but a few helium molecules were withdrawn from the housing and the gyroscopes continued to spin - only 32 microns (0.001 inches) from the housing wall, without any mechanical or fluid support. In the experiment, the near-perfectly spherical, uniform rotor, combined with the highly sophisticated gyroscopic suspension system (GSS), resulted in an average spin-down time constant of about 15,000 years for the four GP-B gyros.

The Gyro Suspension System (GSS) electrostatically suspends scientific gyroscopes within their shell cavities, allowing them to spin freely with minimal friction and torque, which is an engineering marvel in itself. In order to perform successfully, the GSS must meet the following requirements:

1. Operate over 8 orders of magnitude of force. The same system must be able to suspend a gyroscope on Earth in a magnetic field of 1g with minimal interference at the 10-8g level during data collection.
2. Reliable suspension or levitation of the gyroscope. The system must never allow a rotating rotor to touch the housing. In this case, a rotor with a speed of 4000 rpm has enough mechanical energy to effectively damage the rotor and the housing.
3. Operation is compatible with Superconducting Quantum Interferometer (SQUID) readout systems. the SQUID magnetometers are extremely sensitive.

Therefore, the suspension system must not interfere with these sensors, both on the ground and in orbit operation.
4. Minimize static torque during scientific data collection. The suspension system must meet the centering requirements with the absolute minimum control effort and thus with the minimum residual torque to the rotor.
5. Control torque is applied to the rotor for calibration and initial rotor spin axis alignment.
6. As part of the accelerometer, the "drag-free" translational control system further reduces the classical torque on the rotor by another 106 times.

All of these requirements were met during the GP-B mission. In addition, the suspension systems function well even after the Dewar bottles run out of helium because they do not require cryogenic electronics to operate.

11.5 Data Processing Method of Gravity Probe B Experiment Compared with Relativistic Prediction

In this section, a brief overview of what is involved in analyzing the GP-B science data is given first. The data analysis actually starts before the launch. The Stanford University Science Data Analysis Team built prototype computer code to process the simulated science data in preparation for the actual data from the spacecraft. After in-orbit checks, the team used real data to test and refine these routines, taking into account anomalies encountered by the mission (immediate polarization cycles) that were not part of the initial simulation. All of these activities precede the formal data analysis effort.

11.5.1 GP-B Scientific Data Analysis Process

After scientific data are acquired from satellites and archived, the formal analysis of these data is divided into three phases; each subsequent phase builds on the premise of the final scientific results [2].

1. Short-term (day-by-day) analysis

Initially, the data are analyzed mainly in short periods of time, usually a few tracks or a day, with the aim of understanding the limitations of the data analysis practices and improving them if necessary. The main activities in this phase include:

(1) Final instrument calibration was completed based on IOC, science missions, and post-science calibration tests run on the vehicle.
(2) Improvements were made to the subsystem performance estimates (i.e., GSS, ATC, TRE, and SRE subsystems). The performance of these systems determines

a lower bound on the quality of results that the design team expects from the experiment.

(3) The data is analyzed to remove known features, such as aberrations and pointing errors from the scientific signal.

(4) Implement a data grading scheme that flags and selectively ignores patches of data sets that are not of scientific quality due to spacecraft anomalies or unusual instrument configurations.

The overall goal of this phase is to improve the quality of the scientific data analysis package, calibrate the instrument effects, and produce an initial "same-day positioning" estimate based on the adjusted data analysis package. In this phase, it does not attempt to eliminate systematic errors that span many days or months. In addition, this analysis focuses on the performance of each gyroscope and does not attempt to combine the four measurements, nor does it attempt to estimate the gyroscope feed (due to relativistic or classical effects).

2. Mid-term (month-by-month) analysis

In the second phase, the analysis team's attention turns to understanding and compensating for long-term system effects that span many days or months, such as those produced by time-varying gyro polar cycles or drift in on-board instrument calibration.

(1) Identify, model and eliminate systematic errors and improve instrument calibration.

(2) Understand the causes and effects of spacecraft anomalies. Model and remove the effects from these anomalies.

(3) The magnitude of the residual Newtonian moment on the gyroscope is evaluated and compared to the pre-launch estimate.

3. One year forecast

In the final stage, data from all four gyroscopes will be combined with intrinsic motion measurements of guide stars from our partners at the Smithsonian Astrophysical Observatory (SAO) to generate overall estimates of gyroscope feed rates during the mission. These estimates form the basis of the main science result: gyroscope feeds due to the effects of general relativity. This process naturally produces a third, improved "day orientation" for each gyroscope.

11.5.2 Data Processing Methods and Analysis of Results

According to the relevant literature [3], the gyro output signal model used in this experiment is

$$z_G(t) = Cg\{[NS_0 + R_g t + \varepsilon_1(t)]\cos(\omega_r t + \delta\phi) +$$

$$[EW_0 + R_f t + \varepsilon_2(t)] \sin(\omega_r t + \delta\phi)\} + \frac{C_T}{Cg} z_T(t) + b + v(t) \quad (11.40)$$

where

Cg, C_T are the scale factors for SQUID and telescope, respectively;

NS_0, EW_0 are the components of the declination of the initial gyroscopic axis of rotation from the line of aim of the gravitational detector in the north–south and east–west directions, respectively;

R_g, R_f are the drift rates of the gyroscopic rotation axis in the north–south and east–west directions, respectively;

ω_r is the angular velocity of rotation of the gravitational detector;$\delta\phi$ is its initial phase of rotation;

b and v are the zero bias and white noise of the SQUID output signal;

$\varepsilon_1(t)$, $\varepsilon_2(t)$ are the aberrations of starlight in the north–south and east–west directions respectively. Since the satellite orbits the Earth and its orbit is a polar orbit, this movement causes periodic aberration changes in the north–south direction (called Orbital Aberration), whose period is the satellite orbital period, but not in the east–west direction; at the same time, the Earth revolves around the Sun, which leads to aberration changes in the north–south and east–west directions (called Annual Aberration), so the relationship between $\varepsilon_1(t)$, $\varepsilon_2(t)$ and time can be expressed as follows.

$$\varepsilon_1(t) = A_0 \cos(\omega_o t) + L_1 \cos(\omega_a t) + L_2 \sin(\omega_a t) \quad (11.41)$$

$$\varepsilon_1(t) = L_3 \cos(\omega_a t) + L_4 \sin(\omega_a t) \quad (11.42)$$

where ω_o is the angular velocity of the satellite's flight around the Earth;

ω_a is the angular velocity of the Earth's rotation.

A_0, L_1, L_2, L_3, L_4 is a known quantity.

It is necessary to obtain the estimated values of the following eight quantities through the analysis and processing of experimental data, and write them in vector form, i.e.

$$X = [R_g, R_f, C_g, \delta\phi, NS_0, EW_0, b, C_T]^T \quad (11.43)$$

Since it is a nonlinear filtering problem to estimate X directly according to Eq. (11.1), in order to avoid the error caused by linearization of Kalman filtering directly, the following new state quantity is selected for estimation:

$$Y = \begin{bmatrix} x_3 \cos x_4 \\ -x_3 \sin x_4 \\ x_3(x_1 \cos x_4 + x_2 \sin x_4) \\ x_3(-x_1 \sin x_4 + x_2 \cos x_4) \\ x_3(x_5 \cos x_4 + x_6 \sin x_4) \\ x_3(-x_5 \sin x_4 + x_6 \cos x_4) \\ x_7 \\ x_3/x_8 \end{bmatrix} \tag{11.44}$$

In this case, Eq. (11.1) can be expressed as

$$z = H(t)Y + v \tag{11.45}$$

where

$$\begin{aligned} H = [&\varepsilon_1 \cos(\omega_\mathrm{r} t) + \varepsilon_2 \sin(\omega_\mathrm{r} t), \varepsilon_1 \sin(\omega_\mathrm{r} t) - \varepsilon_2 \cos(\omega_\mathrm{r} t), \\ &t\cos(\omega_\mathrm{r} t), t\sin(\omega_\mathrm{r} t), \cos(\omega_\mathrm{r} t), \sin(\omega_\mathrm{r} t), 1, z_T] \end{aligned} \tag{11.46}$$

Then through the Kalman filter can be directly estimated, according to the first estimated, and then according to the following transformation relations to obtain the estimated value

$$X = \begin{bmatrix} (\mathrm{y}_3\mathrm{y}_1 + \mathrm{y}_4\mathrm{y}_2)/x_3^2 \\ (\mathrm{y}_4\mathrm{y}_1 - \mathrm{y}_3\mathrm{y}_2)/x_3^2 \\ x_3 \\ \arctan(-y_2/\mathrm{y}_1) \\ (\mathrm{y}_5\mathrm{y}_1 + \mathrm{y}_6\mathrm{y}_2)/x_3^2 \\ -(\mathrm{y}_5\mathrm{y}_2 - \mathrm{y}_6\mathrm{y}_1)/x_3^2 \\ \mathrm{y}_7 \\ x_3/\mathrm{y}_8 \end{bmatrix} \tag{11.47}$$

It can be seen from the above equation that the estimated value of x_3 has A great influence on the estimated value of other quantities. Therefore, in order to ensure the unbiased estimation, it cannot be calculated directly from the transformation relation $x_3 = \sqrt{\mathrm{y}_1^2 + \mathrm{y}_2^2}$, because the transformation between y_1, y_2 and x_3 and x_4 is essentially a rectangular coordinate polar coordinate transformation, which will cause some deviation to the estimated value of x_3. In order to ensure the unbiased estimation, the deviation needs to be added, so the following estimation formula is presented

$$\hat{x}_3 = \sqrt{\hat{y}_1^2 + \hat{y}_2^2} + \frac{\sigma^2}{2\sqrt{\hat{y}_1^2 + \hat{y}_2^2}} \tag{11.48}$$

where σ^2 is the variance of white noise $v(t)$.

After a full study of level 2 data, it is found that there are a lot of signal data that can be directly used in level 2 data, instead of following the above methods. The provision of some data makes data processing much simpler. Such as the phase of the satellite's rotation around its axis, starlight differences.

In this way, the above estimation method can be changed to only estimate the following seven items:

$$X = [R_g, R_f, C_g, NS_0, EW_0, b, C_T]^T \tag{11.49}$$

As Kalman filtering is a recursive method, its characteristic is to gradually improve the estimation accuracy with the increase of data processing volume. However, in the actual processing, it is found that many abnormal data are still found in level 2 data. Due to the huge experimental data that has been running for a year, there will be a small number of recording errors in the data itself. It is also inevitable that there are some periods of large experimental interference, although it will not affect the normal operation of its core–gyro, but interference such as satellite attitude control will inevitably cause SQUID signal interference, resulting in a lot of data (even a short continuous section) abnormal data. Therefore, if Kalman filter is directly used to estimate the experimental data of one year, it will be found that there are too many places interrupted in the estimation process, that is, the estimated value runs to an unreasonable range, and even leads to the divergence of the filter, which makes all the previous data processing information lost. Later, to solve this problem, the method of segmental estimation of the data of every day is adopted. Avoid local error diffusion. However, since it is difficult to estimate the correct R_g, R_f within one day, the following estimation method is adopted in order to obtain the results.

Suppose the SQUID output model is:

$$\begin{aligned} z_G(t) &= Cg\{[NS + \varepsilon_1(t)]\cos\phi + [EW + \varepsilon_2(t)]\sin\phi\} + b + v(t) \\ &= Hx + v(t) \end{aligned} \tag{11.50}$$

where, $x = [C_g NS, C_g EW, C_g, b]^T$,$H = [\cos\phi, \sin\phi, \varepsilon_1\cos\phi + \varepsilon_2\sin\phi, 1]$.

Firstly, estimate A, store the estimate in a file, and process the data of every day in this way. Finally, read out all the estimated values, calculate B, and make linear fitting processing on it. The theory predicts that C should change linearly.

Figure 11.25 shows the NS drift curve predicted by the first gyroscope. Although it fluctuates greatly under the influence of noise, especially the front and back sections of the curve change irregularly, the middle part of the curve shows obvious linear change characteristics. The slope obtained by fitting the figure is roughly in line with the drift rate of the short-range effect. The short-range effect predicted by general relativity is preliminarily verified.

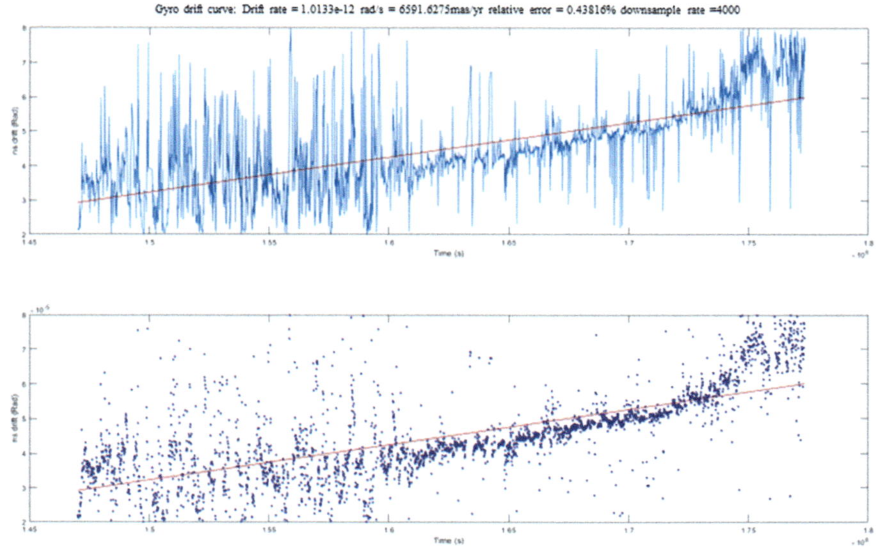

Fig. 11.25 NS drift curve of the first gyro

11.6 Summary

Electrostatic gyro can be used for precision navigation, its hardware can't be realized without the support system, and the integrated technology of suspension support inertial sensor is closely related to its many parameters. This chapter establishes the relationship and the master–slave effect, which has important guiding significance for improving the system performance, eliminating and reducing the source of error.

The research shows that the relationship between carrier and ESS passband directly affects the transmission quality of ESS dynamic control signal. This chapter systematically deduces the relation between carrier frequency and rotor displacement deviation and determines the carrier mechanism of floating support. Carrier frequency has intrinsic physical connection with the design and application of afterforce actuator. This relationship is manifested in the deep high pressure and suspension preload, the relationship between the power and capacity of the afteracting element and the working rate point and the supporting wave, etc. Thus, the material and electromagnetic characteristics of the afterburner are determined. The special electromagnetic characteristics of afteracting element—high voltage transformer are guaranteed by its process, so the mathematical analysis of its process becomes the theoretical basis for standardizing its product index. The combined action of carrier and transformer capacitive load leads to a special electronic circuit phenomenon—the sine wave support of power amplifier in the single tube Class B state. In the first section of this chapter, the object of this image is proved theoretically, and the suspension support of this state is proved to be effective within the whole passband.

This chapter explores the verification of the navigation technology of electrostatic gyroscope and the theory of relativity. Relying on the GP-B project, the linear drift phenomenon is preliminatively observed from the gyro data, which is the preliminary verification of the short thread effect, one of the two predictions of the theory of relativity.

References

1. Yang, W. (1988). *Research on electrostatic gyro bearing system with variable structure and variable preload*. Doctoral Dissertation, Tsinghua University.
2. Post Flight Analysis—Final Report Preface & Executive Summary October 2006. https://www.nasa.gov/pdf/168808main_gp-b_pfar_cvr-pref-execsum.pdf
3. Heifetz, M. I., Keiser, G. M., Krechetov, A. S., Silbergleit, A. S. (2002) Multisensor data integration in the NASA/Stanford gravity probe B relativity mission. In: *International conference on information fusion*. IEEE.

GPSR Compliance

The European Union's (EU) General Product Safety Regulation (GPSR) is a set of rules that requires consumer products to be safe and our obligations to ensure this.

If you have any concerns about our products, you can contact us on ProductSafety@springernature.com

In case Publisher is established outside the EU, the EU authorized representative is:

Springer Nature Customer Service Center GmbH
Europaplatz 3
69115 Heidelberg, Germany

Batch number: 10726133

Printed by Printforce, the Netherlands